PREFACE

For the past 60 years, Building News has been dedicated to providing construction professionals with timely and reliable information. Based on this experience, our staff has researched and compiled thousands of up-to-the-minute costs for the **Building News 2009 Costbooks**. This book is an essential reference for contractors, engineers, architects, facilities managers — any construction professional who must provide an estimate or any type of building project.

Whether working up a preliminary estimate or submitting a formal bid, the costs listed here can quickly and easily be tailored to your needs. All costs are based on national averages, while a table of modifiers is provided for regional adjustments. Overhead and profit are included in all costs.

Complete man-hour tables follow the unit costs to provide data on typical durations of specific tasks. This information can be used to schedule projects as well as to determine specific labor costs based on local labor rates.

All data is categorized according to the MASTERFORMAT of the Construction Specifications Institute (CSI). This industry standard provides an all-inclusive checklist to ensure that no element of a project is overlooked. In addition, to make specific items even easier to locate, there is a complete alphabetical index.

This costbook contains an appendix with reference charts and tables taken from an array of sources. The text explains the costs in certain categories and provides helpful pointers that should be taken into account with every estimate.

A section on square foot costs provides an overview of project costs for different building types — commercial, residential, etc. — with summaries of the actual projects. Square foot costs are invaluable for making budget estimates and checking prices when time is a factor.

The "Features in this Book" section presents a clear overview of the many features of this book. Included is an explanation of the data, sample page layout and discussion of how to best use the information in the book.

Of course, all buildings and construction projects are unique. The costs provided in this book are based on averages from well-managed projects with good labor productivity under normal working conditions (eight hours a day). Other circumstances affecting costs such as overtime, unusual working conditions, savings from buying bulk quantities for large projects, and unusual or hidden costs must be factored in as they arise.

The data provided in this book is for estimating purposes only. Check all applicable federal, state and local codes and regulations for specific requirements.

BNi Building News

TABLE OF CONTENTS

CSI MASTERFORMAT

All data in the Costbook pages and Man-Hour tables is organized according to the CSI MASTERFORMAT — the industry standard numbering/classification system. The data is divided into the 16 divisions as shown below. The five digit numbers within each division correspond to the MASTERFORMAT Broadscope designations. Each section is further broken down into MASTERFORMAT Mediumscope designations. These numbers, in most cases, are the same as those used in architectural and engineering specifications.

The construction estimating information in this book is divided into two main sections: Costbook Pages and Man-Hour Tables. Each is organized to the 16 divisions of the CSI MASTERFORMAT as shown below. In addition, there are extensive Construction Reference Tables, Geographic Cost Modifiers, Square Foot Tables and a detailed Index. Sample pages with graphic explanations are included before the Costbook pages, Man-Hour Tables and Square Foot Tables. These explanations, along with the discussions below, will provide a good understanding of what is included in this book and how it can best be used for construction estimating.

FURNISHINGS *Division 12*

SPECIAL CONSTRUCTION *Division 13*

CONVEYING SYSTEMS *Division 14*

MECHANICAL *Division 15*

ELECTRICAL *Division 16*

FEATURES IN THIS BOOK

The construction estimating information in this book is divided into two main sections: Costbook Pages and Man-Hour Tables. Each section is organized according to the 16 divisions of the MASTERFORMAT as shown on the previous pages. In addition, there are extensive Supporting Construction Reference tables, Geographic Costs Modifiers, Square Foot tables and a detailed Index.

Sample pages with graphic explanations are included before the Costbook pages and Man-Hour tables. These explanations, along with the discussions below, will provide a good understanding of what is included in this book and how it can best be used in construction estimating.

Material Costs

The material costs used in this book represent national averages for prices that a contractor would expect to pay plus an allowance for freight (if applicable), handling and storage. These costs reflect neither the lowest or highest prices, but rather a typical average cost over time. Periodic fluctuations in availability and in certain commodities (e.g. copper, lumber) can significantly affect local material pricing. In the final estimating and bidding stages of a project when the highest degree of accuracy is required, it is best to check local, current prices.

Labor Costs

Labor costs include the basic wage, plus commonly applicable taxes, insurance and markups for overhead and profit. The labor rates used here to develop the costs are typical average prevailing wage rates. Rates for different trades are used where appropriate for each type of work.

Taxes and insurance which are most often applied to labor rates include employer-paid Social Security/Medicare taxes (FICA), Worker's Compensation insurance, state and federal unemployment taxes, and business insurance. Fixed government rates as well as average allowances are included in the labor costs.

However, most of these items vary significantly from state to state and within states. For more specific data, local agencies and sources should be consulted.

Equipment Costs

Costs for various types and pieces of equipment are included in Division 1 - General Requirements and can be included in an estimate when required either as a total "Equipment" category or with specific appropriate trades. Costs for equipment are included when appropriate in the installation costs in the Costbook pages.

Overhead And Profit

Included in the labor costs are allowances for overhead and profit for the contractor/employer whose workers are performing the specific tasks. No cost allowances or fees are included for management of subcontractors by the general contractor or construction manager. These costs, where appropriate, must be added to the costs as listed in the book.

The allowance for overhead is included to account for office overhead, the contractors' typical costs of doing business. These costs normally include in-house office staff salaries and benefits, office rent and operating expenses, professional fees, vehicle costs and other operating costs which are not directly applicable to specific jobs. It should be noted for this book that office overhead as included should be distinguished from project overhead, the General Requirements (CSI Division 1) which are specific to particular projects. Project overhead should be included on an item by item basis for each job.

Depending on the trade, an allowance of 10-15 percent is incorporated into the labor/installation costs to account for typical profit of the installing contractor. See Division 1, General Requirements, for a more detailed review of typical profit allowances.

Adjustments to Costs

The costs as presented in this book attempt to represent national averages. Costs, however, vary among regions, states and even between adjacent localities.

In order to more closely approximate the probable costs for specific locations throughout the U.S., a table of Geographic Cost Modifiers is provided. These adjustment factors are used to modify costs obtained from this book to help account for regional variations of construction costs. Whenever local current costs are known, whether material or equipment prices or labor rates, they should be used if more accuracy is required.

Man-Hour Tables

The man-hour data used to develop the labor costs are listed in the second main section of this book, the "Man-Hour Tables". These productivities represent typical installation labor for thousands of construction items. The data takes into account all activities involved in normal construction under commonly experienced working conditions such as site movement, material handling, start-up, etc. As with the Costbook pages, these items are listed according to the CSI MASTERFORMAT.

Square Foot Tables

Included as an additional reference are Square Foot Tables which list hundreds of actual projects for dozens of building types, each with associated building size and total square foot building cost. This data provides an overview of construction costs by building type. These costs are for actual projects. The variations within similar building types may be due, among other factors, to size, location, quality and specified components, material and processes. Depending upon all such factors, specific building costs can vary significantly and may not necessarily fall within the range of costs as presented.

Editor's Note: The **Building News 2009 Costbooks** are intended to provide accurate, reliable, average costs and typical productivities for thousands of common construction components. The data is developed and compiled from various industry sources, including government, manufacturers, suppliers and working professionals. The intent of the information is to provide assistance and guidelines to construction professionals in estimating. The user should be aware that local conditions, material and labor availability and cost variations, economic considerations, weather, local codes and regulations, etc., all affect the actual cost of construction. These and other such factors must be considered and incorporated into any and all construction estimates.

Sample Costbook Page

In order to best use the information in this book, please review this sample page and read the "Features In This Book" section.

CSI MASTERFORMAT Division

CSI Broadscope Category

CSI Mediumscope Category (First 5 Digits)

Detailed Descriptions

Complete descriptions of items may include information listed above a particular line. Review of the whole category is recommended for a complete description.

Material Cost

Material cost represent average contractor prices plus an allowance for freight, handling and storage.

Installation Cost

Installation cost includes basic wage rates, markups for taxes, insurance overhead and profit and also includes equipment costs where appropriate.

Total Cost

The total cost is the sum of material and installation costs. This total represents typical contractors' costs including overhead and profit, but does not include markups for the general contractor or construction management fees.

Unit of Measurement

Each item (and cost) is defined in terms of the common estimating unit. All costs are listed in dollars per unit.

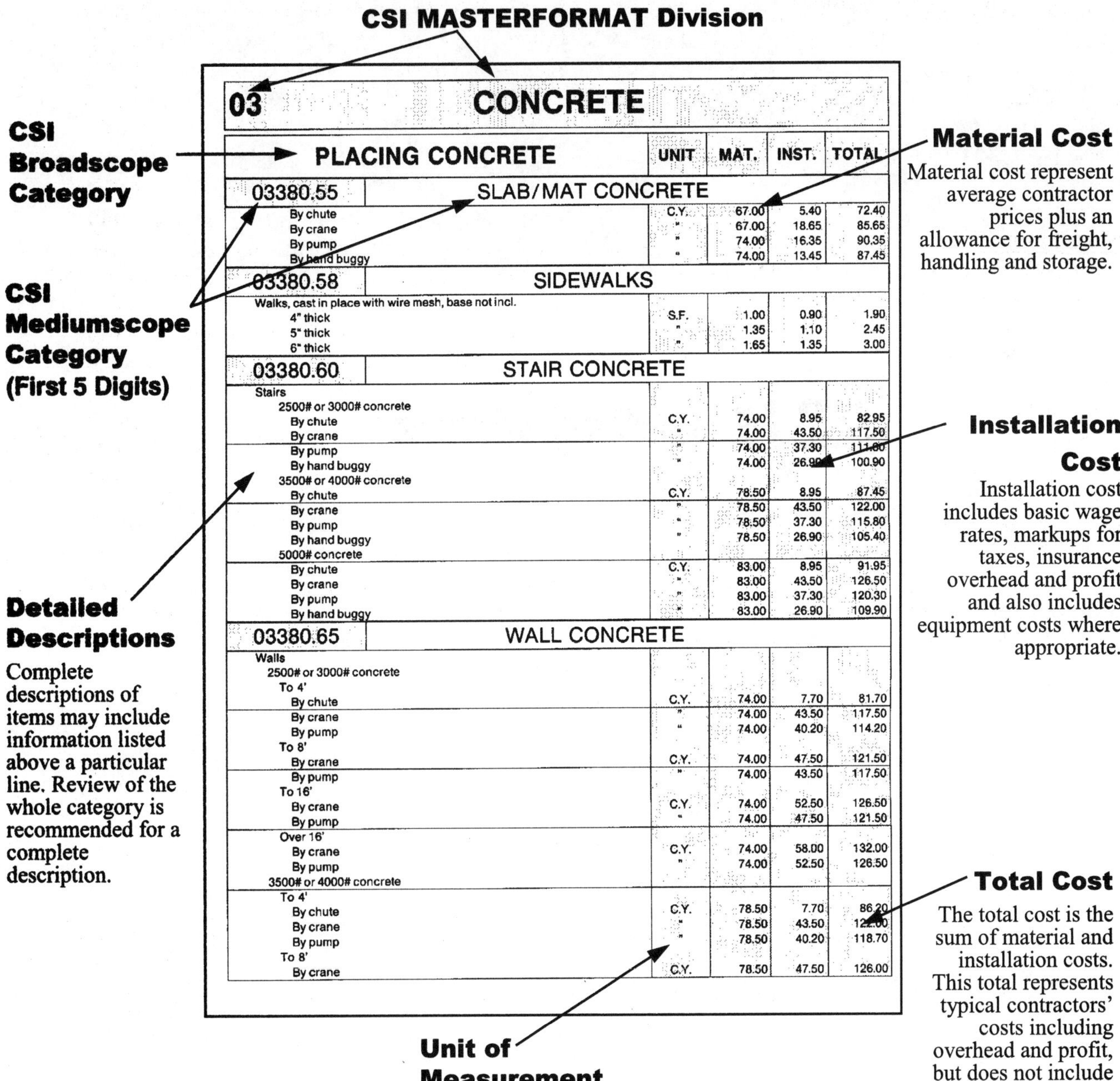

PLACING CONCRETE	UNIT	MAT.	INST.	TOTAL
03380.55 SLAB/MAT CONCRETE				
By chute	C.Y.	67.00	5.40	72.40
By crane	"	67.00	18.65	85.65
By pump	"	74.00	16.35	90.35
By hand buggy	"	74.00	13.45	87.45
03380.58 SIDEWALKS				
Walks, cast in place with wire mesh, base not incl.				
4" thick	S.F.	1.00	0.90	1.90
5" thick	"	1.35	1.10	2.45
6" thick	"	1.65	1.35	3.00
03380.60 STAIR CONCRETE				
Stairs				
2500# or 3000# concrete				
By chute	C.Y.	74.00	8.95	82.95
By crane	"	74.00	43.50	117.50
By pump	"	74.00	37.30	111.80
By hand buggy	"	74.00	26.90	100.90
3500# or 4000# concrete				
By chute	C.Y.	78.50	8.95	87.45
By crane	"	78.50	43.50	122.00
By pump	"	78.50	37.30	115.80
By hand buggy	"	78.50	26.90	105.40
5000# concrete				
By chute	C.Y.	83.00	8.95	91.95
By crane	"	83.00	43.50	126.50
By pump	"	83.00	37.30	120.30
By hand buggy	"	83.00	26.90	109.90
03380.65 WALL CONCRETE				
Walls				
2500# or 3000# concrete				
To 4'				
By chute	C.Y.	74.00	7.70	81.70
By crane	"	74.00	43.50	117.50
By pump	"	74.00	40.20	114.20
To 8'				
By crane	C.Y.	74.00	47.50	121.50
By pump	"	74.00	43.50	117.50
To 16'				
By crane	C.Y.	74.00	52.50	126.50
By pump	"	74.00	47.50	121.50
Over 16'				
By crane	C.Y.	74.00	58.00	132.00
By pump	"	74.00	52.50	126.50
3500# or 4000# concrete				
To 4'				
By chute	C.Y.	78.50	7.70	86.20
By crane	"	78.50	43.50	122.00
By pump	"	78.50	40.20	118.70
To 8'				
By crane	C.Y.	78.50	47.50	126.00

Requirements	UNIT	MAT.	INST.	TOTAL
01020.10 Allowances				
Overhead				
$20,000 project				
Minimum	PCT.			15.00
Average	"			20.00
Maximum	"			40.00
$100,000 project				
Minimum	PCT.			12.00
Average	"			15.00
Maximum	"			25.00
$500,000 project				
Minimum	PCT.			10.00
Average	"			12.00
Maximum	"			20.00
$1,000,000 project				
Minimum	PCT.			6.00
Average	"			10.00
Maximum	"			12.00
Profit				
$20,000 project				
Minimum	PCT.			10.00
Average	"			15.00
Maximum	"			25.00
$100,000 project				
Minimum	PCT.			10.00
Average	"			12.00
Maximum	"			20.00
$500,000 project				
Minimum	PCT.			5.00
Average	"			10.00
Maximum	"			15.00
$1,000,000 project				
Minimum	PCT.			3.00
Average	"			8.00
Maximum	"			15.00
Professional fees				
Architectural				
$100,000 project				
Minimum	PCT.			5.00
Average	"			10.00
Maximum	"			20.00
$500,000 project				
Minimum	PCT.			5.00
Average	"			8.00
Maximum	"			12.00
$1,000,000 project				
Minimum	PCT.			3.50
Average	"			7.00
Maximum	"			10.00
Structural engineering				
Minimum	PCT.			2.00
Average	"			3.00
Maximum	"			5.00
Mechanical engineering				

Requirements	UNIT	MAT.	INST.	TOTAL

01020.10 — Allowances *(Cont.)*

Requirements	UNIT	MAT.	INST.	TOTAL
Minimum	PCT.			4.00
Average	"			5.00
Maximum	"			15.00
Electrical engineering				
Minimum	PCT.			3.00
Average	"			5.00
Maximum	"			12.00
Taxes				
Sales tax				
Minimum	PCT.			4.00
Average	"			5.00
Maximum	"			10.00
Unemployment				
Minimum	PCT.			3.00
Average	"			6.50
Maximum	"			8.00
Social security (FICA)	"			7.85

01050.10 — Field Staff

Requirements	UNIT	MAT.	INST.	TOTAL
Superintendent				
Minimum	YEAR			75,600
Average	"			110,250
Maximum	"			157,500
Field engineer				
Minimum	YEAR			57,750
Average	"			89,250
Maximum	"			131,250
Foreman				
Minimum	YEAR			42,000
Average	"			68,250
Maximum	"			99,750
Bookkeeper/timekeeper				
Minimum	YEAR			24,150
Average	"			31,500
Maximum	"			52,500
Watchman				
Minimum	YEAR			16,800
Average	"			21,000
Maximum	"			33,600

01310.10 — Scheduling

Requirements	UNIT	MAT.	INST.	TOTAL
Scheduling for				
$100,000 project				
Minimum	PCT.			1.05
Average	"			2.10
Maximum	"			4.20
$500,000 project				
Minimum	PCT.			0.53
Average	"			1.05
Maximum	"			2.10
$1,000,000 project				
Minimum	PCT.			0.32

Requirements	UNIT	MAT.	INST.	TOTAL
01310.10 **Scheduling** *(Cont.)*				
Average	PCT.			0.84
Maximum	"			1.58
Scheduling software				
Minimum	EA.			400
Average	"			2,310
Maximum	"			46,200
01330.10 **Surveying**				
Surveying				
Small crew	DAY		830	830
Average crew	"		1,250	1,250
Large crew	"		1,650	1,650
01380.10 **Job Requirements**				
Job photographs, small jobs				
Minimum	EA.			120
Average	"			170
Maximum	"			400
Large projects				
Minimum	EA.			580
Average	"			870
Maximum	"			2,890
01410.10 **Testing**				
Testing concrete, per test				
Minimum	EA.			19.00
Average	"			31.75
Maximum	"			64.00
Soil, per test				
Minimum	EA.			38.25
Average	"			96.00
Maximum	"			250
Welding, per test				
Minimum	EA.			19.00
Average	"			31.75
Maximum	"			130
01500.10 **Temporary Facilities**				
Barricades, temporary				
Highway				
Concrete	L.F.	11.25	4.12	15.37
Wood	"	3.67	1.65	5.32
Steel	"	3.93	1.37	5.30
Pedestrian barricades				
Plywood	S.F.	2.62	1.37	3.99
Chain link fence	"	2.95	1.37	4.32
Trailers, general office type, per month				
Minimum	EA.			200

Requirements	UNIT	MAT.	INST.	TOTAL
01500.10 — **Temporary Facilities** *(Cont.)*				
Average	EA.			330
Maximum	"			660
Crew change trailers, per month				
Minimum	EA.			120
Average	"			130
Maximum	"			200
01505.10 — **Mobilization**				
Equipment mobilization				
Bulldozer				
Minimum	EA.			180
Average	"			380
Maximum	"			630
Backhoe/front-end loader				
Minimum	EA.			110
Average	"			190
Maximum	"			410
Crane, crawler type				
Minimum	EA.			1,990
Average	"			4,880
Maximum	"			10,480
Truck crane				
Minimum	EA.			470
Average	"			700
Maximum	"			1,210
Pile driving rig				
Minimum	EA.			9,030
Average	"			18,070
Maximum	"			32,530
01525.10 — **Construction Aids**				
Scaffolding/staging, rent per month				
Measured by lineal feet of base				
10' high	L.F.			11.50
20' high	"			20.75
30' high	"			29.00
40' high	"			33.50
50' high	"			40.00
Measured by square foot of surface				
Minimum	S.F.			0.51
Average	"			0.87
Maximum	"			1.57
Safety nets, heavy duty, per job				
Minimum	S.F.			0.33
Average	"			0.40
Maximum	"			0.89
Tarpaulins, fabric, per job				
Minimum	S.F.			0.23
Average	"			0.40
Maximum	"			1.03

Requirements	UNIT	MAT.	INST.	TOTAL
01570.10			**Signs**	
Construction signs, temporary				
Signs, 2' x 4'				
Minimum	EA.			33.25
Average	"			80.00
Maximum	"			280
Signs, 4' x 8'				
Minimum	EA.			70.00
Average	"			180
Maximum	"			780
Signs, 8' x 8'				
Minimum	EA.			90.00
Average	"			280
Maximum	"			2,900
01600.10			**Equipment**	
Air compressor				
60 cfm				
By day	EA.			88.00
By week	"			260
By month	"			790
300 cfm				
By day	EA.			180
By week	"			550
By month	"			1,670
600 cfm				
By day	EA.			510
By week	"			1,520
By month	"			4,560
Air tools, per compressor, per day				
Minimum	EA.			33.75
Average	"			42.25
Maximum	"			59.00
Generators, 5 kw				
By day	EA.			85.00
By week	"			250
By month	"			770
Heaters, salamander type, per week				
Minimum	EA.			100
Average	"			140
Maximum	"			300
Pumps, submersible				
50 gpm				
By day	EA.			68.00
By week	"			200
By month	"			610
100 gpm				
By day	EA.			85.00
By week	"			250
By month	"			760
500 gpm				
By day	EA.			140
By week	"			400
By month	"			1,220

Requirements	UNIT	MAT.	INST.	TOTAL
01600.10 **Equipment** *(Cont.)*				
Diaphragm pump, by week				
Minimum	EA.			120
Average	"			200
Maximum	"			420
Pickup truck				
By day	EA.			130
By week	"			370
By month	"			1,150
Dump truck				
6 cy truck				
By day	EA.			340
By week	"			1,010
By month	"			3,040
10 cy truck				
By day	EA.			420
By week	"			1,270
By month	"			3,800
16 cy truck				
By day	EA.			670
By week	"			2,030
By month	"			6,080
Backhoe, track mounted				
1/2 cy capacity				
By day	EA.			690
By week	"			2,110
By month	"			6,250
1 cy capacity				
By day	EA.			1,100
By week	"			3,290
By month	"			9,880
2 cy capacity				
By day	EA.			1,860
By week	"			5,570
By month	"			16,730
3 cy capacity				
By day	EA.			3,550
By week	"			10,640
By month	"			31,940
Backhoe/loader, rubber tired				
1/2 cy capacity				
By day	EA.			420
By week	"			1,270
By month	"			3,800
3/4 cy capacity				
By day	EA.			510
By week	"			1,520
By month	"			4,560
Bulldozer				
75 hp				
By day	EA.			590
By week	"			1,770
By month	"			5,320
Cranes, crawler type				

Requirements	UNIT	MAT.	INST.	TOTAL
01600.10 **Equipment** *(Cont.)*				
15 ton capacity				
By day	EA.			760
By week	"			2,280
By month	"			6,840
25 ton capacity				
By day	EA.			930
By week	"			2,790
By month	"			8,360
50 ton capacity				
By day	EA.			1,690
By week	"			5,070
By month	"			15,210
100 ton capacity				
By day	EA.			2,530
By week	"			7,600
By month	"			22,810
Truck mounted, hydraulic				
15 ton capacity				
By day	EA.			720
By week	"			2,150
By month	"			6,210
Loader, rubber tired				
1 cy capacity				
By day	EA.			510
By week	"			1,520
By month	"			4,560
01740.10 **Bonds**				
Performance bonds				
Minimum	PCT.			0.64
Average	"			2.01
Maximum	"			3.18

Soil Tests	UNIT	MAT.	INST.	TOTAL
02010.10			**Soil Boring**	
Borings, uncased, stable earth				
2-1/2" dia.				
Minimum	L.F.		18.50	18.50
Average	"		27.75	27.75
Maximum	"		44.50	44.50
4" dia.				
Minimum	L.F.		20.25	20.25
Average	"		31.75	31.75
Maximum	"		56.00	56.00
Cased, including samples				
2-1/2" dia.				
Minimum	L.F.		22.25	22.25
Average	"		37.00	37.00
Maximum	"		74.00	74.00
4" dia.				
Minumum	L.F.		44.50	44.50
Average	"		64.00	64.00
Maximum	"		89.00	89.00
Drilling in rock				
No sampling				
Minimum	L.F.		40.50	40.50
Average	"		59.00	59.00
Maximum	"		80.00	80.00
With casing and sampling				
Minimum	L.F.		56.00	56.00
Average	"		74.00	74.00
Maximum	"		110	110
Test pits				
Light soil				
Minimum	EA.		280	280
Average	"		370	370
Maximum	"		740	740
Heavy soil				
Minimum	EA.		450	450
Average	"		560	560
Maximum	"		1,110	1,110

Demolition	UNIT	MAT.	INST.	TOTAL
02060.10			**Building Demolition**	
Building, complete with disposal				
Wood frame	C.F.		0.32	0.32
Concrete	"		0.48	0.48
Steel frame	"		0.64	0.64
Partition removal				
Concrete block partitions				

Demolition	UNIT	MAT.	INST.	TOTAL
02060.10 **Building Demolition** *(Cont.)*				
4" thick	S.F.		2.06	2.06
8" thick	"		2.75	2.75
12" thick	"		3.75	3.75
Brick masonry partitions				
4" thick	S.F.		2.06	2.06
8" thick	"		2.58	2.58
12" thick	"		3.44	3.44
16" thick	"		5.16	5.16
Cast in place concrete partitions				
Unreinforced				
6" thick	S.F.		14.75	14.75
8" thick	"		16.00	16.00
10" thick	"		18.50	18.50
12" thick	"		22.25	22.25
Reinforced				
6" thick	S.F.		17.00	17.00
8" thick	"		22.25	22.25
10" thick	"		24.75	24.75
12" thick	"		29.75	29.75
Terra cotta				
To 6" thick	S.F.		2.06	2.06
Stud partitions				
Metal or wood, with drywall both sides	S.F.		2.06	2.06
Metal studs, both sides, lath and plaster	"		2.75	2.75
Door and frame removal				
Hollow metal in masonry wall				
Single				
2'6"x6'8"	EA.		52.00	52.00
3'x7'	"		69.00	69.00
Double				
3'x7'	EA.		83.00	83.00
4'x8'	"		83.00	83.00
Wood in framed wall				
Single				
2'6"x6'8"	EA.		29.50	29.50
3'x6'8"	"		34.50	34.50
Double				
2'6"x6'8"	EA.		41.25	41.25
3'x6'8"	"		45.75	45.75
Remove for re-use				
Hollow metal	EA.		100	100
Wood	"		69.00	69.00
Floor removal				
Brick flooring	S.F.		1.65	1.65
Ceramic or quarry tile	"		0.91	0.91
Terrazzo	"		1.83	1.83
Heavy wood	"		1.10	1.10
Residential wood	"		1.17	1.17
Resilient tile or linoleum	"		0.41	0.41
Ceiling removal				
Acoustical tile ceiling				
Adhesive fastened	S.F.		0.41	0.41
Furred and glued	"		0.34	0.34

Demolition	UNIT	MAT.	INST.	TOTAL
02060.10 Building Demolition *(Cont.)*				
Suspended grid	S.F.		0.25	0.25
Drywall ceiling				
Furred and nailed	S.F.		0.45	0.45
Nailed to framing	"		0.41	0.41
Plastered ceiling				
Furred on framing	S.F.		1.03	1.03
Suspended system	"		1.37	1.37
Roofing removal				
Steel frame				
Corrugated metal roofing	S.F.		0.82	0.82
Built-up roof on metal deck	"		1.37	1.37
Wood frame				
Built up roof on wood deck	S.F.		1.27	1.27
Roof shingles	"		0.68	0.68
Roof tiles	"		1.37	1.37
Concrete frame	C.F.		2.75	2.75
Concrete plank	S.F.		2.06	2.06
Built-up roof on concrete	"		1.17	1.17
Cut-outs				
Concrete, elevated slabs, mesh reinforcing				
Under 5 cf	C.F.		41.25	41.25
Over 5 cf	"		34.50	34.50
Bar reinforcing				
Under 5 cf	C.F.		69.00	69.00
Over 5 cf	"		52.00	52.00
Window removal				
Metal windows, trim included				
2'x3'	EA.		41.25	41.25
2'x4'	"		45.75	45.75
2'x6'	"		52.00	52.00
3'x4'	"		52.00	52.00
3'x6'	"		59.00	59.00
3'x8'	"		69.00	69.00
4'x4'	"		69.00	69.00
4'x6'	"		83.00	83.00
4'x8'	"		100	100
Wood windows, trim included				
2'x3'	EA.		23.00	23.00
2'x4'	"		24.25	24.25
2'x6'	"		25.75	25.75
3'x4'	"		27.50	27.50
3'x6'	"		29.50	29.50
3'x8'	"		31.75	31.75
6'x4'	"		34.50	34.50
6'x6'	"		37.50	37.50
6'x8'	"		41.25	41.25
Walls, concrete, bar reinforcing				
Small jobs	C.F.		27.50	27.50
Large jobs	"		23.00	23.00
Brick walls, not including toothing				
4" thick	S.F.		2.06	2.06
8" thick	"		2.58	2.58
12" thick	"		3.44	3.44

Demolition		UNIT	MAT.	INST.	TOTAL
02060.10	**Building Demolition** *(Cont.)*				
16" thick		S.F.		5.16	5.16
Concrete block walls, not including toothing					
4" thick		S.F.		2.29	2.29
6" thick		"		2.42	2.42
8" thick		"		2.58	2.58
10" thick		"		2.94	2.94
12" thick		"		3.44	3.44
Rubbish handling					
Load in dumpster or truck					
Minimum		C.F.		0.91	0.91
Maximum		"		1.37	1.37
For use of elevators, add					
Minimum		C.F.		0.20	0.20
Maximum		"		0.41	0.41
Rubbish hauling					
Hand loaded on trucks, 2 mile trip		C.Y.		34.25	34.25
Machine loaded on trucks, 2 mile trip		"		22.25	22.25

Highway Demolition		UNIT	MAT.	INST.	TOTAL
02065.10	**Pavement Demolition**				
Bituminous pavement, up to 3" thick					
On streets		S.Y.		8.90	8.90
On pipe trench		"		11.25	11.25
Concrete pavement, 6" thick					
No reinforcement		S.Y.		14.75	14.75
With wire mesh		"		22.25	22.25
With rebars		"		27.75	27.75
9" thick					
No reinforcement		S.Y.		18.50	18.50
With wire mesh		"		27.75	27.75
With rebars		"		37.00	37.00
12" thick					
No reinforcement		S.Y.		22.25	22.25
With wire mesh		"		31.75	31.75
With rebars		"		44.50	44.50
Sidewalk, 4" thick, with disposal		"		7.42	7.42
Removal of pavement markings by waterblasting		S.F.		0.20	0.20
02065.15	**Saw Cutting Pavement**				
Pavement, bituminous					
2" thick		L.F.		1.71	1.71
3" thick		"		2.14	2.14
4" thick		"		2.64	2.64
5" thick		"		2.86	2.86
6" thick		"		3.06	3.06

Highway Demolition

	UNIT	MAT.	INST.	TOTAL
02065.15 **Saw Cutting Pavement** *(Cont.)*				
Concrete pavement, with wire mesh				
4" thick	L.F.		3.30	3.30
5" thick	"		3.57	3.57
6" thick	"		3.90	3.90
8" thick	"		4.29	4.29
10" thick	"		4.77	4.77
Plain concrete, unreinforced				
4" thick	L.F.		2.86	2.86
5" thick	"		3.30	3.30
6" thick	"		3.57	3.57
8" thick	"		3.90	3.90
10" thick	"		4.29	4.29
02065.80 **Curb & Gutter**				
Removal, plain concrete curb	L.F.		5.56	5.56
Plain concrete curb and 2' gutter	"		7.67	7.67
Curb removal				
Concrete, unreinforced				
Minimum	L.F.		4.45	4.45
Average	"		5.56	5.56
Maximum	"		6.95	6.95
Reinforced				
Minimum	L.F.		7.18	7.18
Average	"		7.95	7.95
Maximum	"		8.90	8.90
Combination curb and 2' gutter				
Unreinforced				
Minimum	L.F.		5.85	5.85
Average	"		7.67	7.67
Maximum	"		11.25	11.25
Reinforced				
Minimum	L.F.		9.27	9.27
Average	"		12.25	12.25
Maximum	"		22.25	22.25
Granite curb				
Minimum	L.F.		6.36	6.36
Average	"		7.42	7.42
Maximum	"		8.56	8.56
Asphalt curb				
Minimum	L.F.		3.71	3.71
Average	"		4.45	4.45
Maximum	"		5.30	5.30
02065.85 **Guardrails**				
Remove standard guardrail				
Steel	L.F.		7.42	7.42
Wood	"		5.70	5.70

Highway Demolition	UNIT	MAT.	INST.	TOTAL
02075.80 — **Core Drilling**				
Concrete				
6" thick				
3" dia.	EA.		38.00	38.00
4" dia.	"		44.50	44.50
6" dia.	"		53.00	53.00
8" dia.	"		89.00	89.00
8" thick				
3" dia.	EA.		53.00	53.00
4" dia.	"		67.00	67.00
6" dia.	"		76.00	76.00
8" dia.	"		110	110
10" thick				
3" dia.	EA.		67.00	67.00
4" dia.	"		76.00	76.00
6" dia.	"		89.00	89.00
8" dia.	"		130	130
12" thick				
3" dia.	EA.		89.00	89.00
4" dia.	"		110	110
6" dia.	"		130	130
8" dia.	"		180	180

Hazardous Waste	UNIT	MAT.	INST.	TOTAL
02080.10 — **Asbestos Removal**				
Enclosure using wood studs & poly, install & remove	S.F.	1.22	1.03	2.25
Trailer (change room)	DAY			110
Disposal suits (4 suits per man day)	"			46.50
Type C respirator mask, includes hose & filters, per man	"			23.25
Respirator mask & filter, light contamination	"			9.08
Air monitoring test, 12 tests per day				
Off job testing	DAY			1,200
On the job testing	"			1,590
Asbestos vacuum with attachments	EA.			700
Hydraspray piston pump	"			890
Negative air pressure system	"			810
Grade D breathing air equipment	"			2,030
Glove bag, 44" x 60" x 6 mil plastic	"			6.19
40 CY asbestos dumpster				
Weekly rental	EA.			770
Pick up/delivery	"			440
Asbestos dump fee	"			190

Hazardous Waste	UNIT	MAT.	INST.	TOTAL
02080.12 — **Duct Insulation Removal**				
Remove duct insulation, duct size				
6" x 12"	L.F.		2.29	2.29
x 18"	"		3.17	3.17
x 24"	"		4.58	4.58
8" x 12"	"		3.44	3.44
x 18"	"		3.75	3.75
x 24"	"		5.16	5.16
12" x 12"	"		3.44	3.44
x 18"	"		4.58	4.58
x 24"	"		5.89	5.89
02080.15 — **Pipe Insulation Removal**				
Removal, asbestos insulation				
2" thick, pipe				
1" to 3" dia.	L.F.		3.44	3.44
4" to 6" dia.	"		3.93	3.93
3" thick				
7" to 8" dia.	L.F.		4.12	4.12
9" to 10" dia.	"		4.34	4.34
11" to 12" dia.	"		4.58	4.58
13" to 14" dia.	"		4.85	4.85
15" to 18" dia.	"		5.16	5.16

Site Demolition	UNIT	MAT.	INST.	TOTAL
02105.10 — **Catch Basins/manholes**				
Abandon catch basin or manhole (fill with sand)	EA.		450	450
Remove and reset frame and cover	"		220	220
Remove catch basin, to 10' deep				
Masonry	EA.		560	560
Concrete	"		740	740
02105.20 — **Fences**				
Remove fencing				
Chain link, 8' high				
For disposal	L.F.		2.06	2.06
For reuse	"		5.16	5.16
Wood				
4' high	S.F.		1.37	1.37
6' high	"		1.65	1.65
8' high	"		2.06	2.06
Masonry				
8" thick				
4' high	S.F.		4.12	4.12
6' high	"		5.16	5.16

Site Demolition	UNIT	MAT.	INST.	TOTAL
02105.20 — **Fences** *(Cont.)*				
8' high	S.F.		5.89	5.89
12" thick				
4' high	S.F.		6.88	6.88
6' high	"		8.25	8.25
8' high	"		10.25	10.25
12' high	"		13.75	13.75
02105.42 — **Drainage Piping**				
Remove drainage pipe, not including excavation				
12" dia.	L.F.		9.27	9.27
18" dia.	"		11.75	11.75
24" dia.	"		14.75	14.75
36" dia.	"		18.50	18.50
02105.43 — **Gas Piping**				
Remove welded steel pipe, not including excavation				
4" dia.	L.F.		14.00	14.00
5" dia.	"		22.25	22.25
6" dia.	"		27.75	27.75
8" dia.	"		44.50	44.50
10" dia.	"		56.00	56.00
02105.45 — **Sanitary Piping**				
Remove sewer pipe, not including excavation				
4" dia.	L.F.		8.90	8.90
6" dia.	"		10.00	10.00
8" dia.	"		11.25	11.25
10" dia.	"		11.75	11.75
12" dia.	"		12.25	12.25
15" dia.	"		13.00	13.00
18" dia.	"		14.75	14.75
24" dia.	"		18.50	18.50
30" dia.	"		22.25	22.25
36" dia.	"		27.75	27.75
02105.48 — **Water Piping**				
Remove water pipe, not including excavation				
4" dia.	L.F.		10.00	10.00
6" dia.	"		10.50	10.50
8" dia.	"		11.75	11.75
10" dia.	"		12.25	12.25
12" dia.	"		13.00	13.00
14" dia.	"		14.00	14.00
16" dia.	"		14.75	14.75
18" dia.	"		16.00	16.00
20" dia.	"		17.00	17.00
Remove valves				
6"	EA.		110	110
10"	"		120	120
14"	"		140	140
18"	"		190	190

Site Demolition

	UNIT	MAT.	INST.	TOTAL

02105.60 — Underground Tanks

	UNIT	MAT.	INST.	TOTAL
Remove underground storage tank, and backfill				
50 to 250 gals	EA.		740	740
600 gals	"		740	740
1000 gals	"		1,110	1,110
4000 gals	"		1,780	1,780
5000 gals	"		1,780	1,780
10,000 gals	"		2,970	2,970
12,000 gals	"		3,710	3,710
15,000 gals	"		4,450	4,450
20,000 gals	"		5,570	5,570

02105.66 — Septic Tanks

	UNIT	MAT.	INST.	TOTAL
Remove septic tank				
1000 gals	EA.		190	190
2000 gals	"		220	220
5000 gals	"		280	280
15,000 gals	"		2,230	2,230
25,000 gals	"		2,970	2,970
40,000 gals	"		4,450	4,450

02105.80 — Walls, Exterior

	UNIT	MAT.	INST.	TOTAL
Concrete wall				
Light reinforcing				
6" thick	S.F.		11.25	11.25
8" thick	"		11.75	11.75
10" thick	"		12.25	12.25
12" thick	"		14.00	14.00
Medium reinforcing				
6" thick	S.F.		11.75	11.75
8" thick	"		12.25	12.25
10" thick	"		14.00	14.00
12" thick	"		16.00	16.00
Heavy reinforcing				
6" thick	S.F.		13.00	13.00
8" thick	"		14.00	14.00
10" thick	"		16.00	16.00
12" thick	"		18.50	18.50
Masonry				
No reinforcing				
8" thick	S.F.		4.94	4.94
12" thick	"		5.56	5.56
16" thick	"		6.36	6.36
Horizontal reinforcing				
8" thick	S.F.		5.56	5.56
12" thick	"		6.01	6.01
16" thick	"		7.18	7.18
Vertical reinforcing				
8" thick	S.F.		7.18	7.18
12" thick	"		8.24	8.24
16" thick	"		10.00	10.00
Remove concrete headwall				
15" pipe	EA.		160	160
24" pipe	"		200	200

Site Demolition	UNIT	MAT.	INST.	TOTAL
02162.10 **Trench Sheeting**				
Closed timber, including pull and salvage, excavation				
8' deep	S.F.	3.24	7.70	10.94
20' deep	"	3.24	11.75	14.99

Earthwork	UNIT	MAT.	INST.	TOTAL
02210.10 **Hauling Material**				
Haul material by 10 cy dump truck, round trip distance				
1 mile	C.Y.		4.77	4.77
2 mile	"		5.72	5.72
5 mile	"		7.80	7.80
10 mile	"		8.58	8.58
20 mile	"		9.54	9.54
30 mile	"		11.50	11.50
Site grading, cut & fill, sandy clay, 200' haul, 75 hp dozer	"		3.43	3.43
Spread topsoil by equipment on site	"		3.81	3.81
Site grading (cut and fill to 6") less than 1 acre				
75 hp dozer	C.Y.		5.72	5.72
1.5 cy backhoe/loader	"		8.58	8.58
02210.30 **Bulk Excavation**				
Excavation, by small dozer				
Small areas	C.Y.		2.86	2.86
Trim banks	"		4.29	4.29
Hydraulic excavator				
1 cy capacity				
Light material	C.Y.		3.71	3.71
Medium material	"		4.45	4.45
Wet material	"		5.56	5.56
Blasted rock	"		6.36	6.36
Wheel mounted front-end loader				
7/8 cy capacity				
Light material	C.Y.		2.95	2.95
Medium material	"		3.37	3.37
Wet material	"		3.93	3.93
Blasted rock	"		4.72	4.72
Track mounted front-end loader				
1-1/2 cy capacity				
Light material	C.Y.		1.96	1.96
Medium material	"		2.14	2.14
Wet material	"		2.36	2.36
Blasted rock	"		2.62	2.62

Earthwork	UNIT	MAT.	INST.	TOTAL
02220.10			**Borrow**	
Borrow fill, F.O.B. at pit				
Sand, haul to site, round trip				
10 mile	C.Y.	15.50	11.75	27.25
20 mile	"	15.50	19.75	35.25
30 mile	"	15.50	29.50	45.00
Place borrow fill and compact				
Less than 1 in 4 slope	C.Y.	15.50	5.90	21.40
Greater than 1 in 4 slope	"	15.50	7.86	23.36
02220.20			**Gravel And Stone**	
F.O.B. PLANT				
No. 21 crusher run stone	C.Y.			30.75
No. 26 crusher run stone	"			30.00
No. 67 gravel	"			15.50
No. 78 gravel, (pea gravel)	"			18.00
02220.40			**Building Excavation**	
Structural excavation, unclassified earth				
3/8 cy backhoe	C.Y.		15.75	15.75
3/4 cy backhoe	"		11.75	11.75
1 cy backhoe	"		9.83	9.83
Foundation backfill and compaction by machine	"		23.50	23.50
02220.50			**Utility Excavation**	
Trencher, sandy clay, 8" wide trench				
18" deep	L.F.		1.90	1.90
24" deep	"		2.14	2.14
36" deep	"		2.45	2.45
Trench backfill, 95% compaction				
Tamp by hand	C.Y.		25.75	25.75
Vibratory compaction	"		20.75	20.75
Trench backfilling, with borrow sand, place & compact	"	12.25	20.75	33.00
02220.60			**Trenching**	
Trenching and continuous footing excavation				
By gradall				
1 cy capacity				
Light soil	C.Y.		3.37	3.37
Medium soil	"		3.63	3.63
Heavy/wet soil	"		3.93	3.93
Loose rock	"		4.29	4.29
Blasted rock	"		4.53	4.53
By hydraulic excavator				
1/2 cy capacity				
Light soil	C.Y.		3.93	3.93
Medium soil	"		4.29	4.29
Heavy/wet soil	"		4.72	4.72
Loose rock	"		5.24	5.24
Blasted rock	"		5.90	5.90
Hand excavation				
Bulk, wheeled 100'				
Normal soil	C.Y.		45.75	45.75
Sand or gravel	"		41.25	41.25

Earthwork	UNIT	MAT.	INST.	TOTAL
02220.60 — **Trenching** *(Cont.)*				
Medium clay	C.Y.		59.00	59.00
Heavy clay	"		83.00	83.00
Loose rock	"		100	100
Trenches, up to 2' deep				
Normal soil	C.Y.		52.00	52.00
Sand or gravel	"		45.75	45.75
Medium clay	"		69.00	69.00
Heavy clay	"		100	100
Loose rock	"		140	140
Trenches, to 6' deep				
Normal soil	C.Y.		59.00	59.00
Sand or gravel	"		52.00	52.00
Medium clay	"		83.00	83.00
Heavy clay	"		140	140
Loose rock	"		210	210
Backfill trenches				
With compaction				
By hand	C.Y.		34.50	34.50
By 60 hp tracked dozer	"		2.14	2.14
By small front-end loader	"		2.45	2.45
Spread dumped fill or gravel, no compaction				
6" layers	S.Y.		1.43	1.43
12" layers	"		1.71	1.71
Compaction in 6" layers				
By hand with air tamper	S.Y.		1.06	1.06
Backfill trenches, sand bedding, no compaction				
By hand	C.Y.	13.00	34.50	47.50
By small front-end loader	"	13.00	3.37	16.37
02220.90 — **Hand Excavation**				
Excavation				
To 2' deep				
Normal soil	C.Y.		45.75	45.75
Sand and gravel	"		41.25	41.25
Medium clay	"		52.00	52.00
Heavy clay	"		59.00	59.00
Loose rock	"		69.00	69.00
To 6' deep				
Normal soil	C.Y.		59.00	59.00
Sand and gravel	"		52.00	52.00
Medium clay	"		69.00	69.00
Heavy clay	"		83.00	83.00
Loose rock	"		100	100
Backfilling foundation without compaction, 6" lifts	"		25.75	25.75
Compaction of backfill around structures or in trench				
By hand with air tamper	C.Y.		29.50	29.50
By hand with vibrating plate tamper	"		27.50	27.50
1 ton roller	"		43.00	43.00
Miscellaneous hand labor				
Trim slopes, sides of excavation	S.F.		0.06	0.06
Trim bottom of excavation	"		0.08	0.08
Excavation around obstructions and services	C.Y.		140	140

Earthwork	UNIT	MAT.	INST.	TOTAL
02240.05 — Soil Stabilization				
Straw bale secured with rebar	L.F.	1.82	1.37	3.19
Filter barrier, 18" high filter fabric	"	1.82	4.12	5.94
Sediment fence, 36" fabric with 6" mesh	"	4.32	5.16	9.48
02280.20 — Soil Treatment				
Soil treatment, termite control pretreatment				
Under slabs	S.F.	0.17	0.22	0.39
By walls	"	0.17	0.27	0.44
02290.30 — Weed Control				
Weed control, bromicil, 15 lb./acre, wettable powder	ACRE	290	210	500
Vegetation control, by application of plant killer	S.Y.	0.02	0.16	0.18
Weed killer, lawns and fields	"	0.24	0.08	0.32

Piles And Caissons	UNIT	MAT.	INST.	TOTAL
02360.50 — Prestressed Piling				
Prestressed concrete piling, less than 60' long				
10" sq.	L.F.	11.25	5.53	16.78
12" sq.	"	15.75	5.77	21.52
14" sq.	"	16.50	5.89	22.39
16" sq.	"	20.25	6.03	26.28
18" sq.	"	28.00	6.47	34.47
20" sq.	"	37.75	6.63	44.38
24" sq.	"	48.25	6.80	55.05
02360.60 — Steel Piles				
H-section piles				
8x8				
36 lb/ft				
30' long	L.F.	16.50	11.00	27.50
40' long	"	16.50	8.84	25.34
50' long	"	16.50	7.37	23.87
10x10				
42 lb/ft				
30' long	L.F.	18.75	11.00	29.75
40' long	"	18.75	8.84	27.59
50' long	"	18.75	7.37	26.12
57 lb/ft				
30' long	L.F.	24.00	11.00	35.00
40' long	"	24.00	8.84	32.84
50' long	"	24.00	7.37	31.37
Splice				
8"	EA.	79.00	69.00	148
10"	"	87.00	83.00	170
Driving cap				

Piles And Caissons

	UNIT	MAT.	INST.	TOTAL
02360.60 **Steel Piles** *(Cont.)*				
8"	EA.	44.75	41.25	86.00
10"	"	44.75	52.00	96.75
Standard point				
8"	EA.	58.00	41.25	99.25
10"	"	72.00	52.00	124
Heavy duty point				
8"	EA.	66.00	45.75	112
10"	"	88.00	59.00	147
02360.65 **Steel Pipe Piles**				
Concrete filled, 3000# concrete, up to 40'				
8" dia.	L.F.	19.75	9.48	29.23
10" dia.	"	24.25	9.83	34.08
12" dia.	"	29.50	10.25	39.75
Pipe piles, non-filled				
8" dia.	L.F.	15.50	7.37	22.87
10" dia.	"	18.25	7.58	25.83
12" dia.	"	22.50	7.80	30.30
Splice				
8" dia.	EA.	53.00	83.00	136
10" dia.	"	56.00	83.00	139
12" dia.	"	63.00	100	163
Standard point				
8" dia.	EA.	68.00	83.00	151
10" dia.	"	77.00	83.00	160
12" dia.	"	120	100	220
Heavy duty point				
8" dia.	EA.	85.00	100	185
10" dia.	"	100	100	200
12" dia.	"	140	140	280
02360.70 **Steel Sheet Piling**				
Steel sheet piling, 12" wide				
20' long	S.F.	12.75	13.25	26.00
35' long	"	12.75	9.48	22.23
02360.80 **Wood And Timber Piles**				
Treated wood piles, 12" butt, 8" tip				
25' long	L.F.	8.94	13.25	22.19
30' long	"	9.55	11.00	20.55
35' long	"	9.55	9.48	19.03
40' long	"	9.55	8.29	17.84
02380.10 **Caissons**				
Caisson, including 3000# concrete, in stable ground				
18" dia.	L.F.	15.50	26.50	42.00
24" dia.	"	25.75	27.75	53.50
Wet ground, casing required but pulled				
18" dia.	L.F.	15.50	33.25	48.75
24" dia.	"	25.75	36.75	62.50
Soft rock				
18" dia.	L.F.	15.50	95.00	111
24" dia.	"	25.75	170	196

Paving And Surfacing

	UNIT	MAT.	INST.	TOTAL

02510.20 — Asphalt Surfaces

	UNIT	MAT.	INST.	TOTAL
Asphalt wearing surface, for flexible pavement				
1" thick	S.Y.	4.90	2.21	7.11
1-1/2" thick	"	7.51	2.65	10.16
2" thick	"	9.97	3.31	13.28
3" thick	"	14.75	4.42	19.17
Binder course				
1-1/2" thick	S.Y.	6.43	2.45	8.88
2" thick	"	8.58	3.01	11.59
3" thick	"	12.75	4.02	16.77
4" thick	"	17.25	4.42	21.67
5" thick	"	22.00	4.91	26.91
6" thick	"	26.25	5.53	31.78
Bituminous sidewalk, no base				
2" thick	S.Y.	10.75	2.61	13.36
3" thick	"	16.00	2.78	18.78

02520.10 — Concrete Paving

	UNIT	MAT.	INST.	TOTAL
Concrete paving, reinforced, 5000 psi concrete				
6" thick	S.Y.	25.00	20.75	45.75
7" thick	"	27.50	22.00	49.50
8" thick	"	31.00	23.75	54.75
Concrete paving, for pipe trench, reinforced				
7" thick	S.Y.	51.00	22.25	73.25
8" thick	"	58.00	24.75	82.75

02545.10 — Asphalt Repair

	UNIT	MAT.	INST.	TOTAL
Coal tar emulsion seal coat, rubber additive, fuel resistant	S.Y.	2.33	0.58	2.91
Bituminous surface treatment, single	"	2.14	0.41	2.55
Double	"	2.85	0.04	2.89
Bituminous prime coat	"	1.43	0.05	1.48
Tack coat	"	0.70	0.04	0.74
Crack sealing, concrete paving	L.F.	1.20	0.27	1.47
Bituminous paving for pipe trench, 4" thick	S.Y.	13.75	14.75	28.50
Polypropylene, nonwoven paving fabric	"	1.76	0.20	1.96
Rubberized asphalt	"	2.85	3.75	6.60
Asphalt slurry seal	"	7.02	2.42	9.44

02580.10 — Pavement Markings

	UNIT	MAT.	INST.	TOTAL
Pavement line marking, paint				
4" wide	L.F.	0.11	0.10	0.21
6" wide	"	0.14	0.22	0.36
8" wide	"	0.16	0.34	0.50
Reflective paint, 4" wide	"	0.42	0.34	0.76
Preformed tape, 4" wide				
Inlaid reflective	L.F.	1.91	0.05	1.96
Reflective paint	"	1.30	0.10	1.40
Thermoplastic				
White	L.F.	0.73	0.20	0.93

Paving And Surfacing

	UNIT	MAT.	INST.	TOTAL
02580.10 **Pavement Markings** *(Cont.)*				
Yellow	L.F.	0.73	0.20	0.93
12" wide, thermoplastic, white	"	1.58	0.58	2.16
Directional arrows, reflective preformed tape	EA.	140	41.25	181
Messages, reflective preformed tape (per letter)	"	69.00	20.75	89.75
Handicap symbol, preformed tape	"	120	41.25	161
Parking stall painting	"	1.82	8.25	10.07

Site Improvements

	UNIT	MAT.	INST.	TOTAL
02810.40 **Lawn Irrigation**				
Residential system, complete				
Minimum	ACRE			19,920
Maximum	"			37,920
Commercial system, complete				
Minimum	ACRE			28,850
Maximum	"			47,820
Components				
Pipe				
Schedule 40, PVC				
1/2"	L.F.	0.51	2.17	2.68
3/4"	"	0.74	2.29	3.03
1"	"	1.10	2.35	3.45
Fittings				
Tee				
1/2"	EA.	0.88	6.88	7.76
3/4"	"	1.01	6.88	7.89
1"	"	1.83	6.88	8.71
El				
1/2"	EA.	0.74	6.35	7.09
3/4"	"	0.83	6.88	7.71
1"	"	1.47	6.88	8.35
Coupling				
1/2"	EA.	0.55	6.35	6.90
3/4"	"	0.74	6.88	7.62
1"	"	1.29	6.88	8.17
45 El				
1/2"	EA.	1.10	6.35	7.45
3/4"	"	1.83	6.88	8.71
1"	"	2.57	6.88	9.45
Riser, 1/2" diameter				
2" (close)	EA.	0.63	10.25	10.88
6"	"	1.29	11.75	13.04
3/4" diameter				
2" (close)	EA.	0.63	10.25	10.88
6"	"	1.19	11.75	12.94
1" diameter				

Site Improvements	UNIT	MAT.	INST.	TOTAL
02810.40 Lawn Irrigation *(Cont.)*				
2" (close)	EA.	0.92	10.25	11.17
6"	"	1.79	11.75	13.54
Valve Box				
Concrete, Square				
12" x 22"	EA.	79.00	52.00	131
Round				
12"	EA.	34.75	41.25	76.00
Plastic				
Square				
12"	EA.	36.50	52.00	88.50
Round				
6"	EA.	9.15	52.00	61.15
Sprinkler, Pop-Up				
Spray				
2" high	EA.	5.86	69.00	74.86
6" high	"	14.75	83.00	97.75
Rotor				
4" high	EA.	27.50	69.00	96.50
Impact				
Brass	EA.	36.50	69.00	106
Plastic	"	18.25	83.00	101
Shrub Head				
Spray	EA.	11.00	69.00	80.00
Rotor	"	27.50	83.00	111
Time Clocks				
Minimum	EA.	180	100	280
Average	"	370	140	510
Maximum	"	3,290	210	3,500
Valves				
Anti-siphon				
Plastic				
3/4"	EA.	45.75	69.00	115
1"	"	55.00	69.00	124
Ball Valve				
Plastic				
1/2"	EA.	10.00	69.00	79.00
3/4"	"	11.00	69.00	80.00
1"	"	14.75	69.00	83.75
Brass				
1/2"	EA.	10.00	69.00	79.00
3/4"	"	15.50	69.00	84.50
1"	"	24.75	69.00	93.75
Gate valves, Brass				
1/2"	EA.	22.00	69.00	91.00
3/4"	"	36.50	69.00	106
1"	"	51.00	69.00	120
Vacuum Breakers				
Brass				
3/4"	EA.	150	140	290
1"	"	200	140	340
Plastic				
3/4"	EA.	110	140	250
1"	"	130	140	270

Site Improvements	UNIT	MAT.	INST.	TOTAL
02810.40 — **Lawn Irrigation** *(Cont.)*				
Backflow Preventors, Brass				
3/4"	EA.	370	1,380	1,750
1"	"	410	1,380	1,790
Pressure Regulators, Brass				
3/4"	EA.	93.00	41.25	134
1"	"	130	41.25	171
02830.10 — **Chain Link Fence**				
Chain link fence, 9 ga., galvanized, with posts 10' o.c.				
4' high	L.F.	6.70	2.94	9.64
5' high	"	8.98	3.75	12.73
6' high	"	10.25	5.16	15.41
7' high	"	11.50	6.35	17.85
8' high	"	13.25	8.25	21.50
For barbed wire with hangers, add				
3 strand	L.F.	2.44	2.06	4.50
6 strand	"	4.14	3.44	7.58
Corner or gate post, 3" post				
4' high	EA.	78.00	13.75	91.75
5' high	"	86.00	15.25	101
6' high	"	96.00	18.00	114
7' high	"	120	20.75	141
8' high	"	120	23.00	143
4" post				
4' high	EA.	130	15.25	145
5' high	"	160	18.00	178
6' high	"	170	20.75	191
7' high	"	190	23.00	213
8' high	"	210	25.75	236
Gate with gate posts, galvanized, 3' wide				
4' high	EA.	87.00	100	187
5' high	"	110	140	250
6' high	"	130	140	270
7' high	"	160	210	370
8' high	"	170	210	380
Fabric, galvanized chain link, 2" mesh, 9 ga.				
4' high	L.F.	3.05	1.37	4.42
5' high	"	3.71	1.65	5.36
6' high	"	3.91	2.06	5.97
8' high	"	5.96	2.75	8.71
Line post, no rail fitting, galvanized, 2-1/2" dia.				
4' high	EA.	23.00	11.75	34.75
5' high	"	25.25	13.00	38.25
6' high	"	27.50	13.75	41.25
7' high	"	31.25	16.50	47.75
8' high	"	35.00	20.75	55.75
1-7/8" H beam				
4' high	EA.	31.75	11.75	43.50
5' high	"	35.25	13.00	48.25
6' high	"	42.25	13.75	56.00
7' high	"	48.00	16.50	64.50
8' high	"	52.00	20.75	72.75
2-1/4" H beam				

Site Improvements	UNIT	MAT.	INST.	TOTAL
02830.10 — **Chain Link Fence** *(Cont.)*				
4' high	EA.	23.00	11.75	34.75
5' high	"	28.75	13.00	41.75
6' high	"	33.00	13.75	46.75
7' high	"	38.25	16.50	54.75
8' high	"	44.50	20.75	65.25
Vinyl coated, 9 ga., with posts 10' o.c.				
4' high	L.F.	7.26	2.94	10.20
5' high	"	8.63	3.75	12.38
6' high	"	10.25	5.16	15.41
7' high	"	11.25	6.35	17.60
8' high	"	12.75	8.25	21.00
For barbed wire w/hangers, add				
3 strand	L.F.	2.63	2.06	4.69
6 Strand	"	4.26	3.44	7.70
Corner, or gate post, 4' high				
3" dia.	EA.	96.00	13.75	110
4" dia.	"	150	13.75	164
6" dia.	"	170	16.50	187
Gate, with posts, 3' wide				
4' high	EA.	110	100	210
5' high	"	120	140	260
6' high	"	140	140	280
7' high	"	170	210	380
8' high	"	180	210	390
Line post, no rail fitting, 2-1/2" dia.				
4' high	EA.	40.50	11.75	52.25
5' high	"	55.00	13.00	68.00
6' high	"	66.00	13.75	79.75
7' high	"	78.00	16.50	94.50
8' high	"	85.00	20.75	106
Corner post, no top rail fitting, 4" dia.				
4' high	EA.	160	13.75	174
5' high	"	180	15.25	195
6' high	"	210	18.00	228
7' high	"	230	20.75	251
8' high	"	240	23.00	263
Fabric, vinyl, chain link, 2" mesh, 9 ga.				
4' high	L.F.	5.54	1.37	6.91
5' high	"	6.70	1.65	8.35
6' high	"	7.86	2.06	9.92
8' high	"	10.50	2.75	13.25
Swing gates, galvanized, 4' high				
Single gate				
3' wide	EA.	200	100	300
4' wide	"	220	100	320
Double gate				
10' wide	EA.	520	170	690
12' wide	"	560	170	730
14' wide	"	570	170	740
16' wide	"	640	170	810
18' wide	"	690	240	930
20' wide	"	720	240	960
22' wide	"	810	240	1,050

Site Improvements	UNIT	MAT.	INST.	TOTAL
02830.10 **Chain Link Fence** *(Cont.)*				
24' wide	EA.	830	280	1,110
26' wide	"	860	280	1,140
28' wide	"	920	330	1,250
30' wide	"	980	330	1,310
5' high				
Single gate				
3' wide	EA.	210	140	350
4' wide	"	240	140	380
Double gate				
10' wide	EA.	560	210	770
12' wide	"	600	210	810
14' wide	"	1,020	210	1,230
16' wide	"	670	210	880
18' wide	"	690	240	930
20' wide	"	780	240	1,020
22' wide	"	820	240	1,060
24' wide	"	850	280	1,130
26' wide	"	890	280	1,170
28' wide	"	990	330	1,320
30' wide	"	1,030	330	1,360
6' high				
Single gate				
3' wide	EA.	240	140	380
4' wide	"	260	140	400
Double gate				
10' wide	EA.	580	210	790
12' wide	"	650	210	860
14' wide	"	690	210	900
16' wide	"	740	210	950
18' wide	"	790	240	1,030
20' wide	"	820	240	1,060
22' wide	"	880	240	1,120
24' wide	"	950	280	1,230
26' wide	"	980	280	1,260
28' wide	"	1,060	330	1,390
30' wide	"	1,120	330	1,450
7' high				
Single gate				
3' wide	EA.	270	210	480
4' wide	"	300	210	510
Double gate				
10' wide	EA.	690	280	970
12' wide	"	750	280	1,030
14' wide	"	810	280	1,090
16' wide	"	860	280	1,140
18' wide	"	920	330	1,250
20' wide	"	980	330	1,310
22' wide	"	1,030	330	1,360
24' wide	"	1,110	410	1,520
26' wide	"	1,180	410	1,590
28' wide	"	1,240	520	1,760
30' wide	"	1,420	520	1,940
8' high				

Site Improvements	UNIT	MAT.	INST.	TOTAL
02830.10 **Chain Link Fence** *(Cont.)*				
Single gate				
3' wide	EA.	290	210	500
4' wide	"	310	210	520
Double gate				
10' wide	EA.	750	280	1,030
12' wide	"	800	280	1,080
14' wide	"	860	280	1,140
16' wide	"	940	280	1,220
18' wide	"	980	330	1,310
20' wide	"	1,010	330	1,340
22' wide	"	1,070	330	1,400
24' wide	"	1,190	410	1,600
26' wide	"	1,230	410	1,640
28' wide	"	1,300	520	1,820
30' wide	"	1,440	520	1,960
Vinyl coated swing gates, 4' high				
Single gate				
3' wide	EA.	310	100	410
4' wide	"	330	100	430
Double gate				
10' wide	EA.	720	170	890
12' wide	"	920	170	1,090
14' wide	"	980	170	1,150
16' wide	"	1,090	170	1,260
18' wide	"	1,220	240	1,460
20' wide	"	1,450	240	1,690
22' wide	"	1,610	240	1,850
24' wide	"	1,730	280	2,010
26' wide	"	1,810	280	2,090
28' wide	"	2,110	330	2,440
30' wide	"	2,210	330	2,540
5' high				
Single gate				
3' wide	EA.	330	140	470
4' wide	"	370	140	510
Double gate				
10' wide	EA.	980	210	1,190
12' wide	"	1,050	210	1,260
14' wide	"	1,180	210	1,390
16' wide	"	1,230	210	1,440
18' wide	"	1,480	240	1,720
20' wide	"	1,610	240	1,850
22' wide	"	1,770	240	2,010
24' wide	"	2,010	280	2,290
26' wide	"	2,100	280	2,380
28' wide	"	2,340	330	2,670
30' wide	"	2,520	330	2,850
6' high				
Single gate				
3' wide	EA.	350	140	490
4' wide	"	360	140	500
Double gate				
10' wide	EA.	880	210	1,090

Site Improvements	UNIT	MAT.	INST.	TOTAL
02830.10 — **Chain Link Fence** *(Cont.)*				
12' wide	EA.	1,000	210	1,210
14' wide	"	1,150	210	1,360
16' wide	"	1,340	210	1,550
18' wide	"	1,490	240	1,730
20' wide	"	1,610	240	1,850
22' wide	"	1,750	240	1,990
24' wide	"	1,950	280	2,230
26' wide	"	2,070	280	2,350
28' wide	"	2,450	330	2,780
30' wide	"	2,530	330	2,860
7' high				
Single gate				
3' wide	EA.	390	210	600
4' wide	"	500	210	710
Double gate				
10' wide	EA.	980	280	1,260
12' wide	"	1,170	280	1,450
14' wide	"	1,320	280	1,600
16' wide	"	1,480	280	1,760
18' wide	"	1,670	330	2,000
20' wide	"	1,910	330	2,240
22' wide	"	2,120	330	2,450
24' wide	"	2,330	410	2,740
26' wide	"	2,520	410	2,930
28' wide	"	2,720	520	3,240
30' wide	"	2,920	520	3,440
8' high				
Single gate				
3' wide	EA.	410	210	620
4' wide	"	500	210	710
Double gate				
10' wide	EA.	1,000	280	1,280
12' wide	"	1,170	280	1,450
14' wide	"	1,340	280	1,620
16' wide	"	950	280	1,230
18' wide	"	1,690	330	2,020
20' wide	"	1,960	330	2,290
22' wide	"	2,250	330	2,580
24' wide	"	2,400	410	2,810
28' wide	"	2,790	410	3,200
30' wide	"	3,000	520	3,520
Motor operator for gates, no wiring	"			5,660
Drilling fence post holes				
In soil				
By hand	EA.		20.75	20.75
By machine auger	"		13.25	13.25
In rock				
By jackhammer	EA.		180	180
By rock drill	"		53.00	53.00
Aluminum privacy slats, installed vertically	S.F.	0.96	1.03	1.99
Post hole, dig by hand	EA.		27.50	27.50
Set fence post in concrete	"	9.45	20.75	30.20

Site Improvements	UNIT	MAT.	INST.	TOTAL
02840.30 **Guardrails**				
Pipe bollard, steel pipe, concrete filled, painted				
6" dia.	EA.	200	34.50	235
8" dia.	"	260	52.00	312
12" dia.	"	420	140	560
Corrugated steel, guardrail, galvanized	L.F.	25.25	3.71	28.96
End section, wrap around or flared	EA.	67.00	41.25	108
Timber guardrail, 4" x 8"	L.F.	29.50	2.78	32.28
Guard rail, 3 cables, 3/4" dia.				
Steel posts	L.F.	13.00	11.25	24.25
Wood posts	"	13.25	8.90	22.15
Steel box beam				
6" x 6"	L.F.	54.00	12.25	66.25
6" x 8"	"	57.00	14.00	71.00
Concrete posts	EA.	35.25	20.75	56.00
Barrel type impact barrier	"	450	41.25	491
Light shield, 6' high	L.F.	29.25	8.25	37.50
02840.40 **Parking Barriers**				
Timber, treated, 4' long				
4" x 4"	EA.	17.25	34.50	51.75
6" x 6"	"	32.25	41.25	73.50
Precast concrete, 6' long, with dowels				
12" x 6"	EA.	33.50	20.75	54.25
12" x 8"	"	36.75	23.00	59.75
02870.10 **Prefabricated Planters**				
Concrete precast, circular				
24" dia., 18" high	EA.	300	41.25	341
42" dia., 30" high	"	400	52.00	452
Fiberglass, circular				
36" dia., 27" high	EA.	560	20.75	581
60" dia., 39" high	"	1,300	23.00	1,323
Tapered, circular				
24" dia., 36" high	EA.	440	18.75	459
40" dia., 36" high	"	730	20.75	751
Square				
2' by 2', 17" high	EA.	370	18.75	389
4' by 4', 39" high	"	1,300	23.00	1,323
Rectangular				
4' by 1', 18" high	EA.	420	20.75	441

Formwork

Formwork	UNIT	MAT.	INST.	TOTAL
03110.05 — Beam Formwork				
Beam forms, job built				
Beam bottoms				
1 use	S.F.	4.32	8.85	13.17
3 uses	"	1.94	8.17	10.11
5 uses	"	1.46	7.58	9.04
Beam sides				
1 use	S.F.	3.09	5.90	8.99
3 uses	"	1.60	5.31	6.91
5 uses	"	1.31	4.82	6.13
03110.15 — Column Formwork				
Column, square forms, job built				
8" x 8" columns				
1 use	S.F.	3.39	10.50	13.89
3 uses	"	1.55	9.83	11.38
5 uses	"	1.20	9.15	10.35
12" x 12" columns				
1 use	S.F.	3.09	9.65	12.74
3 uses	"	1.37	9.00	10.37
5 uses	"	1.01	8.43	9.44
16" x 16" columns				
1 use	S.F.	2.95	8.85	11.80
3 uses	"	1.24	8.30	9.54
5 uses	"	0.92	7.81	8.73
3 uses	"	1.12	7.69	8.81
3 uses	"	1.16	7.17	8.33
Round fiber forms, 1 use				
10" dia.	L.F.	4.08	10.50	14.58
12" dia.	"	5.02	10.75	15.77
18" dia.	"	14.25	12.75	27.00
24" dia.	"	17.25	13.50	30.75
36" dia.	"	32.25	16.00	48.25
03110.18 — Curb Formwork				
Curb forms				
Straight, 6" high				
1 use	L.F.	1.98	5.31	7.29
3 uses	"	0.88	4.82	5.70
5 uses	"	0.72	4.42	5.14
Curved, 6" high				
1 use	L.F.	2.13	6.64	8.77
3 uses	"	1.02	5.90	6.92
5 uses	"	0.88	5.42	6.30
03110.20 — Elevated Slab Formwork				
Elevated slab formwork				
Slab, with drop panels				
1 use	S.F.	3.63	4.24	7.87
3 uses	"	1.62	3.93	5.55
5 uses	"	1.29	3.66	4.95
Floor slab, hung from steel beams				
1 use	S.F.	2.92	4.08	7.00
3 uses	"	1.45	3.79	5.24

Formwork	UNIT	MAT.	INST.	TOTAL
03110.20 — **Elevated Slab Formwork** *(Cont.)*				
5 uses	S.F.	1.08	3.54	4.62
Floor slab, with pans or domes				
1 use	S.F.	5.20	4.82	10.02
3 uses	"	3.14	4.42	7.56
5 uses	"	2.61	4.08	6.69
Equipment curbs, 12" high				
1 use	L.F.	2.64	5.31	7.95
3 uses	"	1.47	4.82	6.29
5 uses	"	1.11	4.42	5.53
03110.25 — **Equipment Pad Formwork**				
Equipment pad, job built				
1 use	S.F.	3.46	6.64	10.10
2 uses	"	2.07	6.24	8.31
3 uses	"	1.66	5.90	7.56
4 uses	"	1.29	5.59	6.88
5 uses	"	1.05	5.31	6.36
03110.35 — **Footing Formwork**				
Wall footings, job built, continuous				
1 use	S.F.	1.62	5.31	6.93
3 uses	"	0.94	4.82	5.76
5 uses	"	0.72	4.42	5.14
Column footings, spread				
1 use	S.F.	1.71	6.64	8.35
3 uses	"	0.91	5.90	6.81
5 uses	"	0.69	5.31	6.00
03110.50 — **Grade Beam Formwork**				
Grade beams, job built				
1 use	S.F.	2.55	5.31	7.86
3 uses	"	1.12	4.82	5.94
5 uses	"	0.77	4.42	5.19
03110.53 — **Pile Cap Formwork**				
Pile cap forms, job built				
Square				
1 use	S.F.	2.91	6.64	9.55
3 uses	"	1.33	5.90	7.23
5 uses	"	0.97	5.31	6.28
Triangular				
1 use	S.F.	3.09	7.58	10.67
3 uses	"	1.62	6.64	8.26
5 uses	"	1.06	5.90	6.96
03110.55 — **Slab/mat Formwork**				
Mat foundations, job built				
1 use	S.F.	2.54	6.64	9.18
3 uses	"	1.07	5.90	6.97
5 uses	"	0.72	5.31	6.03
Edge forms				
6" high				

Formwork	UNIT	MAT.	INST.	TOTAL
03110.55 — **Slab/mat Formwork** *(Cont.)*				
1 use	L.F.	2.54	4.82	7.36
3 uses	"	1.07	4.42	5.49
5 uses	"	0.72	4.08	4.80
12" high				
1 use	L.F.	2.38	5.31	7.69
3 uses	"	1.00	4.82	5.82
5 uses	"	0.68	4.42	5.10
Formwork for openings				
1 use	S.F.	3.47	10.50	13.97
3 uses	"	1.65	8.85	10.50
5 uses	"	1.06	7.58	8.64
03110.60 — **Stair Formwork**				
Stairway forms, job built				
1 use	S.F.	4.33	10.50	14.83
3 uses	"	1.88	8.85	10.73
5 uses	"	1.45	7.58	9.03
Stairs, elevated				
1 use	S.F.	5.22	10.50	15.72
3 uses	"	2.42	7.58	10.00
5 uses	"	1.73	6.64	8.37
03110.65 — **Wall Formwork**				
Wall forms, exterior, job built				
Up to 8' high wall				
1 use	S.F.	2.81	5.31	8.12
3 uses	"	1.37	4.82	6.19
5 uses	"	1.02	4.42	5.44
Over 8' high wall				
1 use	S.F.	3.09	6.64	9.73
3 uses	"	1.60	5.90	7.50
5 uses	"	1.27	5.31	6.58
Column pier and pilaster				
1 use	S.F.	3.09	10.50	13.59
3 uses	"	1.70	8.85	10.55
5 uses	"	1.39	7.58	8.97
Interior wall forms				
Up to 8' high				
1 use	S.F.	2.81	4.82	7.63
3 uses	"	1.39	4.42	5.81
5 uses	"	1.01	4.08	5.09
Over 8' high				
1 use	S.F.	3.09	5.90	8.99
3 uses	"	1.60	5.31	6.91
5 uses	"	1.28	4.82	6.10
PVC form liner, per side, smooth finish				
1 use	S.F.	6.46	4.42	10.88
3 uses	"	2.95	4.08	7.03
5 uses	"	1.84	3.54	5.38

Formwork	UNIT	MAT.	INST.	TOTAL
03110.90 **Miscellaneous Formwork**				
Keyway forms (5 uses)				
2 x 4	L.F.	0.18	2.65	2.83
2 x 6	"	0.27	2.95	3.22
Bulkheads				
Walls, with keyways				
2 piece	L.F.	3.16	4.82	7.98
3 piece	"	3.90	5.31	9.21
Elevated slab, with keyway				
2 piece	L.F.	3.57	4.42	7.99
3 piece	"	5.19	4.82	10.01
Ground slab, with keyway				
2 piece	L.F.	3.90	3.79	7.69
3 piece	"	5.19	4.08	9.27
Chamfer strips				
Wood				
1/2" wide	L.F.	0.23	1.18	1.41
3/4" wide	"	0.30	1.18	1.48
1" wide	"	0.40	1.18	1.58
PVC				
1/2" wide	L.F.	0.86	1.18	2.04
3/4" wide	"	0.95	1.18	2.13
1" wide	"	1.23	1.18	2.41
Radius				
1"	L.F.	1.04	1.26	2.30
1-1/2"	"	1.85	1.26	3.11
Reglets				
Galvanized steel, 24 ga.	L.F.	1.29	2.12	3.41
Metal formwork				
Straight edge forms				
4" high	L.F.	0.18	3.32	3.50
6" high	"	0.20	3.54	3.74
8" high	"	0.23	3.79	4.02
12" high	"	0.29	4.08	4.37
16" high	"	0.36	4.42	4.78
Curb form, S-shape				
12" x				
1'-6"	L.F.	0.36	7.58	7.94
2'	"	0.43	7.08	7.51
2'-6"	"	0.50	6.64	7.14
3'	"	0.56	5.90	6.46

Reinforcement	UNIT	MAT.	INST.	TOTAL
03210.05 — **Beam Reinforcing**				
Beam-girders				
#3 - #4	TON	1,220	1,480	2,700
#5 - #6	"	1,070	1,180	2,250
#7 - #8	"	1,020	980	2,000
Galvanized				
#3 - #4	TON	2,080	1,480	3,560
#5 - #6	"	1,960	1,180	3,140
#7 - #8	"	1,890	980	2,870
Bond Beams				
#3 - #4	TON	1,220	1,970	3,190
#5 - #6	"	1,070	1,480	2,550
#7 - #8	"	1,020	1,310	2,330
Galvanized				
#3 - #4	TON	1,990	1,970	3,960
#5 - #6	"	1,960	1,480	3,440
#7 - #8	"	1,890	1,310	3,200
03210.15 — **Column Reinforcing**				
Columns				
#3 - #4	TON	1,220	1,690	2,910
#5 - #6	"	1,070	1,310	2,380
#7 - #8	"	1,020	1,180	2,200
Galvanized				
#3 - #4	TON	2,080	1,690	3,770
#5 - #6	"	1,960	1,310	3,270
#7 - #8	"	1,890	1,180	3,070
03210.20 — **Elevated Slab Reinforcing**				
Elevated slab				
#3 - #4	TON	1,220	740	1,960
#5 - #6	"	1,070	660	1,730
#7 - #8	"	1,020	590	1,610
Galvanized				
#3 - #4	TON	1,990	740	2,730
#5 - #6	"	1,960	660	2,620
03210.25 — **Equip. Pad Reinforcing**				
Equipment pad				
#3 - #4	TON	1,220	1,180	2,400
#5 - #6	"	1,070	1,070	2,140
#7 - #8	"	1,020	980	2,000
#9 - #10	"	1,020	910	1,930
#11 - #12	"	1,020	840	1,860
03210.35 — **Footing Reinforcing**				
Footings				
Grade 50				
#3 - #4	TON	1,220	980	2,200
#5 - #6	"	1,070	840	1,910
#7 - #8	"	1,020	740	1,760
#9 - #10	"	1,020	660	1,680
Grade 60				
#3 - #4	TON	1,220	980	2,200

Reinforcement

	UNIT	MAT.	INST.	TOTAL
03210.35 **Footing Reinforcing** *(Cont.)*				
#5 - #6	TON	1,070	840	1,910
#7 - #8	"	1,020	740	1,760
Straight dowels, 24" long				
1" dia. (#8)	EA.	3.79	5.90	9.69
3/4" dia. (#6)	"	3.42	5.90	9.32
5/8" dia. (#5)	"	2.95	4.92	7.87
1/2" dia. (#4)	"	2.22	4.22	6.44
03210.45 **Foundation Reinforcing**				
Foundations				
#3 - #4	TON	1,220	980	2,200
#5 - #6	"	1,070	840	1,910
#7 - #8	"	1,020	740	1,760
Galvanized				
#3 - #4	TON	2,080	980	3,060
#5 - #6	"	1,960	840	2,800
#7 - #8	"	1,890	740	2,630
03210.50 **Grade Beam Reinforcing**				
Grade beams				
#3 - #4	TON	1,220	910	2,130
#5 - #6	"	1,070	790	1,860
#7 - #8	"	1,020	700	1,720
Galvanized				
#3 - #4	TON	2,080	910	2,990
#5 - #6	"	1,960	790	2,750
#7 - #8	"	1,890	700	2,590
03210.53 **Pile Cap Reinforcing**				
Pile caps				
#3 - #4	TON	1,220	1,480	2,700
#5 - #6	"	1,070	1,310	2,380
#7 - #8	"	1,020	1,180	2,200
Galvanized				
#3 - #4	TON	2,080	1,480	3,560
#5 - #6	"	1,960	1,310	3,270
#7 - #8	"	1,890	1,180	3,070
03210.55 **Slab/mat Reinforcing**				
Bars, slabs				
#3 - #4	TON	1,220	980	2,200
#5 - #6	"	1,070	840	1,910
#7 - #8	"	1,020	740	1,760
Galvanized				
#3 - #4	TON	2,080	980	3,060
#5 - #6	"	1,960	840	2,800
#7 - #8	"	1,890	740	2,630
Wire mesh, slabs				
Galvanized				
4x4				
W1.4xW1.4	S.F.	0.30	0.39	0.69
W2.0xW2.0	"	0.39	0.42	0.81
W2.9xW2.9	"	0.55	0.45	1.00

Reinforcement	UNIT	MAT.	INST.	TOTAL
03210.55 **Slab/mat Reinforcing** *(Cont.)*				
W4.0xW4.0	S.F.	0.81	0.49	1.30
6x6				
W1.4xW1.4	S.F.	0.28	0.29	0.57
W2.0xW2.0	"	0.39	0.32	0.71
W2.9xW2.9	"	0.53	0.34	0.87
W4.0xW4.0	"	0.57	0.39	0.96
Standard				
2x2				
W.9xW.9	S.F.	0.30	0.39	0.69
4x4				
W1.4xW1.4	S.F.	0.19	0.39	0.58
W4.0xW4.0	"	0.55	0.49	1.04
6x6				
W1.4xW1.4	S.F.	0.12	0.29	0.41
W4.0xW4.0	"	0.37	0.39	0.76
03210.60 **Stair Reinforcing**				
Stairs				
#3 - #4	TON	1,220	1,180	2,400
#5 - #6	"	1,070	980	2,050
Galvanized				
#3 - #4	TON	2,080	1,180	3,260
#5 - #6	"	1,960	980	2,940
03210.65 **Wall Reinforcing**				
Walls				
#3 - #4	TON	1,220	840	2,060
#5 - #6	"	1,070	740	1,810
#7 - #8	"	1,020	660	1,680
Galvanized				
#3 - #4	TON	2,080	840	2,920
#5 - #6	"	1,960	740	2,700
#7 - #8	"	1,890	660	2,550
Masonry wall (horizontal)				
#3 - #4	TON	1,220	2,360	3,580
#5 - #6	"	1,070	1,970	3,040
Galvanized				
#3 - #4	TON	2,080	2,360	4,440
#5 - #6	"	1,960	1,970	3,930
Masonry wall (vertical)				
#3 - #4	TON	1,220	2,950	4,170
#5 - #6	"	1,070	2,360	3,430
Galvanized				
#3 - #4	TON	2,080	2,950	5,030
#5 - #6	"	1,960	2,360	4,320

Accessories	UNIT	MAT.	INST.	TOTAL
03250.40 — **Concrete Accessories**				
Expansion joint, poured				
Asphalt				
1/2" x 1"	L.F.	0.77	0.82	1.59
1" x 2"	"	2.36	0.89	3.25
Liquid neoprene, cold applied				
1/2" x 1"	L.F.	2.66	0.84	3.50
1" x 2"	"	11.00	0.91	11.91
Polyurethane, 2 parts				
1/2" x 1"	L.F.	2.49	1.37	3.86
1" x 2"	"	10.25	1.50	11.75
Rubberized asphalt, cold				
1/2" x 1"	L.F.	0.61	0.82	1.43
1" x 2"	"	1.91	0.89	2.80
Hot, fuel resistant				
1/2" x 1"	L.F.	1.15	0.82	1.97
1" x 2"	"	5.67	0.89	6.56
Expansion joint, premolded, in slabs				
Asphalt				
1/2" x 6"	L.F.	0.81	1.03	1.84
1" x 12"	"	1.37	1.37	2.74
Cork				
1/2" x 6"	L.F.	1.63	1.03	2.66
1" x 12"	"	6.30	1.37	7.67
Neoprene sponge				
1/2" x 6"	L.F.	2.49	1.03	3.52
1" x 12"	"	9.02	1.37	10.39
Polyethylene foam				
1/2" x 6"	L.F.	0.93	1.03	1.96
1" x 12"	"	4.49	1.37	5.86
Polyurethane foam				
1/2" x 6"	L.F.	1.23	1.03	2.26
1" x 12"	"	2.75	1.37	4.12
Polyvinyl chloride foam				
1/2" x 6"	L.F.	2.65	1.03	3.68
1" x 12"	"	9.64	1.37	11.01
Rubber, gray sponge				
1/2" x 6"	L.F.	4.29	1.03	5.32
1" x 12"	"	18.25	1.37	19.62
Asphalt felt control joints or bond breaker, screed joints				
4" slab	L.F.	1.15	0.82	1.97
6" slab	"	1.47	0.91	2.38
8" slab	"	1.97	1.03	3.00
10" slab	"	2.75	1.17	3.92
Keyed cold expansion and control joints, 24 ga.				
4" slab	L.F.	1.01	2.58	3.59
5" slab	"	1.27	2.58	3.85
6" slab	"	1.47	2.75	4.22
8" slab	"	1.83	2.94	4.77
10" slab	"	2.07	3.17	5.24
Waterstops				
Polyvinyl chloride				
Ribbed				
3/16" thick x				

Accessories

	UNIT	MAT.	INST.	TOTAL
03250.40 **Concrete Accessories** *(Cont.)*				
4" wide	L.F.	1.63	2.06	3.69
6" wide	"	2.00	2.29	4.29
1/2" thick x				
9" wide	L.F.	5.50	2.58	8.08
Ribbed with center bulb				
3/16" thick x 9" wide	L.F.	4.66	2.58	7.24
3/8" thick x 9" wide	"	5.73	2.58	8.31
Dumbbell type, 3/8" thick x 6" wide	"	5.73	2.29	8.02
Plain, 3/8" thick x 9" wide	"	7.06	2.58	9.64
Center bulb, 3/8" thick x 9" wide	"	9.48	2.58	12.06
Rubber				
Vapor barrier				
4 mil polyethylene	S.F.	0.03	0.13	0.16
6 mil polyethylene	"	0.05	0.13	0.18
Gravel porous fill, under floor slabs, 3/4" stone	C.Y.	22.50	69.00	91.50

Cast-in-place Concrete

	UNIT	MAT.	INST.	TOTAL
03300.10 **Concrete Admixtures**				
Concrete admixtures				
Water reducing admixture	GAL			17.00
Set retarder	"			30.25
Air entraining agent	"			10.25
03350.10 **Concrete Finishes**				
Floor finishes				
Broom	S.F.		0.58	0.58
Screed	"		0.51	0.51
Darby	"		0.51	0.51
Steel float	"		0.68	0.68
Wall finishes				
Burlap rub, with cement paste	S.F.	0.05	0.68	0.73
Break ties and patch holes	"		0.82	0.82
Carborundum				
Dry rub	S.F.		1.37	1.37
Wet rub	"		2.06	2.06
Floor hardeners				
Metallic				
Light service	S.F.	0.28	0.51	0.79
Heavy service	"	0.92	0.68	1.60
Non-metallic				
Light service	S.F.	0.15	0.51	0.66
Heavy service	"	0.67	0.68	1.35

Cast-in-place Concrete	UNIT	MAT.	INST.	TOTAL
03360.10 **Pneumatic Concrete**				
Pneumatic applied concrete (gunite)				
2" thick	S.F.	4.76	2.78	7.54
3" thick	"	5.91	3.71	9.62
4" thick	"	7.06	4.45	11.51
Finish surface				
Minimum	S.F.		2.65	2.65
Maximum	"		5.31	5.31
03370.10 **Curing Concrete**				
Sprayed membrane				
Slabs	S.F.	0.05	0.08	0.13
Walls	"	0.07	0.10	0.17
Curing paper				
Slabs	S.F.	0.07	0.10	0.17
Walls	"	0.07	0.12	0.19
Burlap				
7.5 oz.	S.F.	0.06	0.13	0.19
12 oz.	"	0.08	0.14	0.22

Placing Concrete	UNIT	MAT.	INST.	TOTAL
03380.05 **Beam Concrete**				
Beams and girders				
2500# or 3000# concrete				
By pump	C.Y.	98.00	73.00	171
By hand buggy	"	98.00	41.25	139
5000# concrete				
By pump	C.Y.	110	73.00	183
By hand buggy	"	110	41.25	151
Bond beam, 3000# concrete				
By pump				
8" high				
4" wide	L.F.	0.26	1.61	1.87
6" wide	"	0.65	1.83	2.48
8" wide	"	0.83	2.01	2.84
10" wide	"	1.10	2.23	3.33
12" wide	"	1.48	2.51	3.99
03380.15 **Column Concrete**				
Columns				
2500# or 3000# concrete				
By pump	C.Y.	98.00	67.00	165
5000# concrete				
By pump	C.Y.	110	67.00	177

Placing Concrete	UNIT	MAT.	INST.	TOTAL
03380.25 **Equipment Pad Concrete**				
Equipment pad				
2500# or 3000# concrete				
By chute	C.Y.	98.00	13.75	112
By pump	"	98.00	58.00	156
3500# or 4000# concrete				
By chute	C.Y.	100	13.75	114
By pump	"	100	58.00	158
5000# concrete				
By chute	C.Y.	110	13.75	124
By pump	"	110	58.00	168
03380.35 **Footing Concrete**				
Continuous footing				
2500# or 3000# concrete				
By chute	C.Y.	98.00	13.75	112
By pump	"	98.00	50.00	148
5000# concrete				
By chute	C.Y.	110	13.75	124
By pump	"	110	50.00	160
Spread footing				
2500# or 3000# concrete				
Under 5 cy				
By chute	C.Y.	98.00	13.75	112
By pump	"	98.00	54.00	152
5000# concrete				
Under 5 c.y.				
By chute	C.Y.	110	13.75	124
By pump	"	110	54.00	164
03380.50 **Grade Beam Concrete**				
Grade beam				
2500# or 3000# concrete				
By chute	C.Y.	98.00	13.75	112
By pump	"	98.00	50.00	148
By hand buggy	"	98.00	41.25	139
5000# concrete				
By chute	C.Y.	110	13.75	124
By pump	"	110	50.00	160
By hand buggy	"	110	41.25	151
03380.53 **Pile Cap Concrete**				
Pile cap				
2500# or 3000 concrete				
By chute	C.Y.	98.00	13.75	112
By pump	"	98.00	58.00	156
By hand buggy	"	98.00	41.25	139
5000# concrete				
By chute	C.Y.	110	13.75	124
By pump	"	110	58.00	168
By hand buggy	"	110	41.25	151

Placing Concrete	UNIT	MAT.	INST.	TOTAL
03380.55 **Slab/mat Concrete**				
Slab on grade				
2500# or 3000# concrete				
By chute	C.Y.	98.00	10.25	108
By pump	"	98.00	28.75	127
By hand buggy	"	98.00	27.50	126
5000# concrete				
By chute	C.Y.	110	10.25	120
By pump	"	110	28.75	139
By hand buggy	"	110	27.50	138
03380.58 **Sidewalks**				
Walks, cast in place with wire mesh, base not incl.				
4" thick	S.F.	1.34	1.37	2.71
5" thick	"	1.80	1.65	3.45
6" thick	"	2.26	2.06	4.32
03380.60 **Stair Concrete**				
Stairs				
2500# or 3000# concrete				
By chute	C.Y.	98.00	13.75	112
By pump	"	98.00	58.00	156
By hand buggy	"	98.00	41.25	139
3500# or 4000# concrete				
By chute	C.Y.	100	13.75	114
By pump	"	100	58.00	158
By hand buggy	"	100	41.25	141
5000# concrete				
By chute	C.Y.	110	13.75	124
By pump	"	110	58.00	168
By hand buggy	"	110	41.25	151
03380.65 **Wall Concrete**				
Walls				
2500# or 3000# concrete				
To 4'				
By chute	C.Y.	98.00	11.75	110
By pump	"	98.00	62.00	160
To 8'				
By pump	C.Y.	98.00	67.00	165
3500# or 4000# concrete				
To 4'				
By chute	C.Y.	100	11.75	112
By pump	"	100	62.00	162
To 8'				
By pump	C.Y.	100	67.00	167
Filled block (CMU)				
3000# concrete, by pump				
4" wide	S.F.	0.36	2.87	3.23

Placing Concrete

	UNIT	MAT.	INST.	TOTAL
03380.65 **Wall Concrete** *(Cont.)*				
6" wide	S.F.	0.85	3.35	4.20
8" wide	"	1.31	4.02	5.33
10" wide	"	1.75	4.73	6.48
12" wide	"	2.23	5.75	7.98
Pilasters, 3000# concrete	C.F.	5.14	81.00	86.14
Wall cavity, 2" thick, 3000# concrete	S.F.	0.94	2.68	3.62

Cementitous Toppings

	UNIT	MAT.	INST.	TOTAL
03550.10 **Concrete Toppings**				
Gypsum fill				
2" thick	S.F.	1.61	0.41	2.02
2-1/2" thick	"	1.84	0.42	2.26
3" thick	"	2.25	0.43	2.68
3-1/2" thick	"	2.58	0.44	3.02
4" thick	"	3.00	0.50	3.50
Formboard				
Mineral fiber board				
1" thick	S.F.	1.44	1.03	2.47
1-1/2" thick	"	3.78	1.17	4.95
Cement fiber board				
1" thick	S.F.	1.12	1.37	2.49
1-1/2" thick	"	1.44	1.58	3.02
Glass fiber board				
1" thick	S.F.	1.77	1.03	2.80
1-1/2" thick	"	2.39	1.17	3.56
Poured deck				
Vermiculite or perlite				
1 to 4 mix	C.Y.	150	67.00	217
1 to 6 mix	"	140	62.00	202
Vermiculite or perlite				
2" thick				
1 to 4 mix	S.F.	1.44	0.42	1.86
1 to 6 mix	"	1.05	0.38	1.43
3" thick				
1 to 4 mix	S.F.	1.96	0.61	2.57
1 to 6 mix	"	1.55	0.57	2.12
Concrete plank, lightweight				
2" thick	S.F.	7.76	3.31	11.07
2-1/2" thick	"	7.95	3.31	11.26
3-1/2" thick	"	8.28	3.68	11.96
4" thick	"	8.61	3.68	12.29
Channel slab, lightweight, straight				
2-3/4" thick	S.F.	6.26	3.31	9.57
3-1/2" thick	"	6.53	3.31	9.84
3-3/4" thick	"	6.93	3.31	10.24

Cementitous Toppings	UNIT	MAT.	INST.	TOTAL
03550.10 — **Concrete Toppings** *(Cont.)*				
4-3/4" thick	S.F.	8.79	3.68	12.47
Gypsum plank				
2" thick	S.F.	2.92	3.31	6.23
3" thick	"	3.05	3.31	6.36
Cement fiber, T and G planks				
1" thick	S.F.	1.63	3.01	4.64
1-1/2" thick	"	1.69	3.01	4.70
2" thick	"	1.98	3.31	5.29
2-1/2" thick	"	2.08	3.31	5.39
3" thick	"	2.73	3.31	6.04
3-1/2" thick	"	3.12	3.68	6.80
4" thick	"	3.54	3.68	7.22

Concrete Restoration	UNIT	MAT.	INST.	TOTAL
03730.10 — **Concrete Repair**				
Epoxy grout floor patch, 1/4" thick	S.F.	6.06	4.12	10.18
Grout, epoxy, 2 component system	C.F.			290
Epoxy sand	BAG			19.75
Epoxy modifier	GAL			130
Epoxy gel grout	S.F.	2.95	41.25	44.20
Injection valve, 1 way, threaded plastic	EA.	8.11	8.25	16.36
Grout crack seal, 2 component	C.F.	680	41.25	721
Grout, non shrink	"	70.00	41.25	111
Concrete, epoxy modified				
Sand mix	C.F.	110	16.50	127
Gravel mix	"	85.00	15.25	100
Concrete repair				
Soffit repair				
16" wide	L.F.	3.45	8.25	11.70
18" wide	"	3.62	8.60	12.22
24" wide	"	4.26	9.17	13.43
30" wide	"	4.89	9.82	14.71
32" wide	"	5.25	10.25	15.50
Edge repair				
2" spall	L.F.	1.62	10.25	11.87
3" spall	"	1.62	10.75	12.37
4" spall	"	1.72	11.25	12.97
6" spall	"	1.81	11.50	13.31
8" spall	"	1.89	12.25	14.14
9" spall	"	1.95	13.75	15.70
Crack repair, 1/8" crack	"	3.20	4.12	7.32
Reinforcing steel repair				
1 bar, 4 ft				
#4 bar	L.F.	0.48	7.38	7.86
#5 bar	"	0.67	7.38	8.05

Concrete Restoration	UNIT	MAT.	INST.	TOTAL
03730.10 Concrete Repair *(Cont.)*				
#6 bar	L.F.	0.83	7.87	8.70
#8 bar	"	1.49	7.87	9.36
#9 bar	"	1.90	8.44	10.34
#11 bar	"	2.96	8.44	11.40
Form fabric, nylon				
18" diameter	L.F.			12.75
20" diameter	"			13.00
24" diameter	"			21.25
30" diameter	"			21.75
36" diameter	"			25.00
Pile repairs				
Polyethylene wrap				
30 mil thick				
60" wide	S.F.	13.25	13.75	27.00
72" wide	"	14.25	16.50	30.75
60 mil thick				
60" wide	S.F.	15.75	13.75	29.50
80" wide	"	18.25	18.75	37.00
Pile spall, average repair 3'				
18" x 18"	EA.	41.00	34.50	75.50
20" x 20"	"	55.00	41.25	96.25

Mortar And Grout	UNIT	MAT.	INST.	TOTAL

04100.10 — Masonry Grout

	UNIT	MAT.	INST.	TOTAL
Grout, non shrink, non-metallic, trowelable	C.F.	4.95	1.48	6.43
Grout door frame, hollow metal				
Single	EA.	12.25	56.00	68.25
Double	"	17.25	59.00	76.25
Grout-filled concrete block (CMU)				
4" wide	S.F.	0.33	1.85	2.18
6" wide	"	0.86	2.02	2.88
8" wide	"	1.27	2.22	3.49
12" wide	"	2.09	2.34	4.43
Grout-filled individual CMU cells				
4" wide	L.F.	0.27	1.11	1.38
6" wide	"	0.37	1.11	1.48
8" wide	"	0.49	1.11	1.60
10" wide	"	0.61	1.27	1.88
12" wide	"	0.74	1.27	2.01
Bond beams or lintels, 8" deep				
6" thick	L.F.	0.74	1.83	2.57
8" thick	"	0.99	2.01	3.00
10" thick	"	1.24	2.23	3.47
12" thick	"	1.48	2.51	3.99
Cavity walls				
2" thick	S.F.	0.82	2.68	3.50
3" thick	"	1.24	2.68	3.92
4" thick	"	1.65	2.87	4.52
6" thick	"	2.47	3.35	5.82

04150.10 — Masonry Accessories

	UNIT	MAT.	INST.	TOTAL
Foundation vents	EA.	31.00	20.50	51.50
Bar reinforcing				
Horizontal				
#3 - #4	Lb.	0.56	2.05	2.61
#5 - #6	"	0.56	1.71	2.27
Vertical				
#3 - #4	Lb.	0.56	2.57	3.13
#5 - #6	"	0.56	2.05	2.61
Horizontal joint reinforcing				
Truss type				
4" wide, 6" wall	L.F.	0.20	0.20	0.40
6" wide, 8" wall	"	0.20	0.21	0.41
8" wide, 10" wall	"	0.25	0.22	0.47
10" wide, 12" wall	"	0.25	0.23	0.48
12" wide, 14" wall	"	0.30	0.24	0.54
Ladder type				
4" wide, 6" wall	L.F.	0.14	0.20	0.34
6" wide, 8" wall	"	0.16	0.21	0.37
8" wide, 10" wall	"	0.17	0.22	0.39
10" wide, 12" wall	"	0.20	0.22	0.42
Rectangular wall ties				
3/16" dia., galvanized				
2" x 6"	EA.	0.22	0.85	1.07
2" x 8"	"	0.23	0.85	1.08
2" x 10"	"	0.26	0.85	1.11
2" x 12"	"	0.30	0.85	1.15

Mortar And Grout

04150.10	Masonry Accessories *(Cont.)*	UNIT	MAT.	INST.	TOTAL
4" x 6"		EA.	0.25	1.02	1.27
4" x 8"		"	0.27	1.02	1.29
4" x 10"		"	0.36	1.02	1.38
4" x 12"		"	0.42	1.02	1.44
1/4" dia., galvanized					
2" x 6"		EA.	0.41	0.85	1.26
2" x 8"		"	0.45	0.85	1.30
2" x 10"		"	0.51	0.85	1.36
2" x 12"		"	0.58	0.85	1.43
4" x 6"		"	0.47	1.02	1.49
4" x 8"		"	0.51	1.02	1.53
4" x 10"		"	0.58	1.02	1.60
4" x 12"		"	0.61	1.02	1.63
"Z" type wall ties, galvanized					
6" long					
1/8" dia.		EA.	0.22	0.85	1.07
3/16" dia.		"	0.23	0.85	1.08
1/4" dia.		"	0.25	0.85	1.10
8" long					
1/8" dia.		EA.	0.23	0.85	1.08
3/16" dia.		"	0.25	0.85	1.10
1/4" dia.		"	0.26	0.85	1.11
10" long					
1/8" dia.		EA.	0.25	0.85	1.10
3/16" dia.		"	0.27	0.85	1.12
1/4" dia.		"	0.32	0.85	1.17
Dovetail anchor slots					
Galvanized steel, filled					
24 ga.		L.F.	0.58	1.28	1.86
20 ga.		"	0.73	1.28	2.01
16 oz. copper, foam filled		"	1.45	1.28	2.73
Dovetail anchors					
16 ga.					
3-1/2" long		EA.	0.17	0.85	1.02
5-1/2" long		"	0.20	0.85	1.05
12 ga.					
3-1/2" long		EA.	0.22	0.85	1.07
5-1/2" long		"	0.44	0.85	1.29
Dovetail, triangular galvanized ties, 12 ga.					
3" x 3"		EA.	0.40	0.85	1.25
5" x 5"		"	0.43	0.85	1.28
7" x 7"		"	0.49	0.85	1.34
7" x 9"		"	0.51	0.85	1.36
Brick anchors					
Corrugated, 3-1/2" long					
16 ga.		EA.	0.15	0.85	1.00
12 ga.		"	0.27	0.85	1.12
Non-corrugated, 3-1/2" long					
16 ga.		EA.	0.21	0.85	1.06
12 ga.		"	0.38	0.85	1.23
Cavity wall anchors, corrugated, galvanized					
5" long					
16 ga.		EA.	0.49	0.85	1.34

Mortar And Grout	UNIT	MAT.	INST.	TOTAL
04150.10 **Masonry Accessories** *(Cont.)*				
12 ga.	EA.	0.72	0.85	1.57
7" long				
28 ga.	EA.	0.53	0.85	1.38
24 ga.	"	0.67	0.85	1.52
22 ga.	"	0.69	0.85	1.54
16 ga.	"	0.77	0.85	1.62
Mesh ties, 16 ga., 3" wide				
8" long	EA.	0.64	0.85	1.49
12" long	"	0.72	0.85	1.57
20" long	"	0.99	0.85	1.84
24" long	"	1.09	0.85	1.94
04150.20 **Masonry Control Joints**				
Control joint, cross shaped PVC	L.F.	2.17	1.28	3.45
Closed cell joint filler				
1/2"	L.F.	0.38	1.28	1.66
3/4"	"	0.77	1.28	2.05
Rubber, for				
4" wall	L.F.	2.50	1.28	3.78
6" wall	"	3.10	1.35	4.45
8" wall	"	3.74	1.42	5.16
PVC, for				
4" wall	L.F.	1.30	1.28	2.58
6" wall	"	2.20	1.35	3.55
8" wall	"	3.33	1.42	4.75
04150.50 **Masonry Flashing**				
Through-wall flashing				
5 oz. coated copper	S.F.	3.35	4.28	7.63
0.030" elastomeric	"	1.10	3.42	4.52

Unit Masonry	UNIT	MAT.	INST.	TOTAL
04210.10 **Brick Masonry**				
Standard size brick, running bond				
Face brick, red (6.4/sf)				
Veneer	S.F.	5.39	8.56	13.95
Cavity wall	"	5.39	7.34	12.73
9" solid wall	"	10.75	14.75	25.50
Common brick (6.4/sf)				
Select common for veneers	S.F.	3.74	8.56	12.30
Back-up				
4" thick	S.F.	3.41	6.42	9.83
8" thick	"	6.82	10.25	17.07
Firewall				
12" thick	S.F.	10.25	17.25	27.50

Unit Masonry	UNIT	MAT.	INST.	TOTAL
04210.10 **Brick Masonry** *(Cont.)*				
16" thick	S.F.	13.75	23.25	37.00
Glazed brick (7.4/sf)				
Veneer	S.F.	12.00	9.34	21.34
Buff or gray face brick (6.4/sf)				
Veneer	S.F.	5.98	8.56	14.54
Cavity wall	"	5.98	7.34	13.32
Jumbo or oversize brick (3/sf)				
4" veneer	S.F.	4.12	5.14	9.26
4" back-up	"	4.12	4.28	8.40
8" back-up	"	8.25	7.34	15.59
12" firewall	"	12.25	12.75	25.00
16" firewall	"	16.50	17.25	33.75
Norman brick, red face, (4.5/sf)				
4" veneer	S.F.	5.94	6.42	12.36
Cavity wall	"	5.94	5.71	11.65
Chimney, standard brick, including flue				
16" x 16"	L.F.	26.50	51.00	77.50
16" x 20"	"	29.50	51.00	80.50
16" x 24"	"	38.00	51.00	89.00
20" x 20"	"	38.25	64.00	102
20" x 24"	"	40.75	64.00	105
20" x 32"	"	49.75	73.00	123
Window sill, face brick on edge	"	2.97	12.75	15.72
04210.20 **Structural Tile**				
Structural glazed tile				
6T series, 5-1/2" x 12"				
Glazed on one side				
2" thick	S.F.	7.26	5.14	12.40
4" thick	"	8.60	5.14	13.74
6" thick	"	13.50	5.71	19.21
8" thick	"	16.50	6.42	22.92
Glazed on two sides				
4" thick	S.F.	15.00	6.42	21.42
6" thick	"	17.25	7.34	24.59
Special shapes				
Group 1	S.F.	11.25	10.25	21.50
Group 2	"	14.25	10.25	24.50
Group 3	"	15.75	10.25	26.00
Group 4	"	27.50	10.25	37.75
Group 5	"	33.00	10.25	43.25
Fire rated				
4" thick, 1 hr rating	S.F.	21.00	5.14	26.14
6" thick, 2 hr rating	"	25.50	5.71	31.21
8W series, 8" x 16"				
Glazed on one side				
2" thick	S.F.	8.10	3.42	11.52
4" thick	"	8.80	3.42	12.22
6" thick	"	12.25	3.95	16.20
8" thick	"	14.25	3.95	18.20
Glazed on two sides				
4" thick	S.F.	15.00	4.28	19.28
6" thick	"	18.00	5.14	23.14

Unit Masonry	UNIT	MAT.	INST.	TOTAL
04210.20 **Structural Tile** *(Cont.)*				
8" thick	S.F.	25.50	5.14	30.64
Special shapes				
Group 1	S.F.	13.50	7.34	20.84
Group 2	"	16.50	7.34	23.84
Group 3	"	18.00	7.34	25.34
Group 4	"	30.00	7.34	37.34
Group 5	"	75.00	7.34	82.34
Fire rated				
4" thick, 1 hr rating	S.F.	22.50	7.34	29.84
6" thick, 2 hr rating	"	27.00	7.34	34.34
04210.60 **Pavers, Masonry**				
Brick walk laid on sand, sand joints				
Laid flat, (4.5 per sf)	S.F.	3.78	5.71	9.49
Laid on edge, (7.2 per sf)	"	6.05	8.56	14.61
Precast concrete patio blocks				
2" thick				
Natural	S.F.	2.68	1.71	4.39
Colors	"	3.74	1.71	5.45
Exposed aggregates, local aggregate				
Natural	S.F.	3.13	1.71	4.84
Colors	"	4.19	1.71	5.90
Granite or limestone aggregate	"	5.74	1.71	7.45
White tumblestone aggregate	"	4.47	1.71	6.18
Stone pavers, set in mortar				
Bluestone				
1" thick				
Irregular	S.F.	3.48	12.75	16.23
Snapped rectangular	"	5.30	10.25	15.55
1-1/2" thick, random rectangular	"	6.08	12.75	18.83
2" thick, random rectangular	"	7.19	14.75	21.94
Slate				
Natural cleft				
Irregular, 3/4" thick	S.F.	3.48	14.75	18.23
Random rectangular				
1-1/4" thick	S.F.	7.54	12.75	20.29
1-1/2" thick	"	8.51	14.25	22.76
Granite blocks				
3" thick, 3" to 6" wide				
4" to 12" long	S.F.	17.00	17.25	34.25
6" to 15" long	"	17.00	14.75	31.75
Crushed stone, white marble, 3" thick	"	1.60	0.82	2.42
04220.10 **Concrete Masonry Units**				
Hollow, load bearing				
4"	S.F.	1.60	3.80	5.40
6"	"	2.01	3.95	5.96
8"	"	2.59	4.28	6.87
10"	"	3.47	4.67	8.14
12"	"	3.66	5.14	8.80
Solid, load bearing				
4"	S.F.	2.15	3.80	5.95
6"	"	2.57	3.95	6.52

Unit Masonry	UNIT	MAT.	INST.	TOTAL

04220.10	**Concrete Masonry Units** *(Cont.)*			
8"	S.F.	2.98	4.28	7.26
10"	"	4.04	4.67	8.71
12"	"	4.49	5.14	9.63
Back-up block, 8" x 16"				
2"	S.F.	1.32	2.93	4.25
4"	"	1.60	3.02	4.62
6"	"	2.01	3.21	5.22
8"	"	2.59	3.42	6.01
10"	"	3.47	3.67	7.14
12"	"	3.66	3.95	7.61
Foundation wall, 8" x 16"				
6"	S.F.	2.01	3.67	5.68
8"	"	2.59	3.95	6.54
10"	"	3.47	4.28	7.75
12"	"	3.66	4.67	8.33
Solid				
6"	S.F.	2.57	3.95	6.52
8"	"	2.98	4.28	7.26
10"	"	4.04	4.67	8.71
12"	"	4.49	5.14	9.63
Exterior, styrofoam inserts, standard weight, 8" x 16"				
6"	S.F.	2.36	3.95	6.31
8"	"	2.47	4.28	6.75
10"	"	2.91	4.67	7.58
12"	"	3.04	5.14	8.18
Lightweight				
6"	S.F.	3.13	3.95	7.08
8"	"	3.80	4.28	8.08
10"	"	4.80	4.67	9.47
12"	"	5.08	5.14	10.22
Acoustical slotted block				
4"	S.F.	3.39	4.67	8.06
6"	"	4.14	4.67	8.81
8"	"	6.21	5.14	11.35
Filled cavities				
4"	S.F.	4.29	5.71	10.00
6"	"	5.31	6.04	11.35
8"	"	7.09	6.42	13.51
Hollow, split face				
4"	S.F.	3.25	3.80	7.05
6"	"	3.77	3.95	7.72
8"	"	3.96	4.28	8.24
10"	"	4.43	4.67	9.10
12"	"	4.73	5.14	9.87
Split rib profile				
4"	S.F.	2.83	4.67	7.50
6"	"	3.27	4.67	7.94
8"	"	3.56	5.14	8.70
10"	"	3.90	5.14	9.04
12"	"	4.25	5.14	9.39
High strength block, 3500 psi				
2"	S.F.	1.54	3.80	5.34
4"	"	1.92	3.95	5.87

Unit Masonry	UNIT	MAT.	INST.	TOTAL
04220.10 Concrete Masonry Units *(Cont.)*				
6"	S.F.	2.29	3.95	6.24
8"	"	2.59	4.28	6.87
10"	"	3.02	4.67	7.69
12"	"	3.58	5.14	8.72
Solar screen concrete block				
4" thick				
6" x 6"	S.F.	4.58	11.50	16.08
8" x 8"	"	5.47	10.25	15.72
12" x 12"	"	5.61	7.90	13.51
8" thick				
8" x 16"	S.F.	5.61	7.34	12.95
Glazed block				
Cove base, glazed 1 side, 2"	L.F.	10.50	5.71	16.21
4"	"	10.75	5.71	16.46
6"	"	11.00	6.42	17.42
8"	"	11.75	6.42	18.17
Single face				
2"	S.F.	8.14	4.28	12.42
4"	"	8.81	4.28	13.09
6"	"	8.93	4.67	13.60
8"	"	9.25	5.14	14.39
10"	"	9.57	5.71	15.28
12"	"	9.87	6.04	15.91
Double face				
4"	S.F.	13.00	5.41	18.41
6"	"	14.00	5.71	19.71
8"	"	14.50	6.42	20.92
Corner or bullnose				
2"	EA.	11.00	6.42	17.42
4"	"	14.00	7.34	21.34
6"	"	17.25	7.34	24.59
8"	"	18.75	8.56	27.31
10"	"	20.25	9.34	29.59
12"	"	22.00	10.25	32.25
Gypsum unit masonry				
Partition blocks (12"x30")				
Solid				
2"	S.F.	1.01	2.05	3.06
Hollow				
3"	S.F.	1.02	2.05	3.07
4"	"	1.16	2.14	3.30
6"	"	1.25	2.33	3.58
Vertical reinforcing				
4' o.c., add 5% to labor				
2'8" o.c., add 15% to labor				
Interior partitions, add 10% to labor				
04220.90 Bond Beams & Lintels				
Bond beam, no grout or reinforcement				
8" x 16" x				
4" thick	L.F.	1.12	3.95	5.07
6" thick	"	1.70	4.11	5.81
8" thick	"	1.95	4.28	6.23

Unit Masonry

	UNIT	MAT.	INST.	TOTAL
04220.90 **Bond Beams & Lintels** *(Cont.)*				
10" thick	L.F.	2.42	4.46	6.88
12" thick	"	2.75	4.67	7.42
Beam lintel, no grout or reinforcement				
8" x 16" x				
10" thick	L.F.	5.58	5.14	10.72
12" thick	"	7.09	5.71	12.80
Precast masonry lintel				
6 lf, 8" high x				
4" thick	L.F.	5.36	8.56	13.92
6" thick	"	6.85	8.56	15.41
8" thick	"	7.75	9.34	17.09
10" thick	"	9.25	9.34	18.59
10 lf, 8" high x				
4" thick	L.F.	6.74	5.14	11.88
6" thick	"	8.31	5.14	13.45
8" thick	"	9.25	5.71	14.96
10" thick	"	12.50	5.71	18.21
Steel angles and plates				
Minimum	Lb.	0.66	0.73	1.39
Maximum	"	0.88	1.28	2.16
Various size angle lintels				
1/4" stock				
3" x 3"	L.F.	3.74	3.21	6.95
3" x 3-1/2"	"	3.89	3.21	7.10
3/8" stock				
3" x 4"	L.F.	5.47	3.21	8.68
3-1/2" x 4"	"	5.75	3.21	8.96
4" x 4"	"	6.20	3.21	9.41
5" x 3-1/2"	"	6.49	3.21	9.70
6" x 3-1/2"	"	6.64	3.21	9.85
1/2" stock				
6" x 4"	L.F.	8.27	3.21	11.48
04240.10 **Clay Tile**				
Hollow clay tile, for back-up, 12" x 12"				
Scored face				
Load bearing				
4" thick	S.F.	4.54	3.67	8.21
6" thick	"	5.56	3.80	9.36
8" thick	"	6.72	3.95	10.67
10" thick	"	7.46	4.11	11.57
12" thick	"	8.77	4.28	13.05
Non-load bearing				
3" thick	S.F.	4.25	3.54	7.79
4" thick	"	4.38	3.67	8.05
6" thick	"	4.54	3.80	8.34
8" thick	"	5.85	3.95	9.80
12" thick	"	6.44	4.28	10.72
Partition, 12" x 12"				
In walls				
3" thick	S.F.	3.67	4.28	7.95
4" thick	"	4.25	4.28	8.53
6" thick	"	4.68	4.46	9.14

Unit Masonry	UNIT	MAT.	INST.	TOTAL
04240.10 **Clay Tile** *(Cont.)*				
8" thick	S.F.	6.14	4.67	10.81
10" thick	"	7.32	4.89	12.21
12" thick	"	7.76	5.14	12.90
Clay tile floors				
4" thick	S.F.	4.54	2.85	7.39
6" thick	"	5.56	3.02	8.58
8" thick	"	6.72	3.21	9.93
10" thick	"	7.46	3.42	10.88
12" thick	"	8.77	3.67	12.44
Terra cotta				
Coping, 10" or 12" wide, 3" thick	L.F.	14.75	10.25	25.00
04270.10 **Glass Block**				
Glass block, 4" thick				
6" x 6"	S.F.	23.50	17.25	40.75
8" x 8"	"	15.00	12.75	27.75
12" x 12"	"	18.75	10.25	29.00
Replacement glass blocks, 4" x 8" x 8"				
Minimum	S.F.	18.25	51.00	69.25
Maximum	"	21.00	100	121
04295.10 **Parging/masonry Plaster**				
Parging				
1/2" thick	S.F.	0.27	3.42	3.69
3/4" thick	"	0.34	4.28	4.62
1" thick	"	0.45	5.14	5.59

Stone	UNIT	MAT.	INST.	TOTAL
04400.10 **Stone**				
Rubble stone				
Walls set in mortar				
8" thick	S.F.	15.00	12.75	27.75
12" thick	"	18.25	20.50	38.75
18" thick	"	24.00	25.75	49.75
24" thick	"	30.25	34.25	64.50
Dry set wall				
8" thick	S.F.	17.00	8.56	25.56
12" thick	"	19.00	12.75	31.75
18" thick	"	26.50	17.25	43.75
24" thick	"	32.25	20.50	52.75
Cut stone				
Imported marble				
Facing panels				
3/4" thick	S.F.	39.00	20.50	59.50
1-1/2" thick	"	59.00	23.25	82.25

Stone	UNIT	MAT.	INST.	TOTAL
04400.10 **Stone** *(Cont.)*				
2-1/4" thick	S.F.	68.00	28.50	96.50
Base				
1" thick				
4" high	L.F.	17.75	25.75	43.50
6" high	"	21.75	25.75	47.50
Columns, solid				
Plain faced	C.F.	120	340	460
Fluted	"	320	340	660
Flooring, travertine, minimum	S.F.	16.00	7.90	23.90
Average	"	21.50	10.25	31.75
Maximum	"	32.25	11.50	43.75
Domestic marble				
Facing panels				
7/8" thick	S.F.	35.25	20.50	55.75
1-1/2" thick	"	53.00	23.25	76.25
2-1/4" thick	"	64.00	28.50	92.50
Stairs				
12" treads	L.F.	31.50	25.75	57.25
6" risers	"	23.50	17.25	40.75
Thresholds, 7/8" thick, 3' long, 4" to 6" wide				
Plain	EA.	30.00	42.75	72.75
Beveled	"	33.25	42.75	76.00
Window sill				
6" wide, 2" thick	L.F.	15.25	20.50	35.75
Stools				
5" wide, 7/8" thick	L.F.	22.25	20.50	42.75
Limestone panels up to 12' x 5', smooth finish				
2" thick	S.F.	23.50	8.90	32.40
3" thick	"	27.50	8.90	36.40
4" thick	"	39.00	8.90	47.90
Miscellaneous limestone items				
Steps, 14" wide, 6" deep	L.F.	75.00	34.25	109
Coping, smooth finish	C.F.	100	17.25	117
Sills, lintels, jambs, smooth finish	"	100	20.50	121
Granite veneer facing panels, polished				
7/8" thick				
Black	S.F.	41.75	20.50	62.25
Gray	"	36.25	20.50	56.75
Base				
4" high	L.F.	19.50	10.25	29.75
6" high	"	23.50	11.50	35.00
Curbing, straight, 6" x 16"	"	21.50	37.00	58.50
Radius curbs, radius over 5'	"	26.25	49.50	75.75
Ashlar veneer				
4" thick, random	S.F.	35.75	20.50	56.25
Pavers, 4" x 4" split				
Gray	S.F.	35.00	10.25	45.25
Pink	"	34.50	10.25	44.75
Black	"	34.00	10.25	44.25
Slate, panels				
1" thick	S.F.	25.00	20.50	45.50
2" thick	"	34.00	23.25	57.25
Sills or stools				

Stone	UNIT	MAT.	INST.	TOTAL
04400.10 **Stone** *(Cont.)*				
1" thick				
6" wide	L.F.	11.75	20.50	32.25
10" wide	"	19.00	22.25	41.25
2" thick				
6" wide	L.F.	19.00	23.25	42.25
10" wide	"	31.50	25.75	57.25

Masonry Restoration	UNIT	MAT.	INST.	TOTAL
04520.10 **Restoration And Cleaning**				
Masonry cleaning				
Washing brick				
Smooth surface	S.F.	0.42	0.85	1.27
Rough surface	"	0.59	1.14	1.73
Steam clean masonry				
Smooth face				
Minimum	S.F.		0.66	0.66
Maximum	"		0.96	0.96
Rough face				
Minimum	S.F.		0.88	0.88
Maximum	"		1.33	1.33
Sandblast masonry				
Minimum	S.F.	0.51	1.06	1.57
Maximum	"	0.77	1.77	2.54
Pointing masonry				
Brick	S.F.	1.35	2.05	3.40
Concrete block	"	0.60	1.46	2.06
Cut and repoint				
Brick				
Minimum	S.F.	0.37	2.57	2.94
Maximum	"	0.63	5.14	5.77
Stone work	L.F.	1.70	3.95	5.65
Cut and recaulk				
Oil base caulks	L.F.	1.32	3.42	4.74
Butyl caulks	"	1.15	3.42	4.57
Polysulfides and acrylics	"	2.24	3.42	5.66
Silicones	"	2.61	3.42	6.03
Cement and sand grout on walls, to 1/8" thick				
Minimum	S.F.	0.64	2.05	2.69
Maximum	"	1.63	2.57	4.20
Brick removal and replacement				
Minimum	EA.	0.68	6.42	7.10
Average	"	0.89	8.56	9.45
Maximum	"	1.80	25.75	27.55

Masonry Restoration	UNIT	MAT.	INST.	TOTAL
04550.10 **Refractories**				
Flue liners				
Rectangular				
8" x 12"	L.F.	6.98	8.56	15.54
12" x 12"	"	8.71	9.34	18.05
12" x 18"	"	15.25	10.25	25.50
16" x 16"	"	17.00	11.50	28.50
18" x 18"	"	20.50	12.25	32.75
20" x 20"	"	34.25	12.75	47.00
24" x 24"	"	41.00	14.75	55.75
Round				
18" dia.	L.F.	34.00	12.25	46.25
24" dia.	"	47.75	14.75	62.50

Metal Fastening	UNIT	MAT.	INST.	TOTAL
05050.10 — Structural Welding				
Welding				
Single pass				
1/8"	L.F.	0.33	2.95	3.28
3/16"	"	0.55	3.93	4.48
1/4"	"	0.77	4.92	5.69
Miscellaneous steel shapes				
Plain	Lb.	1.17	0.11	1.28
Galvanized	"	1.74	0.19	1.93
Plates				
Plain	Lb.	1.06	0.14	1.20
Galvanized	"	1.61	0.23	1.84
05050.90 — Metal Anchors				
Anchor bolts				
3/8" x				
8" long	EA.			0.97
10" long	"			1.06
12" long	"			1.16
1/2" x				
8" long	EA.			1.46
10" long	"			1.55
12" long	"			1.70
18" long	"			1.85
5/8" x				
8" long	EA.			1.36
10" long	"			1.50
12" long	"			1.61
18" long	"			1.70
24" long	"			1.85
3/4" x				
8" long	EA.			1.94
12" long	"			2.19
18" long	"			3.01
24" long	"			3.99
7/8" x				
8" long	EA.			1.94
12" long	"			2.19
18" long	"			3.01
24" long	"			3.99
1" x				
12" long	EA.			3.89
18" long	"			4.87
24" long	"			5.84
36" long	"			8.76
Expansion shield				
1/4"	EA.			1.04
3/8"	"			1.73
1/2"	"			3.41
5/8"	"			4.95
3/4"	"			6.05
1"	"			8.19
Non-drilling anchor				
1/4"	EA.			0.64

Metal Fastening

	UNIT	MAT.	INST.	TOTAL
05050.90 **Metal Anchors** *(Cont.)*				
3/8"	EA.			0.80
1/2"	"			1.23
5/8"	"			2.02
3/4"	"			3.46
Self-drilling anchor				
1/4"	EA.			1.62
5/16"	"			2.03
3/8"	"			2.44
1/2"	"			3.25
5/8"	"			6.18
3/4"	"			8.11
7/8"	"			11.25
Add 25% for galvanized anchor bolts				
Channel door frame, with anchors	Lb.	1.61	0.65	2.26
Corner guard angle, with anchors	"	1.45	0.98	2.43
05050.95 **Metal Lintels**				
Lintels, steel				
Plain	Lb.	1.21	1.47	2.68
Galvanized	"	1.82	1.47	3.29
05300.10 **Metal Decking**				
Roof, 1-1/2" deep, non-composite				
16 ga.				
Primed	S.F.	3.09	1.10	4.19
Galvanized	"	3.30	1.10	4.40
Open type decking, galvanized				
1-1/2" deep				
18 ga.	S.F.	2.48	1.10	3.58
20 ga.	"	1.90	1.10	3.00
3" deep				
16 ga.	S.F.	3.85	1.20	5.05
18 ga.	"	3.63	1.20	4.83
20 ga.	"	2.75	1.20	3.95
Cellular type				
1-1/2" deep, galvanized				
18-18 ga.	S.F.	6.20	1.32	7.52
22-18 ga.	"	5.64	1.32	6.96
3" deep, galvanized				
16-16 ga.	S.F.	8.85	1.47	10.32
18-16 ga.	"	8.31	1.47	9.78
18-18 ga.	"	7.60	1.47	9.07
20-18 ga.	"	6.74	1.47	8.21
Composite deck, non-cellular, galvanized				
1-1/2" deep				
18 ga.	S.F.	2.09	1.20	3.29
20 ga.	"	1.90	1.20	3.10
3" deep				
18 ga.	S.F.	2.93	1.27	4.20
20 ga.	"	1.90	1.27	3.17

Cold Formed Framing	UNIT	MAT.	INST.	TOTAL
05410.10 **Metal Framing**				
Furring channel, galvanized				
Beams and columns, 3/4"				
12" o.c.	S.F.	0.51	5.90	6.41
16" o.c.	"	0.39	5.37	5.76
Walls, 3/4"				
12" o.c.	S.F.	0.51	2.95	3.46
16" o.c.	"	0.39	2.46	2.85
24" o.c.	"	0.26	1.96	2.22
1-1/2"				
12" o.c.	S.F.	0.66	2.95	3.61
16" o.c.	"	0.49	2.46	2.95
24" o.c.	"	0.34	1.96	2.30
Stud, load bearing				
16" o.c.				
16 ga.				
2-1/2"	S.F.	1.21	2.62	3.83
3-5/8"	"	1.43	2.62	4.05
4"	"	1.48	2.62	4.10
6"	"	1.87	2.95	4.82
18 ga.				
2-1/2"	S.F.	0.99	2.62	3.61
3-5/8"	"	1.21	2.62	3.83
4"	"	1.26	2.62	3.88
6"	"	1.59	2.95	4.54
8"	"	1.92	2.95	4.87
20 ga.				
2-1/2"	S.F.	0.55	2.62	3.17
3-5/8"	"	0.66	2.62	3.28
4"	"	0.71	2.62	3.33
6"	"	0.88	2.95	3.83
8"	"	1.04	2.95	3.99
24" o.c.				
16 ga.				
2-1/2"	S.F.	0.82	2.27	3.09
3-5/8"	"	0.99	2.27	3.26
4"	"	1.04	2.27	3.31
6"	"	1.26	2.46	3.72
8"	"	1.59	2.46	4.05
18 ga.				
2-1/2"	S.F.	0.66	2.27	2.93
3-5/8"	"	0.77	2.27	3.04
4"	"	0.82	2.27	3.09
6"	"	1.04	2.46	3.50
8"	"	1.26	2.46	3.72
20 ga.				
2-1/2"	S.F.	0.44	2.27	2.71
3-5/8"	"	0.49	2.27	2.76
4"	"	0.55	2.27	2.82
6"	"	0.71	2.46	3.17
8"	"	0.88	2.46	3.34

Metal Fabrications	UNIT	MAT.	INST.	TOTAL
05510.10		**Stairs**		
Stock unit, steel, complete, per riser				
Tread				
3'-6" wide	EA.	130	74.00	204
4' wide	"	150	84.00	234
5' wide	"	170	98.00	268
Metal pan stair, cement filled, per riser				
3'-6" wide	EA.	140	59.00	199
4' wide	"	160	66.00	226
5' wide	"	180	74.00	254
Landing, steel pan	S.F.	55.00	14.75	69.75
Cast iron tread, steel stringers, stock units, per riser				
Tread				
3'-6" wide	EA.	250	74.00	324
4' wide	"	280	84.00	364
5' wide	"	340	98.00	438
Stair treads, abrasive, 12" x 3'-6"				
Cast iron				
3/8"	EA.	140	29.50	170
1/2"	"	170	29.50	200
Cast aluminum				
5/16"	EA.	160	29.50	190
3/8"	"	170	29.50	200
1/2"	"	200	29.50	230
05515.10		**Ladders**		
Ladder, 18" wide				
With cage	L.F.	94.00	39.50	134
Without cage	"	59.00	29.50	88.50
05520.10		**Railings**		
Railing, pipe				
1-1/4" diameter, welded steel				
2-rail				
Primed	L.F.	19.75	11.75	31.50
Galvanized	"	25.25	11.75	37.00
3-rail				
Primed	L.F.	25.25	14.75	40.00
Galvanized	"	33.00	14.75	47.75
Wall mounted, single rail, welded steel				
Primed	L.F.	13.25	9.08	22.33
Galvanized	"	17.25	9.08	26.33
1-1/2" diameter, welded steel				
2-rail				
Primed	L.F.	21.50	11.75	33.25
Galvanized	"	28.00	11.75	39.75
3-rail				
Primed	L.F.	27.00	14.75	41.75
Galvanized	"	35.00	14.75	49.75
Wall mounted, single rail, welded steel				
Primed	L.F.	14.25	9.08	23.33
Galvanized	"	18.50	9.08	27.58
2" diameter, welded steel				
2-rail				

Metal Fabrications	UNIT	MAT.	INST.	TOTAL
05520.10 **Railings** *(Cont.)*				
Primed	L.F.	27.00	13.00	40.00
Galvanized	"	35.00	13.00	48.00
3-rail				
Primed	L.F.	34.00	16.75	50.75
Galvanized	"	44.25	16.75	61.00
Wall mounted, single rail, welded steel				
Primed	L.F.	15.50	9.84	25.34
Galvanized	"	20.25	9.84	30.09
05530.10 **Metal Grating**				
Floor plate, checkered, steel				
1/4"				
Primed	S.F.	7.75	0.84	8.59
Galvanized	"	10.25	0.84	11.09
3/8"				
Primed	S.F.	10.25	0.90	11.15
Galvanized	"	13.75	0.90	14.65
Aluminum grating, pressure-locked bearing bars				
3/4" x 1/8"	S.F.	16.50	1.47	17.97
1" x 1/8"	"	18.25	1.47	19.72
1-1/4" x 1/8"	"	23.25	1.47	24.72
1-1/4" x 3/16"	"	24.75	1.47	26.22
1-1/2" x 1/8"	"	26.50	1.47	27.97
1-3/4" x 3/16"	"	29.75	1.47	31.22
Miscellaneous expenses				
Cutting				
Minimum	L.F.		3.93	3.93
Maximum	"		5.90	5.90
Banding				
Minimum	L.F.		9.84	9.84
Maximum	"		11.75	11.75
Toe plates				
Minimum	L.F.		11.75	11.75
Maximum	"		14.75	14.75
Steel grating, primed				
3/4" x 1/8"	S.F.	7.37	1.96	9.33
1" x 1/8"	"	7.59	1.96	9.55
1-1/4" x 1/8"	"	8.08	1.96	10.04
1-1/4" x 3/16"	"	9.90	1.96	11.86
1-1/2" x 1/8"	"	9.02	1.96	10.98
1-3/4" x 3/16"	"	12.75	1.96	14.71
Galvanized				
3/4" x 1/8"	S.F.	9.02	1.96	10.98
1" x 1/8"	"	9.62	1.96	11.58
1-1/4" x 1/8"	"	10.50	1.96	12.46
1-1/4" x 3/16"	"	12.50	1.96	14.46
1-1/2" x 1/8"	"	11.25	1.96	13.21
1-3/4" x 3/16"	"	16.50	1.96	18.46
Miscellaneous expenses				
Cutting				
Minimum	L.F.		4.22	4.22
Maximum	"		6.56	6.56

Metal Fabrications	UNIT	MAT.	INST.	TOTAL
05530.10 — **Metal Grating** *(Cont.)*				
Banding				
Minimum	L.F.		10.75	10.75
Maximum	"		13.00	13.00
Toe plates				
Minimum	L.F.		13.00	13.00
Maximum	"		16.75	16.75
05540.10 — **Castings**				
Miscellaneous castings				
Light sections	Lb.	1.76	1.18	2.94
Heavy sections	"	1.39	0.84	2.23

Misc. Fabrications	UNIT	MAT.	INST.	TOTAL
05580.10 — **Metal Specialties**				
Kick plate				
4" high x 1/4" thick				
Primed	L.F.	7.31	11.75	19.06
Galvanized	"	8.30	11.75	20.05
6" high x 1/4" thick				
Primed	L.F.	7.92	13.00	20.92
Galvanized	"	9.40	13.00	22.40
05700.10 — **Ornamental Metal**				
Railings, vertical square bars, 6" o.c., with shaped top rails				
Steel	L.F.	83.00	29.50	113
Aluminum	"	110	29.50	140
Bronze	"	180	39.50	220
Stainless steel	"	160	39.50	200
Laminated metal or wood handrails with metal supports				
2-1/2" round or oval shape	L.F.	250	29.50	280
Grilles and louvers				
Fixed type louvers				
4 through 10 sf	S.F.	28.25	9.84	38.09
Over 10 sf	"	33.25	7.38	40.63
Movable type louvers				
4 through 10 sf	S.F.	33.25	9.84	43.09
Over 10 sf	"	36.75	7.38	44.13
Aluminum louvers				
Residential use, fixed type, with screen				
8" x 8"	EA.	17.25	29.50	46.75
12" x 12"	"	19.00	29.50	48.50
12" x 18"	"	22.75	29.50	52.25
14" x 24"	"	32.75	29.50	62.25
18" x 24"	"	36.75	29.50	66.25
30" x 24"	"	50.00	32.75	82.75

Misc. Fabrications	UNIT	MAT.	INST.	TOTAL
05800.10 **Expansion Control**				
Expansion joints with covers, floor assembly type				
With 1" space				
Aluminum	L.F.	27.25	9.84	37.09
Bronze	"	57.00	9.84	66.84
Stainless steel	"	44.25	9.84	54.09
Aluminum	"	33.75	9.84	43.59
Ceiling and wall assembly type				
With 1" space				
Aluminum	L.F.	17.50	11.75	29.25
Bronze	"	60.00	11.75	71.75
Stainless steel	"	54.00	11.75	65.75
Exterior roof and wall, aluminum				
Roof to roof				
With 1" space	L.F.	51.00	9.84	60.84
With 2" space	"	55.00	9.84	64.84
Roof to wall				
With 1" space	L.F.	39.75	10.75	50.50
With 2" space	"	47.25	10.75	58.00
Flat wall to wall				
With 1" space	L.F.	19.25	9.84	29.09
With 2" space	"	21.00	9.84	30.84
Corner to flat wall				
With 1" space	L.F.	23.25	11.75	35.00
With 2" in space	"	23.50	11.75	35.25

Fasteners And Adhesives

06050.10	Accessories	UNIT	MAT.	INST.	TOTAL
Column/post base, cast aluminum					
4" x 4"		EA.	3.32	13.25	16.57
6" x 6"		"	7.30	13.25	20.55
Bridging, metal, per pair					
12" o.c.		EA.	1.22	5.31	6.53
16" o.c.		"	1.22	4.82	6.04
Anchors					
Bolts, threaded two ends, with nuts and washers					
1/2" dia.					
4" long		EA.	1.64	3.32	4.96
7-1/2" long		"	1.98	3.32	5.30
3/4" dia.					
7-1/2" long		EA.	3.15	3.32	6.47
15" long		"	4.09	3.32	7.41
Framing anchors					
10 gauge		EA.	0.86	4.42	5.28
Bolts, carriage					
1/4 x 4		EA.	1.66	5.31	6.97
5/16 x 6		"	1.76	5.59	7.35
3/8 x 6		"	2.02	5.59	7.61
1/2 x 6		"	2.02	5.59	7.61
Joist and beam hangers					
18 ga.					
2 x 4		EA.	0.79	5.31	6.10
2 x 6		"	1.05	5.31	6.36
2 x 8		"	1.22	5.31	6.53
2 x 10		"	1.40	5.90	7.30
2 x 12		"	1.64	6.64	8.28
16 ga.					
3 x 6		EA.	2.92	5.90	8.82
3 x 8		"	3.00	5.90	8.90
3 x 10		"	3.48	6.24	9.72
3 x 12		"	3.77	7.08	10.85
3 x 14		"	4.04	7.58	11.62
4 x 6		"	3.27	5.90	9.17
4 x 8		"	3.77	5.90	9.67
4 x 10		"	4.73	6.24	10.97
4 x 12		"	5.31	7.08	12.39
4 x 14		"	5.79	7.58	13.37
Rafter anchors, 18 ga., 1-1/2" wide					
5-1/4" long		EA.	0.88	4.42	5.30
10-3/4" long		"	1.37	4.42	5.79
Shear plates					
2-5/8" dia.		EA.	2.51	4.08	6.59
4" dia.		"	5.24	4.42	9.66
Sill anchors					
Embedded in concrete		EA.	2.22	5.31	7.53
Split rings					
2-1/2" dia.		EA.	1.74	5.90	7.64
4" dia.		"	3.10	6.64	9.74
Strap ties, 14 ga., 1-3/8" wide					
12" long		EA.	1.85	4.42	6.27

Fasteners And Adhesives	UNIT	MAT.	INST.	TOTAL
06050.10 **Accessories** *(Cont.)*				
18" long	EA.	2.03	4.82	6.85
24" long	"	2.54	5.31	7.85
36" long	"	3.87	5.90	9.77
Toothed rings				
2-5/8" dia.	EA.	1.52	8.85	10.37
4" dia.	"	1.74	10.50	12.24

Rough Carpentry	UNIT	MAT.	INST.	TOTAL
06110.10 **Blocking**				
Steel construction				
Walls				
2x4	L.F.	0.62	3.54	4.16
2x6	"	0.88	4.08	4.96
2x8	"	1.26	4.42	5.68
2x10	"	1.73	4.82	6.55
2x12	"	2.29	5.31	7.60
Ceilings				
2x4	L.F.	0.62	4.08	4.70
2x6	"	0.88	4.82	5.70
2x8	"	1.26	5.31	6.57
2x10	"	1.73	5.90	7.63
2x12	"	2.29	6.64	8.93
Wood construction				
Walls				
2x4	L.F.	0.62	2.95	3.57
2x6	"	0.88	3.32	4.20
2x8	"	1.26	3.54	4.80
2x10	"	1.73	3.79	5.52
2x12	"	2.29	4.08	6.37
Ceilings				
2x4	L.F.	0.62	3.32	3.94
2x6	"	0.88	3.79	4.67
2x8	"	1.26	4.08	5.34
2x10	"	1.73	4.42	6.15
2x12	"	2.29	4.82	7.11
06110.20 **Ceiling Framing**				
Ceiling joists				
12" o.c.				
2x4	S.F.	0.75	1.26	2.01
2x6	"	1.10	1.32	2.42
2x8	"	1.55	1.39	2.94
2x10	"	2.35	1.47	3.82
2x12	"	3.12	1.56	4.68
16" o.c.				

Rough Carpentry	UNIT	MAT.	INST.	TOTAL
06110.20 — **Ceiling Framing** *(Cont.)*				
2x4	S.F.	0.62	1.02	1.64
2x6	"	0.94	1.06	2.00
2x8	"	1.32	1.10	2.42
2x10	"	1.95	1.15	3.10
2x12	"	2.53	1.20	3.73
24" o.c.				
2x4	S.F.	0.49	0.84	1.33
2x6	"	0.75	0.88	1.63
2x8	"	1.01	0.93	1.94
2x10	"	1.55	0.98	2.53
2x12	"	2.09	1.04	3.13
Headers and nailers				
2x4	L.F.	0.62	1.71	2.33
2x6	"	0.88	1.77	2.65
2x8	"	1.26	1.89	3.15
2x10	"	1.73	2.04	3.77
2x12	"	2.29	2.21	4.50
Sister joists for ceilings				
2x4	L.F.	0.62	3.79	4.41
2x6	"	0.88	4.42	5.30
2x8	"	1.26	5.31	6.57
2x10	"	1.73	6.64	8.37
2x12	"	2.29	8.85	11.14
06110.30 — **Floor Framing**				
Floor joists				
12" o.c.				
2x6	S.F.	1.10	1.06	2.16
2x8	"	1.55	1.08	2.63
2x10	"	2.35	1.10	3.45
2x12	"	3.12	1.15	4.27
2x14	"	3.44	1.10	4.54
3x6	"	2.51	1.13	3.64
3x8	"	3.30	1.15	4.45
3x10	"	4.13	1.20	5.33
3x12	"	4.94	1.26	6.20
3x14	"	5.69	1.32	7.01
4x6	"	3.31	1.10	4.41
4x8	"	4.26	1.15	5.41
4x10	"	5.42	1.20	6.62
4x12	"	6.60	1.26	7.86
4x14	"	7.71	1.32	9.03
16" o.c.				
2x6	S.F.	0.94	0.88	1.82
2x8	"	1.32	0.90	2.22
2x10	"	1.95	0.91	2.86
2x12	"	2.53	0.94	3.47
2x14	"	3.12	0.98	4.10
3x6	"	2.12	0.91	3.03
3x8	"	2.76	0.94	3.70
3x10	"	3.44	0.98	4.42
3x12	"	4.11	1.02	5.13
3x14	"	4.84	1.06	5.90

Rough Carpentry

06110.30	Floor Framing *(Cont.)*	UNIT	MAT.	INST.	TOTAL
4x6		S.F.	2.75	0.91	3.66
4x8		"	3.68	0.94	4.62
4x10		"	4.58	0.98	5.56
4x12		"	5.42	1.02	6.44
4x14		"	6.43	1.06	7.49
Sister joists for floors					
2x4		L.F.	0.62	3.32	3.94
2x6		"	0.88	3.79	4.67
2x8		"	1.26	4.42	5.68
2x10		"	1.73	5.31	7.04
2x12		"	2.29	6.64	8.93
3x6		"	2.12	5.31	7.43
3x8		"	2.76	5.90	8.66
3x10		"	3.44	6.64	10.08
3x12		"	4.11	7.58	11.69
4x6		"	2.67	5.31	7.98
4x8		"	3.68	5.90	9.58
4x10		"	4.58	6.64	11.22
4x12		"	5.42	7.58	13.00

06110.40	Furring	UNIT	MAT.	INST.	TOTAL
Furring, wood strips					
Walls					
On masonry or concrete walls					
1x2 furring					
12" o.c.		S.F.	0.29	1.66	1.95
16" o.c.		"	0.24	1.51	1.75
24" o.c.		"	0.20	1.39	1.59
1x3 furring					
12" o.c.		S.F.	0.42	1.66	2.08
16" o.c.		"	0.36	1.51	1.87
24" o.c.		"	0.29	1.39	1.68
On wood walls					
1x2 furring					
12" o.c.		S.F.	0.29	1.18	1.47
16" o.c.		"	0.24	1.06	1.30
24" o.c.		"	0.20	0.96	1.16
1x3 furring					
12" o.c.		S.F.	0.42	1.18	1.60
16" o.c.		"	0.36	1.06	1.42
24" o.c.		"	0.29	0.96	1.25
Ceilings					
On masonry or concrete ceilings					
1x2 furring					
12" o.c.		S.F.	0.29	2.95	3.24
16" o.c.		"	0.24	2.65	2.89
24" o.c.		"	0.20	2.41	2.61
1x3 furring					
12" o.c.		S.F.	0.42	2.95	3.37
16" o.c.		"	0.36	2.65	3.01
24" o.c.		"	0.29	2.41	2.70
On wood ceilings					
1x2 furring					

Rough Carpentry	UNIT	MAT.	INST.	TOTAL
06110.40 **Furring** *(Cont.)*				
12" o.c.	S.F.	0.29	1.96	2.25
16" o.c.	"	0.24	1.77	2.01
24" o.c.	"	0.20	1.60	1.80
1x3				
12" o.c.	S.F.	0.42	1.96	2.38
16" o.c.	"	0.36	1.77	2.13
24" o.c.	"	0.29	1.60	1.89
06110.50 **Roof Framing**				
Roof framing				
Rafters, gable end				
0-2 pitch (flat to 2-in-12)				
12" o.c.				
2x4	S.F.	0.75	1.10	1.85
2x6	"	1.12	1.15	2.27
2x8	"	1.55	1.20	2.75
2x10	"	2.35	1.26	3.61
2x12	"	3.12	1.32	4.44
16" o.c.				
2x6	S.F.	0.94	0.94	1.88
2x8	"	1.32	0.98	2.30
2x10	"	1.96	1.02	2.98
2x12	"	2.53	1.06	3.59
24" o.c.				
2x6	S.F.	0.75	0.80	1.55
2x8	"	1.01	0.83	1.84
2x10	"	1.55	0.85	2.40
2x12	"	2.09	0.88	2.97
4-6 pitch (4-in-12 to 6-in-12)				
12" o.c.				
2x4	S.F.	0.80	1.15	1.95
2x6	"	1.20	1.20	2.40
2x8	"	1.68	1.26	2.94
2x10	"	2.53	1.32	3.85
2x12	"	3.39	1.39	4.78
16" o.c.				
2x6	S.F.	0.98	0.98	1.96
2x8	"	1.44	1.02	2.46
2x10	"	2.09	1.06	3.15
2x12	"	2.80	1.10	3.90
24" o.c.				
2x6	S.F.	0.80	0.83	1.63
2x8	"	1.10	0.85	1.95
2x10	"	1.65	0.91	2.56
2x12	"	2.24	1.02	3.26
8-12 pitch (8-in-12 to 12-in-12)				
12" o.c.				
2x4	S.F.	0.82	1.20	2.02
2x6	"	1.28	1.26	2.54
2x8	"	1.77	1.32	3.09
2x10	"	2.67	1.39	4.06
2x12	"	3.53	1.47	5.00
16" o.c.				

Rough Carpentry	UNIT	MAT.	INST.	TOTAL
06110.50 **Roof Framing** *(Cont.)*				
2x6	S.F.	1.07	1.02	2.09
2x8	"	1.49	1.06	2.55
2x10	"	2.24	1.10	3.34
2x12	"	2.98	1.15	4.13
24" o.c.				
2x6	S.F.	0.82	0.85	1.67
2x8	"	1.16	0.88	2.04
2x10	"	1.77	0.91	2.68
2x12	"	2.37	0.94	3.31
Ridge boards				
2x6	L.F.	0.88	2.65	3.53
2x8	"	1.26	2.95	4.21
2x10	"	1.73	3.32	5.05
2x12	"	2.29	3.79	6.08
Hip rafters				
2x6	L.F.	0.88	1.89	2.77
2x8	"	1.26	1.96	3.22
2x10	"	1.73	2.04	3.77
2x12	"	2.29	2.12	4.41
Jack rafters				
4-6 pitch (4-in-12 to 6-in-12)				
16" o.c.				
2x6	S.F.	1.00	1.56	2.56
2x8	"	1.42	1.60	3.02
2x10	"	2.09	1.71	3.80
2x12	"	2.82	1.77	4.59
24" o.c.				
2x6	S.F.	0.80	1.20	2.00
2x8	"	1.10	1.23	2.33
2x10	"	1.65	1.29	2.94
2x12	"	2.24	1.32	3.56
8-12 pitch (8-in-12 to 12-in-12)				
16" o.c.				
2x6	S.F.	1.57	1.66	3.23
2x8	"	2.21	1.71	3.92
2x10	"	3.30	1.77	5.07
2x12	"	4.40	1.83	6.23
24" o.c.				
2x6	S.F.	1.28	1.26	2.54
2x8	"	1.77	1.29	3.06
2x10	"	2.67	1.32	3.99
2x12	"	3.53	1.36	4.89
Sister rafters				
2x4	L.F.	0.62	3.79	4.41
2x6	"	0.88	4.42	5.30
2x8	"	1.26	5.31	6.57
2x10	"	1.73	6.64	8.37
2x12	"	2.29	8.85	11.14
Fascia boards				
2x4	L.F.	0.62	2.65	3.27
2x6	"	0.88	2.65	3.53
2x8	"	1.26	2.95	4.21
2x10	"	1.73	2.95	4.68

Rough Carpentry	UNIT	MAT.	INST.	TOTAL
06110.50			**Roof Framing** *(Cont.)*	
2x12	L.F.	2.29	3.32	5.61
Cant strips				
Fiber				
3x3	L.F.	0.30	1.51	1.81
4x4	"	0.40	1.60	2.00
Wood				
3x3	L.F.	1.68	1.60	3.28
06110.60			**Sleepers**	
Sleepers, over concrete				
12" o.c.				
1x2	S.F.	0.26	1.20	1.46
1x3	"	0.36	1.26	1.62
2x4	"	0.75	1.47	2.22
2x6	"	1.10	1.56	2.66
16" o.c.				
1x2	S.F.	0.21	1.06	1.27
1x3	"	0.30	1.06	1.36
2x4	"	0.62	1.26	1.88
2x6	"	0.94	1.32	2.26
06110.65			**Soffits**	
Soffit framing				
2x3	L.F.	0.49	3.79	4.28
2x4	"	0.62	4.08	4.70
2x6	"	0.88	4.42	5.30
2x8	"	1.26	4.82	6.08
06110.70			**Wall Framing**	
Framing wall, studs				
12" o.c.				
2x3	S.F.	0.53	0.98	1.51
2x4	"	0.75	0.98	1.73
2x6	"	1.10	1.06	2.16
2x8	"	1.55	1.10	2.65
16" o.c.				
2x3	S.F.	0.46	0.83	1.29
2x4	"	0.62	0.83	1.45
2x6	"	0.88	0.88	1.76
2x8	"	1.32	0.91	2.23
24" o.c.				
2x3	S.F.	0.34	0.71	1.05
2x4	"	0.50	0.71	1.21
2x6	"	0.75	0.75	1.50
2x8	"	1.01	0.78	1.79
Plates, top or bottom				
2x3	L.F.	0.49	1.56	2.05
2x4	"	0.62	1.66	2.28
2x6	"	0.88	1.77	2.65
2x8	"	1.26	1.89	3.15
Headers, door or window				
2x6				
Single				

Rough Carpentry	UNIT	MAT.	INST.	TOTAL
06110.70 **Wall Framing** (Cont.)				
3' long	EA.	2.83	26.50	29.33
6' long	"	5.69	33.25	38.94
Double				
3' long	EA.	5.69	29.50	35.19
6' long	"	11.25	38.00	49.25
2x8				
Single				
4' long	EA.	5.26	33.25	38.51
8' long	"	10.50	40.75	51.25
Double				
4' long	EA.	10.50	38.00	48.50
8' long	"	21.00	48.25	69.25
2x10				
Single				
5' long	EA.	9.80	40.75	50.55
10' long	"	19.50	53.00	72.50
Double				
5' long	EA.	19.50	44.25	63.75
10' long	"	39.25	53.00	92.25
2x12				
Single				
6' long	EA.	15.25	40.75	56.00
12' long	"	30.50	53.00	83.50
Double				
6' long	EA.	30.50	48.25	78.75
12' long	"	61.00	59.00	120
06115.10 **Floor Sheathing**				
Sub-flooring, plywood, CDX				
1/2" thick	S.F.	0.69	0.66	1.35
5/8" thick	"	0.83	0.75	1.58
3/4" thick	"	0.98	0.88	1.86
Structural plywood				
1/2" thick	S.F.	0.77	0.66	1.43
5/8" thick	"	0.90	0.75	1.65
3/4" thick	"	1.07	0.81	1.88
Board type subflooring				
1x6				
Minimum	S.F.	1.02	1.18	2.20
Maximum	"	1.39	1.32	2.71
1x8				
Minimum	S.F.	1.23	1.11	2.34
Maximum	"	1.39	1.24	2.63
1x10				
Minimum	S.F.	1.45	1.06	2.51
Maximum	"	1.52	1.18	2.70
Underlayment				
Hardboard, 1/4" tempered	S.F.	0.49	0.66	1.15
Plywood, CDX				
3/8" thick	S.F.	0.61	0.66	1.27
1/2" thick	"	0.70	0.70	1.40
5/8" thick	"	0.84	0.75	1.59
3/4" thick	"	1.00	0.81	1.81

Rough Carpentry

	UNIT	MAT.	INST.	TOTAL
06115.20 **Roof Sheathing**				
Sheathing				
Plywood, CDX				
3/8" thick	S.F.	0.61	0.68	1.29
1/2" thick	"	0.70	0.70	1.40
5/8" thick	"	0.84	0.75	1.59
3/4" thick	"	1.00	0.81	1.81
Structural plywood				
3/8" thick	S.F.	0.67	0.68	1.35
1/2" thick	"	0.77	0.70	1.47
5/8" thick	"	0.92	0.75	1.67
3/4" thick	"	1.10	0.81	1.91
06115.30 **Wall Sheathing**				
Sheathing				
Plywood, CDX				
3/8" thick	S.F.	0.61	0.78	1.39
1/2" thick	"	0.70	0.81	1.51
5/8" thick	"	0.84	0.88	1.72
3/4" thick	"	1.00	0.96	1.96
Waferboard				
3/8" thick	S.F.	0.53	0.78	1.31
1/2" thick	"	0.66	0.81	1.47
5/8" thick	"	0.80	0.88	1.68
3/4" thick	"	0.87	0.96	1.83
Structural plywood				
3/8" thick	S.F.	0.67	0.78	1.45
1/2" thick	"	0.77	0.81	1.58
5/8" thick	"	0.92	0.88	1.80
3/4" thick	"	1.10	0.96	2.06
Gypsum, 1/2" thick	"	0.64	0.81	1.45
Asphalt impregnated fiberboard, 1/2" thick	"	0.72	0.81	1.53
06125.10 **Wood Decking**				
Decking, T&G solid				
Cedar				
3" thick	S.F.	8.29	1.32	9.61
4" thick	"	10.00	1.41	11.41
Fir				
3" thick	S.F.	3.46	1.32	4.78
4" thick	"	4.23	1.41	5.64
Southern yellow pine				
3" thick	S.F.	3.40	1.51	4.91
4" thick	"	3.65	1.63	5.28
White pine				
3" thick	S.F.	4.30	1.32	5.62
4" thick	"	5.68	1.41	7.09
06130.10 **Heavy Timber**				
Mill framed structures				
Beams to 20' long				
Douglas fir				
6x8	L.F.	6.12	6.71	12.83
6x10	"	7.22	6.94	14.16

Rough Carpentry

Rough Carpentry	UNIT	MAT.	INST.	TOTAL
06130.10	**Heavy Timber** *(Cont.)*			
6x12	L.F.	8.64	7.45	16.09
6x14	"	10.25	7.74	17.99
6x16	"	11.25	8.05	19.30
8x10	"	9.58	6.94	16.52
8x12	"	11.25	7.45	18.70
8x14	"	13.00	7.74	20.74
8x16	"	14.75	8.05	22.80
Southern yellow pine				
6x8	L.F.	4.85	6.71	11.56
6x10	"	5.89	6.94	12.83
6x12	"	7.54	7.45	14.99
6x14	"	8.64	7.74	16.38
6x16	"	9.58	8.05	17.63
8x10	"	8.00	6.94	14.94
8x12	"	9.73	7.45	17.18
8x14	"	11.25	7.74	18.99
8x16	"	12.75	8.05	20.80
Columns to 12' high				
Douglas fir				
6x6	L.F.	4.40	10.00	14.40
8x8	"	7.54	10.00	17.54
10x10	"	13.25	11.25	24.50
12x12	"	16.25	11.25	27.50
Southern yellow pine				
6x6	L.F.	3.78	10.00	13.78
8x8	"	6.37	10.00	16.37
10x10	"	9.90	11.25	21.15
12x12	"	13.75	11.25	25.00
Posts, treated				
4x4	L.F.	2.04	2.12	4.16
6x6	"	4.79	2.65	7.44
06190.20	**Wood Trusses**			
Truss, fink, 2x4 members				
3-in-12 slope				
24' span	EA.	74.00	58.00	132
26' span	"	86.00	58.00	144
28' span	"	89.00	61.00	150
30' span	"	95.00	61.00	156
34' span	"	100	65.00	165
38' span	"	120	65.00	185
5-in-12 slope				
24' span	EA.	76.00	59.00	135
28' span	"	89.00	61.00	150
30' span	"	100	63.00	163
32' span	"	110	63.00	173
40' span	"	150	67.00	217
Gable, 2x4 members				
5-in-12 slope				
24' span	EA.	100	59.00	159
26' span	"	110	59.00	169
28' span	"	130	61.00	191
30' span	"	130	63.00	193

Rough Carpentry

Rough Carpentry	UNIT	MAT.	INST.	TOTAL
06190.20 **Wood Trusses** *(Cont.)*				
32' span	EA.	140	63.00	203
36' span	"	150	65.00	215
40' span	"	160	67.00	227
King post type, 2x4 members				
4-in-12 slope				
16' span	EA.	63.00	54.00	117
18' span	"	67.00	56.00	123
24' span	"	72.00	59.00	131
26' span	"	77.00	59.00	136
30' span	"	98.00	63.00	161
34' span	"	100	63.00	163
38' span	"	130	65.00	195
42' span	"	150	69.00	219

Finish Carpentry

Finish Carpentry	UNIT	MAT.	INST.	TOTAL
06200.10 **Finish Carpentry**				
Mouldings and trim				
Apron, flat				
9/16 x 2	L.F.	1.30	2.65	3.95
9/16 x 3-1/2	"	2.39	2.79	5.18
Base				
Colonial				
7/16 x 2-1/4	L.F.	1.52	2.65	4.17
7/16 x 3	"	1.81	2.65	4.46
7/16 x 3-1/4	"	1.98	2.65	4.63
9/16 x 3	"	1.95	2.79	4.74
9/16 x 3-1/4	"	2.07	2.79	4.86
11/16 x 2-1/4	"	1.81	2.95	4.76
Ranch				
7/16 x 2-1/4	L.F.	1.57	2.65	4.22
7/16 x 3-1/4	"	2.07	2.65	4.72
9/16 x 2-1/4	"	1.71	2.79	4.50
9/16 x 3	"	2.05	2.79	4.84
9/16 x 3-1/4	"	2.07	2.79	4.86
Casing				
11/16 x 2-1/2	L.F.	1.58	2.41	3.99
11/16 x 3-1/2	"	2.04	2.52	4.56
Chair rail				
9/16 x 2-1/2	L.F.	1.65	2.65	4.30
9/16 x 3-1/2	"	2.48	2.65	5.13
Closet pole				
1-1/8" dia.	L.F.	1.89	3.54	5.43
1-5/8" dia.	"	2.18	3.54	5.72
Cove				
9/16 x 1-3/4	L.F.	1.27	2.65	3.92

Finish Carpentry	UNIT	MAT.	INST.	TOTAL
06200.10 — **Finish Carpentry** *(Cont.)*				
11/16 x 2-3/4	L.F.	1.75	2.65	4.40
Crown				
9/16 x 1-5/8	L.F.	1.90	3.54	5.44
9/16 x 2-5/8	"	2.10	4.08	6.18
11/16 x 3-5/8	"	2.51	4.42	6.93
11/16 x 4-1/4	"	3.52	4.82	8.34
11/16 x 5-1/4	"	4.54	5.31	9.85
Drip cap				
1-1/16 x 1-5/8	L.F.	1.80	2.65	4.45
Glass bead				
3/8 x 3/8	L.F.	0.51	3.32	3.83
1/2 x 9/16	"	0.65	3.32	3.97
5/8 x 5/8	"	0.68	3.32	4.00
3/4 x 3/4	"	0.84	3.32	4.16
Half round				
1/2	L.F.	0.63	2.12	2.75
5/8	"	0.84	2.12	2.96
3/4	"	1.04	2.12	3.16
Lattice				
1/4 x 7/8	L.F.	0.49	2.12	2.61
1/4 x 1-1/8	"	0.52	2.12	2.64
1/4 x 1-3/8	"	0.56	2.12	2.68
1/4 x 1-3/4	"	0.65	2.12	2.77
1/4 x 2	"	0.74	2.12	2.86
Ogee molding				
5/8 x 3/4	L.F.	1.04	2.65	3.69
11/16 x 1-1/8	"	1.57	2.65	4.22
11/16 x 1-3/8	"	1.89	2.65	4.54
Parting bead				
3/8 x 7/8	L.F.	0.81	3.32	4.13
Quarter round				
1/4 x 1/4	L.F.	0.24	2.12	2.36
3/8 x 3/8	"	0.37	2.12	2.49
1/2 x 1/2	"	0.51	2.12	2.63
11/16 x 11/16	"	0.54	2.30	2.84
3/4 x 3/4	"	0.60	2.30	2.90
1-1/16 x 1-1/16	"	0.75	2.41	3.16
Railings, balusters				
1-1/8 x 1-1/8	L.F.	2.75	5.31	8.06
1-1/2 x 1-1/2	"	3.22	4.82	8.04
Screen moldings				
1/4 x 3/4	L.F.	0.66	4.42	5.08
5/8 x 5/16	"	0.83	4.42	5.25
Shoe				
7/16 x 11/16	L.F.	0.83	2.12	2.95
Sash beads				
1/2 x 3/4	L.F.	1.02	4.42	5.44
1/2 x 7/8	"	1.21	4.42	5.63
1/2 x 1-1/8	"	1.28	4.82	6.10
5/8 x 7/8	"	1.28	4.82	6.10
Stop				
5/8 x 1-5/8				
Colonial	L.F.	0.65	3.32	3.97

Finish Carpentry	UNIT	MAT.	INST.	TOTAL

06200.10 — Finish Carpentry *(Cont.)*

	UNIT	MAT.	INST.	TOTAL
Ranch	L.F.	0.63	3.32	3.95
Stools				
11/16 x 2-1/4	L.F.	2.93	5.90	8.83
11/16 x 2-1/2	"	3.11	5.90	9.01
11/16 x 5-1/4	"	3.33	6.64	9.97
Exterior trim, casing, select pine, 1x3	"	2.18	2.65	4.83
Douglas fir				
1x3	L.F.	1.16	2.65	3.81
1x4	"	1.57	2.65	4.22
1x6	"	1.90	2.95	4.85
1x8	"	2.75	3.32	6.07
Cornices, white pine, #2 or better				
1x2	L.F.	0.71	2.65	3.36
1x4	"	1.07	2.65	3.72
1x6	"	1.69	2.95	4.64
1x8	"	1.90	3.12	5.02
1x10	"	2.57	3.32	5.89
1x12	"	3.22	3.54	6.76
Shelving, pine				
1x8	L.F.	2.27	4.08	6.35
1x10	"	2.75	4.24	6.99
1x12	"	3.60	4.42	8.02
Plywood shelf, 3/4", with edge band, 12" wide	"	2.48	5.31	7.79
Adjustable shelf, and rod, 12" wide				
3' to 4' long	EA.	16.75	13.25	30.00
5' to 8' long	"	31.50	17.75	49.25
Prefinished wood shelves with brackets and supports				
8" wide				
3' long	EA.	49.50	13.25	62.75
4' long	"	57.00	13.25	70.25
6' long	"	83.00	13.25	96.25
10" wide				
3' long	EA.	55.00	13.25	68.25
4' long	"	78.00	13.25	91.25
6' long	"	92.00	13.25	105

06220.10 — Millwork

	UNIT	MAT.	INST.	TOTAL
Countertop, laminated plastic				
25" x 7/8" thick				
Minimum	L.F.	18.00	13.25	31.25
Average	"	32.25	17.75	50.00
Maximum	"	54.00	21.25	75.25
25" x 1-1/4" thick				
Minimum	L.F.	21.50	17.75	39.25
Average	"	35.75	21.25	57.00
Maximum	"	57.00	26.50	83.50
Add for cutouts	EA.		33.25	33.25
Backsplash, 4" high, 7/8" thick	L.F.	18.00	10.50	28.50
Plywood, sanded, A-C				
1/4" thick	S.F.	1.26	1.77	3.03
3/8" thick	"	1.39	1.89	3.28
1/2" thick	"	1.53	2.04	3.57
A-D				

Finish Carpentry	UNIT	MAT.	INST.	TOTAL
06220.10 **Millwork** *(Cont.)*				
1/4" thick	S.F.	1.15	1.77	2.92
3/8" thick	"	1.32	1.89	3.21
1/2" thick	"	1.48	2.04	3.52
Base cabinets, 34-1/2" high, 24" deep, hardwood, no tops				
Minimum	L.F.	68.00	21.25	89.25
Average	"	120	26.50	147
Maximum	"	180	35.50	216
Wall cabinets				
Minimum	L.F.	55.00	17.75	72.75
Average	"	88.00	21.25	109
Maximum	"	170	26.50	197

Wood Treatment	UNIT	MAT.	INST.	TOTAL
06300.10 **Wood Treatment**				
Creosote preservative treatment				
8 lb/cf	B.F.			0.50
10 lb/cf	"			0.65
Salt preservative treatment				
Oil borne				
Minimum	B.F.			0.38
Maximum	"			0.61
Water borne				
Minimum	B.F.			0.29
Maximum	"			0.47
Fire retardant treatment				
Minimum	B.F.			0.63
Maximum	"			0.76
Kiln dried, softwood, add to framing costs				
1" thick	B.F.			0.21
2" thick	"			0.32
3" thick	"			0.44
4" thick	"			0.57

Architectural Woodwork	UNIT	MAT.	INST.	TOTAL
06420.10		**Panel Work**		
Hardboard, tempered, 1/4" thick				
Natural faced	S.F.	0.70	1.32	2.02
Plastic faced	"	1.18	1.51	2.69
Pegboard, natural	"	0.81	1.32	2.13
Plastic faced	"	1.07	1.51	2.58
Untempered, 1/4" thick				
Natural faced	S.F.	0.64	1.32	1.96
Plastic faced	"	1.14	1.51	2.65
Pegboard, natural	"	0.67	1.32	1.99
Plastic faced	"	1.03	1.51	2.54
Plywood unfinished, 1/4" thick				
Birch				
Natural	S.F.	1.59	1.77	3.36
Select	"	2.07	1.77	3.84
Knotty pine	"	2.07	1.77	3.84
Cedar (closet lining)				
Standard boards T&G	S.F.	1.99	1.77	3.76
Particle board	"	1.55	1.77	3.32
Plywood, prefinished, 1/4" thick, premium grade				
Birch veneer	S.F.	2.77	2.12	4.89
Cherry veneer	"	3.13	2.12	5.25
Chestnut veneer	"	5.55	2.12	7.67
Lauan veneer	"	0.99	2.12	3.11
Mahogany veneer	"	3.18	2.12	5.30
Oak veneer (red)	"	3.18	2.12	5.30
Pecan veneer	"	3.75	2.12	5.87
Rosewood veneer	"	5.28	2.12	7.40
Teak veneer	"	6.78	2.12	8.90
Walnut veneer	"	4.64	2.12	6.76
06430.10		**Stairwork**		
Risers, 1x8, 42" wide				
White oak	EA.	22.75	26.50	49.25
Pine	"	16.00	26.50	42.50
Treads, 1-1/16" x 9-1/2" x 42"				
White oak	EA.	32.50	33.25	65.75
06440.10		**Columns**		
Column, hollow, round wood				
12" diameter				
10' high	EA.	660	74.00	734
12' high	"	800	80.00	880
14' high	"	970	89.00	1,059
16' high	"	1,200	110	1,310
24" diameter				
16' high	EA.	2,730	110	2,840
18' high	"	3,100	120	3,220
20' high	"	3,830	120	3,950
22' high	"	4,010	120	4,130
24' high	"	4,380	120	4,500

Moisture Protection	UNIT	MAT.	INST.	TOTAL
07100.10 **Waterproofing**				
Membrane waterproofing, elastomeric				
Butyl				
1/32" thick	S.F.	1.05	1.65	2.70
1/16" thick	"	1.37	1.72	3.09
Butyl with nylon				
1/32" thick	S.F.	1.23	1.65	2.88
1/16" thick	"	1.48	1.72	3.20
Neoprene				
1/32" thick	S.F.	1.71	1.65	3.36
1/16" thick	"	2.80	1.72	4.52
Neoprene with nylon				
1/32" thick	S.F.	1.88	1.65	3.53
1/16" thick	"	3.02	1.72	4.74
Bituminous membrane waterproofing, asphalt felt, 15 lb.				
One ply	S.F.	0.62	1.03	1.65
Two ply	"	0.73	1.25	1.98
Three ply	"	0.91	1.47	2.38
Four ply	"	1.03	1.72	2.75
Five ply	"	1.12	2.17	3.29
Modified asphalt membrane waterproofing, fibrous asphalt				
One ply	S.F.	0.71	1.72	2.43
Two ply	"	1.07	2.06	3.13
Three ply	"	1.21	2.29	3.50
Four ply	"	1.47	2.75	4.22
Five ply	"	1.71	3.30	5.01
Asphalt coated protective board				
1/8" thick	S.F.	0.40	1.03	1.43
1/4" thick	"	0.56	1.03	1.59
3/8" thick	"	0.61	1.03	1.64
1/2" thick	"	0.74	1.08	1.82
Cement protective board				
3/8" thick	S.F.	1.38	1.37	2.75
1/2" thick	"	1.93	1.37	3.30
Fluid applied, neoprene				
50 mil	S.F.	1.76	1.37	3.13
90 mil	"	2.91	1.37	4.28
Tab extended polyurethane				
.050" thick	S.F.	1.79	1.03	2.82
Fluid applied rubber based polyurethane				
6 mil	S.F.	0.94	1.29	2.23
15 mil	"	1.79	1.03	2.82
Bentonite waterproofing, panels				
3/16" thick	S.F.	1.60	1.03	2.63
1/4" thick	"	1.82	1.03	2.85
5/8" thick	"	2.71	1.08	3.79
Granular admixtures, trowel on, 3/8" thick	"	1.65	1.03	2.68
Metallic oxide waterproofing, iron compound, troweled				
5/8" thick	S.F.	1.65	1.03	2.68
3/4" thick	"	2.00	1.17	3.17

Moisture Protection	UNIT	MAT.	INST.	TOTAL
07150.10 **Dampproofing**				
Silicone dampproofing, sprayed on				
Concrete surface				
1 coat	S.F.	0.64	0.22	0.86
2 coats	"	1.06	0.31	1.37
Concrete block				
1 coat	S.F.	0.64	0.27	0.91
2 coats	"	1.06	0.37	1.43
Brick				
1 coat	S.F.	0.74	0.31	1.05
2 coats	"	1.15	0.41	1.56
07160.10 **Bituminous Dampproofing**				
Building paper, asphalt felt				
15 lb	S.F.	0.15	1.65	1.80
30 lb	"	0.28	1.72	2.00
Asphalt dampproofing, troweled, cold, primer plus				
1 coat	S.F.	0.68	1.37	2.05
2 coats	"	1.43	2.06	3.49
3 coats	"	2.04	2.58	4.62
Fibrous asphalt dampproofing, hot troweled, primer plus				
1 coat	S.F.	0.68	1.65	2.33
2 coats	"	1.43	2.29	3.72
3 coats	"	2.04	2.94	4.98
Asphaltic paint dampproofing, per coat				
Brush on	S.F.	0.28	0.58	0.86
Spray on	"	0.49	0.45	0.94
07190.10 **Vapor Barriers**				
Vapor barrier, polyethylene				
2 mil	S.F.	0.01	0.20	0.21
6 mil	"	0.04	0.20	0.24
8 mil	"	0.05	0.22	0.27
10 mil	"	0.06	0.22	0.28

Insulation	UNIT	MAT.	INST.	TOTAL
07210.10 **Batt Insulation**				
Ceiling, fiberglass, unfaced				
3-1/2" thick, R11	S.F.	0.38	0.48	0.86
6" thick, R19	"	0.50	0.55	1.05
9" thick, R30	"	1.00	0.63	1.63
Suspended ceiling, unfaced				
3-1/2" thick, R11	S.F.	0.38	0.45	0.83
6" thick, R19	"	0.50	0.51	1.01
9" thick, R30	"	1.00	0.58	1.58
Crawl space, unfaced				

Insulation	UNIT	MAT.	INST.	TOTAL
07210.10 **Batt Insulation** *(Cont.)*				
3-1/2" thick, R11	S.F.	0.38	0.63	1.01
6" thick, R19	"	0.50	0.68	1.18
9" thick, R30	"	1.00	0.75	1.75
Wall, fiberglass				
Paper backed				
2" thick, R7	S.F.	0.25	0.43	0.68
3" thick, R8	"	0.27	0.45	0.72
4" thick, R11	"	0.45	0.48	0.93
6" thick, R19	"	0.67	0.51	1.18
Foil backed, 1 side				
2" thick, R7	S.F.	0.58	0.43	1.01
3" thick, R11	"	0.61	0.45	1.06
4" thick, R14	"	0.64	0.48	1.12
6" thick, R21	"	0.67	0.51	1.18
Foil backed, 2 sides				
2" thick, R7	S.F.	0.66	0.48	1.14
3" thick, R11	"	0.83	0.51	1.34
4" thick, R14	"	0.99	0.55	1.54
6" thick, R21	"	1.06	0.58	1.64
Unfaced				
2" thick, R7	S.F.	0.37	0.43	0.80
3" thick, R9	"	0.41	0.45	0.86
4" thick, R11	"	0.45	0.48	0.93
6" thick, R19	"	0.58	0.51	1.09
Mineral wool batts				
Paper backed				
2" thick, R6	S.F.	0.24	0.43	0.67
4" thick, R12	"	0.52	0.45	0.97
6" thick, R19	"	0.67	0.51	1.18
Fasteners, self adhering, attached to ceiling deck				
2-1/2" long	EA.	0.19	0.68	0.87
4-1/2" long	"	0.22	0.75	0.97
Capped, self-locking washers for fastening insulation	"	0.19	0.41	0.60
07210.20 **Board Insulation**				
Insulation, rigid				
Fiberglass, roof				
0.75" thick, R2.78	S.F.	0.36	0.37	0.73
1.06" thick, R4.17	"	0.55	0.39	0.94
1.31" thick, R5.26	"	0.73	0.41	1.14
1.63" thick, R6.67	"	0.91	0.43	1.34
2.25" thick, R8.33	"	1.00	0.45	1.45
Composite board, roof				
1-1/2" thick, R6.67	S.F.	1.19	0.41	1.60
1-5/8" thick, R7.69	"	1.27	0.43	1.70
2" thick, R10.0	"	2.29	0.45	2.74
2-1/4" thick, R12.50	"	2.55	0.48	3.03
2-1/2" thick, R14.29	"	2.77	0.51	3.28
2-3/4" thick, R16.67	"	3.04	0.55	3.59
3-1/4" thick, R20.00	"	3.91	0.58	4.49
Perlite board, roof				
1.00" thick, R2.78	S.F.	0.77	0.34	1.11
1.50" thick, R4.17	"	1.19	0.35	1.54

Insulation	UNIT	MAT.	INST.	TOTAL
07210.20 **Board Insulation** *(Cont.)*				
2.00" thick, R5.92	S.F.	1.47	0.37	1.84
2.50" thick, R6.67	"	1.80	0.39	2.19
3.00" thick, R8.33	"	2.26	0.41	2.67
4.00" thick, R10.00	"	2.51	0.43	2.94
5.25" thick, R14.29	"	2.77	0.45	3.22
Rigid urethane				
Roof				
1" thick, R6.67	S.F.	1.03	0.34	1.37
1.20" thick, R8.33	"	1.18	0.35	1.53
1.50" thick, R11.11	"	1.40	0.35	1.75
2" thick, R14.29	"	1.82	0.37	2.19
2.25" thick, R16.67	"	2.38	0.39	2.77
Wall				
1" thick, R6.67	S.F.	1.03	0.43	1.46
1.5" thick, R11.11	"	1.40	0.45	1.85
2" thick, R14.29	"	1.87	0.48	2.35
Polystyrene				
Roof				
1.0" thick, R4.17	S.F.	0.41	0.34	0.75
1.5" thick, R6.26	"	0.63	0.35	0.98
2.0" thick, R8.33	"	0.77	0.37	1.14
Wall				
1.0" thick, R4.17	S.F.	0.41	0.43	0.84
1.5" thick, R6.26	"	0.63	0.45	1.08
2.0" thick, R8.33	"	0.77	0.48	1.25
Rigid board insulation, deck				
Mineral fiberboard				
1" thick, R3.0	S.F.	0.60	0.34	0.94
2" thick, R5.26	"	1.32	0.37	1.69
Fiberglass				
1" thick, R4.3	S.F.	1.02	0.34	1.36
2" thick, R8.5	"	1.51	0.37	1.88
Polystyrene				
1" thick, R5.4	S.F.	0.41	0.34	0.75
2" thick, R10.8	"	1.04	0.37	1.41
Urethane				
.75" thick, R5.4	S.F.	0.94	0.34	1.28
1" thick, R6.4	"	1.12	0.34	1.46
1.5" thick, R10.7	"	1.35	0.35	1.70
2" thick, R14.3	"	1.54	0.37	1.91
Foamglass				
1" thick, R1.8	S.F.	1.52	0.34	1.86
2" thick, R5.26	"	1.95	0.37	2.32
Wood fiber				
1" thick, R3.85	S.F.	1.73	0.34	2.07
2" thick, R7.7	"	2.09	0.37	2.46
Particle board				
3/4" thick, R2.08	S.F.	0.91	0.34	1.25
1" thick, R2.77	"	0.95	0.34	1.29
2" thick, R5.50	"	1.26	0.37	1.63

Insulation	UNIT	MAT.	INST.	TOTAL
07210.60 **Loose Fill Insulation**				
Blown-in type				
Fiberglass				
5" thick, R11	S.F.	0.36	0.34	0.70
6" thick, R13	"	0.41	0.41	0.82
9" thick, R19	"	0.50	0.58	1.08
Rockwool, attic application				
6" thick, R13	S.F.	0.33	0.41	0.74
8" thick, R19	"	0.39	0.51	0.90
10" thick, R22	"	0.47	0.63	1.10
12" thick, R26	"	0.59	0.68	1.27
15" thick, R30	"	0.71	0.82	1.53
Poured type				
Fiberglass				
1" thick, R4	S.F.	0.33	0.25	0.58
2" thick, R8	"	0.61	0.29	0.90
3" thick, R12	"	0.90	0.34	1.24
4" thick, R16	"	1.18	0.41	1.59
Mineral wool				
1" thick, R3	S.F.	0.36	0.25	0.61
2" thick, R6	"	0.67	0.29	0.96
3" thick, R9	"	1.02	0.34	1.36
4" thick, R12	"	1.18	0.41	1.59
Vermiculite or perlite				
2" thick, R4.8	S.F.	0.77	0.29	1.06
3" thick, R7.2	"	1.10	0.34	1.44
4" thick, R9.6	"	1.43	0.41	1.84
Masonry, poured vermiculite or perlite				
4" block	S.F.	0.33	0.20	0.53
6" block	"	0.41	0.25	0.66
8" block	"	0.60	0.29	0.89
10" block	"	0.71	0.31	1.02
12" block	"	0.88	0.34	1.22
07210.70 **Sprayed Insulation**				
Foam, sprayed on				
Polystyrene				
1" thick, R4	S.F.	0.44	0.41	0.85
2" thick, R8	"	0.85	0.55	1.40
Urethane				
1" thick, R4	S.F.	0.41	0.41	0.82
2" thick, R8	"	0.79	0.55	1.34
07250.10 **Fireproofing**				
Sprayed on				
1" thick				
On beams	S.F.	0.71	0.91	1.62
On columns	"	0.72	0.82	1.54
On decks				
Flat surface	S.F.	0.72	0.41	1.13
Fluted surface	"	0.92	0.51	1.43

Insulation	UNIT	MAT.	INST.	TOTAL
07250.10 **Fireproofing** *(Cont.)*				
1-1/2" thick				
On beams	S.F.	1.28	1.17	2.45
On columns	"	1.45	1.03	2.48
On decks				
Flat surface	S.F.	1.08	0.51	1.59
Fluted surface	"	1.28	0.68	1.96

Shingles And Tiles	UNIT	MAT.	INST.	TOTAL
07310.10 **Asphalt Shingles**				
Standard asphalt shingles, strip shingles				
210 lb/square	SQ.	68.00	51.00	119
235 lb/square	"	72.00	57.00	129
240 lb/square	"	75.00	64.00	139
260 lb/square	"	110	73.00	183
300 lb/square	"	120	85.00	205
385 lb/square	"	160	100	260
Roll roofing, mineral surface				
90 lb	SQ.	38.50	36.50	75.00
110 lb	"	64.00	42.50	107
140 lb	"	65.00	51.00	116
07310.50 **Metal Shingles**				
Aluminum, .020" thick				
Plain	SQ.	260	100	360
Colors	"	280	100	380
Steel, galvanized				
26 ga.				
Plain	SQ.	260	100	360
Colors	"	340	100	440
24 ga.				
Plain	SQ.	280	100	380
Colors	"	350	100	450
Porcelain enamel, 22 ga.				
Minimum	SQ.	800	130	930
Average	"	920	130	1,050
Maximum	"	1,030	130	1,160
07310.60 **Slate Shingles**				
Slate shingles				
Pennsylvania				
Ribbon	SQ.	630	250	880
Clear	"	810	250	1,060
Vermont				
Black	SQ.	680	250	930

Shingles And Tiles	UNIT	MAT.	INST.	TOTAL

07310.60 — Slate Shingles *(Cont.)*

	UNIT	MAT.	INST.	TOTAL
Gray	SQ.	750	250	1,000
Green	"	760	250	1,010
Red	"	1,380	250	1,630
Replacement shingles				
Small jobs	EA.	12.00	17.00	29.00
Large jobs	S.F.	9.51	8.50	18.01

07310.70 — Wood Shingles

	UNIT	MAT.	INST.	TOTAL
Wood shingles, on roofs				
White cedar, #1 shingles				
4" exposure	SQ.	230	170	400
5" exposure	"	210	130	340
#2 shingles				
4" exposure	SQ.	160	170	330
5" exposure	"	140	130	270
Resquared and rebutted				
4" exposure	SQ.	210	170	380
5" exposure	"	170	130	300
On walls				
White cedar, #1 shingles				
4" exposure	SQ.	230	250	480
5" exposure	"	210	200	410
6" exposure	"	170	170	340
#2 shingles				
4" exposure	SQ.	160	250	410
5" exposure	"	140	200	340
6" exposure	"	120	170	290
Add for fire retarding	"			110

07310.80 — Wood Shakes

	UNIT	MAT.	INST.	TOTAL
Shakes, hand split, 24" red cedar, on roofs				
5" exposure	SQ.	240	250	490
7" exposure	"	230	200	430
9" exposure	"	210	170	380
On walls				
6" exposure	SQ.	220	250	470
8" exposure	"	210	200	410
10" exposure	"	200	170	370
Add for fire retarding	"			75.00

Roofing And Siding	UNIT	MAT.	INST.	TOTAL

07410.10   Manufactured Roofs

	UNIT	MAT.	INST.	TOTAL
Aluminum roof panels, for steel framing				
Corrugated				
Unpainted finish				
.024"	S.F.	1.79	1.27	3.06
.030"	"	2.09	1.27	3.36
Painted finish				
.024"	S.F.	2.26	1.27	3.53
.030"	"	2.77	1.27	4.04
V-beam				
Unpainted finish				
.032"	S.F.	2.33	1.27	3.60
.040"	"	2.80	1.27	4.07
.050"	"	3.53	1.27	4.80
Painted finish				
.032"	S.F.	3.03	1.27	4.30
.040"	"	3.63	1.27	4.90
.050"	"	4.34	1.27	5.61
Steel roof panels, for structural steel framing				
Corrugated, painted				
18 ga.	S.F.	5.61	1.27	6.88
20 ga.	"	5.22	1.27	6.49
22 ga.	"	4.62	1.27	5.89
Box rib, painted				
18 ga.	S.F.	6.54	1.34	7.88
20 ga.	"	5.39	1.34	6.73
22 ga.	"	4.78	1.34	6.12
4" rib, painted				
18 ga.	S.F.	7.53	1.41	8.94
20 ga.	"	6.38	1.41	7.79
22 ga.	"	5.72	1.41	7.13
Standing seam roof				
2" high seam, painted				
22 ga.	S.F.	7.31	2.04	9.35
24 ga.	"	7.15	2.04	9.19
26 ga.	"	6.93	2.04	8.97

07410.30   Manufactured Walls

	UNIT	MAT.	INST.	TOTAL
Sandwich panels with 1-1/2" fiberglass insulation				
Galvanized 18 ga. steel interior panels				
Exterior panels				
16 ga. aluminum	S.F.	6.91	7.87	14.78
18 ga. galvanized steel	"	9.80	7.87	17.67
20 ga. painted steel	"	9.96	7.87	17.83
20 ga. stainless steel	"	10.25	7.87	18.12
Metal liner panels, 1-3/8" thick, 24" wide				
Galvanized				
22 ga.	S.F.	3.64	1.96	5.60
20 ga.	"	4.01	1.96	5.97
18 ga.	"	4.93	1.96	6.89
Primed				
22 ga.	S.F.	2.83	1.96	4.79
20 ga.	"	3.26	1.96	5.22
18 ga.	"	4.01	1.96	5.97

Roofing And Siding	UNIT	MAT.	INST.	TOTAL
07440.10 — Aggregate Coated Panels				
Dryvit type system				
1" thick	S.F.	2.65	1.96	4.61
1-1/2" thick	"	2.76	2.11	4.87
2" thick	"	2.93	2.46	5.39
07460.10 — Metal Siding Panels				
Aluminum siding panels				
Corrugated				
Plain finish				
.024"	S.F.	1.54	2.36	3.90
.032"	"	1.81	2.36	4.17
Painted finish				
.024"	S.F.	1.92	2.36	4.28
.032"	"	2.20	2.36	4.56
V. beam				
Plain finish				
.032"	S.F.	2.22	2.36	4.58
.040"	"	2.55	2.36	4.91
.050"	"	3.23	2.36	5.59
Painted finish				
.032"	S.F.	2.72	2.36	5.08
.040"	"	3.22	2.36	5.58
.050"	"	4.15	2.36	6.51
4" rib				
Plain finish				
.032"	S.F.	2.03	2.68	4.71
.040"	"	2.25	2.68	4.93
.050"	"	2.75	2.68	5.43
Painted finish				
.032"	S.F.	2.64	2.68	5.32
.040"	"	2.75	2.68	5.43
.050"	"	3.13	2.68	5.81
Steel siding panels				
Corrugated				
22 ga.	S.F.	1.95	3.93	5.88
24 ga.	"	1.79	3.93	5.72
26 ga.	"	1.62	3.93	5.55
Box rib				
20 ga.	S.F.	3.37	3.93	7.30
22 ga.	"	2.82	3.93	6.75
24 ga.	"	2.48	3.93	6.41
26 ga.	"	1.92	3.93	5.85
07460.50 — Plastic Siding				
Horizontal vinyl siding, solid				
8" wide				
Standard	S.F.	1.23	2.04	3.27
Insulated	"	1.49	2.04	3.53
10" wide				
Standard	S.F.	1.27	1.89	3.16
Insulated	"	1.52	1.89	3.41
Vinyl moldings for doors and windows	L.F.	0.79	2.12	2.91

Roofing And Siding	UNIT	MAT.	INST.	TOTAL
07460.60 — **Plywood Siding**				
Rough sawn cedar, 3/8" thick	S.F.	1.79	1.77	3.56
Fir, 3/8" thick	"	0.99	1.77	2.76
Texture 1-11, 5/8" thick				
Cedar	S.F.	2.42	1.89	4.31
Fir	"	1.69	1.89	3.58
Redwood	"	2.59	1.82	4.41
Southern Yellow Pine	"	1.37	1.89	3.26
07460.70 — **Steel Siding**				
Ribbed, sheets, galvanized				
22 ga.	S.F.	1.98	2.36	4.34
24 ga.	"	1.79	2.36	4.15
26 ga.	"	1.38	2.36	3.74
28 ga.	"	1.18	2.36	3.54
Primed				
24 ga.	S.F.	2.33	2.36	4.69
26 ga.	"	1.65	2.36	4.01
28 ga.	"	1.38	2.36	3.74
07460.80 — **Wood Siding**				
Beveled siding, cedar				
A grade				
1/2 x 6	S.F.	3.60	2.65	6.25
1/2 x 8	"	3.68	2.12	5.80
3/4 x 10	"	4.74	1.77	6.51
Clear				
1/2 x 6	S.F.	4.01	2.65	6.66
1/2 x 8	"	4.10	2.12	6.22
3/4 x 10	"	5.50	1.77	7.27
B grade				
1/2 x 6	S.F.	3.88	2.65	6.53
1/2 x 8	"	4.37	21.25	25.62
3/4 x 10	"	4.12	1.77	5.89
Board and batten				
Cedar				
1x6	S.F.	4.26	2.65	6.91
1x8	"	3.88	2.12	6.00
1x10	"	3.50	1.89	5.39
1x12	"	3.14	1.71	4.85
Pine				
1x6	S.F.	1.29	2.65	3.94
1x8	"	1.26	2.12	3.38
1x10	"	1.21	1.89	3.10
1x12	"	1.12	1.71	2.83
Redwood				
1x6	S.F.	5.56	2.65	8.21
1x8	"	5.18	2.12	7.30
1x10	"	4.80	1.89	6.69
1x12	"	4.44	1.71	6.15
Tongue and groove				
Cedar				
1x4	S.F.	4.80	2.95	7.75
1x6	"	4.63	2.79	7.42

Roofing And Siding	UNIT	MAT.	INST.	TOTAL
07460.80 **Wood Siding** *(Cont.)*				
1x8	S.F.	4.33	2.65	6.98
1x10	"	4.26	2.52	6.78
Pine				
1x4	S.F.	1.44	2.95	4.39
1x6	"	1.36	2.79	4.15
1x8	"	1.27	2.65	3.92
1x10	"	1.21	2.52	3.73
Redwood				
1x4	S.F.	5.09	2.95	8.04
1x6	"	4.90	2.79	7.69
1x8	"	4.73	2.65	7.38
1x10	"	4.51	2.52	7.03

Membrane Roofing	UNIT	MAT.	INST.	TOTAL
07510.10 **Built-up Asphalt Roofing**				
Built-up roofing, asphalt felt, including gravel				
2 ply	SQ.	56.00	130	186
3 ply	"	76.00	170	246
4 ply	"	110	200	310
Walkway, for built-up roofs				
3' x 3' x				
1/2" thick	S.F.	5.21	1.70	6.91
3/4" thick	"	6.90	1.70	8.60
1" thick	"	7.60	1.70	9.30
Roof bonds				
Asphalt felt				
10 yrs	SQ.			34.00
20 yrs	"			38.75
Cant strip, 4" x 4"				
Treated wood	L.F.	1.92	1.45	3.37
Foamglass	"	1.03	1.27	2.30
Mineral fiber	"	0.33	1.27	1.60
New gravel for built-up roofing, 400 lb/sq	SQ.	32.00	100	132
Roof gravel (ballast)	C.Y.	21.50	250	272
Aluminum coating, top surfacing, for built-up roofing	SQ.	33.75	85.00	119
Remove 4-ply built-up roof (includes gravel)	"		250	250
Remove & replace gravel, includes flood coat	"	48.50	170	219
07530.10 **Single-ply Roofing**				
Elastic sheet roofing				
Neoprene, 1/16" thick	S.F.	2.69	0.63	3.32
EPDM rubber				
45 mil	S.F.	1.47	0.63	2.10
60 mil	"	2.02	0.63	2.65
PVC				

Membrane Roofing	UNIT	MAT.	INST.	TOTAL
07530.10 **Single-ply Roofing** *(Cont.)*				
45 mil	S.F.	2.03	0.63	2.66
60 mil	"	2.42	0.63	3.05
Flashing				
Pipe flashing, 90 mil thick				
1" pipe	EA.	28.50	12.75	41.25
2" pipe	"	30.75	12.75	43.50
3" pipe	"	31.00	13.50	44.50
4" pipe	"	33.50	13.50	47.00
5" pipe	"	35.75	14.25	50.00
6" pipe	"	39.00	14.25	53.25
8" pipe	"	44.50	15.00	59.50
10" pipe	"	51.00	17.00	68.00
12" pipe	"	62.00	17.00	79.00
Neoprene flashing, 60 mil thick strip				
6" wide	L.F.	1.73	4.25	5.98
12" wide	"	3.42	6.37	9.79
18" wide	"	5.02	8.50	13.52
24" wide	"	6.61	12.75	19.36
Adhesives				
Mastic sealer, applied at joints only				
1/4" bead	L.F.	0.11	0.25	0.36
Fluid applied roofing				
Urethane, 2 components, elastomeric top membrane				
1" thick	S.F.	2.97	0.85	3.82
Vinyl liquid roofing, 2 coats, 2 mils per coat	"	4.65	0.72	5.37
Silicone roofing, 2 coats sprayed, 16 mil per coat	"	3.47	0.85	4.32
Inverted roof system				
Insulated membrane with coarse gravel ballast				
3 ply with 2" polystyrene	S.F.	6.69	0.85	7.54
Ballast, 3/4" through 1-1/2" dia. river gravel, 100lb/sf	"	0.38	51.00	51.38
Walkway for membrane roofs, 1/2" thick	"	2.13	1.70	3.83

Flashing And Sheet Metal	UNIT	MAT.	INST.	TOTAL
07610.10 **Metal Roofing**				
Sheet metal roofing, copper, 16 oz, batten seam	SQ.	1,000	340	1,340
Standing seam	"	970	320	1,290
Aluminum roofing, natural finish				
Corrugated, on steel frame				
.0175" thick	SQ.	120	150	270
.0215" thick	"	160	150	310
.024" thick	"	190	150	340
.032" thick	"	240	150	390
V-beam, on steel frame				
.032" thick	SQ.	230	150	380
.040" thick	"	240	150	390

Flashing And Sheet Metal	UNIT	MAT.	INST.	TOTAL
07610.10　**Metal Roofing** *(Cont.)*				
.050" thick	SQ.	310	150	460
Ridge cap				
.019" thick	L.F.	3.85	1.70	5.55
Corrugated galvanized steel roofing, on steel frame				
28 ga.	SQ.	120	150	270
26 ga.	"	140	150	290
24 ga.	"	160	150	310
22 ga.	"	170	150	320
26 ga., factory insulated with 1" polystyrene	"	350	200	550
Ridge roll				
10" wide	L.F.	1.91	1.70	3.61
20" wide	"	3.89	2.04	5.93
07620.10　**Flashing And Trim**				
Counter flashing				
Aluminum, .032"	S.F.	1.44	5.10	6.54
Stainless steel, .015"	"	4.62	5.10	9.72
Copper				
16 oz.	S.F.	7.15	5.10	12.25
20 oz.	"	8.49	5.10	13.59
24 oz.	"	10.25	5.10	15.35
32 oz.	"	12.50	5.10	17.60
Valley flashing				
Aluminum, .032"	S.F.	1.52	3.18	4.70
Stainless steel, .015	"	4.81	3.18	7.99
Copper				
16 oz.	S.F.	6.78	3.18	9.96
20 oz.	"	8.49	4.25	12.74
24 oz.	"	10.25	3.18	13.43
32 oz.	"	12.50	3.18	15.68
Base flashing				
Aluminum, .040"	S.F.	2.60	4.25	6.85
Stainless steel, .018"	"	5.78	4.25	10.03
Copper				
16 oz.	S.F.	6.78	4.25	11.03
20 oz.	"	8.49	3.18	11.67
24 oz.	"	10.25	4.25	14.50
32 oz.	"	12.50	4.25	16.75
Waterstop, "T" section, 22 ga.				
1-1/2" x 3"	L.F.	3.25	2.55	5.80
2" x 2"	"	3.60	2.55	6.15
4" x 3"	"	4.01	2.55	6.56
6" x 4"	"	4.24	2.55	6.79
8" x 4"	"	5.26	2.55	7.81
Scupper outlets				
10" x 10" x 4"	EA.	34.00	12.75	46.75
22" x 4" x 4"	"	42.00	12.75	54.75
8" x 8" x 5"	"	34.00	12.75	46.75
Flashing and trim, aluminum				
.019" thick	S.F.	1.28	3.64	4.92
.032" thick	"	1.57	3.64	5.21
.040" thick	"	2.69	3.92	6.61
Neoprene sheet flashing, .060" thick	"	2.04	3.18	5.22

Flashing And Sheet Metal

07620.10 — Flashing And Trim *(Cont.)*

	UNIT	MAT.	INST.	TOTAL
Copper, paper backed				
2 oz.	S.F.	2.09	5.10	7.19
Drainage boots, roof, cast iron				
2 x 3	L.F.	53.00	6.37	59.37
3 x 4	"	67.00	6.37	73.37
4 x 5	"	96.00	6.80	103
4 x 6	"	91.00	6.80	97.80
5 x 7	"	110	7.28	117
Pitch pocket, copper, 16 oz.				
4 x 4	EA.	120	12.75	133
6 x 6	"	130	12.75	143
8 x 8	"	140	12.75	153
8 x 10	"	150	12.75	163
8 x 12	"	180	12.75	193
Reglets, copper 10 oz.	L.F.	4.65	3.40	8.05
Stainless steel, .020"	"	2.77	3.40	6.17
Gravel stop				
Aluminum, .032"				
4"	L.F.	0.93	1.70	2.63
6"	"	1.37	1.70	3.07
8"	"	1.84	1.96	3.80
10"	"	2.31	1.96	4.27
Copper, 16 oz.				
4"	L.F.	3.13	1.70	4.83
6"	"	4.67	1.70	6.37
8"	"	6.27	1.96	8.23
10"	"	7.81	1.96	9.77

07620.20 — Gutters And Downspouts

	UNIT	MAT.	INST.	TOTAL
Copper gutter and downspout				
Downspouts, 16 oz. copper				
Round				
3" dia.	L.F.	10.50	3.40	13.90
4" dia.	"	13.00	3.40	16.40
Rectangular, corrugated				
2" x 3"	L.F.	10.25	3.18	13.43
3" x 4"	"	12.50	3.18	15.68
Rectangular, flat surface				
2" x 3"	L.F.	11.50	3.40	14.90
3" x 4"	"	16.50	3.40	19.90
Lead-coated copper downspouts				
Round				
3" dia.	L.F.	13.75	3.18	16.93
4" dia.	"	16.50	3.64	20.14
Rectangular, corrugated				
2" x 3"	L.F.	13.75	3.40	17.15
3" x 4"	"	16.50	3.40	19.90
Rectangular, plain				
2" x 3"	L.F.	9.57	3.40	12.97
3" x 4"	"	11.00	3.40	14.40
Gutters, 16 oz. copper				
Half round				
4" wide	L.F.	9.38	5.10	14.48

Flashing And Sheet Metal	UNIT	MAT.	INST.	TOTAL
07620.20 — **Gutters And Downspouts** *(Cont.)*				
5" wide	L.F.	11.50	5.66	17.16
Type K				
4" wide	L.F.	10.50	5.10	15.60
5" wide	"	11.00	5.66	16.66
Lead-coated copper gutters				
Half round				
4" wide	L.F.	11.00	5.10	16.10
6" wide	"	15.00	5.66	20.66
Type K				
4" wide	L.F.	12.50	5.10	17.60
5" wide	"	13.50	5.66	19.16
Aluminum gutter and downspout				
Downspouts				
2" x 3"	L.F.	1.04	3.40	4.44
3" x 4"	"	1.43	3.64	5.07
4" x 5"	"	1.59	3.92	5.51
Round				
3" dia.	L.F.	1.76	3.40	5.16
4" dia.	"	2.25	3.64	5.89
Gutters, stock units				
4" wide	L.F.	1.70	5.36	7.06
5" wide	"	2.03	5.66	7.69
Galvanized steel gutter and downspout				
Downspouts, round corrugated				
3" dia.	L.F.	1.76	3.40	5.16
4" dia.	"	2.36	3.40	5.76
5" dia.	"	3.52	3.64	7.16
6" dia.	"	4.67	3.64	8.31
Rectangular				
2" x 3"	L.F.	1.59	3.40	4.99
3" x 4"	"	2.28	3.18	5.46
4" x 4"	"	2.86	3.18	6.04
Gutters, stock units				
5" wide				
Plain	L.F.	1.61	5.66	7.27
Painted	"	1.76	5.66	7.42
6" wide				
Plain	L.F.	2.25	6.00	8.25
Painted	"	2.53	6.00	8.53

Roofing Specialties	UNIT	MAT.	INST.	TOTAL
07700.10	**Manufactured Specialties**			
Moisture relief vent				
Aluminum	EA.	19.25	7.28	26.53
Copper	"	42.00	7.28	49.28
Expansion joint				
Aluminum				
Opening to 2.5"	L.F.	10.00	3.64	13.64
Opening to 3.5"	"	11.50	3.92	15.42
Copper, 16 oz.				
Opening to 2.5"	L.F.	24.50	3.64	28.14
Opening to 3.5"	"	30.25	3.92	34.17
Butyl or neoprene				
4" wide				
16 oz. copper bellows	L.F.	23.25	4.25	27.50
28 ga. stainless steel bellows	"	18.00	4.25	22.25
6" wide				
Copper bellows	L.F.	23.25	4.63	27.88
Stainless steel				
Opening to 2.5"	L.F.	18.75	3.64	22.39
Opening to 3.5"	"	23.75	3.92	27.67
Smoke vent, 48" x 48"				
Aluminum	EA.	1,810	130	1,940
Galvanized steel	"	1,590	130	1,720
Heat/smoke vent, 48" x 96"				
Aluminum	EA.	2,750	170	2,920
Galvanized steel	"	2,360	170	2,530
Ridge vent strips				
Mill finish	L.F.	3.90	3.40	7.30
Connectors	EA.	3.85	12.75	16.60
End cap	"	1.92	14.50	16.42
Soffit vents				
Mill finish				
2-1/2" wide	L.F.	0.49	2.04	2.53
3" wide	"	0.60	2.04	2.64
6" wide	"	1.04	2.04	3.08
Roof hatches				
Steel, plain, primed				
2'6" x 3'0"	EA.	630	130	760
2'6" x 4'6"	"	930	170	1,100
2'6" x 8'0"	"	1,440	250	1,690
Galvanized steel				
2'6" x 3'0"	EA.	650	130	780
2'6" x 4'6"	"	980	170	1,150
2'6" x 8'0"	"	1,530	250	1,780
Aluminum				
2'6" x 3'0"	EA.	680	130	810
2'6" x 4'6"	"	1,030	170	1,200
2'6" x 8'0"	"	1,910	250	2,160
Ceiling access doors				
Swing up model, metal frame				
Steel door				
2'6" x 2'6"	EA.	470	51.00	521
2'6" x 3'0"	"	510	51.00	561
Aluminum door				

Roofing Specialties	UNIT	MAT.	INST.	TOTAL
07700.10 **Manufactured Specialties** *(Cont.)*				
2'6" x 2'6"	EA.	510	51.00	561
2'6" x 3'0"	"	550	51.00	601
Swing down model, metal frame				
Steel door				
2'6" x 2'6"	EA.	460	51.00	511
2'6" x 3'0"	"	490	51.00	541
Aluminum door				
2'6" x 2'6"	EA.	470	51.00	521
2'6" x 3'0"	"	510	51.00	561
Gravity ventilators, with curb, base, damper and screen				
Stationary siphon				
6" dia.	EA.	36.75	34.00	70.75
12" dia.	"	56.00	34.00	90.00
24" dia.	"	230	51.00	281
36" dia.	"	430	51.00	481
Wind driven spinner				
6" dia.	EA.	56.00	34.00	90.00
12" dia.	"	75.00	34.00	109
24" dia.	"	280	51.00	331
36" dia.	"	580	51.00	631
Stationary mushroom				
16" dia.	EA.	350	51.00	401
30" dia.	"	790	64.00	854
36" dia.	"	1,020	85.00	1,105
42" dia.	"	1,510	100	1,610

Skylights	UNIT	MAT.	INST.	TOTAL
07810.10 **Plastic Skylights**				
Single thickness, not including mounting curb				
2' x 4'	EA.	360	64.00	424
4' x 4'	"	490	85.00	575
5' x 5'	"	650	130	780
6' x 8'	"	1,390	170	1,560
Double thickness, not including mounting curb				
2' x 4'	EA.	480	64.00	544
4' x 4'	"	600	85.00	685
5' x 5'	"	880	130	1,010
6' x 8'	"	1,550	170	1,720
Metal framed skylights				
Translucent panels, 2-1/2" thick	S.F.	43.00	5.10	48.10
Continuous vaults, 8' wide				
Single glazed	S.F.	58.00	6.37	64.37
Double glazed	"	94.00	7.28	101

Joint Sealers	UNIT	MAT.	INST.	TOTAL
07920.10		**Caulking**		
Caulk exterior, two component				
1/4 x 1/2	L.F.	0.39	2.65	3.04
3/8 x 1/2	"	0.60	2.95	3.55
1/2 x 1/2	"	0.82	3.32	4.14
Caulk interior, single component				
1/4 x 1/2	L.F.	0.26	2.52	2.78
3/8 x 1/2	"	0.37	2.79	3.16
1/2 x 1/2	"	0.49	3.12	3.61
Butyl rubber fillers				
1/4" x 1/4"	L.F.	0.75	1.06	1.81
1/2" x 1/2"	"	1.10	1.77	2.87
1/2" x 3/4"	"	1.68	2.12	3.80
3/4" x 3/4"	"	2.22	2.12	4.34
1" x 1"	"	2.60	2.36	4.96
Seals, "O" ring type cord				
1/4" dia.	L.F.	0.83	1.32	2.15
1/2" dia.	"	2.48	1.39	3.87
1" dia.	"	8.78	1.47	10.25
1-1/4" dia.	"	11.25	1.56	12.81
1-1/2" dia.	"	15.00	1.66	16.66
1-3/4" dia.	"	21.75	1.71	23.46
2" dia.	"	28.25	1.77	30.02
Polyvinyl chloride, closed cell				
1/4" x 2"	L.F.	0.57	1.89	2.46
1/4" x 6"	"	1.78	2.41	4.19
Silicon foam penetration seal				
1/4" x 1/2"	L.F.	0.20	0.66	0.86
1/2" x 1/2"	"	0.36	0.88	1.24
1/2" x 3/4"	"	0.56	1.06	1.62
3/4" x 3/4"	"	0.83	1.32	2.15
1/8" x 1"	"	0.20	0.66	0.86
1/8" x 3"	"	0.56	1.06	1.62
1/4" x 3"	"	1.10	1.32	2.42
1/4" x 6"	"	2.27	1.77	4.04
1/2" x 6"	"	4.46	4.08	8.54
1/2" x 9"	"	6.71	6.64	13.35
1/2" x 12"	"	8.88	9.65	18.53
Oil base sealants and caulking				
1/4" x 1/4"	L.F.	0.04	1.32	1.36
1/4" x 3/8"	"	0.09	1.37	1.46
1/4" x 1/2"	"	0.11	1.43	1.54
3/8" x 3/8"	"	0.13	1.51	1.64
3/8" x 1/2"	"	0.15	1.60	1.75
3/8" x 5/8"	"	0.27	1.71	1.98
3/8" x 3/4"	"	0.30	1.83	2.13
1/2" x 1/2"	"	0.27	2.04	2.31
1/2" x 5/8"	"	0.34	2.30	2.64
1/2" x 3/4"	"	0.39	2.65	3.04
1/2" x 7/8"	"	0.46	2.72	3.18
1/2" x 1"	"	0.50	2.79	3.29
3/4" x 3/4"	"	0.59	2.87	3.46
1" x 1"	"	1.03	2.95	3.98
Polyurethane compounds				

Joint Sealers	UNIT	MAT.	INST.	TOTAL
07920.10 **Caulking** *(Cont.)*				
1/4" x 1/4"	L.F.	0.20	1.32	1.52
1/4" x 3/8"	"	0.37	1.37	1.74
1/4" x 1/2"	"	0.47	1.43	1.90
3/8" x 3/8"	"	0.52	1.51	2.03
3/8" x 1/2"	"	0.67	1.60	2.27
3/8" x 5/8"	"	0.83	1.71	2.54
3/8" x 3/4"	"	1.01	1.83	2.84
1/2" x 1/2"	"	0.96	2.04	3.00
1/2" x 5/8"	"	1.05	2.30	3.35
1/2" x 3/4"	"	1.21	2.65	3.86
1/2" x 7/8"	"	1.54	2.72	4.26
1/2" x 1"	"	1.94	2.95	4.89
3/4" x 3/4"	"	1.98	2.87	4.85
3/4" x 1"	"	2.15	2.95	5.10
Backer rod, polyethylene				
1/4"	L.F.	0.05	1.32	1.37
1/2"	"	0.09	1.39	1.48
3/4"	"	0.12	1.47	1.59
1"	"	0.15	1.56	1.71

Metal	UNIT	MAT.	INST.	TOTAL
08110.10 — **Metal Doors**				
Flush hollow metal, std. duty, 20 ga., 1-3/8" thick				
2-6 x 6-8	EA.	220	59.00	279
2-8 x 6-8	"	260	59.00	319
3-0 x 6-8	"	280	59.00	339
1-3/4" thick				
2-6 x 6-8	EA.	260	59.00	319
2-8 x 6-8	"	280	59.00	339
3-0 x 6-8	"	300	59.00	359
2-6 x 7-0	"	290	59.00	349
2-8 x 7-0	"	290	59.00	349
3-0 x 7-0	"	310	59.00	369
Heavy duty, 20 ga., unrated, 1-3/4"				
2-8 x 6-8	EA.	280	59.00	339
3-0 x 6-8	"	300	59.00	359
2-8 x 7-0	"	300	59.00	359
3-0 x 7-0	"	320	59.00	379
3-4 x 7-0	"	390	59.00	449
18 ga., 1-3/4", unrated door				
2-0 x 7-0	EA.	310	59.00	369
2-4 x 7-0	"	320	59.00	379
2-6 x 7-0	"	330	59.00	389
2-8 x 7-0	"	340	59.00	399
3-0 x 7-0	"	360	59.00	419
3-4 x 7-0	"	420	59.00	479
2", unrated door				
2-0 x 7-0	EA.	320	66.00	386
2-4 x 7-0	"	340	66.00	406
2-6 x 7-0	"	350	66.00	416
2-8 x 7-0	"	360	66.00	426
3-0 x 7-0	"	380	66.00	446
3-4 x 7-0	"	460	66.00	526
Galvanized metal door				
3-0 x 7-0	EA.	480	66.00	546
For lead lining in doors	"			970
For sound attenuation	"			45.00
Vision glass				
8" x 8"	EA.	100	66.00	166
8" x 48"	"	150	66.00	216
Fixed metal louver	"	160	53.00	213
For fire rating, add				
3 hr door	EA.			180
1-1/2 hr door	"			76.00
3/4 hr door	"			37.75
1' extra height, add to material, 20%				
1'6" extra height, add to material, 60%				
For dutch doors with shelf, add to material, 100%				
08110.40 — **Metal Door Frames**				
Hollow metal, stock, 18 ga., 4-3/4" x 1-3/4"				
2-0 x 7-0	EA.	130	66.00	196
2-4 x 7-0	"	140	66.00	206
2-6 x 7-0	"	150	66.00	216
2-8 x 7-0	"	150	66.00	216

Metal	UNIT	MAT.	INST.	TOTAL
08110.40 **Metal Door Frames** *(Cont.)*				
3-0 x 7-0	EA.	160	66.00	226
4-0 x 7-0	"	160	89.00	249
5-0 x 7-0	"	170	89.00	259
6-0 x 7-0	"	180	89.00	269
16 ga., 6-3/4" x 1-3/4"				
2-0 x 7-0	EA.	140	74.00	214
2-4 x 7-0	"	150	74.00	224
2-6 x 7-0	"	150	74.00	224
2-8 x 7-0	"	150	74.00	224
3-0 x 7-0	"	160	74.00	234
4-0 x 7-0	"	170	98.00	268
6-0 x 7-0	"	180	98.00	278
Transom frame				
3-4 x 1-6	EA.	91.00	74.00	165
3-8 x 1-6	"	91.00	74.00	165
6-4 x 1-6	"	130	74.00	204
Transom sash				
3-0 x 1-4	EA.	620	74.00	694
3-4 x 1-4	"	620	74.00	694
6-0 x 1-4	"	690	74.00	764
1' extension of frame, add	"			19.25
14 ga. frame, add	"			19.25
For fire rating, add				
3 hour	EA.			53.00
1-1/2 hour	"			42.25
3/4 hour	"			37.00
Lead lining in frame, add	"			110
Sidelights, complete				
1-0 x 7-2	EA.	360	74.00	434
1-4 x 7-2	"	400	74.00	474
1-0 x 8-8	"	420	74.00	494
1-6 x 8-8	"	440	74.00	514
16 ga., 4-3/4" x 1-3/4"				
2-0 x 7-0	EA.	110	74.00	184
2-4 x 7-0	"	110	74.00	184
2-6 x 7-0	"	120	74.00	194
2-8 x 7-0	"	120	74.00	194
3-0 x 7-0	"	120	74.00	194
4-0 x 7-0	"	120	98.00	218
6-0 x 7-0	"	130	98.00	228
Transom frame				
3-4 x 1-6	EA.	91.00	74.00	165
3-8 x 1-6	"	91.00	74.00	165
6-4 x 1-6	"	130	74.00	204
Transom sash				
3-0 x 1-4	EA.	600	74.00	674
3-4 x 1-4	"	600	74.00	674
6-0 x 1-4	"	680	74.00	754
1' extension of door frame, add	"			19.25
14 ga., metal frame, add	"			19.25
For fire rating, add				
3 hour	EA.			53.00
1-1/2 hour	"			42.25

Metal	UNIT	MAT.	INST.	TOTAL
08110.40 **Metal Door Frames** *(Cont.)*				
3/4 hour	EA.			37.00
Lead lining in frame, add	"			110
Sidelights, complete				
1-0 x 7-2	EA.	350	74.00	424
1-4 x 7-2	"	370	74.00	444
1-0 x 8-8	"	400	74.00	474
1-4 x 8-8	"	410	74.00	484
16 ga., 5-3/4" x 1-3/4"				
2-0 x 7-0	EA.	110	66.00	176
2-4 x 7-0	"	120	66.00	186
2-6 x 7-0	"	120	66.00	186
2-8 x 7-0	"	120	66.00	186
3-0 x 7-0	"	130	66.00	196
4-0 x 7-0	"	140	89.00	229
5-0 x 7-0	"	140	89.00	229
6-0 x 7-0	"	150	89.00	239
Mullions, vertical				
5-1/4" x 1-3/4"	L.F.	13.25	6.64	19.89
5-1/4" x 2"	"	16.25	6.64	22.89
Horizontal				
5-1/4" x 1-3/4"	L.F.	13.25	6.64	19.89
5-1/4" x 2"	"	16.25	6.64	22.89
08120.10 **Aluminum Doors**				
Aluminum doors, commercial				
Narrow stile				
2-6 x 7-0	EA.	640	300	940
3-0 x 7-0	"	670	300	970
3-6 x 7-0	"	690	300	990
Pair				
5-0 x 7-0	EA.	1,060	590	1,650
6-0 x 7-0	"	1,080	590	1,670
7-0 x 7-0	"	1,130	590	1,720
Wide stile				
2-6 x 7-0	EA.	980	300	1,280
3-0 x 7-0	"	1,010	300	1,310
3-6 x 7-0	"	1,040	300	1,340
Pair				
5-0 x 7-0	EA.	1,800	590	2,390
6-0 x 7-0	"	1,880	590	2,470
7-0 x 7-0	"	1,910	590	2,500

Wood And Plastic	UNIT	MAT.	INST.	TOTAL

08210.10   Wood Doors

	UNIT	MAT.	INST.	TOTAL
Solid core, 1-3/8" thick				
Birch faced				
2-4 x 7-0	EA.	130	66.00	196
2-8 x 7-0	"	140	66.00	206
3-0 x 7-0	"	140	66.00	206
3-4 x 7-0	"	280	66.00	346
2-4 x 6-8	"	130	66.00	196
2-6 x 6-8	"	130	66.00	196
2-8 x 6-8	"	140	66.00	206
3-0 x 6-8	"	140	66.00	206
Lauan faced				
2-4 x 6-8	EA.	140	66.00	206
2-8 x 6-8	"	150	66.00	216
3-0 x 6-8	"	150	66.00	216
3-4 x 6-8	"	150	66.00	216
Tempered hardboard faced				
2-4 x 7-0	EA.	140	66.00	206
2-8 x 7-0	"	150	66.00	216
3-0 x 7-0	"	160	66.00	226
3-4 x 7-0	"	180	66.00	246
Hollow core, 1-3/8" thick				
Birch faced				
2-4 x 7-0	EA.	130	66.00	196
2-8 x 7-0	"	130	66.00	196
3-0 x 7-0	"	130	66.00	196
3-4 x 7-0	"	140	66.00	206
Lauan faced				
2-4 x 6-8	EA.	53.00	66.00	119
2-6 x 6-8	"	57.00	66.00	123
2-8 x 6-8	"	72.00	66.00	138
3-0 x 6-8	"	75.00	66.00	141
3-4 x 6-8	"	83.00	66.00	149
Tempered hardboard faced				
2-4 x 7-0	EA.	59.00	66.00	125
2-6 x 7-0	"	64.00	66.00	130
2-8 x 7-0	"	70.00	66.00	136
3-0 x 7-0	"	75.00	66.00	141
3-4 x 7-0	"	81.00	66.00	147
Solid core, 1-3/4" thick				
Birch faced				
2-4 x 7-0	EA.	230	66.00	296
2-6 x 7-0	"	240	66.00	306
2-8 x 7-0	"	250	66.00	316
3-0 x 7-0	"	230	66.00	296
3-4 x 7-0	"	240	66.00	306
Lauan faced				
2-4 x 7-0	EA.	160	66.00	226
2-6 x 7-0	"	180	66.00	246
2-8 x 7-0	"	200	66.00	266
3-4 x 7-0	"	200	66.00	266
3-0 x 7-0	"	220	66.00	286
Tempered hardboard faced				
2-4 x 7-0	EA.	160	66.00	226

Wood And Plastic	UNIT	MAT.	INST.	TOTAL
08210.10 — **Wood Doors** *(Cont.)*				
2-6 x 7-0	EA.	180	66.00	246
2-8 x 7-0	"	200	66.00	266
3-0 x 7-0	"	210	66.00	276
3-4 x 7-0	"	220	66.00	286
Hollow core, 1-3/4" thick				
Birch faced				
2-4 x 7-0	EA.	130	66.00	196
2-6 x 7-0	"	130	66.00	196
2-8 x 7-0	"	130	66.00	196
3-0 x 7-0	"	130	66.00	196
3-4 x 7-0	"	140	66.00	206
Lauan faced				
2-4 x 6-8	EA.	75.00	66.00	141
2-6 x 6-8	"	83.00	66.00	149
2-8 x 6-8	"	75.00	66.00	141
3-0 x 6-8	"	79.00	66.00	145
3-4 x 6-8	"	81.00	66.00	147
Tempered hardboard				
2-4 x 7-0	EA.	68.00	66.00	134
2-6 x 7-0	"	72.00	66.00	138
2-8 x 7-0	"	75.00	66.00	141
3-0 x 7-0	"	79.00	66.00	145
3-4 x 7-0	"	82.00	66.00	148
Add-on, louver	"	27.00	53.00	80.00
Glass	"	83.00	53.00	136
Exterior doors, 3-0 x 7-0 x 2-1/2", solid core				
Carved				
One face	EA.	1,160	130	1,290
Two faces	"	1,590	130	1,720
Closet doors, 1-3/4" thick				
Bi-fold or bi-passing, includes frame and trim				
Paneled				
4-0 x 6-8	EA.	360	89.00	449
6-0 x 6-8	"	380	89.00	469
Louvered				
4-0 x 6-8	EA.	260	89.00	349
6-0 x 6-8	"	300	89.00	389
Flush				
4-0 x 6-8	EA.	230	89.00	319
6-0 x 6-8	"	260	89.00	349
Primed				
4-0 x 6-8	EA.	260	89.00	349
6-0 x 6-8	"	290	89.00	379
08210.90 — **Wood Frames**				
Frame, interior, pine				
2-6 x 6-8	EA.	32.25	76.00	108
2-8 x 6-8	"	34.75	76.00	111
3-0 x 6-8	"	35.75	76.00	112
5-0 x 6-8	"	37.50	76.00	114
6-0 x 6-8	"	39.50	76.00	116
2-6 x 7-0	"	36.75	76.00	113
2-8 x 7-0	"	41.75	76.00	118

Wood And Plastic	UNIT	MAT.	INST.	TOTAL
08210.90 — **Wood Frames** *(Cont.)*				
3-0 x 7-0	EA.	43.25	76.00	119
5-0 x 7-0	"	47.50	110	158
6-0 x 7-0	"	49.00	110	159
Exterior, custom, with threshold, including trim				
Walnut				
3-0 x 7-0	EA.	310	130	440
6-0 x 7-0	"	360	130	490
Oak				
3-0 x 7-0	EA.	310	130	440
6-0 x 7-0	"	360	130	490
Pine				
2-4 x 7-0	EA.	170	110	280
2-6 x 7-0	"	180	110	290
2-8 x 7-0	"	180	110	290
3-0 x 7-0	"	190	110	300
3-4 x 7-0	"	210	110	320
6-0 x 7-0	"	230	180	410
08300.10 — **Special Doors**				
Vault door and frame, class 5, steel	EA.	6,160	530	6,690
Overhead door, coiling insulated				
Chain gear, no frame, 12' x 12'	EA.	3,410	660	4,070
Aluminum, bronze glass panels, 12-9 x 13-0	"	3,850	530	4,380
Garage door, flush insulated metal, primed, 9-0 x 7-0	"	1,040	180	1,220
Sliding metal fire doors, motorized, fusible link, 3 hr.				
3-0 x 6-8	EA.	3,900	1,060	4,960
3-8 x 6-8	"	3,900	1,060	4,960
4-0 x 8-0	"	4,240	1,060	5,300
5-0 x 8-0	"	4,280	1,060	5,340
Metal clad doors, including electric motor				
Light duty				
Minimum	S.F.	53.00	8.85	61.85
Maximum	"	83.00	21.25	104
Heavy duty				
Minimum	S.F.	83.00	26.50	110
Maximum	"	140	33.25	173
Hangar doors, based on 150' openings				
To 20' high	S.F.	55.00	8.05	63.05
20' to 40' high	"	61.00	5.03	66.03
40' to 60' high	"	64.00	3.35	67.35
60' to 80' high	"	66.00	2.01	68.01
Over 80' high	"	94.00	1.61	95.61
Counter doors, (roll-up shutters), standard, manual				
Opening, 4' high				
4' wide	EA.	1,300	440	1,740
6' wide	"	1,760	440	2,200
8' wide	"	1,980	480	2,460
10' wide	"	2,200	660	2,860
14' wide	"	2,750	660	3,410
6' high				
4' wide	EA.	1,540	440	1,980
6' wide	"	2,010	480	2,490
8' wide	"	2,200	530	2,730

Wood And Plastic	UNIT	MAT.	INST.	TOTAL
08300.10 — **Special Doors** *(Cont.)*				
10' wide	EA.	2,480	660	3,140
14' wide	"	2,810	760	3,570
For stainless steel, add to material, 40%				
For motor operator, add	EA.			1,520
Service doors, (roll up shutters), standard, manual				
Opening				
8' high x 8' wide	EA.	1,650	300	1,950
10' high x 10' wide	"	2,060	440	2,500
12' high x 12' wide	"	2,310	660	2,970
14' high x 14' wide	"	3,030	890	3,920
16' high x 14' wide	"	4,240	890	5,130
20' high x 14' wide	"	5,720	1,330	7,050
24' high x 16' wide	"	7,810	1,180	8,990
For motor operator				
Up to 12-0 x 12-0, add	EA.			1,550
Over 12-0 x 12-0, add	"			1,980
Roll-up doors				
13-0 high x 14-0 wide	EA.	1,400	760	2,160
12-0 high x 14-0 wide	"	1,790	760	2,550
Top coiling grilles, manually operated, steel or aluminum				
Opening, 4' high x				
4' wide	EA.	1,650	210	1,860
6' wide	"	1,710	210	1,920
8' wide	"	1,970	300	2,270
12' wide	"	2,280	300	2,580
16' wide	"	2,610	440	3,050
6' high x				
4' wide	EA.	1,730	440	2,170
6' wide	"	1,930	480	2,410
8' wide	"	2,000	530	2,530
12' wide	"	2,420	590	3,010
16' wide	"	3,080	760	3,840
Side coiling grilles, manually operated, aluminum				
Opening, 8' high x				
18' wide	EA.	12,100	5,030	17,130
24' wide	"	15,400	5,750	21,150
12' high x				
12' wide	EA.	12,100	5,030	17,130
18' wide	"	15,400	5,750	21,150
24' wide	"	22,000	6,710	28,710
Accordion folding doors, tracks and fittings included				
Vinyl covered, 2 layers	S.F.	13.75	21.25	35.00
Woven mahogany and vinyl	"	17.25	21.25	38.50
Economy vinyl	"	11.50	21.25	32.75
Rigid polyvinyl chloride	"	19.00	21.25	40.25
Sectional wood overhead doors, frames not included				
Commercial grade, heavy duty, 1-3/4" thick, manual				
8' x 8'	EA.	1,030	440	1,470
10' x 10'	"	1,480	480	1,960
12' x 12'	"	1,790	530	2,320
Chain hoist				
12' x 16' high	EA.	2,860	890	3,750
14' x 14' high	"	3,140	660	3,800

Wood And Plastic	UNIT	MAT.	INST.	TOTAL
08300.10 **Special Doors** *(Cont.)*				
20' x 8' high	EA.	2,690	1,060	3,750
16' high	"	6,050	1,330	7,380
Sectional metal overhead doors, complete				
Residential grade, manual				
9' x 7'	EA.	710	210	920
16' x 7'	"	1,270	270	1,540
Commercial grade				
8' x 8'	EA.	820	440	1,260
10' x 10'	"	1,090	480	1,570
12' x 12'	"	1,810	530	2,340
20' x 14', with chain hoist	"	4,350	1,060	5,410
Sliding glass doors				
Tempered plate glass, 1/4" thick				
6' wide				
Economy grade	EA.	1,010	180	1,190
Premium grade	"	1,160	180	1,340
12' wide				
Economy grade	EA.	1,430	270	1,700
Premium grade	"	2,150	270	2,420
Insulating glass, 5/8" thick				
6' wide				
Economy grade	EA.	1,250	180	1,430
Premium grade	"	1,610	180	1,790
12' wide				
Economy grade	EA.	1,560	270	1,830
Premium grade	"	2,500	270	2,770
1" thick				
6' wide				
Economy grade	EA.	1,580	180	1,760
Premium grade	"	1,820	180	2,000
12' wide				
Economy grade	EA.	2,440	270	2,710
Premium grade	"	3,580	270	3,850
Added costs				
Custom quality, add to material, 30%				
Tempered glass, 6' wide, add	S.F.			4.01
Vertical lift doors, channel frame construction				
20' high x				
10' wide	EA.	28,620	890	29,510
15' wide	"	35,780	890	36,670
20' wide	"	39,360	1,590	40,950
25' wide	"	42,930	1,590	44,520
25' high x				
20' wide	EA.	44,720	1,590	46,310
25' wide	"	50,090	1,860	51,950
30' high x				
25' wide	EA.	53,670	1,860	55,530
30' wide	"	59,030	1,860	60,890
35' wide	"	69,770	1,860	71,630
Residential storm door				
Minimum	EA.	160	89.00	249
Average	"	220	89.00	309
Maximum	"	480	130	610

Storefronts

08410.10 Storefronts	UNIT	MAT.	INST.	TOTAL
Storefront, aluminum and glass				
Minimum	S.F.	20.00	7.38	27.38
Average	"	29.75	8.44	38.19
Maximum	"	53.00	9.84	62.84
Entrance doors, premium, incl. glass, closers, panic				
1/2" thick glass				
3' x 7'	EA.	3,190	490	3,680
6' x 7'	"	5,450	740	6,190
3/4" thick glass				
3' x 7'	EA.	3,300	490	3,790
6' x 7'	"	5,500	740	6,240
1" thick glass				
3' x 7'	EA.	3,580	490	4,070
6' x 7'	"	6,330	740	7,070
Revolving doors				
7'diameter, 7' high				
Minimum	EA.	24,920	5,570	30,490
Average	"	31,350	8,910	40,260
Maximum	"	40,460	11,130	51,590

Metal Windows

08510.10 Steel Windows	UNIT	MAT.	INST.	TOTAL
Steel windows, primed				
Casements				
Operable				
Minimum	S.F.	37.00	3.47	40.47
Maximum	"	55.00	3.93	58.93
Fixed sash	"	29.50	2.95	32.45
Double hung	"	55.00	3.28	58.28
Industrial windows				
Horizontally pivoted sash	S.F.	52.00	3.93	55.93
Fixed sash	"	40.50	3.28	43.78
Security sash				
Operable	S.F.	65.00	3.93	68.93
Fixed	"	57.00	3.28	60.28
Picture window	"	27.75	3.28	31.03
Projecting sash				
Minimum	S.F.	48.00	3.69	51.69
Maximum	"	59.00	3.69	62.69
Mullions	L.F.	12.50	2.95	15.45

Metal Windows

Metal Windows	UNIT	MAT.	INST.	TOTAL
08520.10 **Aluminum Windows**				
Jalousie				
3-0 x 4-0	EA.	310	74.00	384
3-0 x 5-0	"	370	74.00	444
Fixed window				
6 sf to 8 sf	S.F.	15.75	8.44	24.19
12 sf to 16 sf	"	14.00	6.56	20.56
Projecting window				
6 sf to 8 sf	S.F.	35.00	14.75	49.75
12 sf to 16 sf	"	31.50	9.84	41.34
Horizontal sliding				
6 sf to 8 sf	S.F.	22.75	7.38	30.13
12 sf to 16 sf	"	21.00	5.90	26.90
Double hung				
6 sf to 8 sf	S.F.	31.50	11.75	43.25
10 sf to 12 sf	"	28.00	9.84	37.84
Storm window, 0.5 cfm, up to				
60 u.i. (united inches)	EA.	74.00	29.50	104
70 u.i.	"	75.00	29.50	105
80 u.i.	"	84.00	29.50	114
90 u.i.	"	86.00	32.75	119
100 u.i.	"	88.00	32.75	121
2.0 cfm, up to				
60 u.i.	EA.	95.00	29.50	125
70 u.i.	"	96.00	29.50	126
80 u.i.	"	98.00	29.50	128
90 u.i.	"	110	32.75	143
100 u.i.	"	110	32.75	143

Wood And Plastic

Wood And Plastic	UNIT	MAT.	INST.	TOTAL
08600.10 **Wood Windows**				
Double hung				
24" x 36"				
Minimum	EA.	200	53.00	253
Average	"	290	66.00	356
Maximum	"	390	89.00	479
24" x 48"				
Minimum	EA.	230	53.00	283
Average	"	330	66.00	396
Maximum	"	470	89.00	559
30" x 48"				
Minimum	EA.	240	59.00	299
Average	"	340	76.00	416
Maximum	"	490	110	600
30" x 60"				
Minimum	EA.	260	59.00	319

Wood And Plastic	UNIT	MAT.	INST.	TOTAL
08600.10	**Wood Windows** *(Cont.)*			
Average	EA.	420	76.00	496
Maximum	"	530	110	640
Casement				
1 leaf, 22" x 38" high				
Minimum	EA.	290	53.00	343
Average	"	350	66.00	416
Maximum	"	410	89.00	499
2 leaf, 50" x 50" high				
Minimum	EA.	780	66.00	846
Average	"	1,010	89.00	1,099
Maximum	"	1,160	130	1,290
3 leaf, 71" x 62" high				
Minimum	EA.	1,280	66.00	1,346
Average	"	1,310	89.00	1,399
Maximum	"	1,560	130	1,690
4 leaf, 95" x 75" high				
Minimum	EA.	1,700	76.00	1,776
Average	"	1,940	110	2,050
Maximum	"	2,480	180	2,660
5 leaf, 119" x 75" high				
Minimum	EA.	2,200	76.00	2,276
Average	"	2,370	110	2,480
Maximum	"	3,040	180	3,220
Picture window, fixed glass, 54" x 54" high				
Minimum	EA.	450	66.00	516
Average	"	510	76.00	586
Maximum	"	910	89.00	999
68" x 55" high				
Minimum	EA.	820	66.00	886
Average	"	940	76.00	1,016
Maximum	"	1,230	89.00	1,319
Sliding, 40" x 31" high				
Minimum	EA.	270	53.00	323
Average	"	410	66.00	476
Maximum	"	490	89.00	579
52" x 39" high				
Minimum	EA.	340	66.00	406
Average	"	500	76.00	576
Maximum	"	560	89.00	649
64" x 72" high				
Minimum	EA.	520	66.00	586
Average	"	830	89.00	919
Maximum	"	920	110	1,030
Awning windows				
34" x 21" high				
Minimum	EA.	280	53.00	333
Average	"	330	66.00	396
Maximum	"	370	89.00	459
40" x 21" high				
Minimum	EA.	340	59.00	399
Average	"	370	76.00	446
Maximum	"	410	110	520
48" x 27" high				

Wood And Plastic

	UNIT	MAT.	INST.	TOTAL

08600.10 — Wood Windows *(Cont.)*

	UNIT	MAT.	INST.	TOTAL
Minimum	EA.	360	59.00	419
Average	"	420	76.00	496
Maximum	"	490	110	600
60" x 36" high				
Minimum	EA.	370	66.00	436
Average	"	660	89.00	749
Maximum	"	740	110	850
Window frame, milled				
Minimum	L.F.	5.28	10.50	15.78
Average	"	5.85	13.25	19.10
Maximum	"	8.83	17.75	26.58

Hardware

	UNIT	MAT.	INST.	TOTAL

08710.10 — Hinges

	UNIT	MAT.	INST.	TOTAL
Hinges				
3 x 3 butts, steel, interior, plain bearing	PAIR			20.75
4 x 4 butts, steel, standard	"			30.50
5 x 4-1/2 butts, bronze/s. steel, heavy duty	"			79.00
Pivot hinges				
Top pivot	EA.			76.00
Intermediate pivot	"			81.00
Bottom pivot	"			160
BHMA specifications				
3-1/2 x 3-1/2, full mortise butts				
Plain bearing	PAIR			24.75
Ball bearing	"			29.75
Half surface butts	"			42.25
4 x 4				
Full mortise butts, plain bearing, standard duty	PAIR			27.00
Full mortise butts, ball bearing	"			32.50
Half surface butts				
Standard duty	PAIR			42.50
Ball bearing	"			42.50
4-1/2 x 4-1/2				
Full mortise butts, plain bearing	PAIR			35.75
Ball bearing, heavy duty	"			73.00
Half mortise and half surface butts				
Plain bearing	PAIR			42.25
Full surface and half surface butts				
Standard duty	PAIR			72.00
Heavy duty	"			180
Full mortise and full slide-in butts, ball bearing	"			33.00
Half mortise butts, ball bearing				
Standard duty	PAIR			150
Heavy duty	"			170

Hardware	UNIT	MAT.	INST.	TOTAL
08710.10 **Hinges** *(Cont.)*				
5 x 5, ball bearing				
Full mortise butts	PAIR			69.00
Half mortise, full & half surface butts	"			150
Full mortise, full surface and half surface butts	"			200
4 x 4				
Full mortise butts, plain bearing, standard duty	PAIR			16.50
5 x 4-1/2				
Full mortise butts, ball bearing, heavy duty	PAIR			76.00
08710.20 **Locksets**				
Latchset, heavy duty				
Cylindrical	EA.	160	33.25	193
Mortise	"	160	53.00	213
Lockset, heavy duty				
Cylindrical	EA.	330	33.25	363
Mortise	"	390	53.00	443
Mortise locks and latchsets, chrome				
Latchset passage or closet latch	EA.	150	44.25	194
Privacy (bath or bedroom)	"	120	44.25	164
Entry lockset	"	230	44.25	274
Classroom lockset (outside key operated)	"	230	44.25	274
Storeroom lock	"	230	44.25	274
Front door lock	"	230	44.25	274
Dormitory or exit lock	"	230	44.25	274
Preassembled locks and latches, brass				
Latchset, passage or closet latch	EA.	250	44.25	294
Lockset				
Privacy (bath or bathroom)	EA.	300	44.25	344
Entry lock	"	430	44.25	474
Classroom lock (outside key, operated)	"	430	44.25	474
Storeroom lock	"	470	44.25	514
Bored locks and latches, satin chrome plated				
Latchset passage or closet latch	EA.	140	44.25	184
Lockset				
Privacy (bath or bedroom)	EA.	190	44.25	234
Entry lock	"	220	44.25	264
Classroom lock	"	220	44.25	264
Corridor lock	"	220	44.25	264
Miscellaneous locks				
Exit lock with alarm, single door	EA.	690	210	900
Electric strike				
Rim mounted wrought steel	EA.	510	130	640
Mortised, wrought steel with bronze plating	"	200	210	410
Dead bolt				
Bored, wrought brass, keyed both sides	EA.	110	89.00	199
Mortised, cast brass	"	290	89.00	379
Lockset, cipher, mechanical	"	1,790	53.00	1,843

Hardware	UNIT	MAT.	INST.	TOTAL
08710.30 — Closers				
Door closers				
Surface mounted, traditional type, parallel arm				
Standard	EA.	220	66.00	286
Heavy duty	"	260	66.00	326
Modern type, parallel arm, standard duty	"	260	66.00	326
Overhead, concealed, pivot hung, single acting				
Interior	EA.	410	66.00	476
Exterior	"	610	66.00	676
Floor concealed, single acting, offset, pivoted				
Interior	EA.	670	180	850
Exterior	"	850	180	1,030
08710.40 — Door Trim				
Door bumper, bronze, wall type	EA.	5.99	10.50	16.49
Wall type, 4" dia. with convex rubber pad, aluminum	"	30.75	10.50	41.25
Floor type				
Aluminum	EA.	4.84	10.50	15.34
Brass	"	6.16	10.50	16.66
Door holders				
Wall type, bronze	EA.	8.36	10.50	18.86
Overhead	"	51.00	26.50	77.50
Floor type	"	51.00	26.50	77.50
Plunger type	"	40.75	26.50	67.25
Wall type, aluminum	"	35.75	26.50	62.25
Surface bolt	"	30.75	10.50	41.25
Panic device				
Rim type with thumb piece	EA.	620	130	750
Mortise	"	780	130	910
Vertical rod	"	1,170	130	1,300
Labeled, rim type	"	810	130	940
Mortise	"	1,040	130	1,170
Vertical rod	"	1,130	130	1,260
Silencers, rubber type	"	2.88	1.06	3.94
Dust proof strike with plate, brass	"	17.75	17.75	35.50
Flush bolt, lever extension, brass, rated	"	32.75	10.50	43.25
Surface bolt with strike, brass, 6" long	"	24.75	10.50	35.25
Door coordinator, labeled, brass, satin chrome	"	120	38.00	158
Door plates				
Kick plate, aluminum, 3 beveled edges				
10" x 28"	EA.	22.00	26.50	48.50
10" x 30"	"	24.25	26.50	50.75
10" x 34"	"	26.50	26.50	53.00
10" x 38"	"	28.50	26.50	55.00
Push plate, 4" x 16"				
Aluminum	EA.	22.00	10.50	32.50
Bronze	"	84.00	10.50	94.50
Stainless steel	"	67.00	10.50	77.50
Armor plate, 40" x 34"	"	77.00	21.25	98.25
Pull handle, 4" x 16"				
Aluminum	EA.	95.00	10.50	106
Bronze	"	180	10.50	191
Stainless steel	"	140	10.50	151
Hasp assembly				

Hardware

	UNIT	MAT.	INST.	TOTAL
08710.40 **Door Trim** *(Cont.)*				
3"	EA.	3.52	8.85	12.37
4-1/2"	"	4.40	11.75	16.15
6"	"	6.98	15.25	22.23
Electro-magnetic door holder				
Wall mounted	EA.	160	180	340
Floor mounted	"	300	180	480
Smoke detector door holder				
Photo electric type	EA.	280	180	460
Ionization type	"	290	180	470
Pneumatic operators, activated by rubber mats				
Swing				
Single	EA.	2,060	440	2,500
Double	"	3,410	660	4,070
Sliding				
Single	EA.	2,280	440	2,720
Double	"	3,960	660	4,620
08710.60 **Weatherstripping**				
Weatherstrip, head and jamb, metal strip, neoprene bulb				
Standard duty	L.F.	4.93	2.95	7.88
Heavy duty	"	5.48	3.32	8.80
Spring type				
Metal doors	EA.	55.00	130	185
Wood doors	"	55.00	180	235
Sponge type with adhesive backing	"	51.00	53.00	104
Astragal				
1-3/4" x 13 ga., aluminum	L.F.	5.91	4.42	10.33
1-3/8" x 5/8", oak	"	5.50	3.54	9.04
Thresholds				
Bronze	L.F.	53.00	13.25	66.25
Aluminum				
Plain	L.F.	24.75	13.25	38.00
Vinyl insert	"	29.75	13.25	43.00
Aluminum with grit	"	28.50	13.25	41.75
Steel				
Plain	L.F.	29.75	13.25	43.00
Interlocking	"	39.50	44.25	83.75

Glazing

	UNIT	MAT.	INST.	TOTAL
08810.10 **Glazing**				
Sheet glass, 1/8" thick	S.F.	7.70	3.28	10.98
Plate glass, bronze or grey, 1/4" thick	"	11.25	5.37	16.62
Clear	"	8.80	5.37	14.17
Polished	"	10.50	5.37	15.87
Plexiglass				

Glazing	UNIT	MAT.	INST.	TOTAL
08810.10 — **Glazing** *(Cont.)*				
1/8" thick	S.F.	3.65	5.37	9.02
1/4" thick	"	6.50	3.28	9.78
Float glass, clear				
3/16" thick	S.F.	5.25	4.92	10.17
1/4" thick	"	6.12	5.37	11.49
5/16" thick	"	11.50	5.90	17.40
3/8" thick	"	12.25	7.38	19.63
1/2" thick	"	20.75	9.84	30.59
5/8" thick	"	23.75	11.75	35.50
3/4" thick	"	26.00	14.75	40.75
1" thick	"	45.50	19.75	65.25
Tinted glass, polished plate, twin ground				
3/16" thick	S.F.	8.22	4.92	13.14
1/4" thick	"	8.22	5.37	13.59
3/8" thick	"	13.25	7.38	20.63
1/2" thick	"	21.50	9.84	31.34
Total, full vision, all glass window system				
To 10' high				
Minimum	S.F.	33.25	14.75	48.00
Average	"	43.75	14.75	58.50
Maximum	"	53.00	14.75	67.75
10' to 20' high				
Minimum	S.F.	40.25	14.75	55.00
Average	"	49.00	14.75	63.75
Maximum	"	61.00	14.75	75.75
Insulated glass, bronze or gray				
1/2" thick	S.F.	15.25	9.84	25.09
1" thick	"	18.25	14.75	33.00
Spandrel glass, polished bronze/grey, 1 side, 1/4" thick	"	12.25	5.37	17.62
Tempered glass (safety)				
Clear sheet glass				
1/8" thick	S.F.	8.41	3.28	11.69
3/16" thick	"	10.25	4.54	14.79
Clear float glass				
1/4" thick	S.F.	8.76	4.92	13.68
5/16" thick	"	15.75	5.90	21.65
3/8" thick	"	19.25	7.38	26.63
1/2" thick	"	26.25	9.84	36.09
5/8" thick	"	29.75	11.75	41.50
3/4" thick	"	36.75	19.75	56.50
Tinted float glass				
3/16" thick	S.F.	10.50	4.54	15.04
1/4" thick	"	11.50	4.92	16.42
3/8" thick	"	21.00	7.38	28.38
1/2" thick	"	28.00	9.84	37.84
Laminated glass				
Float safety glass with polyvinyl plastic interlayer				
1/4", sheet or float				
Two lites, 1/8" thick, clear glass	S.F.	12.25	4.92	17.17
1/2" thick, float glass				
Two lites, 1/4" thick, clear glass	S.F.	18.75	9.84	28.59
Tinted glass	"	22.00	9.84	31.84
Insulating glass, two lites, clear float glass				

Glazing	UNIT	MAT.	INST.	TOTAL
08810.10 — **Glazing** *(Cont.)*				
1/2" thick	S.F.	12.00	9.84	21.84
5/8" thick	"	13.75	11.75	25.50
3/4" thick	"	15.25	14.75	30.00
7/8" thick	"	16.00	16.75	32.75
1" thick	"	17.00	19.75	36.75
Glass seal edge				
3/8" thick	S.F.	10.00	9.84	19.84
Tinted glass				
1/2" thick	S.F.	20.50	9.84	30.34
1" thick	"	22.00	19.75	41.75
Tempered, clear				
1" thick	S.F.	40.25	19.75	60.00
Wire reinforced	"	51.00	19.75	70.75
Plate mirror glass				
1/4" thick				
15 sf	S.F.	10.50	5.90	16.40
Over 15 sf	"	9.65	5.37	15.02
Door type, 1/4" thick	"	11.00	5.90	16.90
Transparent, one way vision, 1/4" thick	"	23.25	5.90	29.15
Sheet mirror glass				
3/16" thick	S.F.	10.25	5.90	16.15
1/4" thick	"	10.75	4.92	15.67
Wall tiles, 12" x 12"				
Clear glass	S.F.	3.46	3.28	6.74
Veined glass	"	4.03	3.28	7.31
Wire glass, 1/4" thick				
Clear	S.F.	14.25	19.75	34.00
Hammered	"	14.50	19.75	34.25
Obscure	"	17.25	19.75	37.00
Bullet resistant glass, plate glass with inter-leaved vinyl				
1-3/16" thick				
To 15 sf	S.F.	64.00	29.50	93.50
Over 15 sf	"	69.00	29.50	98.50
2" thick				
To 15 sf	S.F.	91.00	49.25	140
Over 15 sf	"	100	49.25	149
Glazing accessories				
Neoprene glazing gaskets				
1/4" glass	L.F.	1.98	2.36	4.34
3/8" glass	"	2.20	2.46	4.66
1/2" glass	"	2.31	2.56	4.87
3/4" glass	"	3.30	2.68	5.98
1" glass	"	3.85	2.95	6.80
Mullion section				
1/4" glass	L.F.	0.57	1.18	1.75
3/8" glass	"	0.73	1.47	2.20
1/2" glass	"	1.08	1.68	2.76
3/4" glass	"	1.65	1.96	3.61
1" glass	"	2.21	2.36	4.57
Molded corners	EA.	2.37	39.50	41.87

Glazed Curtain Walls

Glazed Curtain Walls	UNIT	MAT.	INST.	TOTAL
08910.10 **Glazed Curtain Walls**				
Curtain wall, aluminum system, framing sections				
2" x 3"				
Jamb	L.F.	11.00	4.92	15.92
Horizontal	"	11.25	4.92	16.17
Mullion	"	15.00	4.92	19.92
2" x 4"				
Jamb	L.F.	15.00	7.38	22.38
Horizontal	"	15.50	7.38	22.88
Mullion	"	15.00	7.38	22.38
3" x 5-1/2"				
Jamb	L.F.	19.75	7.38	27.13
Horizontal	"	22.00	7.38	29.38
Mullion	"	20.00	7.38	27.38
4" corner mullion	"	26.25	9.84	36.09
Coping sections				
1/8" x 8"	L.F.	27.50	9.84	37.34
1/8" x 9"	"	27.75	9.84	37.59
1/8" x 12-1/2"	"	28.50	11.75	40.25
Sill section				
1/8" x 6"	L.F.	27.25	5.90	33.15
1/8" x 7"	"	27.50	5.90	33.40
1/8" x 8-1/2"	"	28.00	5.90	33.90
Column covers, aluminum				
1/8" x 26"	L.F.	27.25	14.75	42.00
1/8" x 34"	"	27.50	15.50	43.00
1/8" x 38"	"	27.75	15.50	43.25
Doors				
Aluminum framed, standard hardware				
Narrow stile				
2-6 x 7-0	EA.	610	300	910
3-0 x 7-0	"	620	300	920
3-6 x 7-0	"	640	300	940
Wide stile				
2-6 x 7-0	EA.	1,060	300	1,360
3-0 x 7-0	"	1,130	300	1,430
3-6 x 7-0	"	1,220	300	1,520
Flush panel doors, to match adjacent wall panels				
2-6 x 7-0	EA.	890	370	1,260
3-0 x 7-0	"	940	370	1,310
3-6 x 7-0	"	970	370	1,340
Wall panel, insulated				
"U"=.08	S.F.	11.25	4.92	16.17
"U"=.10	"	10.50	4.92	15.42
"U"=.15	"	9.54	4.92	14.46
Window wall system, complete				
Minimum	S.F.	19.75	5.90	25.65
Average	"	31.50	6.56	38.06
Maximum	"	73.00	8.44	81.44
Added costs				
For bronze, add 20% to material				
For stainless steel, add 50% to material				

Support Systems	UNIT	MAT.	INST.	TOTAL
09110.10 **Metal Studs**				
Studs, non load bearing, galvanized				
2-1/2", 20 ga.				
12" o.c.	S.F.	0.74	1.10	1.84
16" o.c.	"	0.58	0.88	1.46
25 ga.				
12" o.c.	S.F.	0.48	1.10	1.58
16" o.c.	"	0.38	0.88	1.26
24" o.c.	"	0.29	0.73	1.02
3-5/8", 20 ga.				
12" o.c.	S.F.	0.85	1.32	2.17
16" o.c.	"	0.68	1.06	1.74
24" o.c.	"	0.49	0.88	1.37
25 ga.				
12" o.c.	S.F.	0.57	1.32	1.89
16" o.c.	"	0.46	1.06	1.52
24" o.c.	"	0.35	0.88	1.23
4", 20 ga.				
12" o.c.	S.F.	0.93	1.32	2.25
16" o.c.	"	0.74	1.06	1.80
24" o.c.	"	0.55	0.88	1.43
25 ga.				
12" o.c.	S.F.	0.63	1.32	1.95
16" o.c.	"	0.49	1.06	1.55
24" o.c.	"	0.37	0.88	1.25
6", 20 ga.				
12" o.c.	S.F.	1.19	1.66	2.85
16" o.c.	"	0.94	1.32	2.26
24" o.c.	"	0.70	1.10	1.80
25 ga.				
12" o.c.	S.F.	0.78	1.66	2.44
16" o.c.	"	0.61	1.32	1.93
24" o.c.	"	0.46	1.10	1.56
Load bearing studs, galvanized				
3-5/8", 16 ga.				
12" o.c.	S.F.	1.55	1.32	2.87
16" o.c.	"	1.43	1.06	2.49
18 ga.				
12" o.c.	S.F.	1.11	0.88	1.99
16" o.c.	"	1.21	1.06	2.27
4", 16 ga.				
12" o.c.	S.F.	1.62	1.32	2.94
16" o.c.	"	1.48	1.06	2.54
6", 16 ga.				
12" o.c.	S.F.	2.07	1.66	3.73
16" o.c.	"	1.87	1.32	3.19
Furring				
On beams and columns				
7/8" channel	L.F.	0.51	3.54	4.05
1-1/2" channel	"	0.60	4.08	4.68
On ceilings				
3/4" furring channels				
12" o.c.	S.F.	0.51	2.21	2.72
16" o.c.	"	0.39	2.12	2.51

Support Systems	UNIT	MAT.	INST.	TOTAL
09110.10 **Metal Studs** *(Cont.)*				
24" o.c.	S.F.	0.26	1.89	2.15
1-1/2" furring channels				
12" o.c.	S.F.	0.66	2.41	3.07
16" o.c.	"	0.49	2.21	2.70
24" o.c.	"	0.34	2.04	2.38
On walls				
3/4" furring channels				
12" o.c.	S.F.	0.51	1.77	2.28
16" o.c.	"	0.39	1.66	2.05
24" o.c.	"	0.26	1.56	1.82
1-1/2" furring channels				
12" o.c.	S.F.	0.66	1.89	2.55
16" o.c.	"	0.49	1.77	2.26
24" o.c.	"	0.34	1.66	2.00

Lath And Plaster	UNIT	MAT.	INST.	TOTAL
09205.10 **Gypsum Lath**				
Gypsum lath, 1/2" thick				
Clipped	S.Y.	7.20	2.95	10.15
Nailed	"	7.20	3.32	10.52
09205.20 **Metal Lath**				
Diamond expanded, galvanized				
2.5 lb., on walls				
Nailed	S.Y.	3.13	6.64	9.77
Wired	"	3.13	7.58	10.71
On ceilings				
Nailed	S.Y.	3.13	7.58	10.71
Wired	"	3.13	8.85	11.98
3.4 lb., on walls				
Nailed	S.Y.	3.79	6.64	10.43
Wired	"	3.79	7.58	11.37
On ceilings				
Nailed	S.Y.	3.79	7.58	11.37
Wired	"	3.79	8.85	12.64
Flat rib				
2.75 lb., on walls				
Nailed	S.Y.	3.08	6.64	9.72
Wired	"	3.08	7.58	10.66
On ceilings				
Nailed	S.Y.	3.08	7.58	10.66
Wired	"	3.08	8.85	11.93
3.4 lb., on walls				
Nailed	S.Y.	3.79	6.64	10.43
Wired	"	3.79	7.58	11.37

Lath And Plaster	UNIT	MAT.	INST.	TOTAL
09205.20 — **Metal Lath** *(Cont.)*				
On ceilings				
Nailed	S.Y.	3.79	7.58	11.37
Wired	"	3.79	8.85	12.64
Stucco lath				
1.8 lb.	S.Y.	4.96	6.64	11.60
3.6 lb.	"	5.56	6.64	12.20
Paper backed				
Minimum	S.Y.	3.85	5.31	9.16
Maximum	"	6.21	7.58	13.79
09205.60 — **Plaster Accessories**				
Expansion joint, 3/4", 26 ga., galvanized, one piece	L.F.	0.99	1.32	2.31
Plaster corner beads, 3/4", galvanized	"	0.56	1.51	2.07
Casing bead, expanded flange, galvanized	"	0.56	1.32	1.88
Expanded wing, 1-1/4" wide, galvanized	"	0.47	1.32	1.79
Joint clips for lath	EA.	0.24	0.26	0.50
Metal base, galvanized, 2-1/2" high	L.F.	0.72	1.77	2.49
Stud clips for gypsum lath	EA.	0.24	0.26	0.50
Tie wire galvanized, 18 ga., 25 lb. hank	"			47.00
Sound deadening board, 1/4", nailed or clipped	S.F.	0.45	0.88	1.33
09210.10 — **Plaster**				
Gypsum plaster, trowel finish, 2 coats				
Ceilings	S.Y.	6.20	16.25	22.45
Walls	"	6.20	15.50	21.70
3 coats				
Ceilings	S.Y.	8.60	22.75	31.35
Walls	"	8.60	20.00	28.60
Vermiculite plaster				
2 coats				
Ceilings	S.Y.	7.06	25.00	32.06
Walls	"	7.06	22.75	29.81
3 coats				
Ceilings	S.Y.	11.00	30.75	41.75
Walls	"	11.00	27.50	38.50
Keenes cement plaster				
2 coats				
Ceilings	S.Y.	9.85	20.00	29.85
Walls	"	9.85	17.50	27.35
3 coats				
Ceilings	S.Y.	12.00	22.75	34.75
Walls	"	12.00	20.00	32.00
On columns, add to installation, 50%	"			
Chases, fascia, and soffits, add to installation, 50%	"			
Beams, add to installation, 50%	"			
Patch holes, average size holes				
1 sf to 5 sf				
Minimum	S.F.	2.53	8.72	11.25
Average	"	3.68	10.50	14.18
Maximum	"	4.84	13.00	17.84
Over 5 sf				
Minimum	S.F.	2.25	5.23	7.48

Lath And Plaster	UNIT	MAT.	INST.	TOTAL
09210.10 **Plaster** *(Cont.)*				
Average	S.F.	3.08	7.47	10.55
Maximum	"	3.96	8.72	12.68
Patch cracks				
Minimum	S.F.	1.32	1.74	3.06
average	"	1.65	2.61	4.26
Maximum	"	1.98	5.23	7.21
09220.10 **Portland Cement Plaster**				
Stucco, portland, gray, 3 coat, 1" thick				
Sand finish	S.Y.	7.75	22.75	30.50
Trowel finish	"	7.75	23.75	31.50
White cement				
Sand finish	S.Y.	8.85	23.75	32.60
Trowel finish	"	8.85	26.25	35.10
Scratch coat				
For ceramic tile	S.Y.	2.81	5.23	8.04
For quarry tile	"	2.81	5.23	8.04
Portland cement plaster				
2 coats, 1/2"	S.Y.	5.58	10.50	16.08
3 coats, 7/8"	"	6.66	13.00	19.66
09250.10 **Gypsum Board**				
Drywall, plasterboard, 3/8" clipped to				
Metal furred ceiling	S.F.	0.34	0.59	0.93
Columns and beams	"	0.34	1.32	1.66
Walls	"	0.34	0.53	0.87
Nailed or screwed to				
Wood framed ceiling	S.F.	0.34	0.53	0.87
Columns and beams	"	0.34	1.18	1.52
Walls	"	0.34	0.48	0.82
1/2", clipped to				
Metal furred ceiling	S.F.	0.36	0.59	0.95
Columns and beams	"	0.36	1.32	1.68
Walls	"	0.36	0.53	0.89
Nailed or screwed to				
Wood framed ceiling	S.F.	0.36	0.53	0.89
Columns and beams	"	0.36	1.18	1.54
Walls	"	0.36	0.48	0.84
5/8", clipped to				
Metal furred ceiling	S.F.	0.38	0.66	1.04
Columns and beams	"	0.38	1.47	1.85
Walls	"	0.38	0.59	0.97
Nailed or screwed to				
Wood framed ceiling	S.F.	0.38	0.66	1.04
Columns and beams	"	0.38	1.47	1.85
Walls	"	0.38	0.59	0.97
Vinyl faced, clipped to metal studs				
1/2"	S.F.	1.76	0.66	2.42
5/8"	"	1.87	0.66	2.53
Add for				
Fire resistant	S.F.			0.10
Water resistant	"			0.17
Water and fire resistant	"			0.21

Lath And Plaster

	UNIT	MAT.	INST.	TOTAL

09250.10 — Gypsum Board *(Cont.)*

	UNIT	MAT.	INST.	TOTAL
Taping and finishing joints				
Minimum	S.F.	0.04	0.35	0.39
Average	"	0.06	0.44	0.50
Maximum	"	0.09	0.53	0.62
Casing bead				
Minimum	L.F.	0.14	1.51	1.65
Average	"	0.16	1.77	1.93
Maximum	"	0.20	2.65	2.85
Corner bead				
Minimum	L.F.	0.16	1.51	1.67
Average	"	0.20	1.77	1.97
Maximum	"	0.25	2.65	2.90

Tile

	UNIT	MAT.	INST.	TOTAL

09310.10 — Ceramic Tile

	UNIT	MAT.	INST.	TOTAL
Glazed wall tile, 4-1/4" x 4-1/4"				
Minimum	S.F.	2.32	3.67	5.99
Average	"	3.64	4.28	7.92
Maximum	"	9.95	5.14	15.09
Base, 4-1/4" high				
Minimum	L.F.	3.56	6.42	9.98
Average	"	3.99	6.42	10.41
Maximum	"	6.33	6.42	12.75
Unglazed floor tile				
Portland cement bed, cushion edge, face mounted				
1" x 1"	S.F.	6.71	4.67	11.38
1" x 2"	"	7.09	4.46	11.55
2" x 2"	"	7.09	4.28	11.37
Adhesive bed, with white grout				
1" x 1"	S.F.	6.71	4.67	11.38
1" x 2"	"	7.09	4.46	11.55
2" x 2"	"	7.09	4.28	11.37
Organic adhesive bed, thin set, back mounted				
1" x 1"	S.F.	6.71	4.67	11.38
1" x 2"	"	7.09	4.46	11.55
2" x 2"	"	7.09	4.28	11.37
For group 2 colors, add to material, 10%				
For group 3 colors, add to material, 20%				
For abrasive surface, add to material, 25%				
Unglazed wall tile				
Organic adhesive, face mounted cushion edge				
1" x 1"				
Minimum	S.F.	3.57	4.28	7.85
Average	"	4.67	4.67	9.34
Maximum	"	6.98	5.14	12.12

Tile	UNIT	MAT.	INST.	TOTAL
09310.10 **Ceramic Tile** *(Cont.)*				
1" x 2"				
Minimum	S.F.	3.93	4.11	8.04
Average	"	5.22	4.46	9.68
Maximum	"	7.64	4.89	12.53
2" x 2"				
Minimum	S.F.	4.12	3.95	8.07
Average	"	4.67	4.28	8.95
Maximum	"	7.64	4.67	12.31
Back mounted				
1" x 1"				
Minimum	S.F.	3.57	4.28	7.85
Average	"	4.67	4.67	9.34
Maximum	"	6.98	5.14	12.12
1" x 2"				
Minimum	S.F.	3.93	4.11	8.04
Average	"	5.22	4.46	9.68
Maximum	"	7.64	4.89	12.53
2" x 2"				
Minimum	S.F.	4.12	3.95	8.07
Average	"	4.67	4.28	8.95
Maximum	"	7.64	4.67	12.31
For glazed finish, add to material, 25%				
For glazed mosaic, add to material, 100%				
For metallic colors, add to material, 125%				
For exterior wall use, add to total, 25%				
For exterior soffit, add to total, 25%				
For portland cement bed, add to total, 25%				
For dry set portland cement bed, add to total, 10%				
Conductive floor tile, unglazed square edged				
Portland cement bed				
1 x 1	S.F.	6.93	6.42	13.35
1-9/16 x 1-9/16	"	6.38	6.42	12.80
Dry set				
1 x 1	S.F.	6.93	6.42	13.35
1-9/16 x 1-9/16	"	6.38	6.42	12.80
Epoxy bed with epoxy joints				
1 x 1	S.F.	6.93	6.42	13.35
1-9/16 x 1-9/16	"	6.38	6.42	12.80
For WWF in bed add to total, 15%				
For abrasive surface, add to material, 40%				
Ceramic accessories				
Towel bar, 24" long				
Minimum	EA.	26.50	20.50	47.00
Average	"	34.75	25.75	60.50
Maximum	"	53.00	34.25	87.25
Soap dish				
Minimum	EA.	23.25	34.25	57.50
Average	"	28.25	42.75	71.00
Maximum	"	39.75	51.00	90.75

Tile	UNIT	MAT.	INST.	TOTAL
09330.10 — **Quarry Tile**				
Floor				
4 x 4 x 1/2"	S.F.	5.40	6.85	12.25
6 x 6 x 1/2"	"	5.47	6.42	11.89
6 x 6 x 3/4"	"	6.45	6.42	12.87
Wall, applied to 3/4" portland cement bed				
4 x 4 x 1/2"	S.F.	4.92	10.25	15.17
6 x 6 x 3/4"	"	6.18	8.56	14.74
Cove base				
5 x 6 x 1/2" straight top	L.F.	4.84	8.56	13.40
6 x 6 x 3/4" round top	"	5.11	8.56	13.67
Stair treads 6 x 6 x 3/4"	"	7.53	12.75	20.28
Window sill 6 x 8 x 3/4"	"	6.87	10.25	17.12
For abrasive surface, add to material, 25%				
09410.10 — **Terrazzo**				
Floors on concrete, 1-3/4" thick, 5/8" topping				
Gray cement	S.F.	4.23	7.47	11.70
White cement	"	4.62	7.47	12.09
Sand cushion, 3" thick, 5/8" top, 1/4"				
Gray cement	S.F.	5.00	8.72	13.72
White cement	"	5.55	8.72	14.27
Monolithic terrazzo, 3-1/2" base slab, 5/8" topping	"	3.97	6.54	10.51
Terrazzo wainscot, cast-in-place, 1/2" thick	"	7.48	13.00	20.48
Base, cast in place, terrazzo cove type, 6" high	L.F.	8.86	7.47	16.33
Curb, cast in place, 6" wide x 6" high, polished top	"	9.84	26.25	36.09
For venetian type terrazzo, add to material, 10%				
For abrasive heavy duty terrazzo, add to material, 15%				
Divider strips				
Zinc	L.F.			1.50
Brass	"			2.80
Stairs, cast-in-place, topping on concrete or metal				
1-1/2" thick treads, 12" wide	L.F.	5.90	26.25	32.15
Combined tread and riser	"	8.86	65.00	73.86
Precast terrazzo, thin set				
Terrazzo tiles, non-slip surface				
9" x 9" x 1" thick	S.F.	18.75	7.47	26.22
12" x 12"				
1" thick	S.F.	20.25	6.97	27.22
1-1/2" thick	"	21.00	7.47	28.47
18" x 18" x 1-1/2" thick	"	27.50	7.47	34.97
24" x 24" x 1-1/2" thick	"	35.50	6.15	41.65
For white cement, add to material, 10%				
For venetian type terrazzo, add to material, 25%				
Terrazzo wainscot				
12" x 12" x 1" thick	S.F.	20.75	13.00	33.75
18" x 18" x 1-1/2" thick	"	32.50	15.00	47.50
Base				
6" high				
Straight	L.F.	13.25	4.02	17.27
Coved	"	15.75	4.02	19.77
8" high				
Straight	L.F.	15.00	4.36	19.36
Coved	"	17.50	4.36	21.86

Tile	UNIT	MAT.	INST.	TOTAL
09410.10 Terrazzo *(Cont.)*				
Terrazzo curbs				
8" wide x 8" high	L.F.	34.75	21.00	55.75
6" wide x 6" high	"	31.25	17.50	48.75
Precast terrazzo stair treads, 12" wide				
1-1/2" thick				
Diamond pattern	L.F.	33.50	9.51	43.01
Non-slip surface	"	37.50	9.51	47.01
2" thick				
Diamond pattern	L.F.	37.50	9.51	47.01
Non-slip surface	"	41.50	10.50	52.00
Stair risers, 1" thick to 6" high				
Straight sections	L.F.	14.00	5.23	19.23
Cove sections	"	16.50	5.23	21.73
Combined tread and riser				
Straight sections				
1-1/2" tread, 3/4" riser	L.F.	69.00	15.00	84.00
3" tread, 1" riser	"	72.00	15.00	87.00
Curved sections				
2" tread, 1" riser	L.F.	77.00	17.50	94.50
3" tread, 1" riser	"	80.00	17.50	97.50
Stair stringers, notched for treads and risers				
1" thick	L.F.	36.50	13.00	49.50
2" thick	"	37.75	17.50	55.25
Landings, structural, nonslip				
1-1/2" thick	S.F.	31.25	8.72	39.97
3" thick	"	30.50	10.50	41.00
Conductive terrazzo, spark proof industrial floor				
Epoxy terrazzo				
Floor	S.F.	13.25	3.27	16.52
Base	"	13.25	4.36	17.61
Polyacrylate				
Floor	S.F.	10.75	3.27	14.02
Base	"	10.75	4.36	15.11
Polyester				
Floor	S.F.	3.79	2.09	5.88
Base	"	3.79	2.61	6.40
Synthetic latex mastic				
Floor	S.F.	6.16	3.27	9.43
Base	"	6.16	4.36	10.52

Acoustical Treatment

09510.10	Ceilings And Walls	UNIT	MAT.	INST.	TOTAL
Acoustical panels, suspension system not included					
Fiberglass panels					
5/8" thick					
2' x 2'		S.F.	0.93	0.75	1.68
2' x 4'		"	0.93	0.59	1.52
3/4" thick					
2' x 2'		S.F.	1.54	0.75	2.29
2' x 4'		"	1.54	0.59	2.13
Glass cloth faced fiberglass panels					
3/4" thick		S.F.	2.55	0.88	3.43
1" thick		"	2.83	0.88	3.71
Mineral fiber panels					
5/8" thick					
2' x 2'		S.F.	1.12	0.75	1.87
2' x 4'		"	1.12	0.59	1.71
3/4" thick					
2' x 2'		S.F.	1.51	0.75	2.26
2' x 4'		"	1.51	0.59	2.10
For aluminum faced panels, add to material, 80%					
For vinyl faced panels, add to total, 125%					
For fire rated panels, add to material, 75%					
Wood fiber panels					
2" thick					
2' x 2'		S.F.	2.64	0.88	3.52
2' x 4'		"	2.64	0.66	3.30
For flameproofing, add to material, 10%					
For sculptured finish, add to material, 15%					
Air distributing panels					
3/4" thick		S.F.	2.64	1.32	3.96
5/8" thick		"	2.26	1.06	3.32
Acoustical tiles, suspension system not included					
Fiberglass tile, 12" x 12"					
5/8" thick		S.F.	0.71	0.96	1.67
3/4" thick		"	0.93	1.18	2.11
Glass cloth faced fiberglass tile					
3/4" thick		S.F.	2.42	1.18	3.60
3" thick		"	2.69	1.32	4.01
Mineral fiber tile, 12" x 12"					
5/8" thick					
Standard		S.F.	0.66	1.06	1.72
Vinyl faced		"	1.65	1.06	2.71
3/4" thick					
Standard		S.F.	1.21	1.06	2.27
Vinyl faced		"	2.11	1.06	3.17
Fire rated		"	2.68	1.06	3.74
Aluminum or mylar faced		"	5.11	1.06	6.17
Wood fiber tile, 12" x 12"					
1/2" thick		S.F.	1.06	1.06	2.12
3/4" thick		"	1.54	1.06	2.60
For flameproofing, add to material, 10%					
For sculptured 3 dimensional, add to material, 50%					
Metal pan units, 24 ga. steel					
12" x 12"		S.F.	5.22	2.12	7.34

Acoustical Treatment	UNIT	MAT.	INST.	TOTAL
09510.10 — **Ceilings And Walls** *(Cont.)*				
12" x 24"	S.F.	5.40	1.77	7.17
Aluminum, .025" thick				
12" x 12"	S.F.	5.55	2.12	7.67
12" x 24"	"	5.73	1.77	7.50
Anodized aluminum, 0.25" thick				
12" x 12"	S.F.	6.40	2.12	8.52
12" x 24"	"	7.59	1.77	9.36
Stainless steel, 24 ga.				
12" x 12"	S.F.	12.75	2.12	14.87
12" x 24"	"	13.25	1.77	15.02
For flameproof sound absorbing pads, add to material	"			1.99
Metal ceiling systems				
.020" thick panels				
10', 12', and 16' lengths	S.F.	4.99	1.51	6.50
Custom lengths, 3' to 20'	"	5.04	1.51	6.55
.025" thick panels				
32 sf, 38 sf, and 52 sf pieces	S.F.	5.01	1.77	6.78
Custom lengths, 10 sf to 65 sf	"	5.85	1.77	7.62
Carriers, black, add	"			3.01
Recess filler strip, add	"			1.01
Custom lengths, add	"			1.50
Sound absorption walls, with fabric cover				
2-6" x 9' x 3/4"	S.F.	6.56	1.77	8.33
2' x 9' x 1"	"	7.16	1.77	8.93
Starter spline	L.F.	1.57	1.32	2.89
Internal spline	"	1.37	1.32	2.69
Acoustical treatment				
Barriers for plenums				
Leaded vinyl				
0.48 lb per sf	S.F.	2.98	2.52	5.50
0.87 lb per sf	"	2.98	2.65	5.63
Aluminum foil, fiberglass reinforcement				
Minimum	S.F.	1.11	1.77	2.88
Maximum	"	1.26	2.65	3.91
Aluminum mesh, paper backed	"	1.04	1.77	2.81
Fibered cement sheet, 3/16" thick	"	2.14	1.89	4.03
Sheet lead, 1/64" thick	"	3.06	1.32	4.38
Sound attenuation blanket				
1" thick	S.F.	0.53	5.31	5.84
1-1/2" thick	"	0.71	5.31	6.02
2" thick	"	0.90	5.31	6.21
3" thick	"	1.06	5.90	6.96
Ceiling suspension systems				
T bar system				
2' x 4'	S.F.	1.15	0.53	1.68
2' x 2'	"	1.25	0.59	1.84
Concealed Z bar suspension system, 12" module	"	1.07	0.88	1.95
For 1-1/2" carrier channels, 4' o.c., add	"			0.38
Carrier channel for recessed light fixtures	"			0.69

Flooring	UNIT	MAT.	INST.	TOTAL
09550.10 — **Wood Flooring**				
Wood strip flooring, unfinished				
Fir floor				
C and better				
Vertical grain	S.F.	3.52	1.77	5.29
Flat grain	"	3.32	1.77	5.09
Oak floor				
Minimum	S.F.	3.98	2.52	6.50
Average	"	4.71	2.52	7.23
Maximum	"	5.94	2.52	8.46
Maple floor				
25/32" x 2-1/4"				
Minimum	S.F.	4.32	2.52	6.84
Maximum	"	5.50	2.52	8.02
33/32" x 3-1/4"				
Minimum	S.F.	5.56	2.52	8.08
Maximum	"	6.29	2.52	8.81
Added costs				
For factory finish, add to material, 10%				
For random width floor, add to total, 20%				
For simulated pegs, add to total, 10%				
Wood block industrial flooring				
Creosoted				
2" thick	S.F.	4.18	1.39	5.57
2-1/2" thick	"	4.34	1.66	6.00
3" thick	"	4.51	1.77	6.28
Parquet, 5/16", white oak				
Finished	S.F.	9.13	2.65	11.78
Unfinished	"	3.30	2.65	5.95
Gym floor, 2 ply felt, 25/32" maple, finished, in mastic	"	7.75	2.95	10.70
Over wood sleepers	"	8.69	3.32	12.01
Finishing, sand, fill, finish, and wax	"	0.66	1.32	1.98
Refinish sand, seal, and 2 coats of polyurethane	"	1.15	1.77	2.92
Clean and wax floors	"	0.19	0.26	0.45
09630.10 — **Unit Masonry Flooring**				
Clay brick				
9 x 4-1/2 x 3" thick				
Glazed	S.F.	8.03	4.42	12.45
Unglazed	"	7.70	4.42	12.12
8 x 4 x 3/4" thick				
Glazed	S.F.	7.26	4.61	11.87
Unglazed	"	7.15	4.61	11.76
For herringbone pattern, add to labor, 15%				
09660.10 — **Resilient Tile Flooring**				
Solid vinyl tile, 1/8" thick, 12" x 12"				
Marble patterns	S.F.	4.45	1.32	5.77
Solid colors	"	5.77	1.32	7.09
Travertine patterns	"	6.49	1.32	7.81
Conductive resilient flooring, vinyl tile				
1/8" thick, 12" x 12"	S.F.	6.71	1.51	8.22

Flooring	UNIT	MAT.	INST.	TOTAL
09665.10 **Resilient Sheet Flooring**				
Vinyl sheet flooring				
Minimum	S.F.	2.86	0.53	3.39
Average	"	4.62	0.64	5.26
Maximum	"	7.81	0.88	8.69
Cove, to 6"	L.F.	1.70	1.06	2.76
Fluid applied resilient flooring				
Polyurethane, poured in place, 3/8" thick	S.F.	10.50	4.42	14.92
Vinyl sheet goods, backed				
0.070" thick	S.F.	3.90	0.66	4.56
0.093" thick	"	6.05	0.66	6.71
0.125" thick	"	6.98	0.66	7.64
0.250" thick	"	8.03	0.66	8.69
09678.10 **Resilient Base And Accessories**				
Wall base, vinyl				
Group 1				
4" high	L.F.	1.04	1.77	2.81
6" high	"	1.43	1.77	3.20
Group 2				
4" high	L.F.	0.93	1.77	2.70
6" high	"	1.10	1.77	2.87
Group 3				
4" high	L.F.	1.04	1.77	2.81
6" high	"	1.21	1.77	2.98
Stair accessories				
Treads, 1/4" x 12", rubber diamond surface				
Marbled	L.F.	8.30	4.42	12.72
Plain	"	8.14	4.42	12.56
Grit strip safety tread, 12" wide, colors				
3/16" thick	L.F.	11.50	4.42	15.92
5/16" thick	"	16.00	4.42	20.42
Risers, 7" high, 1/8" thick, colors				
Flat	L.F.	2.42	2.65	5.07
Coved	"	4.01	2.65	6.66
Nosing, rubber				
3/16" thick, 3" wide				
Black	L.F.	4.01	2.65	6.66
Colors	"	4.18	2.65	6.83
6" wide				
Black	L.F.	5.00	4.42	9.42
Colors	"	5.22	4.42	9.64

Carpet	UNIT	MAT.	INST.	TOTAL
09680.10 **Floor Leveling**				
Repair and level floors to receive new flooring				
Minimum	S.Y.	0.88	1.77	2.65
Average	"	3.41	4.42	7.83
Maximum	"	5.06	5.31	10.37
09682.10 **Carpet Padding**				
Carpet padding				
Foam rubber, waffle type, 0.3" thick	S.Y.	6.16	2.65	8.81
Jute padding				
Minimum	S.Y.	4.18	2.41	6.59
Average	"	5.44	2.65	8.09
Maximum	"	8.19	2.95	11.14
Sponge rubber cushion				
Minimum	S.Y.	4.95	2.41	7.36
Average	"	6.60	2.65	9.25
Maximum	"	8.19	2.95	11.14
Urethane cushion, 3/8" thick				
Minimum	S.Y.	4.95	2.41	7.36
Average	"	5.77	2.65	8.42
Maximum	"	6.60	2.95	9.55
09685.10 **Carpet**				
Carpet, acrylic				
24 oz., light traffic	S.Y.	28.00	5.90	33.90
28 oz., medium traffic	"	36.25	5.90	42.15
Residential				
Nylon				
15 oz., light traffic	S.Y.	20.25	5.90	26.15
28 oz., medium traffic	"	26.50	5.90	32.40
Commercial				
Nylon				
28 oz., medium traffic	S.Y.	25.25	5.90	31.15
35 oz., heavy traffic	"	30.75	5.90	36.65
Wool				
30 oz., medium traffic	S.Y.	41.75	5.90	47.65
36 oz., medium traffic	"	44.00	5.90	49.90
42 oz., heavy traffic	"	58.00	5.90	63.90
Carpet tile				
Foam backed				
Minimum	S.F.	3.21	1.06	4.27
Average	"	3.71	1.18	4.89
Maximum	"	5.88	1.32	7.20
Tufted loop or shag				
Minimum	S.F.	3.47	1.06	4.53
Average	"	4.19	1.18	5.37
Maximum	"	6.74	1.32	8.06
Clean and vacuum carpet				
Minimum	S.Y.	0.29	0.20	0.49
Average	"	0.46	0.35	0.81
Maximum	"	0.63	0.53	1.16

Carpet

	UNIT	MAT.	INST.	TOTAL
09700.10 **Special Flooring**				
Epoxy flooring, marble chips				
Epoxy with colored quartz chips in 1/4" base	S.F.	12.00	2.95	14.95
Heavy duty epoxy topping, 3/16" thick	"	10.50	2.95	13.45
Epoxy terrazzo				
1/4" thick chemical resistant	S.F.	12.25	3.32	15.57

Painting

	UNIT	MAT.	INST.	TOTAL
09905.10 **Painting Preparation**				
Dropcloths				
Minimum	S.F.	0.02	0.03	0.05
Average	"	0.03	0.04	0.07
Maximum	"	0.04	0.05	0.09
Masking				
Paper and tape				
Minimum	L.F.	0.02	0.53	0.55
Average	"	0.03	0.66	0.69
Maximum	"	0.04	0.88	0.92
Doors				
Minimum	EA.	0.04	6.64	6.68
Average	"	0.05	8.85	8.90
Maximum	"	0.06	11.75	11.81
Windows				
Minimum	EA.	0.04	6.64	6.68
Average	"	0.05	8.85	8.90
Maximum	"	0.06	11.75	11.81
Sanding				
Walls and flat surfaces				
Minimum	S.F.		0.35	0.35
Average	"		0.44	0.44
Maximum	"		0.53	0.53
Doors and windows				
Minimum	EA.		8.85	8.85
Average	"		13.25	13.25
Maximum	"		17.75	17.75
Trim				
Minimum	L.F.		0.66	0.66
Average	"		0.88	0.88
Maximum	"		1.18	1.18
Puttying				
Minimum	S.F.	0.01	0.81	0.82
Average	"	0.02	1.06	1.08
Maximum	"	0.03	1.32	1.35
Chemical Preparation				
Concrete floors				
Acid Etch				

Painting	UNIT	MAT.	INST.	TOTAL
09905.10 Painting Preparation *(Cont.)*				
Minimum	S.F.	0.03	0.13	0.16
Average	"	0.04	0.22	0.26
Maximum	"	0.05	0.29	0.34
Chemical Stripping				
Minimum	S.F.	0.03	0.88	0.91
Average	"	0.04	1.32	1.36
Maximum	"	0.05	2.12	2.17
Stone				
Chemical cleaning				
Minimum	S.F.	0.02	2.65	2.67
Average	"	0.03	3.54	3.57
Maximum	"	0.04	4.42	4.46
Wood				
Bleaching				
Minimum	S.F.	0.02	0.59	0.61
Average	"	0.03	0.75	0.78
Maximum	"	0.04	0.96	1.00
Chemical stripping				
Minimum	S.F.	0.02	2.12	2.14
Average	"	0.03	5.31	5.34
Maximum	"	0.04	8.85	8.89
Water cleaning/preparation				
Washing (General)				
Minimum	S.F.		0.03	0.03
Average	"		0.04	0.04
Maximum	"		0.06	0.06
Mildew eradication				
Minimum	S.F.	0.03	0.06	0.09
Average	"	0.04	0.10	0.14
Maximum	"	0.05	0.17	0.22
Remove loose paint				
Minimum	S.F.		0.10	0.10
Average	"		0.17	0.17
Maximum	"		0.26	0.26
Steam clean				
Minimum	S.F.		0.13	0.13
Average	"		0.17	0.17
Maximum	"		0.26	0.26
09910.05 Ext. Painting, Sitework				
Benches				
Brush				
First Coat				
Minimum	S.F.	0.22	0.53	0.75
Average	"	0.22	0.66	0.88
Maximum	"	0.22	0.88	1.10
Second Coat				
Minimum	S.F.	0.19	0.33	0.52
Average	"	0.19	0.37	0.56
Maximum	"	0.19	0.44	0.63
Roller				
First Coat				
Minimum	S.F.	0.22	0.26	0.48

Painting	UNIT	MAT.	INST.	TOTAL
09910.05 **Ext. Painting, Sitework** *(Cont.)*				
Average	S.F.	0.22	0.29	0.51
Maximum	"	0.22	0.33	0.55
Second Coat				
Minimum	S.F.	0.19	0.18	0.37
Average	"	0.19	0.22	0.41
Maximum	"	0.19	0.24	0.43
Brickwork				
Brush				
First Coat				
Minimum	S.F.	0.22	0.33	0.55
Average	"	0.22	0.44	0.66
Maximum	"	0.22	0.66	0.88
Second Coat				
Minimum	S.F.	0.22	0.29	0.51
Average	"	0.22	0.35	0.57
Maximum	"	0.22	0.44	0.66
Roller				
First Coat				
Minimum	S.F.	0.22	0.26	0.48
Average	"	0.22	0.33	0.55
Maximum	"	0.22	0.44	0.66
Second Coat				
Minimum	S.F.	0.22	0.22	0.44
Average	"	0.22	0.26	0.48
Maximum	"	0.22	0.33	0.55
Spray				
First Coat				
Minimum	S.F.	0.17	0.14	0.31
Average	"	0.17	0.18	0.35
Maximum	"	0.17	0.24	0.41
Second Coat				
Minimum	S.F.	0.17	0.13	0.30
Average	"	0.17	0.17	0.34
Maximum	"	0.17	0.22	0.39
Concrete Block				
Roller				
First Coat				
Minimum	S.F.	0.22	0.26	0.48
Average	"	0.22	0.35	0.57
Maximum	"	0.22	0.53	0.75
Second Coat				
Minimum	S.F.	0.22	0.22	0.44
Average	"	0.22	0.29	0.51
Maximum	"	0.22	0.44	0.66
Spray				
First Coat				
Minimum	S.F.	0.17	0.14	0.31
Average	"	0.17	0.17	0.34
Maximum	"	0.17	0.20	0.37
Second Coat				
Minimum	S.F.	0.17	0.09	0.26
Average	"	0.17	0.12	0.29
Maximum	"	0.17	0.16	0.33

Painting	UNIT	MAT.	INST.	TOTAL
09910.05 — Ext. Painting, Sitework *(Cont.)*				
Fences, Chain Link				
Brush				
First Coat				
Minimum	S.F.	0.13	0.53	0.66
Average	"	0.13	0.59	0.72
Maximum	"	0.13	0.66	0.79
Second Coat				
Minimum	S.F.	0.13	0.35	0.48
Average	"	0.13	0.40	0.53
Maximum	"	0.13	0.48	0.61
Roller				
First Coat				
Minimum	S.F.	0.13	0.37	0.50
Average	"	0.13	0.44	0.57
Maximum	"	0.13	0.50	0.63
Second Coat				
Minimum	S.F.	0.13	0.22	0.35
Average	"	0.13	0.26	0.39
Maximum	"	0.13	0.33	0.46
Spray				
First Coat				
Minimum	S.F.	0.11	0.16	0.27
Average	"	0.11	0.18	0.29
Maximum	"	0.11	0.22	0.33
Second Coat				
Minimum	S.F.	0.11	0.12	0.23
Average	"	0.11	0.14	0.25
Maximum	"	0.11	0.16	0.27
Fences, Wood or Masonry				
Brush				
First Coat				
Minimum	S.F.	0.22	0.55	0.77
Average	"	0.22	0.66	0.88
Maximum	"	0.22	0.88	1.10
Second Coat				
Minimum	S.F.	0.22	0.33	0.55
Average	"	0.22	0.40	0.62
Maximum	"	0.22	0.53	0.75
Roller				
First Coat				
Minimum	S.F.	0.22	0.29	0.51
Average	"	0.22	0.35	0.57
Maximum	"	0.22	0.40	0.62
Second Coat				
Minimum	S.F.	0.22	0.20	0.42
Average	"	0.22	0.25	0.47
Maximum	"	0.22	0.33	0.55
Spray				
First Coat				
Minimum	S.F.	0.17	0.18	0.35
Average	"	0.17	0.24	0.41
Maximum	"	0.17	0.33	0.50
Second Coat				

Painting	UNIT	MAT.	INST.	TOTAL
09910.05 **Ext. Painting, Sitework** *(Cont.)*				
Minimum	S.F.	0.17	0.13	0.30
Average	"	0.17	0.16	0.33
Maximum	"	0.17	0.22	0.39
Storage Tanks				
Roller				
First Coat				
Minimum	S.F.	0.17	0.22	0.39
Average	"	0.17	0.26	0.43
Maximum	"	0.17	0.33	0.50
Second Coat				
Minimum	S.F.	0.17	0.17	0.34
Average	"	0.17	0.21	0.38
Maximum	"	0.17	0.26	0.43
Spray				
First Coat				
Minimum	S.F.	0.13	0.13	0.26
Average	"	0.13	0.15	0.28
Maximum	"	0.13	0.18	0.31
Second Coat				
Minimum	S.F.	0.13	0.10	0.23
Average	"	0.13	0.11	0.24
Maximum	"	0.13	0.13	0.26
09910.15 **Ext. Painting, Buildings**				
Decks, Metal				
Spray				
First Coat				
Minimum	S.F.	0.13	0.24	0.37
Average	"	0.13	0.26	0.39
Maximum	"	0.13	0.29	0.42
Second Coat				
Minimum	S.F.	0.13	0.16	0.29
Average	"	0.13	0.18	0.31
Maximum	"	0.13	0.22	0.35
Decks, Wood, Stained				
Brush				
First Coat				
Minimum	S.F.	0.17	0.26	0.43
Average	"	0.17	0.29	0.46
Maximum	"	0.17	0.33	0.50
Second Coat				
Minimum	S.F.	0.17	0.18	0.35
Average	"	0.17	0.20	0.37
Maximum	"	0.17	0.22	0.39
Roller				
First Coat				
Minimum	S.F.	0.17	0.18	0.35
Average	"	0.17	0.20	0.37
Maximum	"	0.17	0.22	0.39
Second Coat				
Minimum	S.F.	0.17	0.16	0.33
Average	"	0.17	0.17	0.34
Maximum	"	0.17	0.20	0.37

Painting	UNIT	MAT.	INST.	TOTAL

09910.15   Ext. Painting, Buildings *(Cont.)*

	UNIT	MAT.	INST.	TOTAL
Spray				
First Coat				
Minimum	S.F.	0.14	0.16	0.30
Average	"	0.14	0.17	0.31
Maximum	"	0.14	0.20	0.34
Second Coat				
Minimum	S.F.	0.13	0.14	0.27
Average	"	0.13	0.16	0.29
Maximum	"	0.13	0.17	0.30
Doors, Metal				
Roller				
First Coat				
Minimum	S.F.	0.17	0.37	0.54
Average	"	0.17	0.44	0.61
Maximum	"	0.17	0.53	0.70
Second Coat				
Minimum	S.F.	0.17	0.26	0.43
Average	"	0.17	0.29	0.46
Maximum	"	0.17	0.33	0.50
Spray				
First Coat				
Minimum	S.F.	0.13	0.33	0.46
Average	"	0.13	0.37	0.50
Maximum	"	0.13	0.44	0.57
Second Coat				
Minimum	S.F.	0.13	0.24	0.37
Average	"	0.13	0.26	0.39
Maximum	"	0.13	0.29	0.42
Door Frames, Metal				
Brush				
First Coat				0.21
Minimum	L.F.	0.22	0.66	0.88
Average	"	0.22	0.83	1.05
Maximum	"	0.22	0.96	1.18
Second Coat				
Minimum	L.F.	0.22	0.37	0.59
Average	"	0.22	0.44	0.66
Maximum	"	0.22	0.53	0.75
Spray				
First Coat				
Minimum	L.F.	0.13	0.29	0.42
Average	"	0.13	0.37	0.50
Maximum	"	0.13	0.53	0.66
Second Coat				
Minimum	L.F.	0.13	0.24	0.37
Average	"	0.13	0.26	0.39
Maximum	"	0.13	0.29	0.42
Doors, Wood				
Brush				
First Coat				
Minimum	S.F.	0.17	0.81	0.98
Average	"	0.17	1.06	1.23
Maximum	"	0.17	1.32	1.49

Painting	UNIT	MAT.	INST.	TOTAL
09910.15 — **Ext. Painting, Buildings** *(Cont.)*				
Second Coat				
Minimum	S.F.	0.17	0.66	0.83
Average	"	0.17	0.75	0.92
Maximum	"	0.17	0.88	1.05
Roller				
First Coat				
Minimum	S.F.	0.17	0.35	0.52
Average	"	0.17	0.44	0.61
Maximum	"	0.17	0.66	0.83
Second Coat				
Minimum	S.F.	0.17	0.26	0.43
Average	"	0.17	0.29	0.46
Maximum	"	0.17	0.44	0.61
Spray				
First Coat				
Minimum	S.F.	0.13	0.16	0.29
Average	"	0.13	0.20	0.33
Maximum	"	0.13	0.26	0.39
Second Coat				
Minimum	S.F.	0.13	0.13	0.26
Average	"	0.13	0.15	0.28
Maximum	"	0.13	0.17	0.30
Gutters and Downspouts				
Brush				
First Coat				
Minimum	L.F.	0.22	0.66	0.88
Average	"	0.22	0.75	0.97
Maximum	"	0.22	0.88	1.10
Second Coat				
Minimum	L.F.	0.22	0.44	0.66
Average	"	0.22	0.53	0.75
Maximum	"	0.22	0.66	0.88
Siding, Metal				
Roller				
First Coat				
Minimum	S.F.	0.17	0.22	0.39
Average	"	0.17	0.24	0.41
Maximum	"	0.17	0.26	0.43
Second Coat				
Minimum	S.F.	0.17	0.20	0.37
Average	"	0.17	0.22	0.39
Maximum	"	0.17	0.24	0.41
Spray				
First Coat				
Minimum	S.F.	0.13	0.16	0.29
Average	"	0.13	0.18	0.31
Maximum	"	0.13	0.22	0.35
Second Coat				
Minimum	S.F.	0.13	0.10	0.23
Average	"	0.13	0.13	0.26
Maximum	"	0.13	0.17	0.30
Siding, Wood				
Roller				

Painting	UNIT	MAT.	INST.	TOTAL
09910.15 **Ext. Painting, Buildings** *(Cont.)*				
First Coat				
Minimum	S.F.	0.15	0.18	0.33
Average	"	0.15	0.22	0.37
Maximum	"	0.15	0.24	0.39
Second Coat				
Minimum	S.F.	0.15	0.22	0.37
Average	"	0.15	0.24	0.39
Maximum	"	0.15	0.26	0.41
Spray				
First Coat				
Minimum	S.F.	0.13	0.17	0.30
Average	"	0.13	0.18	0.31
Maximum	"	0.13	0.20	0.33
Second Coat				
Minimum	S.F.	0.13	0.13	0.26
Average	"	0.13	0.17	0.30
Maximum	"	0.13	0.26	0.39
Stucco				
Roller				
First Coat				
Minimum	S.F.	0.22	0.24	0.46
Average	"	0.22	0.27	0.49
Maximum	"	0.22	0.33	0.55
Second Coat				
Minimum	S.F.	0.22	0.19	0.41
Average	"	0.22	0.22	0.44
Maximum	"	0.22	0.26	0.48
Spray				
First Coat				
Minimum	S.F.	0.17	0.16	0.33
Average	"	0.17	0.18	0.35
Maximum	"	0.17	0.22	0.39
Second Coat				
Minimum	S.F.	0.17	0.13	0.30
Average	"	0.17	0.15	0.32
Maximum	"	0.17	0.17	0.34
Trim				
Brush				
First Coat				
Minimum	L.F.	0.22	0.22	0.44
Average	"	0.22	0.26	0.48
Maximum	"	0.22	0.33	0.55
Second Coat				
Minimum	L.F.	0.22	0.16	0.38
Average	"	0.22	0.22	0.44
Maximum	"	0.22	0.33	0.55
Walls				
Roller				
First Coat				
Minimum	S.F.	0.17	0.18	0.35
Average	"	0.17	0.19	0.36
Maximum	"	0.17	0.21	0.38
Second Coat				

Painting

	UNIT	MAT.	INST.	TOTAL
09910.15 — Ext. Painting, Buildings (Cont.)				
Minimum	S.F.	0.17	0.16	0.33
Average	"	0.17	0.17	0.34
Maximum	"	0.17	0.20	0.37
Spray				
First Coat				
Minimum	S.F.	0.13	0.08	0.21
Average	"	0.13	0.10	0.23
Maximum	"	0.13	0.13	0.26
Second Coat				
Minimum	S.F.	0.13	0.07	0.20
Average	"	0.13	0.08	0.21
Maximum	"	0.13	0.11	0.24
Windows				
Brush				
First Coat				
Minimum	S.F.	0.15	0.88	1.03
Average	"	0.15	1.06	1.21
Maximum	"	0.15	1.32	1.47
Second Coat				
Minimum	S.F.	0.15	0.75	0.90
Average	"	0.15	0.88	1.03
Maximum	"	0.15	1.06	1.21
09910.25 — Ext. Painting, Misc.				
Gratings, Metal				
Roller				
First Coat				
Minimum	S.F.	0.17	1.51	1.68
Average	"	0.17	1.77	1.94
Maximum	"	0.17	2.12	2.29
Second Coat				
Minimum	S.F.	0.17	1.06	1.23
Average	"	0.17	1.32	1.49
Maximum	"	0.17	1.77	1.94
Spray				
First Coat				
Minimum	S.F.	0.15	0.75	0.90
Average	"	0.15	0.88	1.03
Maximum	"	0.15	1.06	1.21
Second Coat				
Minimum	S.F.	0.15	0.59	0.74
Average	"	0.15	0.66	0.81
Maximum	"	0.15	0.75	0.90
Ladders				
Brush				
First Coat				
Minimum	L.F.	0.22	1.32	1.54
Average	"	0.22	1.51	1.73
Maximum	"	0.22	1.77	1.99
Second Coat				
Minimum	L.F.	0.22	1.06	1.28
Average	"	0.22	1.18	1.40
Maximum	"	0.22	1.32	1.54

Painting

09910.25	Ext. Painting, Misc. *(Cont.)*	UNIT	MAT.	INST.	TOTAL
Spray					
First Coat					
Minimum		L.F.	0.15	0.88	1.03
Average		"	0.15	0.96	1.11
Maximum		"	0.15	1.06	1.21
Second Coat					
Minimum		L.F.	0.15	0.75	0.90
Average		"	0.15	0.81	0.96
Maximum		"	0.15	0.88	1.03
Shakes					
Spray					
First Coat					
Minimum		S.F.	0.16	0.22	0.38
Average		"	0.16	0.24	0.40
Maximum		"	0.16	0.26	0.42
Second Coat					
Minimum		S.F.	0.16	0.20	0.36
Average		"	0.16	0.22	0.38
Maximum		"	0.16	0.24	0.40
Shingles, Wood					
Roller					
First Coat					
Minimum		S.F.	0.17	0.29	0.46
Average		"	0.17	0.33	0.50
Maximum		"	0.17	0.37	0.54
Second Coat					
Minimum		S.F.	0.17	0.20	0.37
Average		"	0.17	0.22	0.39
Maximum		"	0.17	0.24	0.41
Spray					
First Coat					
Minimum		L.F.	0.13	0.20	0.33
Average		"	0.13	0.22	0.35
Maximum		"	0.13	0.24	0.37
Second Coat					
Minimum		L.F.	0.13	0.15	0.28
Average		"	0.13	0.16	0.29
Maximum		"	0.13	0.17	0.30
Shutters and Louvres					
Brush					
First Coat					
Minimum		EA.	0.22	10.50	10.72
Average		"	0.22	13.25	13.47
Maximum		"	0.22	17.75	17.97
Second Coat					
Minimum		EA.	0.22	6.64	6.86
Average		"	0.22	8.17	8.39
Maximum		"	0.22	10.50	10.72
Spray					
First Coat					
Minimum		EA.	0.15	3.54	3.69
Average		"	0.15	4.24	4.39
Maximum		"	0.15	5.31	5.46

Painting	UNIT	MAT.	INST.	TOTAL
09910.25 **Ext. Painting, Misc.** *(Cont.)*				
Second Coat				
Minimum	EA.	0.15	2.65	2.80
Average	"	0.15	3.54	3.69
Maximum	"	0.15	4.24	4.39
Stairs, metal				
Brush				
First Coat				
Minimum	S.F.	0.22	0.59	0.81
Average	"	0.22	0.66	0.88
Maximum	"	0.22	0.75	0.97
Second Coat				
Minimum	S.F.	0.22	0.33	0.55
Average	"	0.22	0.37	0.59
Maximum	"	0.22	0.44	0.66
Spray				
First Coat				
Minimum	S.F.	0.15	0.29	0.44
Average	"	0.15	0.37	0.52
Maximum	"	0.15	0.40	0.55
Second Coat				
Minimum	S.F.	0.15	0.22	0.37
Average	"	0.15	0.26	0.41
Maximum	"	0.15	0.33	0.48
Steel, Structural, Light				
Brush				
First Coat				
Minimum	S.F.	0.17	1.06	1.23
Average	"	0.17	1.32	1.49
Maximum	"	0.17	1.77	1.94
Second Coat				
Minimum	S.F.	0.15	0.75	0.90
Average	"	0.15	0.88	1.03
Maximum	"	0.15	1.06	1.21
Roller				
First Coat				
Minimum	S.F.	0.17	0.75	0.92
Average	"	0.17	0.88	1.05
Maximum	"	0.17	1.06	1.23
Second Coat				
Minimum	S.F.	0.15	0.44	0.59
Average	"	0.15	0.53	0.68
Maximum	"	0.15	0.66	0.81
Spray				
First Coat				
Minimum	S.F.	0.15	0.44	0.59
Average	"	0.15	0.53	0.68
Maximum	"	0.15	0.66	0.81
Second Coat				
Minimum	S.F.	0.12	0.37	0.49
Average	"	0.12	0.44	0.56
Maximum	"	0.12	0.53	0.65
Steel, Medium to Heavy				
Brush				

Painting

09910.25 — Ext. Painting, Misc. *(Cont.)*

	UNIT	MAT.	INST.	TOTAL
First Coat				
Minimum	S.F.	0.17	0.53	0.70
Average	"	0.17	0.59	0.76
Maximum	"	0.17	0.66	0.83
Second Coat				
Minimum	S.F.	0.15	0.44	0.59
Average	"	0.15	0.53	0.68
Maximum	"	0.15	0.62	0.77
Roller				
First Coat				
Minimum	S.F.	0.17	0.44	0.61
Average	"	0.17	0.53	0.70
Maximum	"	0.17	0.62	0.79
Second Coat				
Minimum	S.F.	0.15	0.33	0.48
Average	"	0.15	0.37	0.52
Maximum	"	0.15	0.44	0.59
Spray				
First Coat				
Minimum	S.F.	0.15	0.29	0.44
Average	"	0.15	0.33	0.48
Maximum	"	0.15	0.37	0.52
Second Coat				
Minimum	S.F.	0.12	0.24	0.36
Average	"	0.12	0.26	0.38
Maximum	"	0.12	0.29	0.41

09910.35 — Int. Painting, Buildings

	UNIT	MAT.	INST.	TOTAL
Acoustical Ceiling				
Roller				
First Coat				
Minimum	S.F.	0.22	0.33	0.55
Average	"	0.22	0.44	0.66
Maximum	"	0.22	0.66	0.88
Second Coat				
Minimum	S.F.	0.22	0.26	0.48
Average	"	0.22	0.33	0.55
Maximum	"	0.22	0.44	0.66
Spray				
First Coat				
Minimum	S.F.	0.17	0.14	0.31
Average	"	0.17	0.17	0.34
Maximum	"	0.17	0.22	0.39
Second Coat				
Minimum	S.F.	0.17	0.11	0.28
Average	"	0.17	0.13	0.30
Maximum	"	0.17	0.15	0.32
Cabinets and Casework				
Brush				
First Coat				
Minimum	S.F.	0.22	0.53	0.75
Average	"	0.22	0.59	0.81
Maximum	"	0.22	0.66	0.88

Painting	UNIT	MAT.	INST.	TOTAL
09910.35 **Int. Painting, Buildings** *(Cont.)*				
Second Coat				
Minimum	S.F.	0.22	0.44	0.66
Average	"	0.22	0.48	0.70
Maximum	"	0.22	0.53	0.75
Spray				
First Coat				
Minimum	S.F.	0.17	0.26	0.43
Average	"	0.17	0.31	0.48
Maximum	"	0.17	0.37	0.54
Second Coat				
Minimum	S.F.	0.17	0.21	0.38
Average	"	0.17	0.23	0.40
Maximum	"	0.17	0.29	0.46
Ceilings				
Roller				
First Coat				
Minimum	S.F.	0.17	0.22	0.39
Average	"	0.17	0.24	0.41
Maximum	"	0.17	0.26	0.43
Second Coat				
Minimum	S.F.	0.17	0.17	0.34
Average	"	0.17	0.20	0.37
Maximum	"	0.17	0.22	0.39
Spray				
First Coat				
Minimum	S.F.	0.13	0.13	0.26
Average	"	0.13	0.14	0.27
Maximum	"	0.13	0.16	0.29
Second Coat				
Minimum	S.F.	0.13	0.10	0.23
Average	"	0.13	0.11	0.24
Maximum	"	0.13	0.13	0.26
Doors, Metal				
Roller				
First Coat				
Minimum	L.F.	0.22	0.35	0.57
Average	"	0.22	0.40	0.62
Maximum	"	0.22	0.48	0.70
Second Coat				
Minimum	L.F.	0.22	0.25	0.47
Average	"	0.22	0.27	0.49
Maximum	"	0.22	0.31	0.53
Spray				
First Coat				
Minimum	L.F.	0.14	0.29	0.43
Average	"	0.14	0.33	0.47
Maximum	"	0.14	0.37	0.51
Second Coat				
Minimum	L.F.	0.22	0.22	0.44
Average	"	0.22	0.24	0.46
Maximum	"	0.22	0.26	0.48
Doors, Wood				
Brush				

Painting	UNIT	MAT.	INST.	TOTAL
09910.35 **Int. Painting, Buildings** *(Cont.)*				
First Coat				
Minimum	S.F.	0.22	0.75	0.97
Average	"	0.22	0.96	1.18
Maximum	"	0.22	1.18	1.40
Second Coat				
Minimum	S.F.	0.16	0.59	0.75
Average	"	0.16	0.66	0.82
Maximum	"	0.16	0.75	0.91
Spray				
First Coat				
Minimum	S.F.	0.15	0.15	0.30
Average	"	0.15	0.18	0.33
Maximum	"	0.15	0.24	0.39
Second Coat				
Minimum	S.F.	0.15	0.12	0.27
Average	"	0.15	0.14	0.29
Maximum	"	0.15	0.16	0.31
Ductwork				
Brush				
Minimum	L.F.	0.17	0.66	0.83
Average	"	0.17	0.75	0.92
Maximum	"	0.17	0.88	1.05
Roller				
Minimum	S.F.	0.17	0.44	0.61
Average	"	0.17	0.48	0.65
Maximum	"	0.17	0.53	0.70
Spray				
Minimum	S.F.	0.15	0.18	0.33
Average	"	0.15	0.20	0.35
Maximum	"	0.15	0.22	0.37
Floors				
Roller				
First Coat				
Minimum	S.F.	0.17	0.16	0.33
Average	"	0.17	0.19	0.36
Maximum	"	0.17	0.22	0.39
Second Coat				
Minimum	S.F.	0.17	0.13	0.30
Average	"	0.17	0.14	0.31
Maximum	"	0.17	0.15	0.32
Spray				
First Coat				
Minimum	S.F.	0.15	0.11	0.26
Average	"	0.15	0.12	0.27
Maximum	"	0.15	0.13	0.28
Second Coat				
Minimum	S.F.	0.15	0.10	0.25
Average	"	0.15	0.11	0.26
Maximum	"	0.15	0.12	0.27
Pipes to 6" diameter				
Brush				
Minimum	L.F.	0.22	0.66	0.88
Average	"	0.22	0.75	0.97

Painting	UNIT	MAT.	INST.	TOTAL

09910.35 — Int. Painting, Buildings *(Cont.)*

	UNIT	MAT.	INST.	TOTAL
Maximum	L.F.	0.22	0.88	1.10
Spray				
Minimum	L.F.	0.17	0.22	0.39
Average	"	0.17	0.26	0.43
Maximum	"	0.17	0.35	0.52
Pipes to 12" diameter				
Brush				
Minimum	L.F.	0.37	1.32	1.69
Average	"	0.37	1.51	1.88
Maximum	"	0.37	1.77	2.14
Spray				
Minimum	L.F.	0.34	0.44	0.78
Average	"	0.34	0.53	0.87
Maximum	"	0.34	0.66	1.00
Trim				
Brush				
First Coat				
Minimum	L.F.	0.22	0.21	0.43
Average	"	0.22	0.24	0.46
Maximum	"	0.22	0.29	0.51
Second Coat				
Minimum	L.F.	0.22	0.15	0.37
Average	"	0.22	0.20	0.42
Maximum	"	0.22	0.29	0.51
Walls				
Roller				
First Coat				
Minimum	S.F.	0.17	0.18	0.35
Average	"	0.17	0.19	0.36
Maximum	"	0.17	0.22	0.39
Second Coat				
Minimum	S.F.	0.17	0.16	0.33
Average	"	0.17	0.17	0.34
Maximum	"	0.17	0.20	0.37
Spray				
First Coat				
Minimum	S.F.	0.14	0.08	0.22
Average	"	0.14	0.10	0.24
Maximum	"	0.14	0.13	0.27
Second Coat				
Minimum	S.F.	0.14	0.07	0.21
Average	"	0.14	0.09	0.23
Maximum	"	0.14	0.12	0.26

09955.10 — Wall Covering

	UNIT	MAT.	INST.	TOTAL
Vinyl wall covering				
Medium duty	S.F.	1.43	0.75	2.18
Heavy duty	"	2.09	0.88	2.97
Over pipes and irregular shapes				
Lightweight, 13 oz.	S.F.	1.43	1.06	2.49
Medium weight, 25 oz.	"	1.70	1.18	2.88
Heavy weight, 34 oz.	"	2.09	1.32	3.41
Cork wall covering				

Painting	UNIT	MAT.	INST.	TOTAL
09955.10 **Wall Covering** *(Cont.)*				
1' x 1' squares				
1/4" thick	S.F.	4.45	1.32	5.77
1/2" thick	"	5.66	1.32	6.98
3/4" thick	"	6.38	1.32	7.70
Wall fabrics				
Natural fabrics, grass cloths				
Minimum	S.F.	1.43	0.81	2.24
Average	"	1.59	0.88	2.47
Maximum	"	5.33	1.06	6.39
Flexible gypsum coated wall fabric, fire resistant	"	1.60	0.53	2.13
Vinyl corner guards				
3/4" x 3/4" x 8'	EA.	7.53	6.64	14.17
2-3/4" x 2-3/4" x 4'	"	4.45	6.64	11.09
09980.15 **Paint**				
Paint, enamel				
600 sf per gal.	GAL			49.50
550 sf per gal.	"			46.25
500 sf per gal.	"			33.00
450 sf per gal.	"			30.75
350 sf per gal.	"			29.75
Filler, 60 sf per gal.	"			35.25
Latex, 400 sf per gal.	"			33.00
Aluminum				
400 sf per gal.	GAL			44.00
500 sf per gal.	"			70.00
Red lead, 350 sf per gal.	"			62.00
Primer				
400 sf per gal.	GAL			29.75
300 sf per gal.	"			29.75
Latex base, interior, white	"			33.00
Sealer and varnish				
400 sf per gal.	GAL			30.75
425 sf per gal.	"			44.00
600 sf per gal.	"			57.00

Specialties	UNIT	MAT.	INST.	TOTAL

10110.10 Chalkboards

Item	UNIT	MAT.	INST.	TOTAL
Chalkboard, metal frame, 1/4" thick				
48"x60"	EA.	430	53.00	483
48"x96"	"	650	59.00	709
48"x144"	"	720	66.00	786
48"x192"	"	960	76.00	1,036
Liquid chalkboard				
48"x60"	EA.	380	53.00	433
48"x96"	"	480	59.00	539
48"x144"	"	600	66.00	666
48"x192"	"	690	76.00	766
Map rail, deluxe	L.F.	8.80	2.65	11.45

10165.10 Toilet Partitions

Item	UNIT	MAT.	INST.	TOTAL
Toilet partition, plastic laminate				
Ceiling mounted	EA.	930	180	1,110
Floor mounted	"	890	130	1,020
Metal				
Ceiling mounted	EA.	740	180	920
Floor mounted	"	700	130	830
Wheel chair partition, plastic laminate				
Ceiling mounted	EA.	430	180	610
Floor mounted	"	290	130	420
Painted metal				
Ceiling mounted	EA.	390	180	570
Floor mounted	"	280	130	410
Urinal screen, plastic laminate				
Wall hung	EA.	480	66.00	546
Floor mounted	"	400	66.00	466
Porcelain enameled steel, floor mounted	"	560	66.00	626
Painted metal, floor mounted	"	360	66.00	426
Stainless steel, floor mounted	"	940	66.00	1,006
Metal toilet partitions				
Front door and side divider, floor mounted				
Porcelain enameled steel	EA.	1,260	130	1,390
Painted steel	"	420	130	550
Stainless steel	"	1,280	130	1,410

10185.10 Shower Stalls

Item	UNIT	MAT.	INST.	TOTAL
Shower receptors				
Precast, terrazzo				
32" x 32"	EA.	470	48.75	519
32" x 48"	"	530	59.00	589
Concrete				
32" x 32"	EA.	280	48.75	329
48" x 48"	"	330	65.00	395
Shower door, trim and hardware				
Economy, 24" wide, chrome frame, tempered glass	EA.	290	59.00	349
Porcelain enameled steel, flush	"	530	59.00	589
Baked enameled steel, flush	"	370	59.00	429
Aluminum frame, tempered glass, 48" wide, sliding	"	570	73.00	643
Folding	"	540	73.00	613
Aluminum frame and tempered glass, molded plastic				
Complete with receptor and door				

Specialties	UNIT	MAT.	INST.	TOTAL
10185.10 **Shower Stalls** *(Cont.)*				
32" x 32"	EA.	380	150	530
36" x 36"	"	410	150	560
40" x 40"	"	440	170	610
Shower compartment, precast concrete receptor				
Single entry type				
Porcelain enameled steel	EA.	2,620	590	3,210
Baked enameled steel	"	1,540	590	2,130
Stainless steel	"	3,110	590	3,700
Double entry type				
Porcelain enameled steel	EA.	5,490	730	6,220
Baked enameled steel	"	1,520	730	2,250
Stainless steel	"	2,970	730	3,700
10190.10 **Cubicles**				
Hospital track				
Ceiling hung	L.F.	7.70	5.90	13.60
Suspended	"	7.15	7.58	14.73
Hospital metal dividers, galvanized steel				
Baked enamel finish				
54" high				
10" glass light	L.F.	130	26.50	157
14" glass light	"	170	26.50	197
24" glass light	"	210	26.50	237
60" high				
10" glass light	L.F.	160	29.50	190
14" glass light	"	190	29.50	220
24" glass light	"	220	29.50	250
Stainless steel				
54" high				
10" glass light	L.F.	210	29.50	240
14" glass light	"	240	29.50	270
24" glass light	"	290	29.50	320
60" high				
14" glass light	L.F.	250	33.25	283
14" glass light	"	260	33.25	293
24" glass light	"	310	33.25	343
10210.10 **Vents And Wall Louvres**				
Block vent, 8"x16"x4" aluminum, w/screen, mill finish	EA.	150	19.75	170
Standard	"	88.00	18.50	107
Vents w/screen, 4" deep, 8" wide, 5" high				
Modular	EA.	94.00	18.50	113
Aluminum gable louvers	S.F.	21.00	9.84	30.84
Vent screen aluminum, 4" wide, continuous	L.F.	6.60	1.96	8.56
Louvers, aluminum, anodized, fixed blade				
Horizontal line	S.F.	86.00	14.75	101
Vertical line	"	86.00	14.75	101
Wall louvre, aluminum mill finish				
Under, 2 sf	S.F.	49.75	7.38	57.13

Specialties	UNIT	MAT.	INST.	TOTAL
10210.10 **Vents And Wall Louvres** *(Cont.)*				
2 to 4 sf	S.F.	36.00	6.56	42.56
5 to 10 sf	"	31.25	6.56	37.81
Galvanized steel				
Under 2 sf	S.F.	34.00	7.38	41.38
2 to 4 sf	"	29.00	6.56	35.56
5 to 10 sf	"	25.25	6.56	31.81
10225.10 **Door Louvres**				
Fixed, 1" thick, enameled steel				
8"x8"	EA.	53.00	6.64	59.64
12"x8"	"	72.00	6.64	78.64
12"x12"	"	95.00	7.58	103
16"x12"	"	100	8.17	108
18"x12"	"	100	13.25	113
20"x8"	"	120	7.58	128
20"x12"	"	130	15.25	145
20"x16"	"	130	17.75	148
20"x20"	"	140	21.25	161
24"x12"	"	120	17.75	138
24"x16"	"	120	19.00	139
24"x18"	"	140	20.50	161
24"x20"	"	170	22.25	192
24"x24"	"	170	24.25	194
26"x26"	"	200	33.25	233
10270.40 **Access & Pedestal Floor**				
Panels, no covering, 2'x2'				
Plain	S.F.	16.50	0.66	17.16
Perforated	"	21.00	26.50	47.50
Pedestals				13.50
For 6" to 12" clearance	EA.	13.75	5.31	19.06
Stringers				
2'	L.F.	2.20	2.52	4.72
6'	"	2.36	1.77	4.13
Accessories				
Ramp assembly	S.F.	53.00	2.12	55.12
Elevated floor assembly	"	83.00	1.96	84.96
Handrail	L.F.	62.00	26.50	88.50
Fascia plate	"	44.00	13.25	57.25
For carpet tiles, add	S.F.			7.15
For vinyl flooring, add	"			9.07
RF shielding components, floor liner				
Hot rolled steel sheet				
14 ga.	S.F.	4.73	1.32	6.05
11 ga.	"	5.50	4.08	9.58
10290.10 **Pest Control**				
Termite control				
Under slab spraying				
Minimum	S.F.	0.41	0.10	0.51
Average	"	0.55	0.20	0.75
Maximum	"	0.82	0.41	1.23

Specialties	UNIT	MAT.	INST.	TOTAL
10350.10 — **Flagpoles**				
Installed in concrete base				
Fiberglass				
25' high	EA.	1,460	350	1,810
50' high	"	6,480	890	7,370
Aluminum				
25' high	EA.	2,120	350	2,470
50' high	"	6,930	890	7,820
Bonderized steel				
25' high	EA.	3,520	410	3,930
50' high	"	8,580	1,060	9,640
Freestanding tapered, fiberglass				
30' high	EA.	1,460	380	1,840
40' high	"	2,020	480	2,500
50' high	"	6,480	530	7,010
60' high	"	7,770	620	8,390
Wall mounted, with collar, brushed aluminum finish				
15' long	EA.	1,210	270	1,480
18' long	"	1,320	270	1,590
20' long	"	1,480	280	1,760
24' long	"	1,590	310	1,900
Outrigger, wall, including base				
10' long	EA.	680	350	1,030
20' long	"	1,230	440	1,670
10400.10 — **Identifying Devices**				
Directory and bulletin boards				
Open face boards				
Chrome plated steel frame	S.F.	27.50	26.50	54.00
Aluminum framed	"	23.00	26.50	49.50
Bronze framed	"	30.75	26.50	57.25
Stainless steel framed	"	55.00	26.50	81.50
Tack board, aluminum framed	"	38.50	26.50	65.00
Visual aid board, aluminum framed	"	110	26.50	137
Glass encased boards, hinged and keyed				
Aluminum framed	S.F.	120	66.00	186
Bronze framed	"	120	66.00	186
Stainless steel framed	"	210	66.00	276
Chrome plated steel framed	"	200	66.00	266
Metal plaque				
Cast bronze	S.F.	240	44.25	284
Aluminum	"	190	44.25	234
Metal engraved plaque				
Porcelain steel	S.F.	210	44.25	254
Stainless steel	"	190	44.25	234
Brass	"	150	44.25	194
Aluminum	"	130	44.25	174
Metal built-up plaque				
Bronze	S.F.	160	53.00	213
Copper and bronze	"	140	53.00	193
Copper and aluminum	"	120	53.00	173
Metal nameplate plaques				
Cast bronze	S.F.	1,730	33.25	1,763
Cast aluminum	"	1,110	33.25	1,143

Specialties	UNIT	MAT.	INST.	TOTAL
10400.10 **Identifying Devices** *(Cont.)*				
Engraved, 1-1/2" x 6"				
Bronze	EA.	200	33.25	233
Aluminum	"	200	33.25	233
Letters, on masonry or concrete, aluminum, satin finish				
1/2" thick				
2" high	EA.	19.75	21.25	41.00
4" high	"	29.75	26.50	56.25
6" high	"	39.75	29.50	69.25
3/4" thick				
8" high	EA.	60.00	33.25	93.25
10" high	"	69.00	38.00	107
1" thick				
12" high	EA.	99.00	44.25	143
14" high	"	120	53.00	173
16" high	"	150	66.00	216
For polished aluminum add, 15%				
For clear anodized aluminum add, 15%				
For colored anodic aluminum add, 30%				
For profiled and color enameled letters add, 50%				
Cast bronze, satin finish letters				
3/8" thick				
2" high	EA.	52.00	21.25	73.25
4" high	"	67.00	26.50	93.50
1/2" thick, 6" high	"	91.00	29.50	121
5/8" thick, 8" high	"	120	33.25	153
1" thick				
10" high	EA.	180	38.00	218
12" high	"	210	44.25	254
14" high	"	270	53.00	323
16" high	"	350	66.00	416
Interior door signs, adhesive, flexible				
2" x 8"	EA.	35.75	10.25	46.00
4" x 4"	"	38.50	10.25	48.75
6" x 7"	"	42.25	10.25	52.50
6" x 9"	"	42.50	10.25	52.75
10" x 9"	"	49.50	10.25	59.75
10" x 12"	"	61.00	10.25	71.25
Hard plastic type, no frame				
3" x 8"	EA.	8.80	10.25	19.05
4" x 4"	"	13.25	10.25	23.50
4" x 12"	"	25.25	10.25	35.50
Hard plastic type, with frame				
3" x 8"	EA.	14.25	10.25	24.50
4" x 4"	"	20.75	10.25	31.00
4" x 12"	"	27.50	10.25	37.75
10450.10 **Control**				
Access control, 7' high, indoor or outdoor impenetrability				
Remote or card control, type B	EA.	4,390	720	5,110
Free passage, type B	"	3,980	720	4,700
Remote or card control, type AA	"	5,720	720	6,440
Free passage, type AA	"	5,390	720	6,110

Specialties	UNIT	MAT.	INST.	TOTAL
10500.10 **Lockers**				
Locker bench, floor mounted, laminated maple				
4'	EA.	220	44.25	264
6'	"	320	44.25	364
Wardrobe locker, single tier type				
12" x 15" x 72"	EA.	310	53.00	363
18" x 15" x 72"	"	420	56.00	476
12" x 18" x 72"	"	330	59.00	389
18" x 18" x 72"	"	440	62.00	502
Double tier type				
12" x 15" x 36"	EA.	310	26.50	337
18" x 15" x 36"	"	420	26.50	447
12" x 18" x 36"	"	440	26.50	467
18" x 18" x 36"	"	550	26.50	577
Two person unit				
18" x 15" x 72"	EA.	470	89.00	559
18" x 18" x 72"	"	280	110	390
Duplex unit				
15" x 15" x 72"	EA.	460	53.00	513
15" x 21" x 72"	"	490	53.00	543
Basket lockers, basket sets with baskets				
24 basket set	SET	360	270	630
30 basket set	"	380	330	710
36 basket set	"	420	440	860
42 basket set	"	470	530	1,000
10520.10 **Fire Protection**				
Portable fire extinguishers				
Water pump tank type				
2.5 gal.				
Red enameled galvanized	EA.	130	27.50	158
Red enameled copper	"	140	27.50	168
Polished copper	"	240	27.50	268
Carbon dioxide type, red enamel steel				
Squeeze grip with hose and horn				
2.5 lb	EA.	160	27.50	188
5 lb	"	220	31.75	252
10 lb	"	300	41.25	341
15 lb	"	350	52.00	402
20 lb	"	420	52.00	472
Wheeled type				
125 lb	EA.	2,200	83.00	2,283
250 lb	"	3,850	83.00	3,933
500 lb	"	5,720	83.00	5,803
Dry chemical, pressurized type				
Red enameled steel				
2.5 lb	EA.	22.00	27.50	49.50
5 lb	"	44.00	31.75	75.75
10 lb	"	66.00	41.25	107
20 lb	"	120	52.00	172
30 lb	"	160	52.00	212
Chrome plated steel, 2.5 lb	"	220	27.50	248
Other type extinguishers				
2.5 gal, stainless steel, pressurized water tanks	EA.	100	27.50	128

Specialties	UNIT	MAT.	INST.	TOTAL
10520.10 **Fire Protection** *(Cont.)*				
Soda and acid type	EA.	180	27.50	208
Cartridge operated, water type	"	290	27.50	318
Loaded stream, water type	"	130	27.50	158
Foam type	"	110	27.50	138
40 gal, wheeled foam type	"	77.00	83.00	160
Fire extinguisher cabinets				
Enameled steel				
8" x 12" x 27"	EA.	310	83.00	393
8" x 16" x 38"	"	370	83.00	453
Aluminum				
8" x 12" x 27"	EA.	190	83.00	273
8" x 16" x 38"	"	230	83.00	313
8" x 12" x 27"	"	190	83.00	273
Stainless steel				
8" x 16" x 38"	EA.	250	83.00	333
10550.10 **Postal Specialties**				
Mail chutes				
Single mail chute				
Finished aluminum	L.F.	1,290	130	1,420
Bronze	"	1,840	130	1,970
Single mail chute receiving box				
Finished aluminum	EA.	1,240	270	1,510
Bronze	"	2,040	270	2,310
Twin mail chute, double parallel				
Finished aluminum	FLR	1,810	270	2,080
Bronze	"	2,200	270	2,470
Receiving box, 36" x 20" x 12"				
Finished aluminum	EA.	2,270	440	2,710
Bronze	"	3,270	440	3,710
Locked receiving mail box				
Finished aluminum	EA.	1,120	270	1,390
Bronze	"	1,870	270	2,140
Commercial postal accessories for mail chutes				
Letter slot, brass	EA.	500	89.00	589
Bulk mail slot, brass	"	1,100	89.00	1,189
Mail boxes				
Residential postal accessories				
Letter slot	EA.	120	26.50	147
Rural letter box	"	94.00	66.00	160
Apartment house, keyed, 3.5" x 4.5" x 16"	"	120	17.75	138
Ranch style	"	140	26.50	167
Commercial postal accessories				
Letter box, with combination lock	EA.	83.00	19.00	102
Key lock	"	100	19.00	119
Mail box, aluminum w/glass front, 4x5				
Horizontal rear load	EA.	120	15.25	135
Vertical front load	"	120	15.25	135

Specialties	UNIT	MAT.	INST.	TOTAL
10600.10 **Movable Partitions**				
Partition, movable office type, 2-1/2" thick, vinyl-gypsum	S.F.	17.50	2.65	20.15
Enameled steel frame, with 1/4" thick clear glass	"	22.25	2.65	24.90
Door frame and hardware for movable partitions	EA.	540	180	720
Add for acoustic movable partition	S.F.			2.33
Accordion partition, 12' high				
Vinyl	S.F.	22.00	8.85	30.85
Acoustical	"	27.50	8.85	36.35
Standard office cubicles, 8' high, steel framed				
Baked enamel finish				
100% flush	L.F.	170	13.25	183
75% flush and 25% glass	"	190	14.75	205
50% flush and 50% glass	"	280	14.75	295
100% glass	"	300	17.75	318
Natural hardwood panels				
100% flush	L.F.	140	17.75	158
50% flush and 50% glass	"	190	19.00	209
Plastic laminated panels				
100% flush	L.F.	170	17.75	188
75% flush and 25% glass	"	210	19.00	229
50% flush and 50% glass	"	250	19.00	269
Vinyl covered panels				
100% flush	L.F.	150	18.25	168
75% flush and 25% glass	"	200	19.75	220
50% and 50% glass	"	240	19.75	260
Aluminum framed				
Enameled or anodized aluminum panels				
100% flush	L.F.	130	13.25	143
75% flush and 25% glass	"	120	14.75	135
50% flush and 50% glass	"	100	14.75	115
Vinyl covered panels				
100% flush	L.F.	130	18.25	148
75% flush and 25% glass	"	150	19.75	170
50% flush and 50% glass	"	180	19.75	200
60" high partitions, steel framed				
Enameled panels	L.F.	64.00	11.75	75.75
Natural hardwood panels, two sides	"	250	12.25	262
Plastic laminated panels	"	130	12.25	142
Vinyl covered panels	"	170	11.75	182
Aluminum framed				
Anodized or baked enamel panels	L.F.	100	11.75	112
Natural hardwood panels	"	280	12.25	292
Plastic laminated panels	"	140	12.25	152
Vinyl covered panels	"	180	11.75	192
Wire mesh partitions				
Wall panels				
4' x 7'	EA.	160	33.25	193
4' x 8'	"	180	35.50	216
4' x 10'	"	200	40.75	241
Wall filler panels				
1' x 7'	EA.	100	33.25	133
1' x 8'	"	110	35.50	146
1' x 10'	"	150	38.00	188
2' x 7'	"	110	33.25	143

Specialties	UNIT	MAT.	INST.	TOTAL
10600.10 **Movable Partitions** *(Cont.)*				
2' x 8'	EA.	120	35.50	156
2' x 10'	"	160	38.00	198
3' x 7'	"	130	33.25	163
3' x 8'	"	140	35.50	176
3' x 10'	"	180	38.00	218
Ceiling panels				
10' x 2'	EA.	94.00	76.00	170
10' x 4'	"	160	110	270
Wall panel with service window				
5' wide				390
7' high	EA.	410	33.25	443
8' high	"	420	35.50	456
10' high	"	440	38.00	478
Doors				
Sliding				
3' x 7'	EA.	690	130	820
3' x 8'	"	750	150	900
3' x 10'	"	830	210	1,040
4' x 7'	"	700	150	850
4' x 8'	"	800	210	1,010
4' x 10'	"	860	270	1,130
5' x 7'	"	750	210	960
5' x 8'	"	920	270	1,190
5' x 10'	"	1,060	270	1,330
Swing door				
3' x 7'	EA.	330	130	460
4' x 7'	"	370	150	520
Swing door, with 1' transom				
3' x 7'	EA.	370	150	520
4' x 7'	"	420	210	630
Swing door, with 3' transom				
3' x 7'	EA.	420	210	630
4' x 7'	"	460	270	730
10670.10 **Shelving**				
Shelving, enamel, closed side and back, 12" x 36"				
5 shelves	EA.	230	89.00	319
8 shelves	"	260	120	380
Open				
5 shelves	EA.	150	89.00	239
8 shelves	"	170	120	290
Metal storage shelving, baked enamel				
7 shelf unit, 72" or 84" high				
10" shelf	L.F.	60.00	53.00	113
12" shelf	"	63.00	56.00	119
15" shelf	"	72.00	59.00	131
18" shelf	"	73.00	62.00	135
24" shelf	"	83.00	66.00	149
30" shelf	"	90.00	71.00	161
36" shelf	"	110	76.00	186
4 shelf unit, 40" high				
10" shelf	L.F.	52.00	44.25	96.25
12" shelf	"	52.00	48.25	100

Specialties	UNIT	MAT.	INST.	TOTAL
10670.10 — Shelving (Cont.)				
15" shelf	L.F.	66.00	53.00	119
18" shelf	"	76.00	56.00	132
24" shelf	"	76.00	59.00	135
3 shelf unit, 32" high				
10" shelf	L.F.	42.50	26.50	69.00
12" shelf	"	42.50	28.00	70.50
15" shelf	"	46.25	29.50	75.75
18" shelf	"	49.50	31.25	80.75
24" shelf	"	53.00	33.25	86.25
Single shelf unit, attached to masonry				
10" shelf	L.F.	15.00	8.85	23.85
12" shelf	"	17.00	9.65	26.65
15" shelf	"	17.50	10.25	27.75
18" shelf	"	18.75	10.75	29.50
24" shelf	"	21.75	11.50	33.25
For stainless steel, add to material, 120%				
For attachment to gypsum board, add to labor, 50%				
Built-in wood shelves				
Posts and trimmed plywood	L.F.	1.92	7.58	9.50
Solid clear pine	"	1.92	8.17	10.09
Closet shelf, pine with rod	"	2.25	8.17	10.42
For lumber edge band, add to material	"			0.82
For prefinished shelves, add to material, 225%				
10750.10 — Telephone Enclosures				
Telephone enclosure, wall mounted, shelf, 28" x 30" x 15"	EA.	1,480	130	1,610
Directory shelf, stainless steel, 3 binders	"	1,190	89.00	1,279
10800.10 — Bath Accessories				
Ash receiver, wall mounted, aluminum	EA.	99.00	26.50	126
Grab bar, 1-1/2" dia., stainless steel, wall mounted				
24" long	EA.	33.00	26.50	59.50
36" long	"	55.00	28.00	83.00
42" long	"	72.00	29.50	102
48" long	"	88.00	31.25	119
52" long	"	100	33.25	133
1" dia., stainless steel				
12" long	EA.	22.50	23.00	45.50
18" long	"	30.50	24.25	54.75
24" long	"	33.00	26.50	59.50
30" long	"	42.25	28.00	70.25
36" long	"	46.25	29.50	75.75
48" long	"	53.00	31.25	84.25
Hand dryer, surface mounted, 110 volt	"	360	66.00	426
Medicine cabinet, 16 x 22, baked enamel, steel, lighted	"	89.00	21.25	110
With mirror, lighted	"	150	35.50	186
Mirror, 1/4" plate glass, up to 10 sf	S.F.	8.96	5.31	14.27
Mirror, stainless steel frame				
18"x24"	EA.	58.00	17.75	75.75
18"x32"	"	61.00	21.25	82.25
18"x36"	"	67.00	26.50	93.50
24"x30"	"	74.00	26.50	101
24"x36"	"	79.00	29.50	109

Specialties	UNIT	MAT.	INST.	TOTAL
10800.10 **Bath Accessories** *(Cont.)*				
24"x48"	EA.	160	44.25	204
24"x60"	"	240	53.00	293
30"x30"	"	200	53.00	253
30"x72"	"	400	66.00	466
48"x72"	"	450	89.00	539
With shelf, 18"x24"	"	200	21.25	221
Sanitary napkin dispenser, stainless steel, wall mounted	"	510	35.50	546
Shower rod, 1" diameter				
Chrome finish over brass	EA.	34.00	26.50	60.50
Stainless steel	"	77.00	26.50	104
Soap dish, stainless steel, wall mounted	"	220	35.50	256
Toilet tissue dispenser, stainless, wall mounted				
Single roll	EA.	170	13.25	183
Double roll	"	260	15.25	275
Towel dispenser, stainless steel				
Flush mounted	EA.	170	29.50	200
Surface mounted	"	72.00	26.50	98.50
Combination towel dispenser and waste receptacle	"	720	35.50	756
Towel bar, stainless steel				
18" long	EA.	66.00	21.25	87.25
24" long	"	74.00	24.25	98.25
30" long	"	92.00	26.50	119
36" long	"	78.00	29.50	108
Toothbrush and tumbler holder	"	5.50	17.75	23.25
Waste receptacle, stainless steel, wall mounted	"	92.00	44.25	136
10900.10 **Wardrobe Specialties**				
Hospital wardrobe units, 24" x 24" x 76", with door				
Baked enameled steel	EA.	990	300	1,290
Hardwood	"	1,560	300	1,860
Stainless steel	"	1,540	300	1,840
Plastic laminated	"	370	300	670
Dormitory wardrobe units, 24" x 76", with door				
Hardwood	EA.	550	300	850
Plastic laminated	"	200	300	500
Hat and coat rack				
Single tier				
Baked enameled steel	L.F.	150	13.25	163
Stainless steel	"	220	13.25	233
Aluminum	"	180	13.25	193
Double tier				
Baked enameled steel	L.F.	550	15.25	565
Stainless steel	"	690	15.25	705
Aluminum	"	470	15.25	485

Architectural Equipment

	UNIT	MAT.	INST.	TOTAL
11010.10 Maintenance Equipment				
Vacuum cleaning system				
3 valves				
1.5 hp	EA.	2,850	590	3,440
2.5 hp	"	3,290	760	4,050
5 valves	"	3,840	1,060	4,900
7 valves	"	4,390	1,330	5,720
11020.10 Security Equipment				
Bulletproof teller window				
4' x 4'	EA.	3,440	890	4,330
5' x 4'	"	4,240	1,060	5,300
Bulletproof partitions				
Up to 12' high, 2.5" thick	S.F.	86.00	3.54	89.54
Counter for banks				
Minimum	L.F.	690	110	800
Maximum	"	3,130	180	3,310
Drive-up window				
Minimum	EA.	8,630	760	9,390
Maximum	"	4,280	1,770	6,050
Night depository				
Minimum	EA.	11,170	760	11,930
Maximum	"	24,200	1,770	25,970
Office safes, 30" x 20" x 20", 1 hr rating	"	3,250	130	3,380
30" x 16" x 15", 2 hr rating	"	2,470	110	2,580
30" x 28" x 20", H&G rating	"	12,600	66.00	12,666
Service windows, pass through painted steel				
24" x 36"	EA.	5,280	530	5,810
48" x 40"	"	7,430	660	8,090
72" x 40"	"	8,080	1,060	9,140
Special doors and windows				
3' x 7' bulletproof door with frame	EA.	2,330	760	3,090
12" x 12" vision panel	"	2,060	380	2,440
Surveillance system				
Minimum	EA.	5,940	1,060	7,000
Maximum	"	47,850	5,310	53,160
Vault door, 3' wide, 6'6" high				
3-1/2" thick	EA.	24,580	6,640	31,220
7" thick	"	27,720	8,850	36,570
10" thick	"	33,160	10,620	43,780
Insulated vault door				
2 hr rating				
32" wide	EA.	3,620	530	4,150
40" wide	"	4,110	560	4,670
4 hr rating				
32" wide	EA.	3,840	590	4,430
40" wide	"	4,230	660	4,890
6 hr rating				
32" wide	EA.	4,610	590	5,200
40" wide	"	5,190	660	5,850
Insulated file room door				
1 hr rating				
32" wide	EA.	4,260	530	4,790
40" wide	"	4,430	590	5,020

Architectural Equipment	UNIT	MAT.	INST.	TOTAL
11060.10 **Theater Equipment**				
Roll out stage, steel frame, wood floor				
Manual	S.F.	40.25	3.32	43.57
Electric	"	43.00	5.31	48.31
Portable stages				
8" high	S.F.	18.25	2.65	20.90
18" high	"	19.00	2.95	21.95
36" high	"	22.50	3.12	25.62
48" high	"	24.50	3.32	27.82
Band risers				
Minimum	S.F.	35.75	2.65	38.40
Maximum	"	61.00	2.65	63.65
Chairs for risers				
Minimum	EA.	94.00	1.87	95.87
Maximum	"	160	1.87	162
11080.10 **Police Equipment**				
Firing range equipment, rifle				
3 position	EA.	12,890	1,770	14,660
4 position	"	16,920	2,660	19,580
5 position	"	20,950	2,950	23,900
6 position	"	25,790	3,120	28,910
11090.10 **Checkroom Equipment**				
Motorized checkroom equipment				
No shelf system, 6'4" height				
7'6" length	EA.	2,670	530	3,200
14'6" length	"	3,020	530	3,550
28' length	"	4,160	530	4,690
One shelf, 6'8" height				
7'6" length	EA.	3,270	530	3,800
14'6" length	"	3,960	530	4,490
28' length	"	5,860	530	6,390
Two shelves, 7'5" height				
7'6" length	EA.	3,960	530	4,490
14'6" length	"	4,830	530	5,360
28' length	"	6,900	530	7,430
Three shelves, 8' height				
7'6" length	EA.	4,130	1,060	5,190
14'6" length	"	5,000	1,060	6,060
28' length	"	7,070	1,060	8,130
Four shelves, 8'7" height				
7'6" length	EA.	4,130	1,060	5,190
14'6" length	"	5,170	1,060	6,230
28' length	"	7,070	1,060	8,130
11110.10 **Laundry Equipment**				
High capacity, heavy duty				
Washer extractors				
135 lb				
Standard	EA.	31,790	440	32,230
Pass through	"	70,730	440	71,170
200 lb				
Standard	EA.	66,200	440	66,640

Architectural Equipment	UNIT	MAT.	INST.	TOTAL

11110.10   Laundry Equipment *(Cont.)*

	UNIT	MAT.	INST.	TOTAL
Pass through	EA.	72,820	440	73,260
110 lb dryer	"	9,630	440	10,070
Hand operated presser	"	8,910	590	9,500
Mushroom press	"	6,000	590	6,590
Spreader feeders				
2 station	EA.	56,320	590	56,910
4 station	"	71,720	1,060	72,780
Delivery carts				
12 bushel	EA.	510	6.64	517
16 bushel	"	620	7.08	627
18 bushel	"	660	7.58	668
30 bushel	"	920	8.85	929
40 bushel	"	1,120	10.50	1,131
Low capacity				
Pressers				
Air operated	EA.	7,320	210	7,530
Hand operated	"	6,290	210	6,500
Extractor, low capacity	"	10,910	210	11,120
Ironer, 48"	"	3,410	110	3,520
Coin washers				
10 lb capacity	EA.	1,620	110	1,730
20 lb capacity	"	4,350	110	4,460
Coin dryer	"	860	66.00	926
Coin dry cleaner, 20 lb	"	25,410	210	25,620

11161.10   Loading Dock Equipment

	UNIT	MAT.	INST.	TOTAL
Dock leveler, 10 ton capacity				
6' x 8'	EA.	4,250	530	4,780
7' x 8'	"	4,740	530	5,270
Bumpers, laminated rubber				
4-1/2" thick				
6" x 14"	EA.	84.00	10.50	94.50
6" x 36"	"	85.00	11.75	96.75
10" x 14"	"	89.00	13.25	102
10" x 24"	"	100	15.25	115
10" x 36"	"	120	17.75	138
12" x 14"	"	99.00	14.00	113
12" x 24"	"	110	16.50	127
12" x 36"	"	130	19.75	150
6" thick				
10" x 14"	EA.	91.00	15.25	106
10" x 24"	"	100	18.25	118
10" x 36"	"	120	26.50	147
Extruded rubber bumpers				
T-section, 22" x 22" x 3"	EA.	160	10.50	171
Molded rubber bumpers				
24" x 12" x 3" thick	EA.	84.00	26.50	111
Door seal, 12" x 12", vinyl covered	L.F.	44.75	13.25	58.00
Dock boards, heavy duty, 5' x 5'				
5000 lb				
Minimum	EA.	920	440	1,360
Maximum	"	2,250	440	2,690
9000 lb				

Architectural Equipment	UNIT	MAT.	INST.	TOTAL
11161.10 Loading Dock Equipment _(Cont.)_				
Minimum	EA.	1,500	440	1,940
Maximum	"	2,730	480	3,210
15,000 lb	"	2,000	480	2,480
Truck shelters				
Minimum	EA.	1,530	410	1,940
Maximum	"	2,550	760	3,310
11170.10 Waste Handling				
Incinerator, electric				
100 lb/hr				
Minimum	EA.	71,500	540	72,040
Maximum	"	78,980	540	79,520
400 lb/hr				
Minimum	EA.	148,500	1,090	149,590
Maximum	"	170,500	1,090	171,590
1000 lb/hr				
Minimum	EA.	247,500	1,640	249,140
Maximum	"	291,500	1,640	293,140
Incinerator, medical-waste				
25 lb/hr, 2-7 x 4-0	EA.	51,700	1,090	52,790
50 lb/hr, 2-11 x 4-11	"	61,600	1,090	62,690
75 lb/hr, 3-8 x 5-0	"	66,000	2,170	68,170
100 lb/hr, 3-8 x 6-0	"	71,500	2,170	73,670
Industrial compactor				
1 cy	EA.	12,730	600	13,330
3 cy	"	23,760	780	24,540
5 cy	"	30,330	1,090	31,420
Trash chutes steel, including sprinklers				
18" dia.	L.F.	80.00	270	350
24" dia.	"	100	280	380
30" dia.	"	130	300	430
36" dia.	"	170	310	480
Refuse bottom hopper	EA.	2,540	300	2,840
11400.10 Food Service Equipment				
Unit kitchens				
30" compact kitchen				
Refrigerator, with range, sink	EA.	1,390	270	1,660
Sink only	"	770	180	950
Range only	"	660	140	800
Cabinet for upper wall section	"	460	78.00	538
Stainless shield, for rear wall	"	180	21.75	202
Side wall	"	86.00	21.75	108
42" compact kitchen				
Refrigerator with range, sink	EA.	1,920	300	2,220
Sink only	"	1,380	270	1,650
Cabinet for upper wall section	"	660	90.00	750
Stainless shield, for rear wall	"	210	22.50	233
Side wall	"	86.00	22.50	109
54" compact kitchen				
Refrigerator, oven, range, sink	EA.	2,520	390	2,910
Cabinet for upper wall section	"	2,070	110	2,180
Stainless shield, for				

Architectural Equipment

	UNIT	MAT.	INST.	TOTAL
11400.10 **Food Service Equipment** *(Cont.)*				
Rear wall	EA.	290	24.75	315
Side wall	"	86.00	24.75	111
60" compact kitchen				
Refrigerator, oven, range, sink	EA.	3,720	390	4,110
Cabinet for upper wall section	"	930	110	1,040
Stainless shield, for				
Rear wall	EA.	280	24.75	305
Side wall	"	86.00	24.75	111
72" compact kitchen				
Refrigerator, oven, range, sink	EA.	4,670	450	5,120
Cabinet for upper wall section	"	4,290	110	4,400
Stainless shield for				
Rear wall	EA.	340	27.25	367
Side wall	"	86.00	27.25	113
Bake oven				
Single deck				
Minimum	EA.	2,510	68.00	2,578
Maximum	"	3,120	140	3,260
Double deck				
Minimum	EA.	3,100	90.00	3,190
Maximum	"	5,650	140	5,790
Triple deck				
Minimum	EA.	4,130	90.00	4,220
Maximum	"	8,970	180	9,150
Convection type oven, electric, 40" x 45" x 57"				
Minimum	EA.	4,140	68.00	4,208
Maximum	"	8,210	140	8,350
Broiler, without oven, 69" x 26" x 39"				
Minimum	EA.	4,180	68.00	4,248
Maximum	"	6,940	90.00	7,030
Coffee urns, 10 gallons				
Minimum	EA.	2,290	180	2,470
Maximum	"	3,280	270	3,550
Fryer, with submerger				
Single				
Minimum	EA.	1,420	110	1,530
Maximum	"	4,070	180	4,250
Double				
Minimum	EA.	2,710	140	2,850
Maximum	"	6,400	180	6,580
Griddle, counter				
3' long				
Minimum	EA.	1,780	90.00	1,870
Maximum	"	4,460	110	4,570
5' long				
Minimum	EA.	2,740	140	2,880
Maximum	"	7,050	180	7,230
Kettles, steam, jacketed				
20 gallons				
Minimum	EA.	3,190	140	3,330
Maximum	"	6,780	270	7,050
40 gallons				
Minimum	EA.	4,150	140	4,290

Architectural Equipment	UNIT	MAT.	INST.	TOTAL
11400.10 Food Service Equipment *(Cont.)*				
Maximum	EA.	7,590	270	7,860
60 gallons				
Minimum	EA.	4,400	140	4,540
Maximum	"	9,840	270	10,110
Range				
Heavy duty, single oven, open top				
Minimum	EA.	2,200	68.00	2,268
Maximum	"	6,650	180	6,830
Fry top				
Minimum	EA.	3,180	68.00	3,248
Maximum	"	7,220	180	7,400
Steamers, electric				
27 kw				
Minimum	EA.	11,160	140	11,300
Maximum	"	16,030	180	16,210
18 kw				
Minimum	EA.	8,030	140	8,170
Maximum	"	14,850	180	15,030
Dishwasher, rack type				
Single tank, 190 racks/hr	EA.	11,220	270	11,490
Double tank				
234 racks/hr	EA.	20,900	300	21,200
265 racks/hr	"	22,690	360	23,050
Dishwasher, automatic 100 meals/hr	"	65,890	180	66,070
Disposals				
100 gal/hr	EA.	1,930	180	2,110
120 gal/hr	"	2,480	190	2,670
250 gal/hr	"	6,310	190	6,500
Exhaust hood for dishwasher, gutter 4 sides, s-steel				
4'x4'x2'	EA.	2,560	200	2,760
4'x7'x2'	"	3,750	220	3,970
Food preparation machines				
Vertical cutter mixers				
25 quart	EA.	8,980	180	9,160
40 quart	"	13,090	180	13,270
80 quart	"	14,080	270	14,350
130 quart	"	15,540	450	15,990
Choppers				
5 lb	EA.	2,940	140	3,080
16 lb	"	3,110	180	3,290
40 lb	"	5,420	270	5,690
Mixers, floor models				
20 quart	EA.	5,280	68.00	5,348
60 quart	"	12,430	68.00	12,498
80 quart	"	14,630	78.00	14,708
140 quart	"	23,010	110	23,120
Ice cube maker				
50 lb per day				
Minimum	EA.	1,660	540	2,200
Maximum	"	3,520	540	4,060
500 lb per day				
Minimum	EA.	4,500	900	5,400
Maximum	"	6,600	900	7,500

Architectural Equipment	UNIT	MAT.	INST.	TOTAL
11400.10 **Food Service Equipment** *(Cont.)*				
Ice flakers				
300 lb per day	EA.	4,590	540	5,130
600 lb per day	"	6,380	900	7,280
1000 lb per day	"	7,810	1,210	9,020
2000 lb per day	"	14,630	1,360	15,990
Refrigerated cases				
Dairy products				
Multi deck type	L.F.	820	36.25	856
For rear sliding doors, add	"			86.00
Delicatessen case, service deli				
Single deck	L.F.	640	270	910
Multi deck	"	670	340	1,010
Meat case				
Single deck	L.F.	550	320	870
Multi deck	"	600	340	940
Produce case				
Single deck	L.F.	640	320	960
Multi deck	"	680	340	1,020
Bottle coolers				
6' long				
Minimum	EA.	1,670	1,090	2,760
Maximum	"	4,400	1,090	5,490
10' long				
Minimum	EA.	2,080	1,810	3,890
Maximum	"	5,550	1,810	7,360
Frozen food cases				
Chest type	L.F.	750	320	1,070
Reach-in, glass door	"	1,380	340	1,720
Island case, single	"	710	320	1,030
Multi deck	"	1,480	340	1,820
Ice storage bins				
500 lb capacity	EA.	2,290	780	3,070
1000 lb capacity	"	3,200	1,550	4,750
11450.10 **Residential Equipment**				
Compactor, 4 to 1 compaction	EA.	810	140	950
Dishwasher, built-in				
2 cycles	EA.	820	270	1,090
4 or more cycles	"	1,540	270	1,810
Disposal				
Garbage disposer	EA.	120	180	300
Heaters, electric, built-in				
Ceiling type	EA.	210	180	390
Wall type				
Minimum	EA.	77.00	140	217
Maximum	"	240	180	420
Hood for range, 2-speed, vented				
30" wide	EA.	130	180	310
42" wide	"	770	180	950
Ice maker, automatic				
30 lb per day	EA.	1,330	78.00	1,408
50 lb per day	"	1,760	270	2,030
Folding access stairs, disappearing metal stair				

Architectural Equipment	UNIT	MAT.	INST.	TOTAL
11450.10 **Residential Equipment** *(Cont.)*				
8' long	EA.	1,240	78.00	1,318
11' long	"	1,380	78.00	1,458
12' long	"	1,600	78.00	1,678
Wood frame, wood stair				
22" x 54" x 8'9" long	EA.	220	54.00	274
25" x 54" x 10' long	"	280	54.00	334
Ranges electric				
Built-in, 30", 1 oven	EA.	1,670	180	1,850
2 oven	"	2,030	180	2,210
Counter top, 4 burner, standard	"	910	140	1,050
With grill	"	2,500	140	2,640
Free standing, 21", 1 oven	"	830	180	1,010
30", 1 oven	"	1,300	110	1,410
2 oven	"	3,300	110	3,410
Water softener				
30 grains per gallon	EA.	940	180	1,120
70 grains per gallon	"	1,130	270	1,400
11470.10 **Darkroom Equipment**				
Dryers				
36" x 25" x 68"	EA.	9,570	290	9,860
48" x 25" x 68"	"	9,890	290	10,180
Processors, film				
Black and white	EA.	15,250	290	15,540
Color negatives	"	17,270	290	17,560
Prints	"	19,800	290	20,090
Transparencies	"	21,730	290	22,020
Sinks with cabinet and/or stand				
5" sink with stand				
24" x 48"	EA.	1,430	150	1,580
32" x 64"	"	1,540	200	1,740
38" x 52"	"	1,760	200	1,960
42" x 132"	"	2,640	290	2,930
48" x 52"	"	2,640	290	2,930
5" sink with cabinet				
24" x 48"	EA.	2,200	150	2,350
32" x 64"	"	2,310	200	2,510
38" x 52"	"	2,420	200	2,620
42" x 132"	"	3,850	290	4,140
48" x 52"	"	2,640	290	2,930
10" sink with stand				
24" x 48"	EA.	1,650	150	1,800
32" x 64"	"	2,010	200	2,210
38" x 52"	"	2,200	200	2,400
10" sink with cabinet				
24" x 48"	EA.	2,750	150	2,900
38" x 52"	"	3,180	200	3,380

Architectural Equipment	UNIT	MAT.	INST.	TOTAL

11480.10 Athletic Equipment

	UNIT	MAT.	INST.	TOTAL
Basketball backboard				
Fixed	EA.	1,210	660	1,870
Swing-up	"	3,410	1,060	4,470
Portable, hydraulic	"	13,190	270	13,460
Suspended type, standard	"	5,390	1,060	6,450
For glass backboard, add	"			2,940
For electrically operated, add	"			3,290
Bleacher, telescoping, manual				
15 tier, minimum	SEAT	160	10.50	171
Maximum	"	170	10.50	181
20 tier, minimum	"	160	11.75	172
Maximum	"	170	11.75	182
30 tier, minimum	"	160	17.75	178
Maximum	"	180	17.75	198
Boxing ring elevated, complete, 22' x 22'	EA.	7,810	7,590	15,400
Gym divider curtain				
Minimum	S.F.	9.07	0.70	9.77
Maximum	"	15.25	0.70	15.95
Scoreboards, single face				
Minimum	EA.	4,390	530	4,920
Maximum	"	28,760	2,660	31,420
Parallel bars				
Minimum	EA.	1,810	530	2,340
Maximum	"	4,170	890	5,060

11500.10 Industrial Equipment

	UNIT	MAT.	INST.	TOTAL
Vehicular paint spray booth, solid back, 14'4" x 9'6"				
24' deep	EA.	14,070	530	14,600
26'6" deep	"	14,180	530	14,710
28'6" deep	"	14,620	530	15,150
Drive through, 14'9" x 9'6"				
24' deep	EA.	14,620	530	15,150
26'6" deep	"	14,950	530	15,480
28'6" deep	"	15,060	530	15,590
Water wash, paint spray booth				
5' x 11'2" x 10'8"	EA.	7,370	530	7,900
6' x 11'2" x 10'8"	"	7,590	530	8,120
8' x 11'2" x 10'8"	"	9,630	530	10,160
10' x 11'2" x 11'2"	"	10,450	530	10,980
12' x 12'2" x 11'2"	"	12,760	530	13,290
14' x 12'2" x 11'2"	"	14,960	530	15,490
16' x 12'2" x 11'2"	"	17,110	530	17,640
20' x 12'2" x 11'2"	"	20,180	530	20,710
Dry type spray booth, with paint arrestors				
5'4" x 7'2" x 6'8"	EA.	3,250	530	3,780
6'4" x 7'2" x 6'8"	"	4,170	530	4,700
8'4" x 7'2" x 9'2"	"	4,620	530	5,150
10'4" x 7'2" x 9'2"	"	5,250	530	5,780
12'4" x 7'6" x 9'2"	"	4,980	530	5,510
14'4" x 7'6" x 9'8"	"	6,780	530	7,310
16'4" x 7'7" x 9'8"	"	7,690	530	8,220
20'4" x 7'7" x 10'8"	"	8,690	530	9,220
Air compressor, electric				

Architectural Equipment	UNIT	MAT.	INST.	TOTAL

11500.10 — Industrial Equipment *(Cont.)*

Item	UNIT	MAT.	INST.	TOTAL
1 hp				
115 volt	EA.	1,070	350	1,420
7.5 hp				
115 volt	EA.	3,890	530	4,420
230 volt	"	4,350	530	4,880
Hydraulic lifts				
8,000 lb capacity	EA.	6,330	1,330	7,660
11,000 lb capacity	"	8,460	2,120	10,580
24,000 lb capacity	"	11,830	3,540	15,370
Power tools				
Band saws				
10"	EA.	930	44.25	974
14"	"	1,430	53.00	1,483
Motorized shaper	"	830	40.75	871
Motorized lathe	"	1,060	44.25	1,104
Bench saws				
9" saw	EA.	540	35.50	576
10" saw	"	560	38.00	598
12" saw	"	620	44.25	664
Electric grinders				
1/3 hp	EA.	380	21.25	401
1/2 hp	"	540	23.00	563
3/4 hp	"	600	23.00	623

11600.10 — Laboratory Equipment

Item	UNIT	MAT.	INST.	TOTAL
Cabinets, base				
Minimum	L.F.	170	44.25	214
Maximum	"	300	44.25	344
Full storage, 7' high				
Minimum	L.F.	240	44.25	284
Maximum	"	340	44.25	384
Wall				
Minimum	L.F.	100	53.00	153
Maximum	"	190	53.00	243
Counter tops				
Minimum	S.F.	29.75	6.64	36.39
Average	"	59.00	7.58	66.58
Maximum	"	87.00	8.85	95.85
Tables				
Open underneath	S.F.	150	26.50	177
Doors underneath	"	310	33.25	343
Medical laboratory equipment				
Analyzer				
Chloride	EA.	720	27.25	747
Blood	"	4,380	45.25	4,425
Bath, water, utility, countertop unit	"	410	54.00	464
Hot plate, lab, countertop	"	190	49.25	239
Stirrer	"	290	49.25	339
Incubator, anaerobic, 23x23x36"	"	9,890	270	10,160
Dry heat bath	"	650	90.00	740
Incinerator, for sterilizing	"	310	5.42	315
Meter, serum protein	"	1,090	6.78	1,097
Ph analog, general purpose	"	760	7.75	768

Architectural Equipment	UNIT	MAT.	INST.	TOTAL

11600.10   Laboratory Equipment *(Cont.)*

	UNIT	MAT.	INST.	TOTAL
Refrigerator, blood bank, undercounter type 153 litres	EA.	5,590	90.00	5,680
5.4 cf, undercounter type	"	5,290	90.00	5,380
Refrigerator/freezer, 4.4 cf, undercounter type	"	630	90.00	720
Sealer, impulse, free standing, 20x12x4"	"	320	18.00	338
Timer, electric, 1-60 minutes, bench or wall mounted	"	77.00	30.25	107
Glassware washer - dryer, undercounter	"	6,210	680	6,890
Balance, torsion suspension, tabletop, 4.5 lb capacity	"	580	30.25	610
Binocular microscope, with in base illuminator	"	3,240	20.75	3,261
Centrifuge, table model, 19x16x13"	"	1,650	21.75	1,672
Clinical model, with four place head	"	720	12.00	732

11700.10   Medical Equipment

	UNIT	MAT.	INST.	TOTAL
Hospital equipment, lights				
Examination, portable	EA.	1,190	45.25	1,235
Meters				
Air flow meter	EA.	80.00	30.25	110
Oxygen flow meters	"	57.00	22.50	79.50
Racks				
40 chart, revolving open frame; mobile caddy	EA.	1,650	45.25	1,695
Scales.				
Clinical, metric with measure rod, 350 lb	EA.	480	49.25	529
Physical therapy				
Chair, hydrotherapy	EA.	440	8.85	449
Diathermy, shortwave, portable, on casters	"	3,300	21.25	3,321
Exercise bicycle, floor standing, 35" x 15"	"	880	17.75	898
Hydrocollator, 4 pack, portable, 129 x 90 x 160"	"	400	7.58	408
Lamp, infrared, mobile with variable heat control	"	560	40.75	601
Ultra violet, base mounted	"	500	40.75	541
Mirror, posture training, 27" wide and 72" high	"	490	13.25	503
Parallel bars, adjustable	"	2,190	66.00	2,256
Platform mat 10'x6', 1" thick	"	590	13.25	603
Pulley, duplex, wall mounted	"	540	180	720
Rack, crutch, wall mounted, 66 x 16 x 13"	"	910	53.00	963
Stimulator, galvanic-faradic, hand held	"	100	3.54	104
Ultrasound, muscle stimulator, portable, 13x13x8"	"	3,550	4.42	3,554
Sandbag set, velcro straps, saddle bag type	"	120	7.58	128
Whirlpool, 85 gallon	"	3,840	270	4,110
65 gallon capacity	"	3,350	270	3,620
Radiology				
Radiographic table, motor driven tilting table	EA.	30,180	5,310	35,490
Fluoroscope image/tv system	"	10,180	10,620	20,800
Processor for washing and drying radiographs				
Water filter unit, 30" x 48-1/2" x 37-1/2"	EA.	35.75	900	936
Cassette transfer cabinet	"	1,010	45.25	1,055
Base storage cabinets, sectional design				
With back splash, 24" deep and 35" high	L.F.	2,780	45.25	2,825
Wall storage cabinets	"	240	68.00	308
Steam sterilizers				
For heat and moisture stable materials	EA.	4,280	54.00	4,334
For fast drying after sterilization	"	3,300	68.00	3,368
Compact unit	"	1,680	68.00	1,748
Semi-automatic	"	6,940	270	7,210
Floor loading				

Architectural Equipment

11700.10 — Medical Equipment *(Cont.)*

	UNIT	MAT.	INST.	TOTAL
Single door	EA.	126,070	450	126,520
Double door	"	158,960	540	159,500
Utensil washer, sanitizer	"	8,630	420	9,050
Automatic washer/sterilizer	"	27,410	1,090	28,500
16 x 16 x 26", including generator & accessories	"	31,460	1,810	33,270
Steam generator, elec., 10 kw to 180 kw	"	35,750	1,090	36,840
Surgical scrub				
Minimum	EA.	2,700	180	2,880
Maximum	"	5,660	180	5,840
Gas sterilizers				
Automatic, free standing, 21x19x29"	EA.	5,660	540	6,200
Surgical tables				
Minimum	EA.	13,750	780	14,530
Maximum	"	38,500	1,090	39,590
Surgical lights, ceiling mounted				
Minimum	EA.	3,100	900	4,000
Maximum	"	9,440	1,090	10,530
Water stills				
4 liters/hr	EA.	4,180	180	4,360
8 liters/hr	"	4,760	180	4,940
19 liters/hr	"	11,000	450	11,450
X-ray equipment				
Mobile unit				
Minimum	EA.	6,160	270	6,430
Maximum	"	31,840	540	32,380
Film viewers				
Minimum	EA.	190	90.00	280
Maximum	"	960	180	1,140
Autopsy table				
Minimum	EA.	4,730	540	5,270
Maximum	"	9,420	540	9,960
Incubators				
15 cf	EA.	4,640	270	4,910
29 cf	"	7,580	450	8,030
Infant transport, portable	"	11,570	290	11,860
Beds				
Stretcher, with pad, 30" x 78"	EA.	2,040	140	2,180
Transfer, for patient transport	"	3,460	140	3,600
Headwall				
Aluminum, with back frame and console	EA.	4,040	270	4,310
Hospital ground detection system				
Power ground module	EA.	860	160	1,020
Ground slave module	"	600	120	720
Master ground module	"	370	100	470
Remote indicator	"	360	110	470
X-ray indicator	"	400	120	520
Micro ammeter	"	1,220	140	1,360
Supervisory module	"	1,500	120	1,620
Ground cords	"	100	20.00	120
Hospital isolation monitors, 5 ma				
120v	EA.	2,620	240	2,860
208v	"	2,510	240	2,750
240v	"	2,510	240	2,750

Architectural Equipment	UNIT	MAT.	INST.	TOTAL
11700.10 — **Medical Equipment** *(Cont.)*				
Digital clock-timers separate display	EA.	1,090	110	1,200
One display	"	910	110	1,020
Remote control	"	530	85.00	615
Battery pack	"	120	85.00	205
Surgical chronometer clock and 3 timers	"	1,970	170	2,140
Auxilary control	"	710	79.00	789
11700.20 — **Dental Equipment**				
Dental care equipment				
Drill console with accessories	EA.	14,740	900	15,640
Amalgamator	"	470	27.25	497
Lathe	"	460	18.00	478
Finish polisher	"	320	36.25	356
Model trimmer	"	570	24.75	595
Motor, wall mounted	"	690	24.75	715
Cleaner, ultrasonic	"	1,040	54.00	1,094
Curing unit, bench mounted	"	1,720	90.00	1,810
Oral evacuation system, dual pump	"	5,660	68.00	5,728
Sterilizer, table top, self contained	"	840	30.25	870
Dental lights				
Light, floor or ceiling mounted	EA.	3,720	270	3,990
X-ray unit				
Portable	EA.	4,620	140	4,760
Wall mounted with remote control	"	8,490	450	8,940
Illuminator, single panel	"	5,820	780	6,600
X-ray film processor	"	2,470	450	2,920
Shield, portable x-ray, lead lined	"	450	36.25	486

Interior	UNIT	MAT.	INST.	TOTAL
12302.10 **Casework**				
Kitchen base cabinet, prefinished, 24" deep, 35" high				
12"wide	EA.	170	53.00	223
18" wide	"	240	53.00	293
24" wide	"	280	59.00	339
27" wide	"	310	59.00	369
36" wide	"	370	66.00	436
48" wide	"	420	66.00	486
Corner cabinet, 36" wide	"	440	66.00	506
Wall cabinet, 12" deep, 12" high				
30" wide	EA.	170	53.00	223
36" wide	"	190	53.00	243
15" high				
30" wide	EA.	190	59.00	249
36" wide	"	150	59.00	209
24" high				
30" wide	EA.	220	59.00	279
36" wide	"	170	59.00	229
30" high				
12" wide	EA.	140	66.00	206
18" wide	"	170	66.00	236
24" wide	"	220	66.00	286
27" wide	"	240	66.00	306
30" wide	"	250	76.00	326
36" wide	"	280	76.00	356
Corner cabinet, 30" high				
24" wide	EA.	230	89.00	319
30" wide	"	290	89.00	379
36" wide	"	300	89.00	389
Wardrobe	"	960	130	1,090
Vanity with top, laminated plastic				
24" wide	EA.	450	130	580
30" wide	"	540	130	670
36" wide	"	570	180	750
48" wide	"	650	210	860
12390.10 **Counter Tops**				
Stainless steel, counter top, with backsplash	S.F.	140	13.25	153
Acid-proof, kemrock surface	"	46.25	8.85	55.10
12500.10 **Window Treatment**				
Drapery tracks, wall or ceiling mounted				
Basic traverse rod				
50 to 90"	EA.	61.00	26.50	87.50
84 to 156"	"	67.00	29.50	96.50
136 to 250"	"	86.00	29.50	116
165 to 312"	"	140	33.25	173
Traverse rod with stationary curtain rod				
30 to 50"	EA.	55.00	26.50	81.50
50 to 90"	"	68.00	26.50	94.50
84 to 156"	"	99.00	29.50	129

Interior	UNIT	MAT.	INST.	TOTAL
12500.10 **Window Treatment** *(Cont.)*				
136 to 250"	EA.	110	33.25	143
Double traverse rod				
30 to 50"	EA.	66.00	26.50	92.50
50 to 84"	"	86.00	26.50	113
84 to 156"	"	110	29.50	140
136 to 250"	"	130	33.25	163
12510.10 **Blinds**				
Venetian blinds				
2" slats	S.F.	7.97	1.32	9.29
1" slats	"	9.35	1.32	10.67
12690.40 **Floor Mats**				
Recessed entrance mat, 3/8" thick, aluminum link	S.F.	26.50	26.50	53.00
Steel, flexible	"	11.25	26.50	37.75

Construction

Construction	UNIT	MAT.	INST.	TOTAL
13056.10 — **Vaults**				
Floor safes				
Class C				
1.0 cf	EA.	770	44.25	814
1.3 cf	"	900	66.00	966
1.9 cf	"	990	89.00	1,079
5.2 cf	"	2,240	89.00	2,329
13121.10 — **Pre-engineered Buildings**				
Pre-engineered metal building, 40'x100'				
14' eave height	S.F.	13.25	4.42	17.67
16' eave height	"	13.75	5.10	18.85
20' eave height	"	15.00	6.63	21.63
60'x100'				
14' eave height	S.F.	9.84	4.42	14.26
16' eave height	"	10.25	5.10	15.35
20' eave height	"	11.75	6.63	18.38
80'x100'				
14' eave height	S.F.	10.50	4.42	14.92
16' eave height	"	11.50	5.10	16.60
20' eave height	"	12.75	6.63	19.38
100'x100'				
14' eave height	S.F.	13.25	4.42	17.67
16' eave height	"	13.00	5.10	18.10
20' eave height	"	15.25	6.63	21.88
100'x150'				
14' eave height	S.F.	11.25	4.42	15.67
16' eave height	"	11.75	5.10	16.85
20' eave height	"	12.50	6.63	19.13
120'x150'				
14' eave height	S.F.	13.25	4.42	17.67
16' eave height	"	13.75	5.10	18.85
20' eave height	"	15.00	6.63	21.63
140'x150'				
14' eave height	S.F.	11.50	4.42	15.92
16' eave height	"	11.75	5.10	16.85
20' eave height	"	12.50	6.63	19.13
160'x200'				
14' eave height	S.F.	11.00	4.42	15.42
16' eave height	"	11.50	5.10	16.60
20' eave height	"	11.50	6.63	18.13
200'x200'				
14' eave height	S.F.	10.50	4.42	14.92
16' eave height	"	10.75	5.10	15.85
20' eave height	"	11.75	6.63	18.38
Hollow metal door and frame, 6' x 7'	EA.			1,590
Sectional steel overhead door, manually operated				
8' x 8'	EA.			1,720
12' x 12'	"			2,340
Roll-up steel door, manually operated				
10' x 10'	EA.			1,300
12' x 12'	"			1,510
For gravity ridge ventilator with birdscreen	"			1,020
9" throat x 10'	"			1,080

Construction	UNIT	MAT.	INST.	TOTAL
13121.10 **Pre-engineered Buildings** *(Cont.)*				
12" throat x 10'	EA.			1,140
For 20" rotary vent with damper	"			290
For 4' x 3' fixed louver	"			260
For 4' x 3' aluminum sliding window	"			280
For 3' x 9' fiberglass panels	"			300
Liner panel, 26 ga, painted steel	S.F.	2.75	1.47	4.22
Wall panel insulated, 26 ga. steel, foam core	"	8.03	1.47	9.50
Roof panel, 26 ga. painted steel	"	2.56	0.84	3.40
Plastic (sky light)	"	5.44	0.84	6.28
Insulation, 3-1/2" thick blanket, R11	"	1.76	0.39	2.15
13152.10 **Swimming Pool Equipment**				
Diving boards				
14' long				
Aluminum	EA.	2,500	230	2,730
Fiberglass	"	1,430	230	1,660
Ladders, heavy duty				
2 steps				
Minimum	EA.	390	83.00	473
Maximum	"	900	83.00	983
4 steps				
Minimum	EA.	730	100	830
Maximum	"	1,180	100	1,280
Lifeguard chair				
Minimum	EA.	2,750	410	3,160
Maximum	"	3,800	410	4,210
Lights, underwater				
12 volt, with transformer	EA.	260	100	360
110 volt				
Minimum	EA.	380	100	480
Maximum	"	580	100	680
Ground fault interrupter for 110 volt, each light	"	170	34.50	205
Pool covers				
Reinforced polyethylene	S.F.	0.70	3.17	3.87
Vinyl water tube				
Minimum	S.F.	0.88	3.17	4.05
Maximum	"	1.48	3.17	4.65
Slides with water tube				
Minimum	EA.	4,050	340	4,390
Maximum	"	6,820	340	7,160

Elevators	UNIT	MAT.	INST.	TOTAL
14210.10 **Elevators**				
Passenger elevators, electric, geared				
Based on a shaft of 6 stops and 6 openings				
50 fpm, 2000 lb	EA.	105,450	2,230	107,680
100 fpm, 2000 lb	"	109,350	2,470	111,820
150 fpm				
2000 lb	EA.	121,070	2,780	123,850
3000 lb	"	152,310	3,180	155,490
4000 lb	"	158,170	3,710	161,880
200 fpm				
2500 lb	EA.	147,430	3,180	150,610
3000 lb	"	151,340	3,430	154,770
4000 lb	"	160,130	3,710	163,840
250 fpm				
2500 lb	EA.	152,310	3,180	155,490
3000 lb	"	160,130	3,430	163,560
4000 lb	"	164,030	3,710	167,740
300 fpm				
2500 lb	EA.	152,310	3,180	155,490
3000 lb	"	158,170	3,430	161,600
4000 lb	"	164,030	2,230	166,260
For each additional; 50 fpm, add per stop, $3000				
500 lb, add per stop, $4000				
Opening, add per stop, $4500				
Stop, add per stop, $4000				
Bonderized steel door, add per opening, $150				
Colored aluminum door, add per opening, $850				
Stainless steel door, add per opening, $600				
Cast bronze door, add per opening, $1100				
Two speed door, add per opening, $360				
Bi-parting door, add per opening, $850				
Custom cab interior add, $4800				
Based on a shaft of 8 stops and 8 openings				
300 fpm				
3000 lb	EA.	181,610	4,450	186,060
3500 lb	"	185,670	4,450	190,120
4000 lb	"	195,280	4,950	200,230
5000 lb	"	218,710	5,300	224,010
400 fpm				
3000 lb	EA.	191,370	4,450	195,820
3500 lb	"	194,300	4,450	198,750
4000 lb	"	214,810	4,950	219,760
5000 lb	"	234,330	5,300	239,630
600 fpm				
3000 lb	EA.	273,390	4,950	278,340
3500 lb	"	277,290	5,300	282,590
4000 lb	"	279,250	5,430	284,680
5000 lb	"	291,940	5,570	297,510
800 fpm				
3000 lb	EA.	322,210	4,950	327,160
3500 lb	"	326,110	5,300	331,410
4000 lb	"	330,020	5,430	335,450
5000 lb	"	331,770	5,570	337,340
For each additional; 100 fpm add per stop, $13,000				

Elevators	UNIT	MAT.	INST.	TOTAL
14210.10 — **Elevators** *(Cont.)*				
500 lb, add per stop, $6500				
Opening add per stop, $12,000				
Stop add per stop, $4800				
Bypass floor, add per each, $2000				
Bonderized steel door, add per opening, $150				
Colored aluminum door, add per opening, $900				
Stainless steel door, add per opening, $600				
Cast bronze door, add per opening, $600				
Two speed bi-parting door, add per opening, $1000				
Custom cab interior, add $5000				
Hydraulic, based on a shaft of 3 stops, 3 openings				
50 fpm				
2000 lb	EA.	69,010	1,860	70,870
2500 lb	"	73,860	1,860	75,720
3000 lb	"	77,930	1,940	79,870
100 fpm				
2000 lb	EA.	75,800	1,860	77,660
2500 lb	"	79,870	1,940	81,810
3000 lb	"	84,910	2,020	86,930
150 fpm				
2000 lb	EA.	81,810	1,860	83,670
2500 lb	"	89,560	1,940	91,500
3000 lb	"	95,760	2,120	97,880
For each additional; 50 fpm add per stop, $3500				
500 lb, add per stop, $3500				
Opening, add, $4200				
Stop, add per stop, $5300				
Bonderized steel door, add per opening, $400				
Colored aluminum door, add per opening, $1500				
Stainless steel door, add per opening, $650				
Cast bronze door, add per opening, $1200				
Two speed door, add per opening, $400				
Bi-parting door, add per opening, $900				
Custom cab interior, add per cab, $5000				
Small elevators, 4 to 6 passenger capacity				
Electric, push				
2 stops	EA.	25,200	1,860	27,060
3 stops	"	31,010	2,020	33,030
4 stops	"	35,860	2,230	38,090
Freight elevators, electric				
Based on a shaft of 6 stops and 6 openings				
50 fpm				
3500 lb	EA.	190,130	2,470	192,600
4000 lb	"	190,850	2,470	193,320
5000 lb	"	192,660	2,780	195,440
100 fpm				
3500 lb	EA.	200,090	2,780	202,870
4000 lb	"	200,990	2,780	203,770
5000 lb	"	196,100	3,180	199,280
200 fpm				
3500 lb	EA.	197,010	3,180	200,190
4000 lb	"	197,910	3,180	201,090
5000 lb	"	199,720	3,710	203,430

Elevators	UNIT	MAT.	INST.	TOTAL
14210.10 **Elevators** *(Cont.)*				
For elevator with manual door, deduct 15%				
For variable voltage control, add 20%				
Based on shaft of 8 stops and 8 openings				
100 fpm				
4000 lb	EA.	200,630	2,780	203,410
6000 lb	"	193,750	2,850	196,600
8000 lb	"	197,370	3,010	200,380
150 fpm				
4000 lb	EA.	193,750	3,180	196,930
6000 lb	"	197,370	3,270	200,640
8000 lb	"	199,180	3,480	202,660
200 fpm				
4000 lb	EA.	195,560	3,710	199,270
6000 lb	"	199,180	3,840	203,020
8000 lb	"	208,230	4,050	212,280
For each additional; 50 fpm, add per stop, $2000				
500 lb, add per stop, $600				
Opening, add per stop, $7000				
Stop, add per stop, $5500				
For variable voltage, add 20%				
Hydraulic, based on 3 stops and 3 openings				
50 fpm				
3000 lb	EA.	81,230	1,590	82,820
4000 lb	"	88,620	1,650	90,270
6000 lb	"	103,390	1,710	105,100
100 fpm				
3000 lb	EA.	91,570	1,590	93,160
4000 lb	"	97,850	1,650	99,500
6000 lb	"	114,280	1,710	115,990
150 fpm				
3000 lb	EA.	100,620	1,590	102,210
4000 lb	"	108,000	1,650	109,650
6000 lb	"	123,880	1,710	125,590
For each additional; 50 fpm, add per stop, $2000				
500 lb, add per stop, $600				
Opening, add per stop, $5500				
Stop, add per stop, $5500				
For elevator with manual door deduct from total, 15%				
14300.10 **Escalators**				
Escalators				
32" wide, floor to floor				
12' high	EA.	137,320	3,710	141,030
15' high	"	148,930	4,450	153,380
18' high	"	160,540	5,570	166,110
22' high	"	174,070	7,420	181,490
25' high	"	193,420	8,910	202,330
48" wide				
12' high	EA.	152,800	3,840	156,640
15' high	"	166,340	4,640	170,980
18' high	"	177,940	5,860	183,800
22' high	"	199,220	7,950	207,170
25' high	"	212,660	8,910	221,570

Lifts

	UNIT	MAT.	INST.	TOTAL
14410.10 — **Personnel Lifts**				
Electrically operated, 1 or 2 person lift				
With attached foot platforms				
3 stops	EA.			24,840
5 stops	"			30,530
7 stops	"			35,380
For each additional stop, add $1250				
Residential stair climber, per story	EA.	6,250	450	6,700
14450.10 — **Vehicle Lifts**				
Automotive hoist, one post, semi-hydraulic, 8,000 lb	EA.	4,900	2,230	7,130
Full hydraulic, 8,000 lb	"	4,620	2,230	6,850
2 post, semi-hydraulic, 10,000 lb	"	4,950	3,180	8,130
Full hydraulic				
10,000 lb	EA.	3,410	3,180	6,590
13,000 lb	"	6,640	5,570	12,210
18,500 lb	"	8,460	5,570	14,030
24,000 lb	"	12,780	5,570	18,350
26,000 lb	"	14,300	5,570	19,870
Pneumatic hoist, fully hydraulic				
11,000 lb	EA.	8,990	7,420	16,410
24,000 lb	"	13,970	7,420	21,390

Material Handling

	UNIT	MAT.	INST.	TOTAL
14560.10 — **Chutes**				
Linen chutes, stainless steel, with supports				
18" dia.	L.F.	130	4.22	134
24" dia.	"	170	4.54	175
30" dia.	"	180	4.92	185
Hopper	EA.	2,100	39.50	2,140
Skylight	"	1,270	59.00	1,329
Sprinkler unit at top	"	460	66.00	526
For galvanized metal, deduct from material cost, 35%				
For aluminum, deduct from material cost, 25%				
14580.10 — **Pneumatic Systems**				
Pneumatic message tube system				
Average, 20 station job				
3" round system	E.A.	332,520	4,930	337,450
4" round system	"	415,640	5,430	421,070
6" round system	"	628,080	6,030	634,110

Material Handling

	UNIT	MAT.	INST.	TOTAL

14580.10 — Pneumatic Systems *(Cont.)*

	UNIT	MAT.	INST.	TOTAL
4" x 7" oval system	E.A.	655,790	10,860	666,650
Trash and linen tube system				
10 stations	EA.	212,440	11,130	223,570
15 stations	"	267,860	14,840	282,700
20 stations	"	359,330	17,130	376,460
30 stations	"	471,060	20,240	491,300

Hoists And Cranes

	UNIT	MAT.	INST.	TOTAL

14600.10 — Industrial Hoists

	UNIT	MAT.	INST.	TOTAL
Industrial hoists, electric, light to medium duty				
500 lb	EA.	6,490	270	6,760
1000 lb	"	6,670	290	6,960
2000 lb	"	6,880	300	7,180
3000 lb	"	7,280	320	7,600
4000 lb	"	7,750	340	8,090
5000 lb	"	9,760	360	10,120
6000 lb	"	10,660	370	11,030
7500 lb	"	11,960	390	12,350
10,000 lb	"	30,090	400	30,490
15,000 lb	"	38,230	420	38,650
20,000 lb	"	46,200	450	46,650
25,000 lb	"	47,350	490	47,840
30,000 lb	"	50,000	540	50,540
Heavy duty				
500 lb	EA.	10,340	270	10,610
1000 lb	"	15,300	290	15,590
2000 lb	"	15,520	300	15,820
3000 lb	"	16,070	320	16,390
4000 lb	"	16,390	340	16,730
5000 lb	"	16,610	360	16,970
6000 lb	"	21,890	370	22,260
7500 lb	"	23,380	390	23,770
10,000 lb	"	30,090	400	30,490
15,000 lb	"	32,370	420	32,790
20,000 lb	"	36,960	450	37,410
25,000 lb	"	39,930	490	40,420
30,000 lb	"	41,690	540	42,230
Air powered hoists				
500 lb	EA.	6,060	270	6,330
1000 lb	"	6,380	270	6,650
2000 lb	"	6,710	290	7,000
4000 lb	"	7,150	320	7,470
6000 lb	"	7,750	420	8,170
Overhead traveling bridge crane				
Single girder, 20' span				

Hoists And Cranes	UNIT	MAT.	INST.	TOTAL
14600.10 **Industrial Hoists** *(Cont.)*				
3 ton	EA.	24,030	1,110	25,140
5 ton	"	25,100	1,110	26,210
7.5 ton	"	31,510	1,110	32,620
10 ton	"	36,720	1,390	38,110
15 ton	"	39,930	1,390	41,320
30' span				
3 ton	EA.	25,720	1,110	26,830
5 ton	"	27,700	1,110	28,810
10 ton	"	35,620	1,390	37,010
15 ton	"	40,860	1,390	42,250
Double girder, 40' span				
3 ton	EA.	42,980	2,470	45,450
5 ton	"	44,590	2,470	47,060
7.5 ton	"	47,050	2,470	49,520
10 ton	"	56,380	3,180	59,560
15 ton	"	60,190	3,180	63,370
25 ton	"	84,920	3,180	88,100
50' span				
3 ton	EA.	50,740	2,470	53,210
5 ton	"	52,530	2,470	55,000
7.5 ton	"	53,850	2,470	56,320
10 ton	"	59,510	3,180	62,690
15 ton	"	68,590	3,180	71,770
25 ton	"	93,150	3,180	96,330
14650.10 **Jib Cranes**				
Self supporting, swinging 8' boom, 200 deg rotation				
1000 lb	EA.	2,300	490	2,790
2000 lb	"	2,700	490	3,190
3000 lb	"	3,220	980	4,200
4000 lb	"	2,330	980	3,310
6000 lb	"	3,900	980	4,880
10,000 lb	"	6,480	980	7,460
Wall mounted, 180 deg rotation				
2000 lb	EA.	1,400	490	1,890
3000 lb	"	1,680	490	2,170
4000 lb	"	2,170	980	3,150
6000 lb	"	2,310	980	3,290
10,000 lb	"	4,620	980	5,600

Basic Materials	UNIT	MAT.	INST.	TOTAL
15100.10 — Specialties				
Wall penetration				
Concrete wall, 6" thick				
2" dia.	EA.		13.75	13.75
4" dia.	"		20.75	20.75
8" dia.	"		29.50	29.50
12" thick				
2" dia.	EA.		18.75	18.75
4" dia.	"		29.50	29.50
8" dia.	"		45.75	45.75
Non-destructive testing, piping systems				
X-ray of welds				
3" dia. pipe	EA.	65.00	59.00	124
4" dia. pipe	"	56.00	59.00	115
6" dia. pipe	"	48.50	59.00	108
8" dia. pipe	"	52.00	73.00	125
10" dia. pipe	"	55.00	73.00	128
Liquid penetration of welds				
2" dia. pipe	EA.	33.00	36.75	69.75
3" dia. pipe	"	35.25	36.75	72.00
4" dia. pipe	"	38.50	36.75	75.25
6" dia. pipe	"	40.75	36.75	77.50
8" dia. pipe	"	45.00	36.75	81.75
10" dia. pipe	"	49.50	36.75	86.25
15120.10 — Backflow Preventers				
Backflow preventer, flanged, cast iron, with valves				
3" pipe	EA.	2,760	290	3,050
4" pipe	"	3,540	330	3,870
6" pipe	"	6,040	490	6,530
8" pipe	"	9,910	590	10,500
Threaded				
3/4" pipe	EA.	520	36.75	557
2" pipe	"	1,000	59.00	1,059
Reduced pressure assembly, bronze, threaded				
3/4"	EA.	600	36.75	637
1"	"	630	42.00	672
1-1/4"	"	910	48.75	959
1-1/2"	"	930	59.00	989
15140.10 — Pipe Hangers, Heavy				
Hangers				
1/2" pipe, clevis pipe hanger				
Black steel	EA.	1.92	19.50	21.42
Galvanized	"	2.91	19.50	22.41
U bolt	"	1.54	5.86	7.40
3/4" pipe, clevis pipe hanger				
Black steel	EA.	1.98	19.50	21.48
Galvanized	"	2.97	19.50	22.47
U bolt	"	1.65	5.86	7.51
1" pipe, clevis pipe hanger				
Black steel	EA.	2.03	19.50	21.53
Galvanized	"	3.19	19.50	22.69
U bolt	"	1.81	5.86	7.67

Basic Materials	UNIT	MAT.	INST.	TOTAL
15140.10 **Pipe Hangers, Heavy** *(Cont.)*				
1-1/4" pipe, clevis pipe hanger				
Black steel	EA.	2.20	19.50	21.70
Galvanized	"	3.52	19.50	23.02
U bolt	"	1.81	5.86	7.67
1-1/2" pipe, clevis pipe hanger				
Black steel	EA.	2.31	19.50	21.81
Galvanized	"	3.79	19.50	23.29
U bolt	"	1.87	5.86	7.73
2" pipe, clevis pipe hanger				
Black steel	EA.	2.75	19.50	22.25
Galvanized	"	4.51	19.50	24.01
U bolt	"	2.03	5.86	7.89
2-1/2" pipe, clevis pipe hanger				
Black steel	EA.	4.34	19.50	23.84
Galvanized	"	7.81	19.50	27.31
3" pipe, clevis pipe hanger				
Black steel	EA.	5.39	19.50	24.89
Galvanized	"	9.40	19.50	28.90
U bolt	"	3.74	5.86	9.60
3-1/2" pipe, clevis pipe hanger				
Black steel	EA.	5.72	19.50	25.22
Galvanized	"	10.00	19.50	29.50
U bolt	"	4.07	5.86	9.93
4" pipe, clevis pipe hanger				
Black steel	EA.	6.60	19.50	26.10
Galvanized	"	11.75	19.50	31.25
U bolt	"	4.95	5.86	10.81
5" pipe, clevis pipe hanger				
Black steel	EA.	9.02	23.50	32.52
Galvanized	"	16.50	23.50	40.00
U bolt	"	9.46	5.86	15.32
6" pipe, clevis pipe hanger				
Black steel	EA.	10.50	23.50	34.00
Galvanized	"	21.25	23.50	44.75
U bolt	"	11.00	7.33	18.33
8" pipe, clevis pipe hanger				
Black steel	EA.	17.25	23.50	40.75
Galvanized	"	32.50	23.50	56.00
U bolt	"	18.25	7.33	25.58
10" clevis pipe hanger				
Black steel	EA.	31.25	23.50	54.75
Galvanized	"	58.00	23.50	81.50
12" pipe, clevis pipe hanger				
Black steel	EA.	37.25	23.50	60.75
Galvanized	"	76.00	23.50	99.50
Threaded rod, galvanized				
3/8"	L.F.			0.33
1/2"	"			0.58
5/8"	"			0.86
3/4"	"			1.38
7/8"	"			1.87
1"	"			2.17
Hex nuts, galvanized				

Basic Materials	UNIT	MAT.	INST.	TOTAL
15140.10 **Pipe Hangers, Heavy** *(Cont.)*				
3/8"	EA.			0.31
1/2"	"			0.67
5/8"	"			1.44
3/4"	"			1.91
7/8"	"			3.03
1"	"			4.22
C-clamp, steel, with lock nut				
3/8"	EA.	2.64	7.33	9.97
1/2"	"	2.95	7.33	10.28
5/8"	"	4.93	7.33	12.26
3/4"	"	6.73	7.33	14.06
7/8"	"	12.50	7.33	19.83
Angle support, medium, welded steel				
12"x18"	EA.	120	59.00	179
18"x24"	"	140	59.00	199
24"x30"	"	190	59.00	249
Heavy, welded steel				
12"x18"	EA.	170	59.00	229
18"x24"	"	240	59.00	299
24"x30"	"	270	59.00	329
15140.11 **Pipe Hangers, Light**				
A band, black iron				
1/2"	EA.	1.05	4.18	5.23
1"	"	1.11	4.34	5.45
1-1/4"	"	1.19	4.51	5.70
1-1/2"	"	1.25	4.88	6.13
2"	"	1.32	5.33	6.65
2-1/2"	"	2.71	5.86	8.57
3"	"	2.82	6.51	9.33
4"	"	4.01	7.33	11.34
5"	"	4.51	7.81	12.32
6"	"	7.64	8.37	16.01
8"	"	9.32	9.77	19.09
Copper				
1/2"	EA.	1.48	4.18	5.66
3/4"	"	1.54	4.34	5.88
1"	"	1.77	4.34	6.11
1-1/4"	"	2.05	4.51	6.56
1-1/2"	"	2.17	4.88	7.05
2"	"	2.75	5.33	8.08
2-1/2"	"	3.89	5.86	9.75
3"	"	4.23	6.51	10.74
4"	"	4.57	7.33	11.90
Black riser friction hangers				
3/4"	EA.	3.96	4.88	8.84
1"	"	4.01	5.09	9.10
1-1/4"	"	5.00	5.33	10.33
1-1/2"	"	5.44	5.58	11.02
2"	"	5.55	5.86	11.41
2-1/2"	"	5.99	6.51	12.50
3"	"	6.16	7.33	13.49
4"	"	7.86	8.37	16.23

Basic Materials	UNIT	MAT.	INST.	TOTAL
15140.11 **Pipe Hangers, Light** *(Cont.)*				
5"	EA.	11.50	9.02	20.52
6"	"	13.25	9.77	23.02
8"	"	22.50	10.75	33.25
10"	"	29.00	11.75	40.75
2 hole clips, galvanized				
3/4"	EA.	0.26	3.90	4.16
1"	"	0.29	4.04	4.33
1-1/4"	"	0.38	4.18	4.56
1-1/2"	"	0.47	4.34	4.81
2"	"	0.61	4.51	5.12
2-1/2"	"	1.11	4.69	5.80
3"	"	1.61	4.88	6.49
4"	"	3.46	5.33	8.79
Perforated strap				
3/4"				
Galvanized, 20 ga.	L.F.	0.40	2.93	3.33
Copper, 22 ga.	"	0.63	2.93	3.56
J-Hooks				
1/2"	EA.	0.41	2.66	3.07
3/4"	"	0.45	2.66	3.11
1"	"	0.46	2.79	3.25
1-1/4"	"	0.48	2.86	3.34
1-1/2"	"	0.49	2.93	3.42
2"	"	0.51	2.93	3.44
3"	"	0.59	3.08	3.67
4"	"	0.63	3.08	3.71
PVC coated hangers, galvanized, 28 ga.				
1-1/2" x 12"	EA.	1.26	3.90	5.16
2" x 12"	"	1.37	4.18	5.55
3" x 12"	"	1.54	4.51	6.05
4" x 12"	"	1.70	4.88	6.58
Copper, 30 ga.				
1-1/2" x 12"	EA.	1.76	3.90	5.66
2" x 12"	"	2.09	4.18	6.27
3" x 12"	"	2.31	4.51	6.82
4" x 12"	"	2.53	4.88	7.41
2" x 24"	"	4.07	4.51	8.58
3" x 24"	"	4.67	4.88	9.55
4" x 24"	"	7.31	5.33	12.64
Wire hook hangers				
Black wire, 1/2" x				
4"	EA.	0.42	2.93	3.35
6"	"	0.49	3.08	3.57
8"	"	0.53	3.25	3.78
10"	"	0.68	3.25	3.93
12"	"	0.81	3.44	4.25
3/4" x				
4"	EA.	0.52	3.08	3.60
6"	"	0.56	3.25	3.81
8"	"	0.57	3.44	4.01
10"	"	0.79	3.66	4.45
12"	"	0.80	3.90	4.70
1" x				

Basic Materials	UNIT	MAT.	INST.	TOTAL
15140.11 — **Pipe Hangers, Light** *(Cont.)*				
4"	EA.	0.52	3.25	3.77
6"	"	0.53	3.44	3.97
8"	"	0.57	3.66	4.23
10"	"	0.78	3.90	4.68
12"	"	0.84	4.18	5.02
1-1/4" x				
4"	EA.	0.56	3.44	4.00
6"	"	0.57	3.66	4.23
8"	"	0.64	3.90	4.54
10"	"	0.74	4.18	4.92
12"	"	0.84	4.51	5.35
1-1/2" x				
6"	EA.	0.63	3.90	4.53
8"	"	0.64	4.18	4.82
10"	"	0.81	4.51	5.32
12"	"	0.88	4.88	5.76
2" x				
6"	EA.	0.64	4.18	4.82
8"	"	0.78	4.51	5.29
10"	"	0.80	4.88	5.68
12"	"	0.90	5.33	6.23
Copper wire hooks				
1/2" x				
4"	EA.	0.59	2.93	3.52
6"	"	0.67	3.08	3.75
8"	"	0.75	3.25	4.00
10"	"	0.95	3.44	4.39
12"	"	1.08	3.66	4.74
3/4" x				
4"	EA.	0.59	3.08	3.67
6"	"	0.73	3.25	3.98
8"	"	0.85	3.44	4.29
10"	"	0.96	3.66	4.62
12"	"	1.15	3.90	5.05
1" x				
4"	EA.	0.63	3.25	3.88
6"	"	0.71	3.44	4.15
8"	"	0.91	3.66	4.57
10"	"	1.08	3.90	4.98
12"	"	1.26	4.18	5.44
1-1/4" x				
6"	EA.	0.83	3.44	4.27
8"	"	0.95	3.66	4.61
10"	"	1.21	3.90	5.11
12"	"	1.26	4.18	5.44
1-1/2" x				
6"	EA.	0.99	3.90	4.89
8"	"	1.07	4.18	5.25
10"	"	1.21	4.51	5.72

Basic Materials	UNIT	MAT.	INST.	TOTAL
15140.11 **Pipe Hangers, Light** *(Cont.)*				
12"	EA.	1.32	4.88	6.20
2" x				
6"	EA.	1.03	4.18	5.21
8"	"	1.15	4.51	5.66
10"	"	1.21	4.88	6.09
12"	"	1.32	5.33	6.65
15175.60 **Expansion Tanks**				
Expansion tank, 125 psi, steel				
20 gallon	EA.	580	73.00	653
30 gallon	"	630	98.00	728
65 gallon	"	780	150	930
80 gallon	"	880	170	1,050
15240.10 **Vibration Control**				
Vibration isolator, in-line, stainless connector, screwed				
1/2"	EA.	93.00	32.50	126
3/4"	"	110	34.50	145
1"	"	110	36.75	147
1-1/4"	"	150	39.00	189
1-1/2"	"	170	42.00	212
2"	"	200	45.00	245
2-1/2"	"	310	48.75	359
3"	"	360	53.00	413
4"	"	460	59.00	519
6"	"	780	65.00	845
Flanged				
8"	EA.	1,490	73.00	1,563
10"	"	1,980	84.00	2,064
12"	"	3,070	98.00	3,168

Insulation	UNIT	MAT.	INST.	TOTAL
15260.10 **Fiberglass Pipe Insulation**				
Fiberglass insulation on 1/2" pipe				
1" thick	L.F.	0.99	1.95	2.94
1-1/2" thick	"	2.09	2.44	4.53
3/4" pipe				
1" thick	L.F.	1.21	1.95	3.16
1-1/2" thick	"	2.20	2.44	4.64
1" pipe				
1" thick	L.F.	1.21	1.95	3.16
1-1/2" thick	"	2.31	2.44	4.75
2" pipe				
1" thick	L.F.	1.65	2.44	4.09
1-1/2" thick	"	2.86	2.66	5.52

Insulation	UNIT	MAT.	INST.	TOTAL
15260.10 **Fiberglass Pipe Insulation** *(Cont.)*				
3" pipe				
1" thick	L.F.	1.98	2.79	4.77
1-1/2" thick	"	3.19	2.93	6.12
6" pipe				
1" thick	L.F.	3.30	3.08	6.38
2" thick	"	6.82	3.25	10.07
10" pipe				
2" thick	L.F.	10.50	3.08	13.58
3" thick	"	15.50	3.25	18.75
15260.20 **Calcium Silicate**				
Calcium silicate insulation, 6" pipe				
2" thick	L.F.	8.25	4.18	12.43
2-1/2" thick	"	11.75	4.51	16.26
3" thick	"	15.50	4.88	20.38
4" thick	"	19.25	5.33	24.58
6" thick	"	28.25	5.86	34.11
12" pipe				
2" thick	L.F.	13.75	4.51	18.26
2-1/2" thick	"	19.00	4.88	23.88
3" thick	"	25.75	5.33	31.08
4" thick	"	30.00	5.86	35.86
6" thick	"	47.75	6.51	54.26
15260.60 **Exterior Pipe Insulation**				
Fiberglass insulation, aluminum jacket				
1/2" pipe				
1" thick	L.F.	1.49	4.51	6.00
1-1/2" thick	"	2.80	4.88	7.68
3/4" pipe				
1" thick	L.F.	1.76	4.51	6.27
1-1/2" thick	"	2.97	4.88	7.85
1" pipe				
1" thick	L.F.	1.81	4.51	6.32
1-1/2" thick	"	3.13	4.88	8.01
2" pipe				
1" thick	L.F.	2.47	5.33	7.80
1-1/2" thick	"	3.68	5.58	9.26
3" pipe				
1" thick	L.F.	2.97	5.86	8.83
1-1/2" thick	"	4.40	6.17	10.57
6" pipe				
1" thick	L.F.	4.95	6.51	11.46
2" thick	"	8.80	6.89	15.69
10" pipe				
2" thick	L.F.	12.50	6.51	19.01
3" thick	"	18.00	6.89	24.89

Insulation

	UNIT	MAT.	INST.	TOTAL
15260.90 **Pipe Insulation Fittings**				
Insulation protection saddle				
1" thick covering				
1/2" pipe	EA.	3.30	23.50	26.80
3/4" pipe	"	3.30	23.50	26.80
1" pipe	"	3.30	23.50	26.80
2" pipe	"	4.73	23.50	28.23
3" pipe	"	5.39	26.75	32.14
6" pipe	"	6.65	36.75	43.40
1-1/2" thick covering				
3/4" pipe	EA.	5.39	23.50	28.89
1" pipe	"	5.39	23.50	28.89
2" pipe	"	6.05	23.50	29.55
3" pipe	"	6.82	23.50	30.32
6" pipe	"	10.00	36.75	46.75
10" pipe	"	12.75	48.75	61.50
15280.10 **Equipment Insulation**				
Equipment insulation, 2" thick, cellular glass	S.F.	3.24	3.66	6.90
Urethane, rigid, field applied jacket, plastered finish	"	3.46	7.33	10.79
Fiberglass, rigid, with vapor barrier	"	3.24	3.25	6.49
15290.10 **Ductwork Insulation**				
Fiberglass duct insulation, plain blanket				
1-1/2" thick	S.F.	1.10	0.73	1.83
2" thick	"	1.32	0.97	2.29
With vapor barrier				
1-1/2" thick	S.F.	0.66	0.73	1.39
2" thick	"	0.73	0.97	1.70
Rigid with vapor barrier				
2" thick	S.F.	2.14	1.95	4.09
3" thick	"	3.52	2.34	5.86
4" thick	"	4.56	2.93	7.49
6" thick	"	6.32	3.90	10.22
Weatherproof, polystyrene, 3" thick, w/vapor barrier	"	4.67	5.86	10.53
Urethane board with vapor barrier	"	6.60	7.33	13.93

Fire Protection

	UNIT	MAT.	INST.	TOTAL
15330.10 **Wet Sprinkler System**				
Sprinkler head, 212 deg, brass, exposed piping	EA.	10.00	23.50	33.50
Chrome, concealed piping	"	11.75	32.50	44.25
Water motor alarm	"	280	98.00	378
Fire department inlet connection	"	210	120	330
Wall plate for fire dept connection	"	99.00	48.75	148
Swing check valve flanged iron body, 4"	"	280	200	480
Check valve, 6"	"	900	290	1,190

Fire Protection	UNIT	MAT.	INST.	TOTAL
15330.10 **Wet Sprinkler System** *(Cont.)*				
Wet pipe valve, flange to groove, 4"	EA.	780	65.00	845
Flange to flange				
6"	EA.	1,160	98.00	1,258
8"	"	2,040	200	2,240
Alarm valve, flange to flange, (wet valve)				
4"	EA.	950	65.00	1,015
8"	"	2,090	490	2,580
Inspector's test connection	"	61.00	48.75	110
Wall hydrant, polished brass, 2-1/2" x 2-1/2", single	"	480	42.00	522
2-way	"	1,080	42.00	1,122
3-way	"	2,210	42.00	2,252
Wet valve trim, includes retard chamber & gauges, 4"-6"	"	760	48.75	809
Retard pressure switch for wet systems	"	1,320	120	1,440
Air maintenance device	"	400	48.75	449
Wall hydrant non-freeze, 8" thick wall, vacuum breaker	"	44.25	29.25	73.50
12" thick wall	"	48.25	29.25	77.50
15330.50 **Dry Sprinkler System**				
Dry pipe valve, flange to flange				
4"	EA.	1,680	120	1,800
6"	"	2,110	150	2,260
Trim, 4" and 6", includes gauges	"	720	48.75	769
Field testing and flushing	"		490	490
Disinfection	"		490	490
Pressure switch double circuit, open/close contacts	"	360	150	510
Low air				
Supervisory unit	EA.	1,180	98.00	1,278
Pressure switch	"	370	48.75	419
15330.70 **Co2 System**				
CO2 system, high pressure, 75# cylinder with				
Valve assemblies	EA.	2,410	120	2,530
Storage rack	"	1,290	84.00	1,374
Manifold	"	930	420	1,350
Flexible loops	"	76.00	7.33	83.33
Beam scale for cylinders	"	640	98.00	738
Mechanically control head	"	540	39.00	579
Electrically control head	"	540	39.00	579
Stop valves	"	1,360	59.00	1,419
Check valves	"	620	73.00	693
Activation station	"	720	59.00	779
Nozzles	"	120	48.75	169
Hose reel with 75' of 3/4" hose	"	4,310	290	4,600
Main/reserve transfer switch	"	5,020	98.00	5,118
Pressure switch	"	440	59.00	499
Heat responsive device	"	700	98.00	798
Battery and charger	"	4,010	290	4,300
Low pressure				
Battery and charger	EA.	4,010	290	4,300
Pressure switch	"	400	65.00	465
Nozzles	"	120	53.00	173

Fire Protection

Fire Protection	UNIT	MAT.	INST.	TOTAL
15330.70 Co2 System *(Cont.)*				
Master selector valve	EA.	340	98.00	438
Selector valve	"	5,020	98.00	5,118
Low pressure hose reel with 75' of 3/4" hose	"	6,230	290	6,520
Tank fill lines	"	1,410	73.00	1,483
Activation stations	"	700	48.75	749
Electro manual pilot panels	"	1,410	73.00	1,483

Plumbing

Plumbing	UNIT	MAT.	INST.	TOTAL
15410.05 C.i. Pipe, Above Ground				
No hub pipe				
1-1/2" pipe	L.F.	6.82	4.18	11.00
2" pipe	"	6.98	4.88	11.86
3" pipe	"	9.62	5.86	15.48
4" pipe	"	12.50	9.77	22.27
6" pipe	"	18.75	11.75	30.50
8" pipe	"	33.50	19.50	53.00
10" pipe	"	56.00	23.50	79.50
No hub fittings, 1-1/2" pipe				
1/4 bend	EA.	10.50	19.50	30.00
1/8 bend	"	7.53	19.50	27.03
Sanitary tee	"	12.75	29.25	42.00
Sanitary cross	"	16.25	29.25	45.50
Plug	"			4.46
Coupling	"			14.00
Wye	"	14.00	29.25	43.25
2" pipe				
1/4 bend	EA.	10.75	23.50	34.25
1/8 bend	"	7.93	23.50	31.43
Sanitary tee	"	13.50	39.00	52.50
Sanitary cross	"	21.50	39.00	60.50
Plug	"			4.46
Coupling	"			14.00
Wye	"	15.00	48.75	63.75
3" pipe				
1/4 bend	EA.	13.75	29.25	43.00
1/8 bend	"	12.25	29.25	41.50
Sanitary tee	"	16.00	36.75	52.75
3"x2" sanitary tee	"	15.50	36.75	52.25
3"x1-1/2" sanitary tee	"	15.75	36.75	52.50
Sanitary cross	"	26.00	48.75	74.75
3x2" sanitary cross	"	27.75	48.75	76.50
Plug	"			6.62
Coupling	"			16.50
Wye	"	17.00	48.75	65.75
4" pipe				

Plumbing	UNIT	MAT.	INST.	TOTAL
15410.05 **C.i. Pipe, Above Ground** *(Cont.)*				
1/4 bend	EA.	18.75	29.25	48.00
1/8 bend	"	16.00	29.25	45.25
Sanitary tee	"	24.25	48.75	73.00
4x3" sanitary tee	"	22.25	48.75	71.00
4x2" sanitary tee	"	18.75	48.75	67.50
Sanitary cross	"	45.75	59.00	105
4x3" sanitary cross	"	39.75	59.00	98.75
4x2" sanitary cross	"	35.25	59.00	94.25
Plug	"			6.87
Coupling	"			18.25
Wye	"	23.50	48.75	72.25
8" deep	"	33.00	29.25	62.25
6" pipe				
1/4 bend	EA.	44.50	48.75	93.25
1/8 bend	"	33.00	48.75	81.75
Sanitary tee	"	70.00	59.00	129
6x4" sanitary tee	"	56.00	59.00	115
Coupling	"			37.50
Wye	"	56.00	59.00	115
8" pipe				
1/4 bend	EA.	92.00	48.75	141
1/8 bend	"	82.00	48.75	131
Sanitary tee	"	160	73.00	233
8x6" sanitary tee	"	120	73.00	193
Plug	"			25.25
Coupling	"			42.25
Wye	"	110	59.00	169
10" pipe				
1/4 bend	EA.	170	48.75	219
1/8 bend	"	160	48.75	209
Plug	"			42.25
Coupling	"			59.00
Wye	"	220	98.00	318
15410.06 **C.i. Pipe, Below Ground**				
No hub pipe				
1-1/2" pipe	L.F.	6.18	2.93	9.11
2" pipe	"	6.34	3.25	9.59
3" pipe	"	8.75	3.66	12.41
4" pipe	"	11.25	4.88	16.13
6" pipe	"	19.50	5.33	24.83
8" pipe	"	30.50	6.51	37.01
10" pipe	"	51.00	7.33	58.33
Fittings, 1-1/2"				
1/4 bend	EA.	10.50	16.75	27.25
1/8 bend	"	7.53	16.75	24.28
Plug	"			2.36
Wye	"	14.00	23.50	37.50
Wye & 1/8 bend	"	14.00	16.75	30.75
P-trap	"	20.50	16.75	37.25
2"				
1/4 bend	EA.	10.75	19.50	30.25
1/8 bend	"	7.93	19.50	27.43

Plumbing	UNIT	MAT.	INST.	TOTAL
15410.06	**C.i. Pipe, Below Ground** *(Cont.)*			
Plug	EA.			2.86
Double wye	"	19.25	36.75	56.00
Wye & 1/8 bend	"	19.00	29.25	48.25
Double wye & 1/8 bend	"	28.25	36.75	65.00
P-trap	"	20.50	19.50	40.00
3"				
1/4 bend	EA.	13.75	23.50	37.25
1/8 bend	"	12.25	23.50	35.75
Plug	"			4.23
Wye	"	16.50	36.75	53.25
3x2" wye	"	17.00	36.75	53.75
Wye & 1/8 bend	"	21.00	36.75	57.75
Double wye & 1/8 bend	"	38.75	36.75	75.50
3x2" double wye & 1/8 bend	"	29.25	36.75	66.00
3x2" reducer	"	8.96	23.50	32.46
P-trap	"	24.75	23.50	48.25
4"				
1/4 bend	EA.	18.75	23.50	42.25
1/8 bend	"	16.00	23.50	39.50
Plug	"			6.87
Wye	"	23.50	36.75	60.25
4x3" wye	"	20.00	36.75	56.75
4x2" wye	"	14.75	36.75	51.50
Double wye	"	55.00	48.75	104
4x3" double wye	"	35.25	48.75	84.00
4x2" double wye	"	28.25	48.75	77.00
Wye & 1/8 bend	"	30.50	36.75	67.25
Double wye & 1/8 bend	"	70.00	48.75	119
6"				
1/4 bend	EA.	44.50	36.75	81.25
1/8 bend	"	33.00	36.75	69.75
Wye & 1/8 bend	"	73.00	48.75	122
P-trap	"	89.00	29.25	118
8"				
1/4 bend	EA.	92.00	36.75	129
1/8 bend	"	82.00	36.75	119
Plug	"			25.25
Wye	"	110	48.75	159
10"				
1/4 bend	EA.	170	36.75	207
1/8 bend	"	160	36.75	197
Plug	"			42.25
Wye	"	220	73.00	293
15410.09	**Service Weight Pipe**			
Service weight pipe, single hub				
3" x 5'	EA.	66.00	12.50	78.50
4" x 5'	"	77.00	13.00	90.00
6" x 5'	"	150	14.75	165
1/8 bend				
3"	EA.	16.75	23.50	40.25
4"	"	24.75	26.75	51.50
6"	"	41.75	29.25	71.00

Plumbing		UNIT	MAT.	INST.	TOTAL
15410.09	**Service Weight Pipe** *(Cont.)*				
1/4 bend					
3"		EA.	20.25	23.50	43.75
4"		"	31.50	26.75	58.25
6"		"	55.00	29.25	84.25
Sweep					
3"		EA.	26.25	23.50	49.75
4"		"	40.50	26.75	67.25
6"		"	82.00	29.25	111
Sanitary T					
3"		EA.	34.00	42.00	76.00
4"		"	41.75	48.75	90.50
6"		"	94.00	53.00	147
Wye					
3"		EA.	35.75	32.50	68.25
4"		"	47.75	34.50	82.25
6"		"	110	42.00	152
15410.10	**Copper Pipe**				
Type "K" copper					
1/2"		L.F.	5.50	1.83	7.33
3/4"		"	10.00	1.95	11.95
1"		"	12.75	2.09	14.84
1-1/4"		"	16.00	2.25	18.25
1-1/2"		"	20.25	2.44	22.69
2"		"	27.75	2.66	30.41
2-1/2"		"	36.00	2.93	38.93
3"		"	48.75	3.08	51.83
4"		"	79.00	3.25	82.25
DWV, copper					
1-1/4"		L.F.	10.25	2.44	12.69
1-1/2"		"	13.00	2.66	15.66
2"		"	14.75	2.93	17.68
3"		"	27.75	3.25	31.00
4"		"	45.75	3.66	49.41
6"		"	180	4.18	184
Refrigeration tubing, copper, sealed					
1/8"		L.F.	1.37	2.34	3.71
3/16"		"	1.43	2.44	3.87
1/4"		"	1.65	2.54	4.19
5/16"		"	2.14	2.66	4.80
3/8"		"	2.42	2.79	5.21
1/2"		"	3.19	2.93	6.12
7/8"		"	5.39	3.35	8.74
1-1/8"		"	8.69	3.90	12.59
1-3/8"		"	10.50	4.51	15.01
Type "L" copper					
1/4"		L.F.	3.35	1.72	5.07
3/8"		"	3.63	1.72	5.35
1/2"		"	5.50	1.83	7.33
3/4"		"	6.27	1.95	8.22
1"		"	8.47	2.09	10.56
1-1/4"		"	11.50	2.25	13.75
1-1/2"		"	13.75	2.44	16.19

Plumbing	UNIT	MAT.	INST.	TOTAL
15410.10 **Copper Pipe** *(Cont.)*				
2"	L.F.	20.50	2.66	23.16
Type "M" copper				
1/2"	L.F.	3.57	1.83	5.40
3/4"	"	5.00	1.95	6.95
1"	"	6.49	2.09	8.58
1-1/4"	"	9.46	2.25	11.71
2"	"	18.25	2.66	20.91
15410.11 **Copper Fittings**				
Coupling, with stop				
1/4"	EA.	1.13	19.50	20.63
3/8"	"	1.27	23.50	24.77
1/2"	"	0.96	25.50	26.46
5/8"	"	4.20	29.25	33.45
3/4"	"	1.87	32.50	34.37
1"	"	3.81	34.50	38.31
3"	"	50.00	59.00	109
4"	"	110	73.00	183
Reducing coupling				
1/4" x 1/8"	EA.	2.69	23.50	26.19
3/8" x 1/4"	"	2.97	25.50	28.47
1/2" x				
3/8"	EA.	2.11	29.25	31.36
1/4"	"	2.56	29.25	31.81
1/8"	"	2.81	29.25	32.06
3/4" x				
3/8"	EA.	4.53	32.50	37.03
1/2"	"	3.58	32.50	36.08
1" x				
3/8"	EA.	8.12	36.75	44.87
1" x 1/2"	"	7.87	36.75	44.62
1" x 3/4"	"	6.63	36.75	43.38
1-1/4" x				
1/2"	EA.	10.50	39.00	49.50
3/4"	"	11.00	39.00	50.00
1"	"	12.25	39.00	51.25
1-1/2" x				
1/2"	EA.	16.50	42.00	58.50
3/4"	"	15.50	42.00	57.50
1"	"	15.50	42.00	57.50
1-1/4"	"	15.50	42.00	57.50
2" x				
1/2"	EA.	26.75	48.75	75.50
3/4"	"	25.75	48.75	74.50
1"	"	25.25	48.75	74.00
1-1/4"	"	24.00	48.75	72.75
1-1/2"	"	24.00	48.75	72.75
2-1/2" x				
1"	EA.	61.00	59.00	120
1-1/4"	"	61.00	59.00	120
1-1/2"	"	54.00	59.00	113
2"	"	53.00	59.00	112
3" x				

Plumbing	UNIT	MAT.	INST.	TOTAL
15410.11		**Copper Fittings** *(Cont.)*		
1-1/2"	EA.	76.00	73.00	149
2"	"	68.00	73.00	141
2-1/2"	"	70.00	73.00	143
4" x				
2"	EA.	160	84.00	244
2-1/2"	"	160	84.00	244
3"	"	140	84.00	224
Slip coupling				
1/4"	EA.	0.79	19.50	20.29
1/2"	"	1.33	23.50	24.83
3/4"	"	2.77	29.25	32.02
1"	"	5.84	32.50	38.34
1-1/4"	"	8.82	36.75	45.57
1-1/2"	"	12.00	39.00	51.00
2"	"	20.25	48.75	69.00
2-1/2"	"	26.25	48.75	75.00
3"	"	52.00	59.00	111
4"	"	99.00	73.00	172
Coupling with drain				
1/2"	EA.	10.75	29.25	40.00
3/4"	"	15.50	32.50	48.00
1"	"	19.50	36.75	56.25
Reducer				
3/8" x 1/4"	EA.	3.57	23.50	27.07
1/2" x 3/8"	"	2.89	23.50	26.39
3/4" x				
1/4"	EA.	4.95	26.75	31.70
3/8"	"	5.17	26.75	31.92
1/2"	"	5.39	26.75	32.14
1" x				
1/2"	EA.	8.75	29.25	38.00
3/4"	"	6.71	29.25	35.96
1-1/4" x				
1/2"	EA.	10.25	32.50	42.75
3/4"	"	10.25	32.50	42.75
1"	"	10.25	32.50	42.75
1-1/2" x				
1/2"	EA.	13.25	36.75	50.00
3/4"	"	13.25	36.75	50.00
1"	"	13.25	36.75	50.00
1-1/4"	"	13.25	36.75	50.00
2" x				
1/2"	EA.	26.50	42.00	68.50
3/4"	"	26.50	42.00	68.50
1"	"	26.50	42.00	68.50
1-1/4"	"	25.00	42.00	67.00
1-1/2"	"	25.00	42.00	67.00
2-1/2" x				
1"	EA.	58.00	48.75	107
1-1/4"	"	52.00	48.75	101
1-1/2"	"	51.00	48.75	99.75
2"	"	49.75	48.75	98.50
3" x				

Plumbing	UNIT	MAT.	INST.	TOTAL
15410.11	**Copper Fittings** *(Cont.)*			
1-1/4"	EA.	68.00	59.00	127
1-1/2"	"	70.00	59.00	129
2"	"	63.00	59.00	122
2-1/2"	"	64.00	59.00	123
4" x				
2"	EA.	150	73.00	223
3"	"	130	73.00	203
Female adapters				
1/4"	EA.	9.04	23.50	32.54
3/8"	"	9.25	26.75	36.00
1/2"	"	4.43	29.25	33.68
3/4"	"	6.07	32.50	38.57
1"	"	17.00	32.50	49.50
1-1/4"	"	37.00	36.75	73.75
1-1/2"	"	25.75	36.75	62.50
2"	"	43.75	39.00	82.75
2-1/2"	"	160	42.00	202
3"	"	240	48.75	289
4"	"	300	59.00	359
Increasing female adapters				
1/8" x				
3/8"	EA.	9.07	23.50	32.57
1/2"	"	9.07	23.50	32.57
1/4" x 1/2"	"	9.07	25.50	34.57
3/8" x 1/2"	"	9.07	26.75	35.82
1/2" X				
3/4"	EA.	9.59	29.25	38.84
1"	"	17.50	29.25	46.75
3/4" X				
1"	EA.	20.50	32.50	53.00
1-1/4"	"	34.00	32.50	66.50
1" x				
1-1/4"	EA.	37.00	32.50	69.50
1-1/2"	"	39.50	32.50	72.00
1-1/4" x				
1-1/2"	EA.	44.00	36.75	80.75
2"	"	57.00	36.75	93.75
1-1/2" x 2"	"	82.00	39.00	121
Reducing female adapters				
3/8" x 1/4"	EA.	7.99	26.75	34.74
1/2" x				
1/4"	EA.	6.88	29.25	36.13
3/8"	"	6.88	29.25	36.13
3/4" x 1/2"	"	9.59	32.50	42.09
1" x				
1/2"	EA.	25.75	32.50	58.25
3/4"	"	20.50	32.50	53.00
1-1/4" x				
1/2"	EA.	35.25	36.75	72.00
3/4"	"	43.25	36.75	80.00
1"	"	43.50	36.75	80.25
1-1/2" x				
1"	EA.	52.00	39.00	91.00

Plumbing	UNIT	MAT.	INST.	TOTAL
15410.11 **Copper Fittings** *(Cont.)*				
1-1/4"	EA.	62.00	39.00	101
2" x				
1"	EA.	54.00	42.00	96.00
1-1/4"	"	84.00	42.00	126
1-1/2"	"	71.00	42.00	113
Female fitting adapters				
1/2"	EA.	6.86	29.25	36.11
3/4"	"	8.86	29.25	38.11
3/4" x 1/2"	"	9.59	30.75	40.34
1"	"	14.00	32.50	46.50
1-1/4"	"	20.50	34.50	55.00
1-1/2"	"	32.25	36.75	69.00
2"	"	43.75	39.00	82.75
Male adapters				
1/4"	EA.	13.75	26.75	40.50
3/8"	"	6.88	26.75	33.63
3"	"	180	48.75	229
4"	"	250	59.00	309
Increasing male adapters				
3/8" x 1/2"	EA.	9.37	26.75	36.12
1/2" x				
3/4"	EA.	8.12	29.25	37.37
1"	"	18.25	29.25	47.50
3/4" x				
1"	EA.	18.00	30.75	48.75
1-1/4"	"	23.50	30.75	54.25
1" x 1-1/4"	"	23.50	32.50	56.00
1-1/2" x				
3/4"	EA.	30.50	34.50	65.00
1"	"	41.50	34.50	76.00
1-1/4"	"	49.25	34.50	83.75
2" x				
1"	EA.	89.00	36.75	126
1-1/4"	"	89.00	36.75	126
1-1/2"	"	80.00	36.75	117
2" x 2-1/2"	"	180	39.00	219
Copper pipe fittings				
1/2"				
90 deg ell	EA.	1.67	13.00	14.67
45 deg ell	"	2.12	13.00	15.12
Tee	"	2.80	16.75	19.55
Cap	"	1.14	6.51	7.65
Coupling	"	1.23	13.00	14.23
Union	"	8.49	14.75	23.24
3/4"				
90 deg ell	EA.	3.65	14.75	18.40
45 deg ell	"	4.26	14.75	19.01
Tee	"	6.11	19.50	25.61
Cap	"	2.23	6.89	9.12
Coupling	"	2.48	14.75	17.23
Union	"	12.50	16.75	29.25
1"				
90 deg ell	EA.	8.49	19.50	27.99

Plumbing	UNIT	MAT.	INST.	TOTAL
15410.11 **Copper Fittings** *(Cont.)*				
45 deg ell	EA.	11.00	19.50	30.50
Tee	"	14.00	23.50	37.50
Cap	"	4.13	9.77	13.90
Coupling	"	6.11	19.50	25.61
Union	"	16.25	19.50	35.75
1-1/4"				
90 deg ell	EA.	11.50	16.75	28.25
45 deg ell	"	14.25	16.75	31.00
Tee	"	18.75	29.25	48.00
Cap	"	3.31	9.77	13.08
Union	"	27.00	21.00	48.00
1-1/2"				
90 deg ell	EA.	15.00	21.00	36.00
45 deg ell	"	18.00	21.00	39.00
Tee	"	24.75	32.50	57.25
Cap	"	3.31	9.77	13.08
Coupling	"	11.00	19.50	30.50
Union	"	41.25	26.75	68.00
2"				
90 deg ell	EA.	29.50	23.50	53.00
45 deg ell	"	27.00	36.75	63.75
Tee	"	42.25	36.75	79.00
Cap	"	6.83	11.75	18.58
Coupling	"	18.00	23.50	41.50
Union	"	44.75	29.25	74.00
2-1/2"				
90 deg ell	EA.	57.00	29.25	86.25
45 deg ell	"	49.50	29.25	78.75
Tee	"	57.00	42.00	99.00
Cap	"	14.00	14.75	28.75
Coupling	"	27.00	29.25	56.25
Union	"	82.00	32.50	115
15410.15 **Brass Fittings**				
Compression fittings, union				
3/8"	EA.	4.40	9.77	14.17
1/2"	"	6.60	9.77	16.37
5/8"	"	8.08	9.77	17.85
Union elbow				
3/8"	EA.	8.14	9.77	17.91
1/2"	"	12.75	9.77	22.52
5/8"	"	22.25	9.77	32.02
Union tee				
3/8"	EA.	10.00	9.77	19.77
1/2"	"	14.25	9.77	24.02
5/8"	"	34.00	9.77	43.77
Male connector				
3/8"	EA.	4.73	9.77	14.50
1/2"	"	6.93	9.77	16.70
5/8"	"	7.42	9.77	17.19
Female connector				
3/8"	EA.	8.47	9.77	18.24
1/2"	"	11.25	9.77	21.02

Plumbing	UNIT	MAT.	INST.	TOTAL
15410.15	**Brass Fittings** *(Cont.)*			
5/8"	EA.	11.25	9.77	21.02
Brass flare fittings, union				
3/8"	EA.	2.75	9.45	12.20
1/2"	"	3.96	9.45	13.41
5/8"	"	5.50	9.45	14.95
90 deg elbow union				
3/8"	EA.	11.75	9.45	21.20
1/2"	"	16.50	9.45	25.95
5/8"	"	23.75	9.45	33.20
Three way tee				
3/8"	EA.	15.75	15.75	31.50
1/2"	"	20.50	15.75	36.25
5/8"	"	34.00	15.75	49.75
Cross				
3/8"	EA.	27.25	21.00	48.25
1/2"	"	27.25	21.00	48.25
5/8"	"	36.50	21.00	57.50
Male connector, half union				
3/8"	EA.	8.29	9.45	17.74
1/2"	"	8.29	9.45	17.74
5/8"	"	12.00	9.45	21.45
Female connector, half union				
3/8"	EA.	6.82	9.45	16.27
1/2"	"	10.75	9.45	20.20
5/8"	"	15.50	9.45	24.95
Long forged nut				
3/8"	EA.	1.43	9.45	10.88
1/2"	"	1.59	9.45	11.04
5/8"	"	3.39	9.45	12.84
Short forged nut				
3/8"	EA.	0.53	9.45	9.98
1/2"	"	0.74	9.45	10.19
5/8"	"	0.74	9.45	10.19
15410.18	**Glass Pipe**			
Glass pipe				
1-1/2" dia.	L.F.	11.50	11.75	23.25
2" dia.	"	15.50	13.00	28.50
3" dia.	"	20.75	14.75	35.50
4" dia.	"	38.00	16.75	54.75
6" dia.	"	69.00	19.50	88.50
15410.30	**Pvc/cpvc Pipe**			
PVC schedule 40				
1/2" pipe	L.F.	0.44	2.44	2.88
3/4" pipe	"	0.60	2.66	3.26
1" pipe	"	0.88	2.93	3.81
1-1/4" pipe	"	1.21	3.25	4.46
1-1/2" pipe	"	1.37	3.66	5.03
2" pipe	"	1.87	4.18	6.05
2-1/2" pipe	"	3.02	4.88	7.90
3" pipe	"	3.85	5.86	9.71
4" pipe	"	5.50	7.33	12.83

Plumbing	UNIT	MAT.	INST.	TOTAL
15410.30	**Pvc/cpvc Pipe** *(Cont.)*			
6" pipe	L.F.	9.62	14.75	24.37
8" pipe	"	14.50	19.50	34.00
Fittings, 1/2"				
90 deg ell	EA.	0.48	7.33	7.81
45 deg ell	"	0.66	7.33	7.99
Tee	"	0.49	8.37	8.86
3/4"				
90 deg elbow	EA.	0.47	9.77	10.24
45 deg elbow	"	1.12	9.77	10.89
Tee	"	0.64	11.75	12.39
1"				
90 deg elbow	EA.	0.82	11.75	12.57
45 deg elbow	"	1.21	11.75	12.96
Tee	"	1.10	13.00	14.10
1-1/4"				
90 deg elbow	EA.	1.43	16.75	18.18
45 deg elbow	"	1.70	16.75	18.45
Tee	"	1.65	19.50	21.15
1-1/2"				
90 deg elbow	EA.	1.59	16.75	18.34
45 deg elbow	"	2.36	16.75	19.11
Tee	"	2.20	19.50	21.70
2"				
90 deg elbow	EA.	2.53	19.50	22.03
45 deg elbow	"	3.19	19.50	22.69
Tee	"	3.35	23.50	26.85
2-1/2"				
90 deg elbow	EA.	7.64	36.75	44.39
45 deg elbow	"	11.00	36.75	47.75
Tee	"	9.84	39.00	48.84
3"				
90 deg elbow	EA.	8.25	48.75	57.00
45 deg elbow	"	10.75	48.75	59.50
Tee	"	13.25	53.00	66.25
4"				
90 deg elbow	EA.	14.75	59.00	73.75
45 deg elbow	"	19.25	59.00	78.25
Tee	"	22.00	65.00	87.00
PVC schedule 80 pipe				
1-1/2" pipe	L.F.	1.48	3.66	5.14
2" pipe	"	1.94	4.18	6.12
3" pipe	"	3.91	5.86	9.77
4" pipe	"	5.63	7.33	12.96
Fittings, 1-1/2"				
90 deg elbow	EA.	4.40	19.50	23.90
45 deg elbow	"	7.75	19.50	27.25
Tee	"	12.25	29.25	41.50
2"				
90 deg elbow	EA.	5.28	23.50	28.78
45 deg elbow	"	10.00	23.50	33.50
Tee	"	15.25	36.75	52.00
2-1/2"				
90 deg elbow	EA.	9.95	36.75	46.70

Plumbing	UNIT	MAT.	INST.	TOTAL
15410.30 — **Pvc/cpvc Pipe** *(Cont.)*				
45 deg elbow	EA.	21.00	36.75	57.75
Tee	"	16.50	48.75	65.25
3"				
90 deg elbow	EA.	13.50	48.75	62.25
45 deg elbow	"	25.75	48.75	74.50
Tee	"	20.50	59.00	79.50
4"				
90 deg elbow	EA.	17.25	59.00	76.25
45 deg elbow	"	46.50	59.00	106
Tee	"	22.75	73.00	95.75
CPVC schedule 40				
1/2" pipe	L.F.	0.82	2.44	3.26
3/4" pipe	"	1.10	2.66	3.76
1" pipe	"	1.59	2.93	4.52
1-1/4" pipe	"	2.09	3.25	5.34
1-1/2" pipe	"	2.53	3.66	6.19
2" pipe	"	3.35	4.18	7.53
Fittings, CPVC, schedule 80				
1/2", 90 deg ell	EA.	6.87	5.86	12.73
Tee	"	8.85	9.77	18.62
3/4", 90 deg ell	"	8.14	5.86	14.00
Tee	"	7.20	9.77	16.97
1", 90 deg ell	"	8.74	6.51	15.25
Tee	"	13.25	10.75	24.00
1-1/4", 90 deg ell	"	10.00	6.51	16.51
Tee	"	14.25	10.75	25.00
1-1/2", 90 deg ell	"	12.00	11.75	23.75
Tee	"	16.75	14.75	31.50
2", 90 deg ell	"	15.50	11.75	27.25
Tee	"	22.00	14.75	36.75
Polypropylene, acid resistant, DWV pipe				
Schedule 40				
1-1/2" pipe	L.F.	5.17	4.18	9.35
2" pipe	"	6.93	4.88	11.81
3" pipe	"	12.75	5.86	18.61
4" pipe	"	18.00	7.33	25.33
6" pipe	"	32.25	14.75	47.00
Fittings				
1-1/2"				
1/4 bend	EA.	21.75	14.75	36.50
1/8 bend	"	21.25	14.75	36.00
Sanitary tee	"	27.00	29.25	56.25
Cleanout with plug	"	55.00	29.25	84.25
Wye	"	50.00	29.25	79.25
2"				
1/4 bend	EA.	27.00	16.75	43.75
1/8 bend	"	27.00	16.75	43.75
Sanitary tee	"	32.50	34.50	67.00
Cleanout with plug	"	65.00	34.50	99.50
Wye	"	68.00	34.50	103
3"				
1/4 bend	EA.	45.75	19.50	65.25
1/8 bend	"	47.75	19.50	67.25

Plumbing	UNIT	MAT.	INST.	TOTAL
15410.30	**Pvc/cpvc Pipe** *(Cont.)*			
Sanitary tee	EA.	65.00	39.00	104
Cleanout with plug	"	72.00	39.00	111
Wye	"	69.00	39.00	108
4"				
1/4 bend	EA.	73.00	29.25	102
1/8 bend	"	54.00	29.25	83.25
Sanitary tee	"	99.00	59.00	158
Cleanout with plug	"	100	59.00	159
Wye	"	100	59.00	159
6"				
1/4 bend	EA.	180	48.75	229
1/8 bend	"	150	48.75	199
Sanitary tee	"	180	98.00	278
Cleanout with plug	"	190	98.00	288
Wye	"	250	98.00	348
Polyethylene pipe and fittings				
SDR-21				
3" pipe	L.F.	3.41	7.33	10.74
4" pipe	"	5.39	9.77	15.16
6" pipe	"	9.35	14.75	24.10
8" pipe	"	13.50	16.75	30.25
10" pipe	"	15.25	19.50	34.75
12" pipe	"	23.75	23.50	47.25
14" pipe	"	30.75	29.25	60.00
16" pipe	"	37.25	36.75	74.00
18" pipe	"	41.00	45.00	86.00
20" pipe	"	51.00	59.00	110
22" pipe	"	59.00	65.00	124
24" pipe	"	73.00	73.00	146
Fittings, 3"				
90 deg elbow	EA.	92.00	29.25	121
45 deg elbow	"	57.00	29.25	86.25
Tee	"	52.00	48.75	101
4"				
90 deg elbow	EA.	130	36.75	167
45 deg elbow	"	79.00	36.75	116
Tee	"	110	59.00	169
8"				
90 deg elbow	EA.	360	73.00	433
45 deg elbow	"	200	73.00	273
Tee	"	340	120	460
10"				
90 deg elbow	EA.	500	98.00	598
45 deg elbow	"	270	98.00	368
Tee	"	450	150	600
12"				
90 deg elbow	EA.	810	120	930
45 deg elbow	"	500	120	620
Tee	"	630	200	830
14"				
90 deg elbow	EA.	1,100	150	1,250
45 deg elbow	"	630	150	780
Tee	"	810	230	1,040

Plumbing	UNIT	MAT.	INST.	TOTAL
15410.30 — **Pvc/cpvc Pipe** *(Cont.)*				
16"				
90 deg elbow	EA.	1,380	150	1,530
45 deg elbow	"	820	150	970
Tee	"	1,000	230	1,230
18"				
90 deg elbow	EA.	2,110	200	2,310
45 deg elbow	"	1,360	200	1,560
Tee	"	1,620	290	1,910
20"				
90 deg elbow	EA.	1,700	200	1,900
45 deg elbow	"	1,010	200	1,210
15410.33 — **Abs Dwv Pipe**				
Schedule 40 ABS				
1-1/2" pipe	L.F.	0.82	2.93	3.75
2" pipe	"	1.10	3.25	4.35
3" pipe	"	2.25	4.18	6.43
4" pipe	"	3.19	5.86	9.05
6" pipe	"	6.54	7.33	13.87
Fittings				
1/8 bend				
1-1/2"	EA.	1.81	11.75	13.56
2"	"	2.64	14.75	17.39
3"	"	6.43	19.50	25.93
4"	"	11.50	23.50	35.00
6"	"	47.00	29.25	76.25
Tee, sanitary				
1-1/2"	EA.	2.64	19.50	22.14
2"	"	4.07	23.50	27.57
3"	"	11.00	29.25	40.25
4"	"	20.25	36.75	57.00
6"	"	87.00	48.75	136
Tee, sanitary reducing				
2 x 1-1/2 x 1-1/2	EA.	3.74	23.50	27.24
2 x 1-1/2 x 2	"	3.85	24.50	28.35
2 x 2 x 1-1/2	"	3.57	26.75	30.32
3 x 3 x 1-1/2	"	6.38	29.25	35.63
3 x 3 x 2	"	8.08	32.50	40.58
4 x 4 x 1-1/2	"	20.25	36.75	57.00
4 x 4 x 2	"	18.75	42.00	60.75
4 x 4 x 3	"	16.50	45.00	61.50
6 x 6 x 4	"	85.00	48.75	134
Wye				
1-1/2"	EA.	3.85	16.75	20.60
2"	"	5.39	23.50	28.89
3"	"	12.25	29.25	41.50
4"	"	26.50	36.75	63.25
6"	"	81.00	48.75	130
Reducer				
2 x 1-1/2	EA.	2.58	14.75	17.33
3 x 1-1/2	"	5.94	19.50	25.44
3 x 2	"	5.39	19.50	24.89
4 x 2	"	11.50	23.50	35.00

Plumbing	UNIT	MAT.	INST.	TOTAL

15410.33 — Abs Dwv Pipe (Cont.)

	UNIT	MAT.	INST.	TOTAL
4 x 3	EA.	11.25	23.50	34.75
6 x 4	"	17.00	29.25	46.25
P-trap				
1-1/2"	EA.	5.99	19.50	25.49
2"	"	8.08	21.75	29.83
3"	"	31.00	25.50	56.50
4"	"	63.00	29.25	92.25
6"	"	100	36.75	137
Double sanitary, tee				
1-1/2"	EA.	5.83	23.50	29.33
2"	"	8.47	29.25	37.72
3"	"	23.25	36.75	60.00
4"	"	37.00	48.75	85.75
Long sweep, 1/4 bend				
1-1/2"	EA.	3.02	11.75	14.77
2"	"	3.85	14.75	18.60
3"	"	9.13	19.50	28.63
4"	"	17.00	29.25	46.25
Wye, standard				
1-1/2"	EA.	3.85	19.50	23.35
2"	"	5.39	23.50	28.89
3"	"	12.25	29.25	41.50
4"	"	26.50	36.75	63.25

15410.35 — Plastic Pipe

	UNIT	MAT.	INST.	TOTAL
Fiberglass reinforced pipe				
2" pipe	L.F.	2.80	4.51	7.31
3" pipe	"	3.85	4.88	8.73
4" pipe	"	5.11	5.33	10.44
6" pipe	"	8.25	5.86	14.11
8" pipe	"	13.25	9.77	23.02
10" pipe	"	18.50	11.75	30.25
12" pipe	"	24.25	14.75	39.00
Fittings				
90 deg elbow, flanged				
2"	EA.	140	59.00	199
3"	"	180	65.00	245
4"	"	220	73.00	293
6"	"	410	98.00	508
8"	"	740	120	860
10"	"	980	150	1,130
12"	"	1,310	200	1,510
45 deg elbow, flanged				
2"	EA.	140	48.75	189
3"	"	180	59.00	239
4"	"	220	73.00	293
6"	"	410	98.00	508
8"	"	620	120	740
10"	"	830	150	980
12"	"	1,040	200	1,240
Tee, flanged				
2"	EA.	180	73.00	253
3"	"	260	84.00	344

Plumbing	UNIT	MAT.	INST.	TOTAL
15410.35 — Plastic Pipe *(Cont.)*				
4"	EA.	290	98.00	388
6"	"	500	120	620
8"	"	870	150	1,020
10"	"	1,420	200	1,620
12"	"	1,960	290	2,250
Wye, flanged				
2"	EA.	360	73.00	433
3"	"	500	84.00	584
4"	"	650	98.00	748
6"	"	830	120	950
8"	"	1,400	150	1,550
10"	"	2,390	200	2,590
12"	"	3,060	290	3,350
15410.70 — Stainless Steel Pipe				
Stainless steel, schedule 40, threaded				
1/2" pipe	L.F.	9.35	8.37	17.72
1" pipe	"	15.25	9.02	24.27
1-1/2" pipe	"	20.75	9.77	30.52
2" pipe	"	31.00	10.75	41.75
2-1/2" pipe	"	43.50	11.75	55.25
3" pipe	"	61.00	13.00	74.00
4" pipe	"	79.00	14.75	93.75
Fittings, 1/2"				
90 deg ell	EA.	14.25	73.00	87.25
45 deg ell	"	16.75	73.00	89.75
Tee	"	19.00	98.00	117
3/4"				
90 deg ell	EA.	19.00	73.00	92.00
45 deg ell	"	22.00	73.00	95.00
Tee	"	25.75	98.00	124
1"				
90 deg ell	EA.	23.75	73.00	96.75
45 deg ell	"	23.75	73.00	96.75
Tee	"	28.25	98.00	126
1-1/4"				
90 deg ell	EA.	33.25	73.00	106
45 deg ell	"	33.25	73.00	106
Tee	"	47.50	98.00	146
1-1/2"				
90 deg ell	EA.	41.25	98.00	139
45 deg ell	"	43.00	98.00	141
Tee	"	60.00	120	180
2"				
90 deg ell	EA.	62.00	120	182
45 deg ell	"	62.00	120	182
Tee	"	73.00	200	273
Type 304, sch 10 pipe				
1" pipe	L.F.	9.02	7.33	16.35
1-1/4" pipe	"	11.50	9.77	21.27
1-1/2" pipe	"	12.50	10.75	23.25
2" pipe	"	13.25	13.00	26.25
2-1/2" pipe	"	17.75	14.75	32.50

Plumbing	UNIT	MAT.	INST.	TOTAL
15410.70 **Stainless Steel Pipe** *(Cont.)*				
3" pipe	L.F.	21.25	16.75	38.00
4" pipe	"	28.00	19.50	47.50
6" pipe	"	44.00	23.50	67.50
Fittings, 1"				
90 deg elbow	EA.	11.00	120	131
45 deg elbow	"	15.50	120	136
Tee	"	34.00	200	234
1-1/4"				
90 deg elbow	EA.	18.00	120	138
45 deg elbow	"	21.50	120	142
Tee	"	65.00	200	265
1-1/2"				
90 deg elbow	EA.	20.00	98.00	118
45 deg elbow	"	22.25	98.00	120
Tee	"	46.50	200	247
2"				
90 deg elbow	EA.	19.00	120	139
45 deg elbow	"	17.25	120	137
Tee	"	46.50	290	337
2-1/2"				
90 deg elbow	EA.	31.75	120	152
45 deg elbow	"	31.75	120	152
Tee	"	100	290	390
3"				
90 deg elbow	EA.	32.00	150	182
45 deg elbow	"	26.50	150	177
Tee	"	80.00	390	470
4"				
90 deg elbow	EA.	68.00	150	218
45 deg elbow	"	51.00	150	201
6"				
90 deg elbow	EA.	140	200	340
45 deg elbow	"	95.00	200	295
Type 304 tubing				
.035 wall				
1/4"	L.F.	3.46	3.25	6.71
3/8"	"	4.29	3.66	7.95
1/2"	"	4.84	4.18	9.02
5/8"	"	7.04	4.88	11.92
3/4"	"	8.14	5.86	14.00
7/8"	"	8.41	6.51	14.92
1"	"	9.02	7.33	16.35
.049 wall				
1/4"	L.F.	4.56	3.44	8.00
3/8"	"	4.84	3.90	8.74
1/2"	"	5.22	4.51	9.73
5/8"	"	7.04	5.33	12.37
3/4"	"	8.25	6.51	14.76
7/8"	"	10.25	7.33	17.58
1"	"	11.00	8.37	19.37
.065 wall				
1/4"	L.F.	6.60	3.90	10.50
3/8"	"	8.14	4.88	13.02

Plumbing	UNIT	MAT.	INST.	TOTAL
15410.70 **Stainless Steel Pipe** (Cont.)				
1/2"	L.F.	8.80	5.33	14.13
5/8"	"	9.51	6.51	16.02
3/4"	"	11.50	8.37	19.87
7/8"	"	13.00	9.77	22.77
1"	"	14.25	11.75	26.00
Type 316 tubing				
.035 wall				
1/4"	L.F.	3.30	3.25	6.55
3/8"	"	4.29	3.66	7.95
1/2"	"	4.40	4.18	8.58
5/8"	"	8.41	4.88	13.29
3/4"	"	9.51	5.86	15.37
7/8"	"	12.75	6.51	19.26
1"	"	14.50	7.33	21.83
.049 wall				
1/4"	L.F.	5.77	3.90	9.67
3/8"	"	6.32	4.88	11.20
1/2"	"	7.97	5.33	13.30
5/8"	"	10.00	6.51	16.51
3/4"	"	10.50	8.37	18.87
7/8"	"	13.25	9.77	23.02
1"	"	14.75	11.75	26.50
.065 wall				
1/4"	L.F.	8.41	3.90	12.31
3/8"	"	11.25	4.88	16.13
1/2"	"	11.25	5.33	16.58
5/8"	"	11.50	6.51	18.01
3/4"	"	13.25	8.37	21.62
7/8"	"	14.25	9.77	24.02
1"	"	15.75	11.75	27.50
Fittings, 1/4"				
90 deg elbow	EA.	13.25	11.75	25.00
Union tee	"	30.00	19.50	49.50
Union	"	7.97	19.50	27.47
Male connector	"	5.22	14.75	19.97
3/8"				
90 deg elbow	EA.	16.00	14.75	30.75
Union tee	"	33.75	22.50	56.25
Union	"	11.00	22.50	33.50
Male connector	"	7.97	14.75	22.72
1/2"				
90 deg elbow	EA.	22.50	15.50	38.00
Union tee	"	45.00	24.50	69.50
Union	"	17.00	24.50	41.50
Male connector	"	11.75	14.75	26.50
5/8"				
90 deg elbow	EA.	24.75	19.50	44.25
Union tee	"	47.75	29.25	77.00
Union	"	26.25	29.25	55.50
Male connector	"	14.75	19.50	34.25
3/4"				
90 deg elbow	EA.	35.00	19.50	54.50
Union tee	"	56.00	29.25	85.25

Plumbing	UNIT	MAT.	INST.	TOTAL
15410.70 — **Stainless Steel Pipe** *(Cont.)*				
Union	EA.	37.50	29.25	66.75
Male connector	"	20.25	19.50	39.75
7/8"				
90 deg elbow	EA.	49.25	21.00	70.25
Union tee	"	76.00	32.50	109
Union	"	48.75	32.50	81.25
Male connector	"	31.00	21.00	52.00
1"				
90 deg elbow	EA.	60.00	26.75	86.75
Union tee	"	93.00	36.75	130
Union	"	60.00	36.75	96.75
Male connector	"	40.25	29.25	69.50
Type 316 valves				
Gate valves				
1/4"	EA.	290	19.50	310
3/8"	"	290	23.50	314
1/2"	"	320	25.50	346
3/4"	"	390	29.25	419
1"	"	450	39.00	489
Globe valves				
1/4"	EA.	190	19.50	210
3/8"	"	300	23.50	324
1/2"	"	370	25.50	396
3/4"	"	420	29.25	449
1"	"	460	39.00	499
Check valves				
1/4"	EA.	130	19.50	150
3/8"	"	130	23.50	154
1/2"	"	130	25.50	156
3/4"	"	150	29.25	179
1"	"	160	39.00	199
Test and balance	"	49.50	48.75	98.25
Pipe identification	"	0.27	11.75	12.02
Disinfect	"	49.50	48.75	98.25
15410.80 — **Steel Pipe**				
Black steel, extra heavy pipe, threaded				
1/2" pipe	L.F.	2.25	2.34	4.59
3/4" pipe	"	2.69	2.34	5.03
1" pipe	"	3.74	2.93	6.67
1-1/2" pipe	"	5.83	3.25	9.08
2-1/2" pipe	"	12.00	7.33	19.33
3" pipe	"	15.50	9.77	25.27
4" pipe	"	22.75	11.75	34.50
5" pipe	"	30.50	14.75	45.25
6" pipe	"	38.00	14.75	52.75
8" pipe	"	56.00	19.50	75.50
10" pipe	"	88.00	23.50	112
12" pipe	"	120	29.25	149
Fittings, malleable iron, threaded, 1/2" pipe				
90 deg ell	EA.	2.31	19.50	21.81
45 deg ell	"	3.63	19.50	23.13
Tee	"	3.02	29.25	32.27

Plumbing		UNIT	MAT.	INST.	TOTAL
15410.80	**Steel Pipe** *(Cont.)*				
Reducing tee		EA.	6.65	29.25	35.90
Cap		"	2.31	11.75	14.06
Coupling		"	3.02	23.50	26.52
Union		"	13.00	19.50	32.50
Nipple, 4" long		"	2.42	19.50	21.92
3/4" pipe					
90 deg ell		EA.	3.19	19.50	22.69
45 deg ell		"	5.06	29.25	34.31
Tee		"	4.23	29.25	33.48
Reducing tee		"	7.42	19.50	26.92
Cap		"	3.08	11.75	14.83
Coupling		"	3.57	19.50	23.07
Union		"	14.75	19.50	34.25
Nipple, 4" long		"	2.80	19.50	22.30
1" pipe					
90 deg ell		EA.	4.95	23.50	28.45
45 deg ell		"	5.66	23.50	29.16
Tee		"	7.42	32.50	39.92
Reducing tee		"	9.51	32.50	42.01
Cap		"	4.18	11.75	15.93
Coupling		"	5.33	23.50	28.83
Union		"	17.50	23.50	41.00
Nipple, 4" long		"	3.96	23.50	27.46
1-1/2" pipe					
90 deg ell		EA.	9.46	29.25	38.71
45 deg ell		"	12.25	29.25	41.50
Tee		"	14.50	42.00	56.50
Reducing tee		"	21.00	42.00	63.00
Cap		"	6.10	14.75	20.85
Coupling		"	8.41	29.25	37.66
Union		"	22.75	29.25	52.00
Nipple, 4" long		"	6.21	29.25	35.46
2-1/2" pipe					
90 deg ell		EA.	38.50	73.00	112
45 deg ell		"	54.00	73.00	127
Tee		"	54.00	98.00	152
Reducing tee		"	67.00	98.00	165
Cap		"	20.00	36.75	56.75
Coupling		"	38.25	98.00	136
Union		"	86.00	98.00	184
Nipple, 4" long		"	22.50	98.00	121
3" pipe					
90 deg ell		EA.	56.00	98.00	154
45 deg ell		"	70.00	98.00	168
Tee		"	79.00	150	229
Reducing tee		"	110	150	260
Cap		"	30.50	48.75	79.25
Coupling		"	52.00	98.00	150
Union		"	140	98.00	238
Nipple, 4" long		"	29.25	98.00	127
4" pipe					
90 deg ell		EA.	110	120	230
45 deg ell		"	140	120	260

Plumbing	UNIT	MAT.	INST.	TOTAL

15410.80   Steel Pipe *(Cont.)*

	UNIT	MAT.	INST.	TOTAL
Tee	EA.	190	200	390
Reducing tee	"	160	200	360
Cap	"	45.75	200	246
Coupling	"	100	59.00	159
Union	"	140	200	340
Nipple, 4" long	"	38.75	200	239
6" pipe				
90 deg ell	EA.	310	120	430
45 deg ell	"	400	120	520
Tee	"	500	200	700
Reducing tee	"	430	200	630
Cap	"	44.00	59.00	103
8" pipe				
90 deg ell	EA.	680	230	910
45 deg ell	"	770	230	1,000
Tee	"	470	370	840
Reducing tee	"	440	310	750
Cap	"	53.00	120	173
10" pipe				
90 deg ell	EA.	750	290	1,040
45 deg ell	"	850	290	1,140
Tee	"	540	370	910
Reducing tee	"	520	150	670
Cap	"	55.00	150	205
12" pipe				
90 deg ell	EA.	830	370	1,200
45 deg ell	"	930	370	1,300
Tee	"	630	490	1,120
Reducing tee	"	580	490	1,070
Cap	"	61.00	200	261

15410.82   Galvanized Steel Pipe

	UNIT	MAT.	INST.	TOTAL
Galvanized pipe				
1/2" pipe	L.F.	2.53	5.86	8.39
3/4" pipe	"	3.30	7.33	10.63
1" pipe	"	5.06	8.37	13.43
1-1/4" pipe	"	5.88	9.77	15.65
1-1/2" pipe	"	6.27	11.75	18.02
2" pipe	"	9.02	14.75	23.77
2-1/2" pipe	"	13.00	19.50	32.50
3" pipe	"	18.00	21.00	39.00
4" pipe	"	24.50	24.50	49.00
6" pipe	"	49.00	48.75	97.75
90 degree ell, 150 lb malleable iron, galvanized				
1/2"	EA.	2.69	11.75	14.44
3/4"	"	3.57	14.75	18.32
1"	"	5.83	15.50	21.33
1-1/4"	"	8.58	17.25	25.83
1-1/2"	"	11.25	19.50	30.75
2"	"	17.50	23.50	41.00
2-1/2"	"	45.00	36.75	81.75
3"	"	65.00	45.00	110
4"	"	130	48.75	179

Plumbing	UNIT	MAT.	INST.	TOTAL
15410.82 **Galvanized Steel Pipe** *(Cont.)*				
5"	EA.	320	59.00	379
6"	"	360	59.00	419
45 degree ell, 150 lb m.i., galv.				
1/2"	EA.	4.29	11.75	16.04
3/4"	"	5.83	14.75	20.58
1"	"	6.54	15.50	22.04
1-1/4"	"	10.25	17.25	27.50
1-1/2"	"	13.00	19.50	32.50
2"	"	20.25	23.50	43.75
2-1/2"	"	56.00	36.75	92.75
3"	"	74.00	45.00	119
4"	"	130	59.00	189
5"	"	320	59.00	379
6"	"	450	73.00	523
Tees, straight, 150 lb m.i., galv.				
1/2"	EA.	3.57	14.75	18.32
3/4"	"	5.94	16.75	22.69
1"	"	8.74	19.50	28.24
1-1/4"	"	11.50	23.50	35.00
1-1/2"	"	16.00	29.25	45.25
2"	"	25.00	36.75	61.75
2-1/2"	"	61.00	48.75	110
3"	"	83.00	59.00	142
4"	"	190	73.00	263
5"	"	290	84.00	374
6"	"	520	98.00	618
Couplings, straight, 150 lb m.i., galv.				
1/2"	EA.	3.30	11.75	15.05
3/4"	"	3.96	13.00	16.96
1"	"	6.76	14.75	21.51
1-1/4"	"	7.70	16.75	24.45
1-1/2"	"	9.95	19.50	29.45
2"	"	13.75	23.50	37.25
2-1/2"	"	37.00	36.75	73.75
3"	"	52.00	48.75	101
4"	"	100	53.00	153
5"	"	300	59.00	359
6"	"	300	59.00	359
Caps, 150 lb m.i., galv.				
1/2"	EA.	2.75	5.86	8.61
3/4"	"	3.63	6.17	9.80
1"	"	4.95	6.51	11.46
1-1/4"	"	5.33	6.89	12.22
1-1/2"	"	7.20	7.33	14.53
2"	"	9.73	8.37	18.10
2-1/2"	"	23.50	10.75	34.25
3"	"	36.00	14.75	50.75
4"	"	60.00	18.25	78.25
5"	"	67.00	22.50	89.50
6"	"	71.00	29.25	100
Unions, 150 lb m.i., galv.				
1/2"	EA.	15.25	14.75	30.00
3/4"	"	17.25	16.75	34.00

Plumbing	UNIT	MAT.	INST.	TOTAL
15410.82 — **Galvanized Steel Pipe** *(Cont.)*				
1"	EA.	20.50	19.50	40.00
1-1/4"	"	23.50	23.50	47.00
1-1/2"	"	27.00	29.25	56.25
2"	"	33.50	32.50	66.00
2-1/2"	"	100	39.00	139
3"	"	170	48.75	219
Nipples, galvanized steel, 4" long				
1/2"	EA.	2.64	7.33	9.97
3/4"	"	3.52	7.81	11.33
1"	"	4.84	8.37	13.21
1-1/4"	"	5.50	9.02	14.52
1-1/2"	"	6.60	9.77	16.37
2"	"	7.70	10.75	18.45
2-1/2"	"	9.24	11.75	20.99
3"	"	12.00	14.75	26.75
4"	"	15.25	19.50	34.75
Square head plug (C.I.)				
1/2"	EA.	1.98	6.51	8.49
3/4"	"	4.40	7.33	11.73
1"	"	4.62	7.81	12.43
1-1/4"	"	4.84	8.37	13.21
1-1/2"	"	6.05	9.02	15.07
2"	"	7.70	9.77	17.47
2-1/2"	"	10.00	13.00	23.00
3"	"	14.75	14.75	29.50
4"	"	21.50	19.50	41.00
5"	"	35.00	23.50	58.50
6"	"	48.25	29.25	77.50
Screwed flanges, galv.				
1"	EA.	18.00	29.25	47.25
1-1/4"	"	19.75	32.50	52.25
1-1/2"	"	22.50	36.75	59.25
2"	"	24.00	36.75	60.75
2-1/2"	"	25.25	39.00	64.25
3"	"	33.00	53.00	86.00
4"	"	44.00	73.00	117
5"	"	55.00	98.00	153
6"	"	70.00	98.00	168
15430.23 — **Cleanouts**				
Cleanout, wall				
2"	EA.	160	39.00	199
3"	"	230	39.00	269
4"	"	240	48.75	289
6"	"	390	59.00	449
8"	"	540	73.00	613
Floor				
2"	EA.	150	48.75	199
3"	"	190	48.75	239
4"	"	200	59.00	259
6"	"	280	73.00	353
8"	"	530	84.00	614

Plumbing	UNIT	MAT.	INST.	TOTAL
15430.24 **Grease Traps**				
Grease traps, cast iron, 3" pipe				
35 gpm, 70 lb capacity	EA.	3,200	590	3,790
50 gpm, 100 lb capacity	"	4,080	730	4,810
15430.25 **Hose Bibbs**				
Hose bibb				
1/2"	EA.	9.07	19.50	28.57
3/4"	"	9.62	19.50	29.12
15430.60 **Valves**				
Gate valve, 125 lb, bronze, soldered				
1/2"	EA.	25.75	14.75	40.50
3/4"	"	51.00	14.75	65.75
1"	"	65.00	19.50	84.50
1-1/2"	"	110	23.50	134
2"	"	150	29.25	179
2-1/2"	"	330	36.75	367
Threaded				
1/4", 125 lb	EA.	25.75	23.50	49.25
1/2"				
125 lb	EA.	28.50	23.50	52.00
150 lb	"	38.00	23.50	61.50
300 lb	"	72.00	23.50	95.50
3/4"				
125 lb	EA.	33.25	23.50	56.75
150 lb	"	45.25	23.50	68.75
300 lb	"	86.00	23.50	110
1"				
125 lb	EA.	42.75	23.50	66.25
150 lb	"	67.00	23.50	90.50
300 lb	"	130	29.25	159
1-1/2"				
125 lb	EA.	95.00	29.25	124
150 lb	"	110	29.25	139
300 lb	"	230	32.50	263
2"				
125 lb	EA.	94.00	42.00	136
150 lb	"	150	42.00	192
300 lb	"	330	48.75	379
Cast iron, flanged				
2", 150 lb	EA.	350	48.75	399
2-1/2"				
125 lb	EA.	220	48.75	269
150 lb	"	320	48.75	369
250 lb	"	590	48.75	639
3"				
125 lb	EA.	240	59.00	299
150 lb	"	340	59.00	399
250 lb	"	650	59.00	709
4"				
125 lb	EA.	500	84.00	584
150 lb	"	660	84.00	744
250 lb	"	1,160	84.00	1,244

Plumbing	UNIT	MAT.	INST.	TOTAL
15430.60 **Valves** *(Cont.)*				
6"				
125 lb	EA.	750	120	870
250 lb	"	1,930	120	2,050
8"				
125 lb	EA.	1,500	150	1,650
250 lb	"	3,640	150	3,790
OS&Y, flanged				
2"				
125 lb	EA.	300	48.75	349
250 lb	"	810	48.75	859
2-1/2"				
125 lb	EA.	320	48.75	369
250 lb	"	1,010	59.00	1,069
3"				
125 lb	EA.	320	59.00	379
250 lb	"	1,040	59.00	1,099
4"				
125 lb	EA.	460	98.00	558
250 lb	"	1,590	98.00	1,688
6"				
125 lb	EA.	790	120	910
250 lb	"	2,420	120	2,540
Check valve, bronze, soldered, 125 lb				
1/2"	EA.	36.75	14.75	51.50
3/4"	"	45.75	14.75	60.50
1"	"	57.00	19.50	76.50
1-1/4"	"	74.00	23.50	97.50
1-1/2"	"	87.00	23.50	111
2"	"	120	29.25	149
Threaded				
1/2"				
125 lb	EA.	25.75	19.50	45.25
150 lb	"	46.25	19.50	65.75
200 lb	"	53.00	19.50	72.50
3/4"				
125 lb	EA.	32.75	23.50	56.25
150 lb	"	56.00	23.50	79.50
200 lb	"	73.00	23.50	96.50
1"				
125 lb	EA.	41.50	29.25	70.75
150 lb	"	74.00	29.25	103
200 lb	"	110	29.25	139
Flow check valve, cast iron, threaded				
1"	EA.	60.00	23.50	83.50
1-1/4"	"	60.00	29.25	89.25
1-1/2"				
125 lb	EA.	58.00	29.25	87.25
150 lb	"	97.00	29.25	126
200 lb	"	97.00	32.50	130
2"				
125 lb	EA.	86.00	32.50	119
150 lb	"	140	32.50	173
200 lb	"	140	36.75	177

Plumbing	UNIT	MAT.	INST.	TOTAL
15430.60 — **Valves** *(Cont.)*				
2-1/2"				
125 lb	EA.	200	48.75	249
250 lb	"	630	59.00	689
3"				
125 lb	EA.	220	59.00	279
250 lb	"	780	73.00	853
4"				
125 lb	EA.	340	84.00	424
250 lb	"	1,000	98.00	1,098
6"				
125 lb	EA.	460	120	580
250 lb	"	1,690	120	1,810
Vertical check valve, bronze, 125 lb, threaded				
1/2"	EA.	50.00	23.50	73.50
3/4"	"	68.00	26.75	94.75
1"	"	82.00	29.25	111
1-1/4"	"	100	32.50	133
1-1/2"	"	110	36.75	147
2"	"	190	42.00	232
Cast iron, flanged				
2-1/2"	EA.	280	59.00	339
3"	"	310	73.00	383
4"	"	480	98.00	578
6	"	820	120	940
8"	"	1,540	150	1,690
10"	"	2,630	200	2,830
12"	"	4,090	230	4,320
Globe valve, bronze, soldered, 125 lb				
1/2"	EA.	55.00	16.75	71.75
3/4"	"	66.00	18.25	84.25
1"	"	90.00	19.50	110
1-1/4"	"	120	21.00	141
1-1/2"	"	150	24.50	175
2"	"	240	29.25	269
Threaded				
1/2"				
125 lb	EA.	55.00	19.50	74.50
150 lb	"	66.00	19.50	85.50
300 lb	"	120	19.50	140
3/4"				
125 lb	EA.	66.00	23.50	89.50
150 lb	"	88.00	23.50	112
300 lb	"	160	23.50	184
1"				
125 lb	EA.	88.00	29.25	117
150 lb	"	140	29.25	169
300 lb	"	200	29.25	229
1-1/4"				
125 lb	EA.	120	29.25	149
150 lb	"	220	29.25	249
300 lb	"	280	29.25	309
1-1/2"				
125 lb	EA.	150	32.50	183

Plumbing	UNIT	MAT.	INST.	TOTAL
15430.60		**Valves** *(Cont.)*		
150 lb	EA.	280	32.50	313
300 lb	"	300	32.50	333
2"				
125 lb	EA.	240	39.00	279
150 lb	"	430	39.00	469
300 lb	"	480	39.00	519
Cast iron flanged				
2-1/2"				
125 lb	EA.	680	59.00	739
250 lb	"	1,180	59.00	1,239
3"				
125 lb	EA.	820	73.00	893
250 lb	"	1,210	73.00	1,283
4"				
125 lb	EA.	1,180	98.00	1,278
250 lb	"	1,350	98.00	1,448
6"				
125 lb	EA.	2,160	120	2,280
250 lb	"	3,180	120	3,300
8"				
125 lb	EA.	4,250	150	4,400
250 lb	"	5,230	150	5,380
Butterfly valve, cast iron, wafer type				
2"				
150 lb	EA.	120	42.00	162
200 lb	"	160	48.75	209
2-1/2"				
150 lb	EA.	160	48.75	209
200 lb	"	170	53.00	223
3"				
150 lb	EA.	160	59.00	219
200 lb	"	170	65.00	235
4"				
150 lb	EA.	180	84.00	264
200 lb	"	200	98.00	298
6"				
150 lb	EA.	220	120	340
200 lb	"	270	120	390
8"				
150 lb	EA.	270	130	400
200 lb	"	340	150	490
10"				
150 lb	EA.	330	150	480
200 lb	"	480	200	680
Ball valve, bronze, 250 lb, threaded				
1/2"	EA.	14.25	23.50	37.75
3/4"	"	21.25	23.50	44.75
1"	"	24.25	29.25	53.50
1-1/4"	"	39.50	32.50	72.00
1-1/2"	"	63.00	36.75	99.75
2"	"	71.00	42.00	113
Angle valve, bronze, 150 lb, threaded				
1/2"	EA.	85.00	21.00	106

Plumbing	UNIT	MAT.	INST.	TOTAL
15430.60 — **Valves** *(Cont.)*				
3/4"	EA.	110	23.50	134
1"	"	160	23.50	184
1-1/4"	"	210	29.25	239
1-1/2"	"	270	32.50	303
Balancing valve, with meter connections, circuit setter				
1/2"	EA.	82.00	23.50	106
3/4"	"	86.00	26.75	113
1"	"	110	29.25	139
1-1/4"	"	150	32.50	183
1-1/2"	"	190	39.00	229
2"	"	260	48.75	309
2-1/2"	"	520	59.00	579
3"	"	760	73.00	833
4"	"	1,070	98.00	1,168
Balancing valve, straight type				
1/2"	EA.	22.25	23.50	45.75
3/4"	"	27.00	23.50	50.50
Angle type				
1/2"	EA.	30.00	23.50	53.50
3/4"	"	41.50	23.50	65.00
Square head cock, 125 lb, bronze body				
1/2"	EA.	17.50	19.50	37.00
3/4"	"	21.00	23.50	44.50
1"	"	29.25	26.75	56.00
1-1/4"	"	39.75	29.25	69.00
Radiator temp control valve, with control and sensor				
1/2" valve	EA.	120	36.75	157
1" valve	"	130	36.75	167
Pressure relief valve, 1/2", bronze				
Low pressure	EA.	28.00	23.50	51.50
High pressure	"	32.75	23.50	56.25
Pressure and temperature relief valve				
Bronze, 3/4"	EA.	100	23.50	124
Cast iron, 3/4"				
High pressure	EA.	48.50	23.50	72.00
Temperature relief	"	65.00	23.50	88.50
Pressure & temp relief valve	"	67.00	23.50	90.50
Pressure reducing valve, bronze, threaded, 250 lb				
1/2"	EA.	290	36.75	327
3/4"	"	290	36.75	327
1"	"	320	36.75	357
1-1/4"	"	620	42.00	662
1-1/2"	"	720	48.75	769
Pressure regulating valve, bronze, class 300				
1"	EA.	620	36.75	657
1-1/2"	"	830	45.00	875
2"	"	930	59.00	989
3"	"	1,060	84.00	1,144
4"	"	1,320	120	1,440
5"	"	2,000	150	2,150
6"	"	2,040	200	2,240
Solar water temperature regulating valve				
3/4"	EA.	640	48.75	689

Plumbing	UNIT	MAT.	INST.	TOTAL
15430.60 Valves *(Cont.)*				
1"	EA.	650	59.00	709
1-1/4"	"	700	65.00	765
1-1/2"	"	780	73.00	853
2"	"	970	84.00	1,054
2-1/2"	"	1,830	150	1,980
Tempering valve, threaded				
3/4"	EA.	330	19.50	350
1"	"	400	23.50	424
1-1/4"	"	580	29.25	609
1-1/2"	"	690	29.25	719
2"	"	910	36.75	947
2-1/2"	"	940	48.75	989
3"	"	1,170	59.00	1,229
4"	"	2,570	84.00	2,654
Thermostatic mixing valve, threaded				
1/2"	EA.	120	21.00	141
3/4"	"	120	23.50	144
1"	"	440	25.50	466
1-1/2"	"	500	29.25	529
2"	"	620	36.75	657
Sweat connection				
1/2"	EA.	130	21.00	151
3/4"	"	160	23.50	184
Mixing valve, sweat connection				
1/2"	EA.	55.00	21.00	76.00
3/4"	"	62.00	23.50	85.50
Liquid level gauge, aluminum body				
3/4"	EA.	360	23.50	384
4125 psi, pvc body				
3/4"	EA.	430	23.50	454
150 psi, crs body				
3/4"	EA.	340	23.50	364
1"	"	370	23.50	394
175 psi, bronze body, 1/2"	"	690	21.00	711
15430.65 Vacuum Breakers				
Vacuum breaker, atmospheric, threaded connection				
3/4"	EA.	45.75	23.50	69.25
1"	"	67.00	23.50	90.50
Anti-siphon, brass				
3/4"	EA.	52.00	23.50	75.50
1"	"	73.00	23.50	96.50
1-1/4"	"	110	29.25	139
1-1/2"	"	130	32.50	163
2"	"	210	36.75	247
15430.68 Strainers				
Strainer, Y pattern, 125 psi, cast iron body, threaded				
3/4"	EA.	10.75	21.00	31.75
1"	"	13.00	23.50	36.50
1-1/4"	"	17.25	29.25	46.50
1-1/2"	"	21.50	29.25	50.75
2"	"	26.50	36.75	63.25

Plumbing	UNIT	MAT.	INST.	TOTAL
15430.68 **Strainers** *(Cont.)*				
250 psi, brass body, threaded				
3/4"	EA.	29.75	23.50	53.25
1"	"	41.50	23.50	65.00
1-1/4"	"	52.00	29.25	81.25
1-1/2"	"	73.00	29.25	102
2"	"	130	36.75	167
Cast iron body, threaded				
3/4"	EA.	17.50	23.50	41.00
1"	"	22.00	23.50	45.50
1-1/4"	"	29.50	29.25	58.75
1-1/2"	"	39.00	29.25	68.25
2"	"	49.50	36.75	86.25
15430.70 **Drains, Roof & Floor**				
Floor drain, cast iron, with cast iron top				
2"	EA.	120	48.75	169
3"	"	140	48.75	189
4"	"	290	48.75	339
6"	"	370	59.00	429
Roof drain, cast iron				
2"	EA.	210	48.75	259
3"	"	260	48.75	309
4"	"	270	48.75	319
5"	"	410	59.00	469
6"	"	410	59.00	469
15430.80 **Traps**				
Bucket trap, threaded				
3/4"	EA.	190	36.75	227
1"	"	540	39.00	579
1-1/4"	"	640	45.00	685
1-1/2"	"	960	53.00	1,013
Inverted bucket steam trap, threaded				
3/4"	EA.	230	36.75	267
1"	"	450	36.75	487
1-1/4"	"	680	32.50	713
1-1/2"	"	730	48.75	779
Float trap, 15 psi				
3/4"	EA.	160	36.75	197
1"	"	250	39.00	289
1-1/4"	"	330	42.00	372
1-1/2"	"	420	48.75	469
2"	"	730	59.00	789
Float and thermostatic trap, 15 psi				
3/4"	EA.	180	36.75	217
1"	"	200	39.00	239
1-1/4"	"	310	42.00	352
1-1/2"	"	400	48.75	449
2"	"	730	59.00	789
Steam trap, cast iron body, threaded, 125 psi				
3/4"	EA.	210	36.75	247
1"	"	240	39.00	279
1-1/4"	"	350	42.00	392

Plumbing	UNIT	MAT.	INST.	TOTAL
15430.80		**Traps** *(Cont.)*		
1-1/2"	EA.	570	48.75	619
Thermostatic trap, low pressure, angle type, 25 psi				
1/2"	EA.	64.00	36.75	101
3/4"	"	110	36.75	147
1"	"	140	39.00	179
50 psi				
1/2"	EA.	100	36.75	137
3/4"	"	130	36.75	167
1"	"	150	39.00	189
Cast iron body, threaded, 125 psi				
3/4"	EA.	140	36.75	177
1"	"	180	42.00	222
1-1/4"	"	230	45.00	275
1-1/2"	"	340	48.75	389

Plumbing Fixtures	UNIT	MAT.	INST.	TOTAL
15440.10		**Baths**		
Bath tub, 5' long				
Minimum	EA.	530	200	730
Average	"	1,160	290	1,450
Maximum	"	2,640	590	3,230
6' long				
Minimum	EA.	590	200	790
Average	"	1,210	290	1,500
Maximum	"	3,420	590	4,010
Square tub, whirlpool, 4'x4'				
Minimum	EA.	1,810	290	2,100
Average	"	2,570	590	3,160
Maximum	"	7,850	730	8,580
5'x5'				
Minimum	EA.	1,810	290	2,100
Average	"	2,570	590	3,160
Maximum	"	8,000	730	8,730
6'x6'				
Minimum	EA.	2,210	290	2,500
Average	"	3,230	590	3,820
Maximum	"	9,270	730	10,000
For trim and rough-in				
Minimum	EA.	190	200	390
Average	"	280	290	570
Maximum	"	780	590	1,370

Plumbing Fixtures	UNIT	MAT.	INST.	TOTAL
15440.12 **Disposals & Accessories**				
Continuous feed				
Minimum	EA.	72.00	120	192
Average	"	200	150	350
Maximum	"	390	200	590
Batch feed, 1/2 hp				
Minimum	EA.	280	120	400
Average	"	550	150	700
Maximum	"	950	200	1,150
Hot water dispenser				
Minimum	EA.	200	120	320
Average	"	320	150	470
Maximum	"	510	200	710
Epoxy finish faucet	"	290	120	410
Lock stop assembly	"	61.00	73.00	134
Mounting gasket	"	7.04	48.75	55.79
Tailpipe gasket	"	1.03	48.75	49.78
Stopper assembly	"	24.00	59.00	83.00
Switch assembly, on/off	"	27.50	98.00	126
Tailpipe gasket washer	"	1.10	29.25	30.35
Stop gasket	"	2.42	32.50	34.92
Tailpipe flange	"	0.27	29.25	29.52
Tailpipe	"	3.13	36.75	39.88
15440.15 **Faucets**				
Kitchen				
Minimum	EA.	130	98.00	228
Average	"	230	120	350
Maximum	"	290	150	440
Bath				
Minimum	EA.	130	98.00	228
Average	"	240	120	360
Maximum	"	370	150	520
Lavatory, domestic				
Minimum	EA.	140	98.00	238
Average	"	280	120	400
Maximum	"	460	150	610
Hospital, patient rooms				
Minimum	EA.	200	150	350
Average	"	390	200	590
Maximum	"	670	290	960
Operating room				
Minimum	EA.	250	150	400
Average	"	550	200	750
Maximum	"	800	290	1,090
Washroom				
Minimum	EA.	160	98.00	258
Average	"	280	120	400
Maximum	"	510	150	660
Handicapped				
Minimum	EA.	200	120	320
Average	"	360	150	510
Maximum	"	560	200	760
Shower				

Plumbing Fixtures	UNIT	MAT.	INST.	TOTAL
15440.15 **Faucets** *(Cont.)*				
Minimum	EA.	160	98.00	258
Average	"	320	120	440
Maximum	"	510	150	660
For trim and rough-in				
Minimum	EA.	79.00	120	199
Average	"	120	150	270
Maximum	"	200	290	490
15440.18 **Hydrants**				
Wall hydrant				
8" thick	EA.	340	98.00	438
12" thick	"	400	120	520
18" thick	"	460	130	590
24" thick	"	510	150	660
Ground hydrant				
2' deep	EA.	670	73.00	743
4' deep	"	710	84.00	794
6' deep	"	770	98.00	868
8' deep	"	970	150	1,120
15440.20 **Lavatories**				
Lavatory, counter top, porcelain enamel on cast iron				
Minimum	EA.	190	120	310
Average	"	290	150	440
Maximum	"	520	200	720
Wall hung, china				
Minimum	EA.	260	120	380
Average	"	310	150	460
Maximum	"	770	200	970
Handicapped				
Minimum	EA.	430	150	580
Average	"	500	200	700
Maximum	"	830	290	1,120
For trim and rough-in				
Minimum	EA.	220	150	370
Average	"	370	200	570
Maximum	"	460	290	750
15440.30 **Showers**				
Shower, fiberglass, 36"x34"x84"				
Minimum	EA.	570	420	990
Average	"	800	590	1,390
Maximum	"	1,160	590	1,750
Steel, 1 piece, 36"x36"				
Minimum	EA.	530	420	950
Average	"	800	590	1,390
Maximum	"	950	590	1,540
Receptor, molded stone, 36"x36"				
Minimum	EA.	220	200	420

Plumbing Fixtures	UNIT	MAT.	INST.	TOTAL
15440.30 *Showers (Cont.)*				
Average	EA.	370	290	660
Maximum	"	570	490	1,060
For trim and rough-in				
Minimum	EA.	220	270	490
Average	"	370	330	700
Maximum	"	460	590	1,050
15440.40 **Sinks**				
Service sink, 24"x29"				
Minimum	EA.	640	150	790
Average	"	790	200	990
Maximum	"	1,170	290	1,460
Kitchen sink, single, stainless steel, single bowl				
Minimum	EA.	280	120	400
Average	"	320	150	470
Maximum	"	580	200	780
Double bowl				
Minimum	EA.	320	150	470
Average	"	360	200	560
Maximum	"	620	290	910
Porcelain enamel, cast iron, single bowl				
Minimum	EA.	200	120	320
Average	"	260	150	410
Maximum	"	410	200	610
Double bowl				
Minimum	EA.	280	150	430
Average	"	390	200	590
Maximum	"	550	290	840
Mop sink, 24"x36"x10"				
Minimum	EA.	480	120	600
Average	"	580	150	730
Maximum	"	780	200	980
Washing machine box				
Minimum	EA.	180	150	330
Average	"	250	200	450
Maximum	"	310	290	600
For trim and rough-in				
Minimum	EA.	290	200	490
Average	"	440	290	730
Maximum	"	560	390	950
15440.50 **Urinals**				
Urinal, flush valve, floor mounted				
Minimum	EA.	500	150	650
Average	"	580	200	780
Maximum	"	680	290	970
Wall mounted				
Minimum	EA.	410	150	560

Plumbing Fixtures	UNIT	MAT.	INST.	TOTAL
15440.50 Urinals *(Cont.)*				
Average	EA.	560	200	760
Maximum	"	730	290	1,020
For trim and rough-in				
Minimum	EA.	180	150	330
Average	"	260	290	550
Maximum	"	360	390	750
15440.60 Water Closets				
Water closet flush tank, floor mounted				
Minimum	EA.	330	150	480
Average	"	650	200	850
Maximum	"	1,020	290	1,310
Handicapped				
Minimum	EA.	370	200	570
Average	"	670	290	960
Maximum	"	1,280	590	1,870
Bowl, with flush valve, floor mounted				
Minimum	EA.	460	150	610
Average	"	510	200	710
Maximum	"	990	290	1,280
Wall mounted				
Minimum	EA.	460	150	610
Average	"	540	200	740
Maximum	"	1,030	290	1,320
For trim and rough-in				
Minimum	EA.	210	150	360
Average	"	250	200	450
Maximum	"	330	290	620
15440.70 Water Heaters				
Water heater, electric				
6 gal	EA.	330	98.00	428
10 gal	"	350	98.00	448
15 gal	"	360	98.00	458
20 gal	"	410	120	530
30 gal	"	440	120	560
40 gal	"	560	120	680
52 gal	"	630	150	780
66 gal	"	760	150	910
80 gal	"	830	150	980
100 gal	"	1,030	200	1,230
120 gal	"	1,320	200	1,520
Oil fired				
20 gal	EA.	1,300	290	1,590
50 gal	"	2,020	420	2,440
15440.90 Miscellaneous Fixtures				
Electric water cooler				

Plumbing Fixtures

Plumbing Fixtures	UNIT	MAT.	INST.	TOTAL

15440.90 — Miscellaneous Fixtures (Cont.)

	UNIT	MAT.	INST.	TOTAL
Floor mounted	EA.	1,010	200	1,210
Wall mounted	"	950	200	1,150
Wash fountain				
Wall mounted	EA.	2,420	290	2,710
Circular, floor supported	"	4,240	590	4,830
Deluge shower and eye wash	"	1,010	290	1,300

15440.95 — Fixture Carriers

	UNIT	MAT.	INST.	TOTAL
Water fountain, wall carrier				
Minimum	EA.	64.00	59.00	123
Average	"	86.00	73.00	159
Maximum	"	110	98.00	208
Lavatory, wall carrier				
Minimum	EA.	140	59.00	199
Average	"	210	73.00	283
Maximum	"	260	98.00	358
Sink, industrial, wall carrier				
Minimum	EA.	190	59.00	249
Average	"	220	73.00	293
Maximum	"	280	98.00	378
Toilets, water closets, wall carrier				
Minimum	EA.	280	59.00	339
Average	"	330	73.00	403
Maximum	"	430	98.00	528
Floor support				
Minimum	EA.	140	48.75	189
Average	"	160	59.00	219
Maximum	"	180	73.00	253
Urinals, wall carrier				
Minimum	EA.	150	59.00	209
Average	"	190	73.00	263
Maximum	"	230	98.00	328
Floor support				
Minimum	EA.	120	48.75	169
Average	"	180	59.00	239
Maximum	"	200	73.00	273

15450.30 — Pumps

	UNIT	MAT.	INST.	TOTAL
In-line pump, bronze, centrifugal				
5 gpm, 20' head	EA.	490	36.75	527
20 gpm, 40' head	"	870	36.75	907
50 gpm				
50' head	EA.	1,150	73.00	1,223
100' head	"	1,320	73.00	1,393
70 gpm, 100' head	"	1,640	98.00	1,738
100 gpm, 80' head	"	1,750	98.00	1,848
250 gpm, 150' head	"	5,490	150	5,640
Cast iron, centrifugal				
50 gpm, 200' head	EA.	770	73.00	843
100 gpm				
100' head	EA.	1,860	98.00	1,958
200' head	"	2,170	98.00	2,268
200 gpm				

Plumbing Fixtures	UNIT	MAT.	INST.	TOTAL

15450.30 — Pumps *(Cont.)*

	UNIT	MAT.	INST.	TOTAL
100' head	EA.	3,240	150	3,390
200' head	"	4,370	150	4,520
Centrifugal, close coupled, c.i., single stage				
50 gpm, 100' head	EA.	1,300	73.00	1,373
100 gpm, 100' head	"	1,580	98.00	1,678
Base mounted				
50 gpm, 100' head	EA.	2,660	73.00	2,733
100 gpm, 50' head	"	3,030	98.00	3,128
200 gpm, 100' head	"	3,870	150	4,020
300 gpm, 175' head	"	5,030	150	5,180
Sump pump, bronze, 1750 rpm, 25 gpm				
20' head	EA.	5,360	730	6,090
150' head	"	7,460	980	8,440
50 gpm				
100' head	EA.	6,370	730	7,100
100 gpm				
50' head	EA.	5,540	730	6,270
Condensate pump, simplex				
1000 sf EDR, 2 gpm	EA.	1,360	490	1,850
2000 sf EDR, 3 gpm	"	1,390	490	1,880
4000 sf EDR, 6 gpm	"	1,400	530	1,930
6000 sf EDR, 9 gpm	"	1,420	530	1,950
Duplex, bronze				
8000 sf EDR, 12 gpm	EA.	1,950	530	2,480
10,000 sf EDR, 15 gpm	"	2,020	730	2,750
15,000 sf EDR, 23 gpm	"	2,430	840	3,270
20,000 sf EDR, 30 gpm	"	2,830	1,170	4,000
25,000 sf EDR, 38 gpm	"	2,930	1,170	4,100

15450.40 — Storage Tanks

	UNIT	MAT.	INST.	TOTAL
Hot water storage tank, cement lined				
10 gallon	EA.	390	200	590
70 gallon	"	1,210	290	1,500
200 gallon	"	2,310	420	2,730
900 gallon	"	5,320	730	6,050
1100 gallon	"	6,550	730	7,280
2000 gallon	"	11,850	730	12,580

15480.10 — Special Systems

	UNIT	MAT.	INST.	TOTAL
Air compressor, air cooled, two stage				
5.0 cfm, 175 psi	EA.	2,350	1,170	3,520
10 cfm, 175 psi	"	2,870	1,300	4,170
20 cfm, 175 psi	"	3,960	1,400	5,360
50 cfm, 125 psi	"	5,780	1,540	7,320
80 cfm, 125 psi	"	8,260	1,680	9,940
Single stage, 125 psi				
1.0 cfm	EA.	2,270	840	3,110
1.5 cfm	"	2,310	840	3,150
2.0 cfm	"	2,360	840	3,200
Automotive, hose reel, air and water, 50' hose	"	1,060	490	1,550
Lube equipment, 3 reel, with pumps	"	5,530	2,350	7,880
Tire changer				
Truck	EA.	14,430	840	15,270

Plumbing Fixtures

	UNIT	MAT.	INST.	TOTAL

15480.10 — Special Systems *(Cont.)*

	UNIT	MAT.	INST.	TOTAL
Passenger car	EA.	3,410	450	3,860
Air hose reel, includes, 50' hose	"	900	450	1,350
Hose reel, 5 reel, motor oil, gear oil, lube, air & water	"	8,420	2,350	10,770
Water hose reel, 50' hose	"	900	450	1,350
Pump, air operated, for motor or gear oil, fits 55 gal drum	"	1,180	59.00	1,239
For chassis lube	"	1,930	59.00	1,989
Fuel dispensing pump, lighted dial, one product				
One hose	EA.	4,200	490	4,690
Two hose	"	7,410	490	7,900
Two products, two hose	"	7,810	490	8,300

Heating & Ventilating

	UNIT	MAT.	INST.	TOTAL

15555.10 — Boilers

	UNIT	MAT.	INST.	TOTAL
Cast iron, gas fired, hot water				
115 mbh	EA.	2,160	1,860	4,020
175 mbh	"	2,530	2,020	4,550
235 mbh	"	3,310	2,230	5,540
940 mbh	"	11,200	4,450	15,650
1600 mbh	"	14,760	5,570	20,330
3000 mbh	"	23,390	7,420	30,810
6000 mbh	"	76,450	11,130	87,580
Steam				
115 mbh	EA.	2,390	1,860	4,250
175 mbh	"	2,830	2,020	4,850
235 mbh	"	3,410	2,230	5,640
940 mbh	"	11,840	4,450	16,290
1600 mbh	"	15,400	5,570	20,970
3000 mbh	"	22,930	7,420	30,350
6000 mbh	"	72,600	11,130	83,730
Electric, hot water				
115 mbh	EA.	3,880	1,110	4,990
175 mbh	"	4,290	1,110	5,400
235 mbh	"	4,900	1,110	6,010
940 mbh	"	11,910	2,230	14,140
1600 mbh	"	16,860	4,450	21,310
3000 mbh	"	25,190	5,570	30,760
6000 mbh	"	43,840	7,420	51,260
Steam				
115 mbh	EA.	4,770	1,110	5,880
175 mbh	"	5,890	1,110	7,000
235 mbh	"	6,420	1,110	7,530
940 mbh	"	13,020	2,230	15,250
1600 mbh	"	21,710	4,450	26,160
3000 mbh	"	30,220	5,570	35,790
6000 mbh	"	46,890	7,420	54,310

Heating & Ventilating	UNIT	MAT.	INST.	TOTAL
15555.10 **Boilers** *(Cont.)*				
Oil fired, hot water				
115 mbh	EA.	2,860	1,480	4,340
175 mbh	"	3,630	1,710	5,340
235 mbh	"	4,630	2,020	6,650
940 mbh	"	9,240	3,710	12,950
1600 mbh	"	15,080	4,450	19,530
3000 mbh	"	20,350	5,570	25,920
6000 mbh	"	75,870	11,130	87,000
Steam				
115 mbh	EA.	2,860	1,480	4,340
175 mbh	"	3,630	1,710	5,340
235 mbh	"	4,630	2,020	6,650
940 mbh	"	9,240	3,710	12,950
1600 mbh	"	15,080	4,450	19,530
3000 mbh	"	20,350	5,570	25,920
6000 mbh	"	75,870	11,130	87,000
15610.10 **Furnaces**				
Electric, hot air				
40 mbh	EA.	810	290	1,100
60 mbh	"	880	310	1,190
80 mbh	"	960	330	1,290
100 mbh	"	1,080	340	1,420
125 mbh	"	1,320	360	1,680
160 mbh	"	1,810	370	2,180
200 mbh	"	2,640	380	3,020
400 mbh	"	4,680	390	5,070
Gas fired hot air				
40 mbh	EA.	810	290	1,100
60 mbh	"	870	310	1,180
80 mbh	"	1,000	330	1,330
100 mbh	"	1,040	340	1,380
125 mbh	"	1,140	360	1,500
160 mbh	"	1,360	370	1,730
200 mbh	"	2,430	380	2,810
400 mbh	"	4,350	390	4,740
Oil fired hot air				
40 mbh	EA.	1,090	290	1,380
60 mbh	"	1,480	310	1,790
80 mbh	"	1,590	330	1,920
100 mbh	"	1,810	340	2,150
125 mbh	"	1,870	360	2,230
160 mbh	"	2,010	370	2,380
200 mbh	"	2,590	380	2,970
400 mbh	"	4,290	390	4,680

Refrigeration	UNIT	MAT.	INST.	TOTAL
15670.10 — Condensing Units				
Air cooled condenser, single circuit				
3 ton	EA.	1,400	98.00	1,498
5 ton	"	2,340	98.00	2,438
7.5 ton	"	3,740	280	4,020
20 ton	"	9,570	290	9,860
25 ton	"	15,320	290	15,610
30 ton	"	17,640	290	17,930
With low ambient dampers				
3 ton	EA.	1,640	150	1,790
5 ton	"	2,590	150	2,740
7.5 ton	"	3,970	290	4,260
20 ton	"	9,850	390	10,240
25 ton	"	15,760	390	16,150
30 ton	"	18,110	390	18,500
Dual circuit				
10 ton	EA.	3,510	290	3,800
15 ton	"	5,140	420	5,560
20 ton	"	10,500	420	10,920
25 ton	"	16,730	420	17,150
30 ton	"	19,460	420	19,880
With low ambient dampers				
15 ton	EA.	5,720	420	6,140
20 ton	"	11,170	420	11,590
25 ton	"	17,710	420	18,130
30 ton	"	19,970	420	20,390
15680.10 — Chillers				
Chiller, reciprocal				
Air cooled, remote condenser, starter				
20 ton	EA.	29,960	740	30,700
25 ton	"	33,810	740	34,550
30 ton	"	35,750	740	36,490
40 ton	"	52,870	1,110	53,980
Water cooled, with starter				
20 ton	EA.	25,690	740	26,430
25 ton	"	28,470	740	29,210
30 ton	"	34,240	1,110	35,350
40 ton	"	47,090	1,110	48,200
Packaged, air cooled, with starter				
20 ton	EA.	28,690	560	29,250
25 ton	"	31,040	560	31,600
30 ton	"	36,180	560	36,740
40 ton	"	41,530	560	42,090
15710.10 — Cooling Towers				
Cooling tower, propeller type				
100 ton	EA.	11,660	740	12,400
200 ton	"	19,360	1,110	20,470
300 ton	"	29,260	1,860	31,120
400 ton	"	38,940	2,230	41,170
600 ton	"	58,410	3,180	61,590
800 ton	"	77,880	4,450	82,330
1000 ton	"	93,500	5,570	99,070

Refrigeration

	UNIT	MAT.	INST.	TOTAL

15710.10 — Cooling Towers *(Cont.)*

	UNIT	MAT.	INST.	TOTAL
Centrifugal				
100 ton	EA.	16,170	740	16,910
200 ton	"	23,430	1,110	24,540
300 ton	"	34,980	1,860	36,840
400 ton	"	46,640	2,230	48,870
600 ton	"	66,760	3,180	69,940
800 ton	"	88,990	4,450	93,440
1000 ton	"	109,120	5,570	114,690

Heat Transfer

	UNIT	MAT.	INST.	TOTAL

15780.10 — Computer Room A/c

	UNIT	MAT.	INST.	TOTAL
Air cooled, alarm, high efficiency filter, elec. heat				
3 ton	EA.	14,960	450	15,410
5 ton	"	16,060	490	16,550
7.5 ton	"	27,280	590	27,870
10 ton	"	28,160	730	28,890
15 ton	"	31,240	840	32,080
Steam heat				
3 ton	EA.	17,270	450	17,720
5 ton	"	18,370	490	18,860
7.5 ton	"	29,260	590	29,850
10 ton	"	30,140	730	30,870
15 ton	"	33,330	840	34,170
Hot water heat				
3 ton	EA.	17,270	450	17,720
5 ton	"	18,370	490	18,860
7.5 ton	"	29,260	590	29,850
10 ton	"	30,140	730	30,870
15 ton	"	33,440	840	34,280
Air cooled condenser, low ambient damper				
3 ton	EA.	1,640	120	1,760
5 ton	"	2,590	150	2,740
7.5 ton	"	3,970	290	4,260
10 ton	"	5,810	420	6,230
15 ton	"	6,410	340	6,750
Water cooled, high efficiency filter, alarm, elec. heat				
3 ton	EA.	17,050	420	17,470
5 ton	"	18,370	490	18,860
7.5 ton	"	29,260	730	29,990
10 ton	"	30,360	840	31,200
15 ton	"	35,530	980	36,510
Steam heat				
3 ton	EA.	19,470	420	19,890
5 ton	"	22,220	490	22,710
7.5 ton	"	31,350	730	32,080

Heat Transfer	UNIT	MAT.	INST.	TOTAL
15780.10 **Computer Room A/c** *(Cont.)*				
10 ton	EA.	32,450	840	33,290
15 ton	"	37,730	980	38,710
Hot water heat				
3 ton	EA.	19,470	420	19,890
5 ton	"	20,790	490	21,280
7.5 ton	"	31,350	730	32,080
10 ton	"	32,450	840	33,290
15 ton	"	37,730	980	38,710
Chilled water, alarm, high eff. filter, elec. heat				
7.5 ton	EA.	13,640	530	14,170
10 ton	"	14,300	650	14,950
15 ton	"	16,280	730	17,010
Steam heat				
7.5 ton	EA.	15,620	530	16,150
10 ton	"	16,280	650	16,930
15 ton	"	18,260	730	18,990
Hot water heat				
7.5 ton	EA.	15,620	530	16,150
10 ton	"	16,280	650	16,930
15 ton	"	18,260	730	18,990
15780.20 **Rooftop Units**				
Packaged, single zone rooftop unit, with roof curb				
2 ton	EA.	3,400	590	3,990
3 ton	"	3,580	590	4,170
4 ton	"	3,900	730	4,630
5 ton	"	4,240	980	5,220
7.5 ton	"	6,160	1,170	7,330
15820.10 **Dehumidifiers**				
Dessicant dehumidifier, 1125 cfm	EA.			23,320
15830.10 **Radiation Units**				
Baseboard radiation unit				
1.7 mbh/lf	L.F.	73.00	23.50	96.50
2.1 mbh/lf	"	97.00	29.25	126
Enclosure only				
Two tier	L.F.	37.50	9.77	47.27
Three tier	"	48.50	9.77	58.27
Copper element only, 3/4" dia.				
Two tier	L.F.	54.00	14.75	68.75
Three tier	"	85.00	19.50	105
Fin-tube, 16 ga, sloping cover, 1-1/4" steel				
One tier	L.F.	52.00	19.50	71.50
Two tier	"	83.00	23.50	107
2" steel				
Two tier	L.F.	85.00	23.50	109
Three tier	"	120	29.25	149
1-1/4" copper				
Two tier	L.F.	110	19.50	130
18 ga flat cover, 1-1/4" steel				
One tier	L.F.	34.75	19.50	54.25
Two tier	"	56.00	23.50	79.50

Heat Transfer	UNIT	MAT.	INST.	TOTAL
15830.10 — **Radiation Units** *(Cont.)*				
Three tier	L.F.	90.00	29.25	119
2" steel				
One tier	L.F.	41.25	19.50	60.75
Two tier	"	65.00	23.50	88.50
Three tier	"	90.00	29.25	119
1-1/4" copper				
One tier	L.F.	40.00	19.50	59.50
Two tier	"	73.00	23.50	96.50
Three tier	"	97.00	29.25	126
15830.20 — **Fan Coil Units**				
Fan coil unit, 2 pipe, complete				
200 cfm ceiling hung	EA.	920	200	1,120
Floor mounted	"	870	150	1,020
300 cfm, ceiling hung	"	980	230	1,210
Floor mounted	"	930	200	1,130
400 cfm, ceiling hung	"	1,030	280	1,310
Floor mounted	"	990	200	1,190
500 cfm, ceiling hung	"	1,200	290	1,490
Floor mounted	"	1,160	230	1,390
600 cfm, ceiling hung	"	1,520	320	1,840
Floor mounted	"	1,410	270	1,680
800 cfm, ceiling hung	"	1,770	370	2,140
Floor mounted	"	1,410	280	1,690
1000 cfm, ceiling hung	"	2,020	420	2,440
Floor mounted	"	2,220	310	2,530
1200 cfm ceiling hung	"	2,300	490	2,790
Floor mounted	"	2,410	370	2,780
15830.70 — **Unit Heaters**				
Steam unit heater, horizontal				
12,500 btuh, 200 cfm	EA.	510	98.00	608
17,000 btuh, 300 cfm	"	570	98.00	668
40,000 btuh, 500 cfm	"	720	98.00	818
60,000 btuh, 700 cfm	"	800	98.00	898
70,000 btuh, 1000 cfm	"	1,000	150	1,150
Vertical				
12,500 btuh, 200 cfm	EA.	430	98.00	528
17,000 btuh, 300 cfm	"	720	98.00	818
40,000 btuh, 500 cfm	"	850	98.00	948
60,000 btuh, 700 cfm	"	1,000	98.00	1,098
70,000 btuh, 1000 cfm	"	1,140	98.00	1,238
Gas unit heater, horizontal				
27,400 btuh	EA.	720	230	950
38,000 btuh	"	780	230	1,010
56,000 btuh	"	850	230	1,080
82,200 btuh	"	940	230	1,170
103,900 btuh	"	990	370	1,360
125,700 btuh	"	1,010	370	1,380
133,200 btuh	"	1,070	370	1,440
149,000 btuh	"	1,250	370	1,620
172,000 btuh	"	1,350	370	1,720
190,000 btuh	"	1,420	370	1,790

Heat Transfer

	UNIT	MAT.	INST.	TOTAL
15830.70 **Unit Heaters** *(Cont.)*				
225,000 btuh	EA.	1,570	370	1,940
Hot water unit heater, horizontal				
12,500 btuh, 200 cfm	EA.	430	98.00	528
17,000 btuh, 300 cfm	"	500	98.00	598
25,000 btuh, 500 cfm	"	540	98.00	638
30,000 btuh, 700 cfm	"	640	98.00	738
50,000 btuh, 1000 cfm	"	720	150	870
60,000 btuh, 1300 cfm	"	1,000	150	1,150
Vertical				
12,500 btuh, 200 cfm	EA.	530	98.00	628
17,000 btuh, 300 cfm	"	570	98.00	668
25,000 btuh, 500 cfm	"	600	98.00	698
30,000 btuh, 700 cfm	"	650	98.00	748
50,000 btuh, 1000 cfm	"	720	98.00	818
60,000 btuh, 1300 cfm	"	780	98.00	878
Cabinet unit heaters, ceiling, exposed, hot water				
200 cfm	EA.	1,000	200	1,200
300 cfm	"	1,070	230	1,300
400 cfm	"	1,110	280	1,390
600 cfm	"	1,140	310	1,450
800 cfm	"	1,420	370	1,790
1000 cfm	"	1,860	420	2,280
1200 cfm	"	2,000	490	2,490
2000 cfm	"	3,420	650	4,070

Air Handling

	UNIT	MAT.	INST.	TOTAL
15855.10 **Air Handling Units**				
Air handling unit, medium pressure, single zone				
1500 cfm	EA.	4,000	370	4,370
3000 cfm	"	5,260	650	5,910
4000 cfm	"	6,730	730	7,460
5000 cfm	"	8,490	780	9,270
6000 cfm	"	10,110	840	10,950
7000 cfm	"	11,740	900	12,640
8500 cfm	"	12,940	980	13,920
10,500 cfm	"	15,520	1,170	16,690
12,500 cfm	"	18,480	1,300	19,780
15,500 cfm	"	25,160	1,680	26,840
17,500 cfm	"	27,350	1,950	29,300
20,500 cfm	"	31,560	2,350	33,910
25,000 cfm	"	35,770	2,930	38,700
31,500 cfm	"	44,180	3,910	48,090
Rooftop air handling units				

Air Handling

	UNIT	MAT.	INST.	TOTAL
15855.10 **Air Handling Units** *(Cont.)*				
4950 cfm	EA.	11,500	650	12,150
7370 cfm	"	14,580	840	15,420
9790 cfm	"	15,510	980	16,490
14,300 cfm	"	21,910	840	22,750
21,725 cfm	"	31,040	840	31,880
33,000 cfm	"	43,820	980	44,800
15870.20 **Exhaust Fans**				
Belt drive roof exhaust fans				
640 cfm, 2618 fpm	EA.	1,030	73.00	1,103
940 cfm, 2604 fpm	"	1,340	73.00	1,413
1050 cfm, 3325 fpm	"	1,200	73.00	1,273
1170 cfm, 2373 fpm	"	1,740	73.00	1,813
2440 cfm, 4501 fpm	"	1,360	73.00	1,433
2760 cfm, 4950 fpm	"	1,510	73.00	1,583
3890 cfm, 6769 fpm	"	1,720	73.00	1,793
2380 cfm, 3382 fpm	"	1,900	73.00	1,973
2880 cfm, 3859 fpm	"	1,990	73.00	2,063
3200 cfm, 4173 fpm	"	2,010	98.00	2,108
3660 cfm, 3437 fpm	"	2,050	98.00	2,148
Direct drive fans				
60 to 390 cfm	EA.	850	73.00	923
145 to 590 cfm	"	1,020	73.00	1,093
295 to 860 cfm	"	1,240	73.00	1,313
235 to 1300 cfm	"	1,330	73.00	1,403
415 to 1630 cfm	"	1,510	73.00	1,583
590 to 2045 cfm	"	1,740	73.00	1,813

Air Distribution

	UNIT	MAT.	INST.	TOTAL
15890.10 **Metal Ductwork**				
Rectangular duct				
Galvanized steel				
Minimum	Lb.	0.88	5.33	6.21
Average	"	1.10	6.51	7.61
Maximum	"	1.68	9.77	11.45
Aluminum				
Minimum	Lb.	2.29	11.75	14.04
Average	"	3.05	14.75	17.80
Maximum	"	3.79	19.50	23.29
Fittings				

Air Distribution	UNIT	MAT.	INST.	TOTAL

15890.10 — Metal Ductwork *(Cont.)*

	UNIT	MAT.	INST.	TOTAL
Minimum	EA.	7.26	19.50	26.76
Average	"	11.00	29.25	40.25
Maximum	"	16.00	59.00	75.00
For work				
10-20' high, add per pound, $.30				
30-50', add per pound, $.50				

15890.30 — Flexible Ductwork

	UNIT	MAT.	INST.	TOTAL
Flexible duct, 1.25" fiberglass				
5" dia.	L.F.	2.58	2.93	5.51
6" dia.	"	3.13	3.25	6.38
7" dia.	"	3.46	3.44	6.90
8" dia.	"	3.96	3.66	7.62
10" dia.	"	4.45	4.18	8.63
12" dia.	"	5.50	4.51	10.01
14" dia.	"	6.71	4.88	11.59
16" dia.	"	8.25	5.33	13.58
Flexible duct connector, 3" wide fabric	"	2.31	9.77	12.08

15895.10 — Roof Curbs

	UNIT	MAT.	INST.	TOTAL
8" high, insulated, with liner and raised can				
15" x 15"	EA.	97.00	29.25	126
17" x 17"	"	100	29.25	129
19" x 19"	"	110	29.25	139
21" x 21"	"	120	29.25	149
25" x 25"	"	130	36.75	167
28" x 28"	"	140	39.00	179
32" x 32"	"	150	42.00	192
36" x 36"	"	180	42.00	222
40" x 40"	"	200	42.00	242
44" x 44"	"	220	45.00	265
48" x 48"	"	500	45.00	545
52" x 52"	"	600	48.75	649
56" x 56"	"	760	48.75	809
60" x 60"	"	910	59.00	969
64" x 64"	"	1,100	59.00	1,159
68" x 68"	"	1,270	65.00	1,335
72" x 72"	"	1,450	73.00	1,523

15910.10 — Dampers

	UNIT	MAT.	INST.	TOTAL
Horizontal parallel aluminum backdraft damper				
12" x 12"	EA.	69.00	14.75	83.75
16" x 16"	"	71.00	16.75	87.75
20" x 20"	"	91.00	21.00	112
24" x 24"	"	110	29.25	139
28" x 28"	"	150	32.50	183
32" x 32"	"	210	36.75	247
36" x 36"	"	250	42.00	292
40" x 40"	"	320	48.75	369
44" x 44"	"	370	53.00	423
48" x 48"	"	450	59.00	509
"Up", parallel dampers				
12" x 12"	EA.	89.00	14.75	104

Air Distribution	UNIT	MAT.	INST.	TOTAL
15910.10 **Dampers** *(Cont.)*				
16" x 16"	EA.	120	16.75	137
20" x 20"	"	140	21.00	161
24" x 24"	"	150	29.25	179
28" x 28"	"	220	32.50	253
32" x 32"	"	250	36.75	287
36" x 36"	"	260	42.00	302
40" x 40"	"	350	48.75	399
44" x 44"	"	420	53.00	473
48" x 48"	"	520	59.00	579
"Down", parallel dampers				
12" x 12"	EA.	89.00	14.75	104
16" x 16"	"	120	16.75	137
20" x 20"	"	140	21.00	161
24" x 24"	"	150	29.25	179
28" x 28"	"	220	32.50	253
32" x 32"	"	250	36.75	287
36" x 36"	"	260	42.00	302
40" x 40"	"	350	48.75	399
44" x 44"	"	420	53.00	473
48" x 48"	"	520	59.00	579
Fire damper, 1.5 hr rating				
12" x 12"	EA.	30.25	29.25	59.50
16" x 16"	"	48.50	29.25	77.75
20" x 20"	"	53.00	29.25	82.25
24" x 24"	"	61.00	29.25	90.25
28" x 28"	"	73.00	42.00	115
32" x 32"	"	85.00	48.75	134
36" x 36"	"	100	59.00	159
40" x 40"	"	120	65.00	185
44" x 44"	"	140	73.00	213
48" x 48"	"	200	84.00	284
15940.10 **Diffusers**				
Ceiling diffusers, round, baked enamel finish				
6" dia.	EA.	54.00	19.50	73.50
8" dia.	"	65.00	24.50	89.50
10" dia.	"	72.00	24.50	96.50
12" dia.	"	92.00	24.50	117
14" dia.	"	110	26.75	137
16" dia.	"	140	26.75	167
18" dia.	"	160	29.25	189
20" dia.	"	190	29.25	219
Rectangular				
6x6"	EA.	58.00	19.50	77.50
9x9"	"	70.00	29.25	99.25
12x12"	"	100	29.25	129
15x15"	"	130	29.25	159
18x18"	"	160	29.25	189
21x21"	"	200	36.75	237
24x24"	"	230	36.75	267
Lay in, flush mounted, perforated face, with grid				
6x6/24x24	EA.	83.00	23.50	107
8x8/24x24	"	83.00	23.50	107

Air Distribution	UNIT	MAT.	INST.	TOTAL
15940.10 **Diffusers** *(Cont.)*				
9x9/24x24	EA.	83.00	23.50	107
10x10/24x24	"	90.00	23.50	114
12x12/24x24	"	96.00	23.50	120
15x15/24x24	"	130	23.50	154
18x6/24x24	"	96.00	23.50	120
18x18/24x24	"	150	23.50	174
Two-way slot diffuser with balancing damper, 4'	"	92.00	59.00	151
15940.20 **Relief Ventilators**				
Intake ventilator, aluminum, with screen, no curbs				
12" x 12"	EA.	200	48.75	249
16" x 16"	"	270	59.00	329
20" x 20"	"	440	59.00	499
30" x 30"	"	690	84.00	774
36" x 36"	"	1,050	98.00	1,148
42" x 42"	"	1,430	98.00	1,528
48" x 48"	"	1,720	120	1,840
15940.40 **Registers And Grilles**				
Lay in flush mounted, perforated face, return				
6x6/24x24	EA.	48.50	23.50	72.00
8x8/24x24	"	48.50	23.50	72.00
9x9/24x24	"	53.00	23.50	76.50
10x10/24x24	"	57.00	23.50	80.50
12x12/24x24	"	57.00	23.50	80.50
Rectangular, ceiling return, single deflection				
10x10	EA.	29.25	29.25	58.50
12x12	"	34.00	29.25	63.25
14x14	"	41.50	29.25	70.75
16x8	"	34.00	29.25	63.25
16x16	"	34.00	29.25	63.25
18x8	"	39.00	29.25	68.25
20x20	"	63.00	29.25	92.25
24x12	"	92.00	29.25	121
24x18	"	120	29.25	149
36x24	"	220	32.50	253
36x30	"	330	32.50	363
Wall, return air register				
12x12	EA.	48.50	14.75	63.25
16x16	"	71.00	14.75	85.75
18x18	"	85.00	14.75	99.75
20x20	"	100	14.75	115
24x24	"	140	14.75	155
Ceiling, return air grille				
6x6	EA.	18.00	19.50	37.50
8x8	"	20.50	23.50	44.00
10x10	"	23.00	23.50	46.50
Ceiling, exhaust grille, aluminum egg crate				
6x6	EA.	19.25	19.50	38.75

Air Distribution	UNIT	MAT.	INST.	TOTAL
15940.40 **Registers And Grilles** *(Cont.)*				
8x8	EA.	19.25	23.50	42.75
10x10	"	21.25	23.50	44.75
12x12	"	26.25	29.25	55.50
14x14	"	34.50	29.25	63.75
16x16	"	40.50	29.25	69.75
18x18	"	48.50	29.25	77.75
15940.80 **Penthouse Louvers**				
Penthouse louvers				
12" high, extruded aluminum, 4" louver				
6' perimeter	EA.	450	150	600
8' perimeter	"	600	150	750
10' perimeter	"	760	150	910
12' perimeter	"	1,100	150	1,250
14' perimeter	"	1,340	200	1,540
16' perimeter	"	1,550	230	1,780
18' perimeter	"	1,810	330	2,140
20' perimeter	"	2,270	390	2,660
16" high x 4' perimeter	"	370	150	520
6' perimeter	"	530	150	680
8' perimeter	"	680	150	830
10' perimeter	"	840	150	990
12' perimeter	"	1,230	150	1,380
14' perimeter	"	1,520	200	1,720
16' perimeter	"	1,710	230	1,940
18' perimeter	"	2,080	330	2,410
20' perimeter	"	2,650	390	3,040
22' perimeter	"	2,840	490	3,330
24' perimeter	"	3,030	650	3,680
20" high x 4' perimeter	"	530	150	680
6' perimeter	"	560	150	710
8' perimeter	"	760	150	910
10' perimeter	"	950	150	1,100
12' perimeter	"	1,390	150	1,540
14' perimeter	"	1,710	200	1,910
16' perimeter	"	2,080	230	2,310
18' perimeter	"	2,360	330	2,690
20' perimeter	"	2,840	390	3,230
22' perimeter	"	3,050	490	3,540
24' perimeter	"	3,390	650	4,040
24" high x 4' perimeter	"	530	150	680
6' perimeter	"	640	150	790
8' perimeter	"	840	150	990
10' perimeter	"	1,040	150	1,190
12' perimeter	"	1,540	150	1,690
16' perimeter	"	2,270	230	2,500
18' perimeter	"	2,650	330	2,980
20' perimeter	"	3,200	390	3,590
22' perimeter	"	3,540	490	4,030
24' perimeter	"	3,780	650	4,430

Controls

15950.10   Hvac Controls

	UNIT	MAT.	INST.	TOTAL
Pressure gauge, direct reading gage cock and siphon	EA.	120	36.75	157
Control valve, 1", modulating				
2-way	EA.	910	48.75	959
3-way	"	1,030	73.00	1,103
Self contained control valve w/ sensing elmnt, 3/4"	"	180	36.75	217
Inst air syst 2-1/2 hp comp, rcvr refrg dryer	"			7,700
Thermostat primary control device	"			170
Humidistat primary control device	"			140
Timers primary control device, indoor/outdoor, 24 hour	"			270
Thermometer, dir. reading, 3 dial	"			140
Control dampers, round				
6" dia.	EA.	93.00	23.50	117
8" dia	"	130	23.50	154
10" dia	"	180	23.50	204
12" dia	"	230	23.50	254
12" dia	"	340	29.25	369
18" dia	"	360	29.25	389
20" dia	"	480	29.25	509
Rectangular, parallel blade standard leakage				
12" x 12"	EA.	69.00	29.25	98.25
16" x 16"	"	100	29.25	129
20" x 20"	"	120	29.25	149
28" x 28"	"	150	36.75	187
32" x 32"	"	180	36.75	217
36" x 36"	"	260	48.75	309
40" x 40"	"	260	59.00	319
44" x 44"	"	320	73.00	393
48" x 48"	"	360	84.00	444
48" x 52"	"	390	98.00	488
48" x 56"	"	430	98.00	528
48" x 60"	"	460	98.00	558
48" x 64"	"	470	98.00	568
48" x 68"	"	520	98.00	618
48" x 72"	"	560	98.00	658
Low leakage				
12" x 12"	EA.	130	29.25	159
16" x 16"	"	160	29.25	189
20" x 20"	"	220	29.25	249
24" x 24"	"	300	29.25	329
28" x 28"	"	340	36.75	377
32" x 32"	"	390	42.00	432
36" x 36"	"	430	48.75	479
40" x 40"	"	690	59.00	749
44" x 44"	"	730	73.00	803
48" x 48"	"	770	84.00	854
48" x 56"	"	890	98.00	988
48" x 60"	"	930	98.00	1,028
48" x 64"	"	970	98.00	1,068
48" x 68"	"	1,010	98.00	1,108
48" x 72"	"	1,200	98.00	1,298
Rectangular, opposed horizontal blade				
12" x 12"	EA.	89.00	29.25	118
16" x 16"	"	120	29.25	149

Controls	UNIT	MAT.	INST.	TOTAL
15950.10	**Hvac Controls** *(Cont.)*			
20" x 20"	EA.	140	29.25	169
24" x 24"	"	160	29.25	189
28" x 28"	"	230	36.75	267
32" x 32"	"	250	39.00	289
36" x 36"	"	280	48.75	329
40" x 40"	"	350	59.00	409
44" x 44"	"	440	73.00	513
48" x 48"	"	550	84.00	634
48" x 52"	"	560	84.00	644
48" x 56"	"	580	98.00	678
48" x 60"	"	620	98.00	718
48" x 64"	"	650	98.00	748
48" x 68"	"	720	98.00	818
48" x 72"	"	770	98.00	868

Basic Materials	UNIT	MAT.	INST.	TOTAL
16110.20 **Conduit Specialties**				
Rod beam clamp, 1/2"	EA.	5.72	3.39	9.11
Hanger rod				
3/8"	L.F.	1.21	2.71	3.92
1/2"	"	2.26	3.39	5.65
All thread rod				
1/4"	L.F.	0.73	2.04	2.77
3/8"	"	1.21	2.71	3.92
1/2"	"	2.26	3.39	5.65
5/8"	"	3.13	5.42	8.55
Hanger channel, 1-1/2"				
No holes	EA.	3.94	2.04	5.98
Holes	"	3.94	2.04	5.98
Channel strap				
1/2"	EA.	0.89	3.39	4.28
3/4"	"	1.38	3.39	4.77
1"	"	1.77	3.39	5.16
1-1/4"	"	3.77	5.42	9.19
1-1/2"	"	4.66	5.42	10.08
2"	"	5.90	5.42	11.32
2-1/2"	"	2.97	8.35	11.32
3"	"	3.23	8.35	11.58
3-1/2"	"	4.02	8.35	12.37
4"	"	4.55	9.86	14.41
5"	"	6.40	9.86	16.26
6"	"	7.59	9.86	17.45
Conduit penetrations, roof and wall, 8" thick				
1/2"	EA.		41.75	41.75
3/4"	"		41.75	41.75
1"	"		54.00	54.00
1-1/4"	"		54.00	54.00
1-1/2"	"		54.00	54.00
2"	"		110	110
2-1/2"	"		110	110
3"	"		110	110
3-1/2"	"		140	140
4"	"		140	140
Fireproofing, for conduit penetrations				
1/2"	EA.	3.36	34.00	37.36
3/4"	"	3.49	34.00	37.49
1"	"	3.56	34.00	37.56
1-1/4"	"	8.38	53.00	61.38
1-1/2"	"	4.95	49.25	54.20
2"	"	5.08	49.25	54.33
2-1/2"	"	9.70	61.00	70.70
3"	"	10.00	66.00	76.00
3-1/2"	"	11.75	85.00	96.75
4"	"	14.25	100	114

Basic Materials	UNIT	MAT.	INST.	TOTAL
16110.21 **Aluminum Conduit**				
Aluminum conduit				
1/2"	L.F.	1.32	2.04	3.36
3/4"	"	1.70	2.71	4.41
1"	"	2.38	3.39	5.77
1-1/4"	"	3.19	4.02	7.21
1-1/2"	"	3.96	5.42	9.38
2"	"	5.28	6.03	11.31
2-1/2"	"	8.36	6.78	15.14
3"	"	11.00	7.23	18.23
3-1/2"	"	13.25	8.35	21.60
4"	"	15.50	9.86	25.36
5"	"	22.25	12.25	34.50
6"	"	29.25	13.50	42.75
90 deg. elbow				
1/2"	EA.	12.75	13.00	25.75
3/4"	"	17.50	17.00	34.50
1"	"	24.25	20.75	45.00
1-1/4"	"	38.50	25.75	64.25
1-1/2"	"	51.00	27.25	78.25
2"	"	76.00	30.25	106
2-1/2"	"	130	38.75	169
3"	"	200	45.25	245
3-1/2"	"	300	54.00	354
4"	"	400	60.00	460
5"	"	1,000	78.00	1,078
6"	"	1,360	150	1,510
Coupling				
1/2"	EA.	4.23	3.39	7.62
3/4"	"	6.41	4.02	10.43
1"	"	8.47	5.42	13.89
1-1/4"	"	10.25	6.03	16.28
1-1/2"	"	12.00	6.78	18.78
2"	"	17.00	7.23	24.23
2-1/2"	"	38.25	8.35	46.60
3"	"	49.75	8.35	58.10
3-1/2"	"	68.00	9.86	77.86
4"	"	83.00	10.75	93.75
5"	"	210	10.75	221
6"	"	340	13.00	353
16110.22 **Emt Conduit**				
EMT conduit				
1/2"	L.F.	0.36	2.04	2.40
3/4"	"	0.68	2.71	3.39
1"	"	1.07	3.39	4.46
1-1/4"	"	1.67	4.02	5.69
1-1/2"	"	2.03	5.42	7.45
2"	"	2.42	6.03	8.45
2-1/2"	"	5.20	6.78	11.98
3"	"	7.38	8.35	15.73
3-1/2"	"	9.74	9.86	19.60
4"	"	11.00	12.25	23.25
90 deg. elbow				

Basic Materials	UNIT	MAT.	INST.	TOTAL
16110.22 **Emt Conduit** *(Cont.)*				
1/2"	EA.	5.22	6.03	11.25
3/4"	"	5.72	6.78	12.50
1"	"	8.69	7.23	15.92
1-1/4"	"	10.75	8.35	19.10
1-1/2"	"	12.50	9.86	22.36
2"	"	18.50	13.00	31.50
2-1/2"	"	38.00	14.25	52.25
3"	"	75.00	16.50	91.50
3-1/2"	"	82.00	19.00	101
4"	"	100	19.50	120
Connector, die cast set screw				
1/2"	EA.	0.66	4.02	4.68
3/4"	"	1.12	4.02	5.14
1"	"	2.11	4.02	6.13
1-1/4"	"	3.69	6.03	9.72
1-1/2"	"	5.01	7.23	12.24
2"	"	6.73	9.86	16.59
2-1/2"	"	21.75	13.50	35.25
3"	"	26.75	15.00	41.75
3-1/2"	"	37.50	17.00	54.50
4"	"	43.25	19.50	62.75
Coupling, steel set screw				
1/2"	EA.	1.65	2.71	4.36
3/4"	"	2.50	2.71	5.21
1"	"	3.89	2.71	6.60
1-1/4"	"	8.51	3.39	11.90
1-1/2"	"	12.50	5.42	17.92
2"	"	16.75	7.23	23.98
2-1/2"	"	36.50	10.75	47.25
3"	"	40.75	13.00	53.75
3-1/2"	"	48.00	15.00	63.00
4"	"	55.00	17.00	72.00
16110.23 **Flexible Conduit**				
Flexible conduit, steel				
3/8"	L.F.	0.39	2.04	2.43
1/2	"	0.45	2.04	2.49
3/4"	"	0.60	2.71	3.31
1"	"	1.18	2.71	3.89
1-1/4"	"	1.45	3.39	4.84
1-1/2"	"	2.40	4.02	6.42
2"	"	3.01	5.42	8.43
2-1/2"	"	3.64	6.03	9.67
3"	"	6.38	7.23	13.61
Flexible conduit, liquid tight				
3/8"	L.F.	1.07	2.04	3.11
1/2"	"	1.37	2.04	3.41
3/4"	"	1.89	2.71	4.60
1"	"	3.10	2.71	5.81
1-1/4"	"	4.11	3.39	7.50
1-1/2"	"	4.80	4.02	8.82
2"	EA.	5.89	5.42	11.31
2-1/2"	"	11.00	6.03	17.03

Basic Materials	UNIT	MAT.	INST.	TOTAL
16110.23 **Flexible Conduit** *(Cont.)*				
3"	EA.	14.50	7.23	21.73
4"	"	21.75	9.86	31.61
Connector, straight				
3/8"	EA.	3.10	5.42	8.52
1/2"	"	3.33	5.42	8.75
3/4"	"	4.22	6.03	10.25
1"	"	6.28	6.78	13.06
1-1/4"	"	10.75	7.23	17.98
1-1/2"	"	15.00	8.35	23.35
2"	"	27.75	9.86	37.61
2-1/2"	"	29.75	12.25	42.00
3"	"	41.50	13.00	54.50
Flexible aluminum conduit				
3/8"	L.F.	0.35	2.04	2.39
1/2"	"	0.41	2.04	2.45
3/4"	"	0.57	2.71	3.28
1"	"	1.07	2.71	3.78
1-1/4"	"	1.46	3.39	4.85
1-1/2"	"	2.40	4.02	6.42
2"	"	2.92	5.42	8.34
2-1/2"	"	3.86	6.03	9.89
3"	"	6.57	7.23	13.80
3-1/2"	"	7.55	8.35	15.90
4"	"	8.31	9.86	18.17
16110.24 **Galvanized Conduit**				
Galvanized rigid steel conduit				
1/2"	L.F.	1.30	2.71	4.01
3/4"	"	1.51	3.39	4.90
1"	"	2.35	4.02	6.37
1-1/4"	"	3.35	5.42	8.77
1-1/2"	"	3.92	6.03	9.95
2"	"	4.65	6.78	11.43
2-1/2"	"	8.98	9.86	18.84
3"	"	10.75	12.25	23.00
3-1/2"	"	12.75	13.00	25.75
4"	"	15.00	14.25	29.25
5"	"	30.50	19.50	50.00
6"	"	44.25	25.75	70.00
90 degree ell				
1/2"	EA.	10.25	17.00	27.25
3/4"	"	10.75	20.75	31.50
1"	"	16.50	25.75	42.25
1-1/4"	"	22.75	30.25	53.00
1-1/2"	"	28.00	34.00	62.00
2"	"	40.75	36.25	77.00
2-1/2"	"	76.00	45.25	121
3"	"	110	60.00	170
3-1/2"	"	170	68.00	238
4"	"	190	90.00	280
5"	"	520	150	670
6"	"	790	230	1,020
Couplings, with set screws				

Basic Materials	UNIT	MAT.	INST.	TOTAL
16110.24 — **Galvanized Conduit** *(Cont.)*				
1/2"	EA.	5.17	3.39	8.56
3/4"	"	6.82	4.02	10.84
1"	"	11.00	5.42	16.42
1-1/4"	"	18.50	6.78	25.28
1-1/2"	"	24.00	8.35	32.35
2"	"	54.00	9.86	63.86
2-1/2"	"	130	13.00	143
3"	"	160	17.00	177
3-1/2"	"	220	19.50	240
4"	"	290	20.75	311
5"	"	420	30.25	450
6"	"	550	34.00	584
Split couplings				
1/2"	EA.	4.40	13.00	17.40
3/4"	"	5.72	17.00	22.72
1"	"	8.03	18.75	26.78
1-1/4"	"	15.75	20.75	36.50
1-1/2"	"	20.50	25.75	46.25
2"	"	47.25	38.75	86.00
2-1/2"	"	96.00	38.75	135
3"	"	140	49.25	189
3-1/2"	"	220	68.00	288
4"	"	260	90.00	350
5"	"	460	110	570
6"	"	620	140	760
Erickson couplings				
1/2"	EA.	5.22	30.25	35.47
3/4"	"	6.38	34.00	40.38
1"	"	12.75	41.75	54.50
1-1/4"	"	23.25	60.00	83.25
1-1/2"	"	30.00	68.00	98.00
2"	"	58.00	90.00	148
2-1/2"	"	120	130	250
3"	"	170	140	310
3-1/2"	"	310	170	480
4"	"	370	180	550
5"	"	740	200	940
6"	"	1,080	220	1,300
Seal fittings				
1/2"	EA.	16.50	45.25	61.75
3/4"	"	18.25	54.00	72.25
1"	"	23.00	68.00	91.00
1-1/4"	"	27.50	78.00	106
1-1/2"	"	41.75	90.00	132
2"	"	53.00	110	163
2-1/2"	"	83.00	130	213
3"	"	99.00	140	239
3-1/2"	"	260	170	430
4"	"	400	200	600
5"	"	620	300	920
6"	"	680	340	1,020
Entrance fitting, (weather head), threaded				
1/2"	EA.	8.80	30.25	39.05

Basic Materials	UNIT	MAT.	INST.	TOTAL
16110.24 **Galvanized Conduit** *(Cont.)*				
3/4"	EA.	10.75	34.00	44.75
1"	"	13.75	38.75	52.50
1-1/4"	"	18.00	49.25	67.25
1-1/2"	"	31.75	54.00	85.75
2"	"	48.50	60.00	109
2-1/2"	"	170	68.00	238
3"	"	240	90.00	330
3-1/2"	"	310	120	430
4"	"	400	170	570
5"	"	420	240	660
6"	"	520	300	820
Locknuts				
1/2"	EA.	0.19	3.39	3.58
3/4"	"	0.24	3.39	3.63
1"	"	0.40	3.39	3.79
1-1/4"	"	0.53	3.39	3.92
1-1/2"	"	0.77	4.02	4.79
2"	"	1.07	4.02	5.09
2-1/2"	"	2.83	5.42	8.25
3"	"	3.56	5.42	8.98
3-1/2"	"	6.13	5.42	11.55
4"	"	7.65	6.03	13.68
5"	"	16.25	6.03	22.28
6"	"	28.75	6.03	34.78
Plastic conduit bushings				
1/2"	EA.	0.33	8.35	8.68
3/4"	"	0.50	9.86	10.36
1"	"	0.71	13.00	13.71
1-1/4"	"	0.93	15.00	15.93
1-1/2"	"	1.26	17.00	18.26
2"	"	2.20	20.75	22.95
2-1/2"	"	5.50	34.00	39.50
3"	"	6.38	45.25	51.63
3-1/2"	"	7.15	54.00	61.15
4"	"	9.79	60.00	69.79
5"	"	18.75	78.00	96.75
6"	"	36.00	110	146
Conduit bushings, steel				
1/2"	EA.	0.50	8.35	8.85
3/4"	"	0.63	9.86	10.49
1"	"	0.96	13.00	13.96
1-1/4"	"	1.38	15.00	16.38
1-1/2"	"	1.98	17.00	18.98
2"	"	3.08	20.75	23.83
2-1/2"	"	7.04	34.00	41.04
3"	"	8.69	45.25	53.94
3-1/2"	"	18.00	54.00	72.00
4"	"	22.00	60.00	82.00
5"	"	45.00	78.00	123
6"	"	83.00	110	193
Pipe cap				
1/2"	EA.	0.57	3.39	3.96
3/4"	"	0.61	3.39	4.00

Basic Materials	UNIT	MAT.	INST.	TOTAL
16110.24 **Galvanized Conduit** *(Cont.)*				
1"	EA.	0.99	3.39	4.38
1-1/4"	"	1.69	5.42	7.11
1-1/2"	"	2.64	5.42	8.06
2"	"	2.97	5.42	8.39
2-1/2"	"	5.06	6.03	11.09
3"	"	6.27	6.03	12.30
3-1/2"	"	8.58	6.03	14.61
4"	"	11.00	7.23	18.23
5"	"	14.75	9.86	24.61
6"	"	18.25	13.50	31.75
16110.25 **Plastic Conduit**				
PVC conduit, schedule 40				
1/2"	L.F.	0.29	2.04	2.33
3/4"	"	0.33	2.04	2.37
1"	"	0.49	2.71	3.20
1-1/4"	"	0.69	2.71	3.40
1-1/2"	"	0.80	3.39	4.19
2"	"	0.96	3.39	4.35
2-1/2"	"	1.46	4.02	5.48
3"	"	1.87	4.02	5.89
3-1/2"	"	2.57	5.42	7.99
4"	"	2.79	5.42	8.21
5"	"	5.17	6.03	11.20
6"	"	6.83	6.78	13.61
Couplings				
1/2"	EA.	0.47	3.39	3.86
3/4"	"	0.49	3.39	3.88
1"	"	0.79	3.39	4.18
1-1/4"	"	1.00	4.02	5.02
1-1/2"	"	1.32	4.02	5.34
2"	"	1.87	4.02	5.89
2-1/2"	"	3.35	4.02	7.37
3"	"	5.33	5.42	10.75
3-1/2"	"	6.05	5.42	11.47
4"	"	8.25	6.78	15.03
5"	"	20.50	6.78	27.28
6"	"	26.50	6.78	33.28
90 degree elbows				
1/2"	EA.	1.87	6.78	8.65
3/4"	"	1.81	8.35	10.16
1"	"	2.82	8.35	11.17
1-1/4"	"	4.07	9.86	13.93
1-1/2"	"	5.39	13.00	18.39
2"	"	7.40	15.00	22.40
2-1/2"	"	13.00	17.00	30.00
3"	"	22.75	20.75	43.50
3-1/2"	"	31.25	25.75	57.00
4"	"	39.25	34.00	73.25
5"	"	69.00	41.75	111
6"	"	120	49.25	169
Terminal adapters				
1/2"	EA.	0.62	6.78	7.40

Basic Materials	UNIT	MAT.	INST.	TOTAL
16110.25 **Plastic Conduit** *(Cont.)*				
3/4"	EA.	1.01	6.78	7.79
1"	"	1.26	6.78	8.04
1-1/4"	"	1.59	10.75"	12.34
1-1/2"	"	2.03	10.75	12.78
2"	"	2.80	10.75	13.55
2-1/2"	"	4.78	15.00	19.78
3"	"	6.76	15.00	21.76
3-1/2"	"	8.69	15.00	23.69
4"	"	11.25	25.75	37.00
5"	"	22.50	25.75	48.25
6"	"	27.00	25.75	52.75
End bells				
1"	EA.	3.38	6.78	10.16
1-1/4"	"	4.01	10.75	14.76
1-1/2"	"	4.18	10.75	14.93
2"	"	6.21	10.75	16.96
2-1/2"	"	6.87	15.00	21.87
3"	"	7.26	15.00	22.26
3-1/2"	"	7.97	15.00	22.97
4"	"	8.69	25.75	34.44
5"	"	13.50	25.75	39.25
6"	"	15.00	25.75	40.75
LB conduit body				
1/2"	EA.	4.73	13.00	17.73
3/4"	"	6.09	13.00	19.09
1	"	6.71	13.00	19.71
1-1/4"	"	10.25	20.75	31.00
1-1/2"	"	12.25	20.75	33.00
2"	"	21.75	20.75	42.50
Direct burial, conduit				
2"	L.F.	1.01	3.39	4.40
3"	"	1.87	4.02	5.89
4"	"	3.74	5.42	9.16
5"	"	5.33	6.03	11.36
6"	"	7.59	6.78	14.37
Encased burial conduit				
2"	L.F.	1.05	3.39	4.44
3"	"	1.76	4.02	5.78
4"	"	2.69	5.42	8.11
5"	"	3.85	6.03	9.88
6"	"	5.00	6.78	11.78
PVC cement				
1 pint	EA.			15.00
1 quart	"			22.00
1 gallon	"			72.00
16110.27 **Plastic Coated Conduit**				
Rigid steel conduit, plastic coated				
1/2"	L.F.	4.29	3.39	7.68
3/4"	"	5.06	4.02	9.08
1"	"	6.54	5.42	11.96
1-1/4"	"	8.19	6.78	14.97
1-1/2"	"	9.95	8.35	18.30

Basic Materials	UNIT	MAT.	INST.	TOTAL

16110.27 — Plastic Coated Conduit (Cont.)

	UNIT	MAT.	INST.	TOTAL
2"	L.F.	13.25	9.86	23.11
2-1/2"	"	20.00	13.00	33.00
3"	"	25.25	15.00	40.25
3-1/2"	"	30.50	17.00	47.50
4"	"	37.50	20.75	58.25
5"	"	64.00	25.75	89.75
90 degree elbows				
1/2"	EA.	16.50	20.75	37.25
3/4"	"	17.00	25.75	42.75
1"	"	19.75	30.25	50.00
1-1/4"	"	24.50	34.00	58.50
1-1/2"	"	29.75	41.75	71.50
2"	"	41.25	54.00	95.25
2-1/2"	"	80.00	78.00	158
3"	"	81.00	90.00	171
3-1/2"	"	180	110	290
4"	"	190	140	330
5"	"	420	170	590
Couplings				
1/2"	EA.	4.84	4.02	8.86
3/4"	"	5.11	5.42	10.53
1"	"	6.82	6.03	12.85
1-1/4"	"	7.97	7.23	15.20
1-1/2"	"	11.00	8.35	19.35
2"	"	14.00	9.86	23.86
2-1/2"	"	34.75	12.25	47.00
3"	"	40.50	13.00	53.50
3-1/2"	"	56.00	13.50	69.50
4"	"	69.00	15.00	84.00
5"	"	200	17.00	217

16110.28 — Steel Conduit

	UNIT	MAT.	INST.	TOTAL
Intermediate metal conduit (IMC)				
1/2"	L.F.	1.43	2.04	3.47
3/4"	"	1.70	2.71	4.41
1"	"	2.57	3.39	5.96
1-1/4"	"	3.28	4.02	7.30
1-1/2"	"	4.12	5.42	9.54
2"	"	5.35	6.03	11.38
2-1/2"	"	10.50	8.10	18.60
3"	"	13.75	9.86	23.61
3-1/2"	"	16.00	12.25	28.25
4"	"	17.75	13.00	30.75
90 degree ell				
1/2"	EA.	9.18	17.00	26.18
3/4"	"	11.25	20.75	32.00
1"	"	16.25	25.75	42.00
1-1/4"	"	26.25	30.25	56.50

Basic Materials	UNIT	MAT.	INST.	TOTAL

16110.28 — Steel Conduit *(Cont.)*

Basic Materials	UNIT	MAT.	INST.	TOTAL
1-1/2"	EA.	29.50	34.00	63.50
2"	"	42.25	38.75	81.00
2-1/2"	"	73.00	45.25	118
3"	"	110	60.00	170
3-1/2"	"	200	78.00	278
4"	"	230	90.00	320

16110.35 — Surface Mounted Raceway

Basic Materials	UNIT	MAT.	INST.	TOTAL
Single Raceway				
3/4" x 17/32" Conduit	L.F.	1.67	2.71	4.38
Mounting Strap	EA.	0.45	3.61	4.06
Connector	"	0.60	3.61	4.21
Elbow				
45 degree	EA.	7.62	3.39	11.01
90 degree	"	2.43	3.39	5.82
internal	"	3.05	3.39	6.44
external	"	2.82	3.39	6.21
Switch	"	19.75	27.25	47.00
Utility Box	"	13.25	27.25	40.50
Receptacle	"	23.50	27.25	50.75
3/4" x 21/32" Conduit	L.F.	1.90	2.71	4.61
Mounting Strap	EA.	0.70	3.61	4.31
Connector	"	0.72	3.61	4.33
Elbow				
45 degree	EA.	9.41	3.39	12.80
90 degree	"	2.59	3.39	5.98
internal	"	3.52	3.39	6.91
external	"	3.52	3.39	6.91
Switch	"	19.75	27.25	47.00
Utility Box	"	13.25	27.25	40.50
Receptacle	"	23.50	27.25	50.75

16110.80 — Wireways

Basic Materials	UNIT	MAT.	INST.	TOTAL
Wireway, hinge cover type				
2-1/2" x 2-1/2"				
1' section	EA.	18.50	10.50	29.00
2'	"	26.50	13.00	39.50
3'	"	35.75	17.00	52.75
5'	"	61.00	25.75	86.75
10'	"	120	45.25	165
4" x 4"				
1'	EA.	20.25	17.00	37.25
2'	"	29.75	17.00	46.75
3'	"	44.00	20.75	64.75
4'	"	61.00	20.75	81.75
10'	"	180	54.00	234
6" x 6"				
1'	EA.	39.00	25.75	64.75
2'	"	47.50	25.75	73.25
3'	"	68.00	30.25	98.25
4'	"	90.00	30.25	120
5'	"	97.00	38.75	136
10'	"	190	60.00	250

Basic Materials	UNIT	MAT.	INST.	TOTAL

16110.80 — Wireways (Cont.)

	UNIT	MAT.	INST.	TOTAL
8" x 8"				
1'	EA.	63.00	30.25	93.25
2'	"	96.00	30.25	126
3'	"	110	34.00	144
4'	"	130	34.00	164
5'	"	200	41.75	242
12" x 12"				
1'	EA.	87.00	41.75	129
2'	"	130	41.75	172
3'	"	190	49.25	239
4'	"	220	49.25	269
5'	"	260	60.00	320

16120.43 — Copper Conductors

	UNIT	MAT.	INST.	TOTAL
Copper conductors, type THW, solid				
#14	L.F.	0.12	0.27	0.39
#12	"	0.18	0.33	0.51
#10	"	0.28	0.40	0.68
Stranded				
#14	L.F.	0.13	0.27	0.40
#12	"	0.16	0.33	0.49
#10	"	0.25	0.40	0.65
#8	"	0.41	0.54	0.95
#6	"	0.67	0.60	1.27
#4	"	1.05	0.67	1.72
#3	"	1.33	0.67	2.00
#2	"	1.67	0.81	2.48
#1	"	2.11	0.95	3.06
1/0	"	2.53	1.08	3.61
2/0	"	3.16	1.35	4.51
3/0	"	3.99	1.69	5.68
4/0	"	4.98	1.90	6.88
250 MCM	"	6.14	2.04	8.18
300 MCM	"	7.23	2.26	9.49
350 MCM	"	8.47	2.71	11.18
400 MCM	"	9.63	3.01	12.64
500 MCM	"	12.00	3.50	15.50
600 MCM	"	15.75	4.02	19.77
750 MCM	"	20.00	4.52	24.52
1000 MCM	"	25.00	5.16	30.16
THHN-THWN, solid				
#14	L.F.	0.12	0.27	0.39
#12	"	0.18	0.33	0.51
#10	"	0.28	0.40	0.68
Stranded				
#14	L.F.	0.11	0.27	0.38
#12	"	0.16	0.33	0.49
#10	"	0.24	0.40	0.64
#8	"	0.49	0.54	1.03
#6	"	0.67	0.60	1.27
#4	"	1.05	0.67	1.72
#2	"	1.67	0.81	2.48
#1	"	2.11	0.95	3.06

Basic Materials	UNIT	MAT.	INST.	TOTAL
16120.43 — **Copper Conductors** *(Cont.)*				
1/0	L.F.	2.53	1.08	3.61
2/0	"	3.16	1.35	4.51
3/0	"	3.99	1.69	5.68
4/0	"	4.98	1.90	6.88
250 MCM	"	6.05	2.04	8.09
350 MCM	"	7.86	2.71	10.57
XHHW				
#14	L.F.	0.18	0.27	0.45
#10	"	0.39	0.40	0.79
#8	"	0.55	0.54	1.09
#6	"	0.86	0.60	1.46
#4	"	1.35	0.60	1.95
#2	"	2.11	0.74	2.85
#1	"	2.67	0.95	3.62
1/0	"	2.91	1.08	3.99
2/0	"	3.66	1.29	4.95
3/0	"	4.55	1.69	6.24
XLP, 600v				
#12	L.F.	0.31	0.33	0.64
#10	"	0.46	0.40	0.86
#8	"	0.60	0.54	1.14
#6	"	0.93	0.60	1.53
#4	"	1.44	0.67	2.11
#3	"	1.79	0.74	2.53
#2	"	2.22	0.81	3.03
#1	"	2.86	0.95	3.81
1/0	"	3.22	1.08	4.30
2/0	"	4.02	1.35	5.37
3/0	"	5.04	1.75	6.79
4/0	"	6.31	1.90	8.21
250 MCM	"	7.37	2.04	9.41
300 MCM	"	8.78	2.26	11.04
350 MCM	"	10.25	2.64	12.89
400 MCM	"	11.75	3.01	14.76
500 MCM	"	14.50	3.50	18.00
600 MCM	"	17.50	4.02	21.52
750 MCM	"	27.00	4.52	31.52
1000 MCM	"	35.75	5.16	40.91
16120.47 — **Sheathed Cable**				
Non-metallic sheathed cable				
Type NM cable with ground				
#14/2	L.F.	0.35	1.01	1.36
#12/2	"	0.53	1.08	1.61
#10/2	"	0.85	1.20	2.05
#8/2	"	1.39	1.35	2.74
#6/2	"	2.20	1.69	3.89
#14/3	"	0.49	1.75	2.24
#12/3	"	0.77	1.80	2.57
#10/3	"	1.22	1.84	3.06
#8/3	"	2.05	1.87	3.92
#6/3	"	3.32	1.90	5.22
#4/3	"	5.84	2.17	8.01

Basic Materials	UNIT	MAT.	INST.	TOTAL
16120.47 **Sheathed Cable** *(Cont.)*				
#2/3	L.F.	8.75	2.36	11.11
Type U.F. cable with ground				
#14/2	L.F.	0.40	1.08	1.48
#12/2	"	0.61	1.29	1.90
#10/2	"	0.97	1.35	2.32
#8/2	"	1.68	1.55	3.23
#6/2	"	2.62	1.84	4.46
#14/3	"	0.57	1.35	1.92
#12/3	"	0.86	1.48	2.34
#10/3	"	1.35	1.69	3.04
#8/3	"	2.55	1.90	4.45
#6/3	"	4.12	2.17	6.29
16130.60 **Pull And Junction Boxes**				
4"				
Octagon box	EA.	2.79	7.75	10.54
Box extension	"	4.64	4.02	8.66
Plaster ring	"	3.08	4.02	7.10
Cover blank	"	1.17	4.02	5.19
Square box	"	3.38	7.75	11.13
Box extension	"	4.69	4.02	8.71
Plaster ring	"	2.57	4.02	6.59
Cover blank	"	1.32	4.02	5.34
4-11/16"				
Square box	EA.	9.70	7.75	17.45
Box extension	"	10.50	4.02	14.52
Plaster ring	"	6.40	4.02	10.42
Cover blank	"	2.37	4.02	6.39
Switch and device boxes				
2 gang	EA.	13.25	7.75	21.00
3 gang	"	16.25	7.75	24.00
4 gang	"	23.00	10.75	33.75
Device covers				
2 gang	EA.	11.50	4.02	15.52
3 gang	"	12.00	4.02	16.02
4 gang	"	16.25	4.02	20.27
Handy box	"	3.57	7.75	11.32
Extension	"	3.36	4.02	7.38
Switch cover	"	1.78	4.02	5.80
Switch box with knockout	"	5.36	9.86	15.22
Weatherproof cover, spring type	"	9.94	5.42	15.36
Cover plate, dryer receptacle 1 gang plastic	"	5.97	6.78	12.75
For 4" receptacle, 2 gang	"	6.10	6.78	12.88
Duplex receptacle cover plate, plastic	"	0.90	4.02	4.92
4", vertical bracket box, 1-1/2" with				
RMX clamps	EA.	6.90	9.86	16.76
BX clamps	"	7.41	9.86	17.27
4", octagon device cover				
1 switch	EA.	4.05	4.02	8.07
1 duplex recept	"	4.05	4.02	8.07
4", square face bracket boxes, 1-1/2"				
RMX	EA.	8.25	9.86	18.11
BX	"	8.96	9.86	18.82

Basic Materials	UNIT	MAT.	INST.	TOTAL
16130.60 **Pull And Junction Boxes** *(Cont.)*				
4" square to round plaster rings	EA.	2.75	4.02	6.77
2 gang device plaster rings	"	2.83	4.02	6.85
Surface covers				
1 gang switch	EA.	2.47	4.02	6.49
2 gang switch	"	2.53	4.02	6.55
1 single recept	"	3.72	4.02	7.74
1 20a twist lock recept	"	4.66	4.02	8.68
1 30a twist lock recept	"	5.97	4.02	9.99
1 duplex recept	"	2.31	4.02	6.33
2 duplex recept	"	2.31	4.02	6.33
Switch and duplex recept	"	3.85	4.02	7.87
4-11/16" square to round plaster rings	"	6.40	4.02	10.42
2 gang device plaster rings	"	5.28	4.02	9.30
Surface covers				
1 gang switch	EA.	7.10	4.02	11.12
2 gang switch	"	11.00	4.02	15.02
1 single recept	"	9.86	4.02	13.88
1 20a twist lock recept	"	9.69	4.02	13.71
1 30a twist lock recept	"	12.25	4.02	16.27
1 duplex recept	"	10.50	4.02	14.52
2 duplex recept	"	9.25	4.02	13.27
Switch and duplex recept	"	16.00	4.02	20.02
4" plastic round boxes, ground straps				
Box only	EA.	1.59	9.86	11.45
Box w/clamps	"	1.85	13.50	15.35
Box w/16" bar	"	3.94	15.50	19.44
Box w/24" bar	"	3.93	17.00	20.93
4" plastic round box covers				
Blank cover	EA.	1.04	4.02	5.06
Plaster ring	"	1.70	4.02	5.72
4" plastic square boxes				
Box only	EA.	1.23	9.86	11.09
Box w/clamps	"	2.14	13.50	15.64
Box w/hanger	"	1.88	17.00	18.88
Box w/nails and clamp	"	4.40	17.00	21.40
4" plastic square box covers				
Blank cover	EA.	1.01	4.02	5.03
1 gang ring	"	1.23	4.02	5.25
2 gang ring	"	1.72	4.02	5.74
Round ring	"	1.37	4.02	5.39
16130.65 **Pull Boxes And Cabinets**				
Galvanized pull boxes, screw cover				
4x4x4	EA.	7.70	13.00	20.70
4x6x4	"	9.62	13.00	22.62
6x6x4	"	12.50	13.00	25.50
6x8x4	"	15.00	13.00	28.00
8x8x4	"	18.75	17.00	35.75
8x10x4	"	21.50	16.50	38.00
8x12x4	"	23.75	17.00	40.75
Screw cover				
10x10x4	EA.	8.27	20.75	29.02
12x12x6	"	11.50	30.25	41.75

Basic Materials	UNIT	MAT.	INST.	TOTAL
16130.65 — Pull Boxes And Cabinets *(Cont.)*				
12x15x6	EA.	13.25	30.25	43.50
12x18x6	"	14.75	34.00	48.75
15x18x6	"	18.75	38.75	57.50
18x24x6	"	33.25	41.75	75.00
18x30x6	"	37.50	49.25	86.75
24x36x6	"	58.00	49.25	107
Cast iron junction box, unflanged				
6x6x4				
3/4" tap	EA.	79.00	34.00	113
1" tap	"	86.00	34.00	120
Two 1/2" taps	"	86.00	34.00	120
3/4" taps	"	86.00	34.00	120
6" adapter plate	"	34.50	23.50	58.00
6" exterior collar	"	56.00	23.50	79.50
Screw cover cabinet				
12x12x4	EA.	66.00	41.75	108
12x16x4	"	86.00	41.75	128
12x16x6	"	100	41.75	142
12x18x4	"	93.00	45.25	138
12x18x6	"	110	45.25	155
18x18x4	"	110	68.00	178
18x18x6	"	130	68.00	198
18x24x6	"	180	78.00	258
24x24x6	"	210	90.00	300
24x36x6	"	300	110	410
36x48x6	"	610	170	780
16130.80 — Receptacles				
Contractor grade duplex receptacles, 15a 120v				
Duplex	EA.	1.04	13.50	14.54
125 volt, 20a, duplex, grounding type, standard grade	"	9.07	13.50	22.57
Ground fault interrupter type	"	12.25	20.00	32.25
250 volt, 20a, 2 pole, single receptacle, ground type	"	6.60	13.50	20.10
120/208v, 4 pole, single receptacle, twist lock				
20a	EA.	21.50	23.50	45.00
50a	"	41.00	23.50	64.50
125/250v, 3 pole, flush receptacle				
30a	EA.	6.82	20.00	26.82
50a	"	7.89	20.00	27.89
60a	"	32.00	23.50	55.50
277v, 20a, 2 pole, grounding type, twist lock	"	11.75	13.50	25.25
Dryer receptacle, 250v, 30a/50a, 3 wire	"	16.25	20.00	36.25
Clock receptacle, 2 pole, grounding type	"	5.70	13.50	19.20
125v, 20a single recept. grounding type				
Standard grade	EA.	11.75	13.50	25.25
Specification	"	14.25	13.50	27.75
Hospital	"	14.75	13.50	28.25
Isolated ground orange	"	25.75	17.00	42.75
Duplex				
Specification grade	EA.	9.05	13.50	22.55
Hospital	"	11.75	13.50	25.25
Isolated ground orange	"	17.50	17.00	34.50
250v, 20a, duplex, 2 pole, grounding, spec. grade	"	11.75	13.50	25.25

Basic Materials	UNIT	MAT.	INST.	TOTAL

16130.80 — Receptacles *(Cont.)*

	UNIT	MAT.	INST.	TOTAL
Combination recepts, 20a, 125v and 250v, duplex	EA.	25.50	13.50	39.00
GFI hospital grade recepts, 20a, 125v, duplex	"	53.00	20.00	73.00
125/250v, 3 pole, 3 wire surface recepts				
30a	EA.	18.50	20.00	38.50
50a	"	20.50	20.00	40.50
60a	"	45.00	23.50	68.50
Cord set, 3 wire, 6' cord				
30a	EA.	16.50	20.00	36.50
50a	"	23.25	20.00	43.25

16198.10 — Electric Manholes

	UNIT	MAT.	INST.	TOTAL
Precast, handhole, 4' deep				
2'x2'	EA.	350	240	590
3'x3'	"	460	380	840
4'x4'	"	1,000	700	1,700
Power manhole, complete, precast, 8' deep				
4'x4'	EA.	1,420	950	2,370
6'x6'	"	1,900	1,360	3,260
8'x8'	"	2,260	1,430	3,690
6' deep, 9' x 12'	"	2,490	1,700	4,190
Cast in place, power manhole, 8' deep				
4'x4'	EA.	1,690	950	2,640
6'x6'	"	2,180	1,360	3,540
8'x8'	"	2,420	1,430	3,850

16199.10 — Utility Poles & Fittings

	UNIT	MAT.	INST.	TOTAL
Wood pole, creosoted				
25'	EA.	470	160	630
30'	"	570	200	770
35'	"	750	240	990
40'	"	900	260	1,160
45'	"	1,040	470	1,510
50'	"	1,220	490	1,710
55'	"	1,410	510	1,920
Treated, wood preservative, 6"x6"				
8'	EA.	95.00	34.00	129
10'	"	140	54.00	194
12'	"	150	60.00	210
14'	"	180	90.00	270
16'	"	220	110	330
18'	"	250	140	390
20'	"	320	140	460
Aluminum, brushed, no base				
8'	EA.	590	140	730
10'	"	680	180	860
15'	"	760	190	950
20'	"	920	220	1,140
25'	"	1,230	260	1,490
30'	"	1,850	300	2,150
35'	"	2,170	340	2,510
40'	"	2,780	420	3,200
Steel, no base				
10'	EA.	690	170	860

Basic Materials	UNIT	MAT.	INST.	TOTAL
16199.10 **Utility Poles & Fittings** *(Cont.)*				
15'	EA.	760	200	960
20'	"	1,010	260	1,270
25'	"	1,140	310	1,450
30'	"	1,460	350	1,810
35'	"	1,710	420	2,130
Concrete, no base				
13'	EA.	890	370	1,260
16'	"	1,240	490	1,730
18'	"	1,490	600	2,090
25'	"	1,820	680	2,500
30'	"	2,430	820	3,250
35'	"	3,130	950	4,080
40'	"	3,650	1,090	4,740
45'	"	4,340	1,150	5,490
50'	"	5,380	1,230	6,610
55'	"	5,990	1,290	7,280
60'	"	6,860	1,360	8,220
Pole line hardware				
Wood crossarm				
4'	EA.	72.00	90.00	162
8'	"	140	110	250
10'	"	280	140	420
Angle steel brace				
1 piece	EA.	10.50	17.00	27.50
2 piece	"	22.25	23.50	45.75
Eye nut, 5/8"	"	2.57	3.39	5.96
Bolt (14-16"), 5/8"	"	19.25	13.50	32.75
Transformer, ground connection	"	6.10	17.00	23.10
Stirrup	"	13.50	20.75	34.25
Secondary lead support	"	20.75	27.25	48.00
Spool insulator	"	4.44	13.50	17.94
Guy grip, preformed				
7/16"	EA.	2.64	9.86	12.50
1/2"	"	4.56	9.86	14.42
Hook	"	3.30	17.00	20.30
Strain insulator	"	26.75	24.75	51.50
Wire				
5/16"	L.F.	1.47	0.33	1.80
7/16"	"	1.46	0.40	1.86
1/2"	"	3.52	0.54	4.06
Soft drawn ground, copper, #8	"	0.45	0.54	0.99
Ground clamp	EA.	3.67	20.75	24.42
Perforated strapping for conduit, 1-1/2"	L.F.	2.75	9.86	12.61
Hot line clamp	EA.	15.00	54.00	69.00
Lightning arrester				
3kv	EA.	240	68.00	308
10kv	"	370	110	480
30kv	"	680	140	820
36kv	"	1,390	170	1,560

Power Generation	UNIT	MAT.	INST.	TOTAL

16210.10 Generators

Description	UNIT	MAT.	INST.	TOTAL
Diesel generator, with auto transfer switch				
30kw	EA.	27,180	2,090	29,270
50kw	"	34,420	2,090	36,510
75kw	"	43,700	2,860	46,560
100kw	"	48,500	3,190	51,690
125kw	"	51,830	3,390	55,220
150kw	"	59,750	3,880	63,630
175kw	"	62,270	4,520	66,790
200kw	"	64,590	5,430	70,020
250kw	"	70,590	6,030	76,620
300kw	"	84,900	6,780	91,680
350kw	"	90,890	7,750	98,640
400kw	"	111,580	9,050	120,630
450kw	"	119,520	9,870	129,390
500kw	"	129,470	10,860	140,330
600kw	"	177,330	13,570	190,900
750kw	"	247,530	13,570	261,100

16320.10 Transformers

Description	UNIT	MAT.	INST.	TOTAL
Floor mounted, single phase, int. dry, 480v-120/240v				
3 kva	EA.	420	120	540
5 kva	"	580	210	790
7.5 kva	"	870	240	1,110
10 kva	"	1,010	260	1,270
15 kva	"	1,350	290	1,640
25 kva	"	1,730	510	2,240
37.5 kva	"	2,360	640	3,000
50 kva	"	2,940	700	3,640
75 kva	"	3,900	720	4,620
100 kva	"	4,700	790	5,490
Three phase, 480v-120/208v				
15 kva	EA.	1,630	410	2,040
30 kva	"	1,960	640	2,600
45 kva	"	2,600	730	3,330
75 kva	"	3,920	740	4,660
112.5 kva	"	5,130	860	5,990
150 kva	"	6,100	920	7,020
225 kva	"	8,330	1,040	9,370

16350.10 Circuit Breakers

Description	UNIT	MAT.	INST.	TOTAL
Molded case, 240v, 15-60a, bolt-on				
1 pole	EA.	16.50	17.00	33.50
2 pole	"	35.25	23.50	58.75
70-100a, 2 pole	"	100	36.25	136
15-60a, 3 pole	"	120	27.25	147
70-100a, 3 pole	"	200	41.75	242
480v, 2 pole				
15-60a	EA.	250	20.00	270
70-100a	"	330	27.25	357
3 pole				
15-60a	EA.	330	27.25	357
70-100a	"	390	30.25	420
70-225a	"	800	41.75	842

Power Generation	UNIT	MAT.	INST.	TOTAL

16350.10 Circuit Breakers *(Cont.)*

	UNIT	MAT.	INST.	TOTAL
Draw out air circuit breakers				
600a	EA.	12,720	1,090	13,810
800a	"	16,430	1,230	17,660
1600a	"	26,390	1,640	28,030
2000a	"	35,360	1,870	37,230
3000a	"	61,370	2,170	63,540
4000a	"	94,260	2,580	96,840
Load center circuit breakers, 240v				
1 pole, 10-60a	EA.	16.50	17.00	33.50
2 pole				
10-60a	EA.	35.25	27.25	62.50
70-100a	"	100	45.25	145
110-150a	"	200	49.25	249
3 pole				
10-60a	EA.	110	34.00	144
70-100a	"	160	49.25	209
Load center, G.F.I. breakers, 240v				
1 pole, 15-30a	EA.	110	20.00	130
2 pole, 15-30a	"	200	27.25	227
Key operated breakers, 240v, 1 pole, 10-30a	"	19.50	20.00	39.50
Tandem breakers, 240v				
1 pole, 15-30a	EA.	31.25	27.25	58.50
2 pole, 15-30a	"	57.00	36.25	93.25
Bolt-on, G.F.I. breakers, 240v, 1 pole, 15-30a	"	130	23.50	154

16360.10 Safety Switches

	UNIT	MAT.	INST.	TOTAL
Fused, 3 phase, 30 amp, 600v, heavy duty				
NEMA 1	EA.	200	78.00	278
NEMA 3r	"	350	78.00	428
NEMA 4	"	910	110	1,020
NEMA 12	"	330	120	450
60a				
NEMA 1	EA.	280	78.00	358
NEMA 3r	"	420	78.00	498
NEMA 4	"	1,090	110	1,200
NEMA 12	"	340	120	460
100a				
NEMA 1	EA.	540	120	660
NEMA 3r	"	700	120	820
NEMA 4	"	2,180	140	2,320
NEMA 12	"	600	170	770
200a				
NEMA 1	EA.	900	170	1,070
NEMA 3r	"	790	170	960
NEMA 4	"	3,090	190	3,280
NEMA 12	"	920	240	1,160
400a				
NEMA 1	EA.	1,810	370	2,180
NEMA 3r	"	2,360	370	2,730
NEMA 4	"	6,140	390	6,530
NEMA 12	"	2,930	480	3,410
NEMA 4	"	7,280	610	7,890
NEMA 12	"	3,450	840	4,290

Power Generation	UNIT	MAT.	INST.	TOTAL

16360.10 — Safety Switches *(Cont.)*

	UNIT	MAT.	INST.	TOTAL
Non-fused, 240-600v, heavy duty, 3 phase, 30 amp				
NEMA 1	EA.	140	78.00	218
NEMA 3r	"	230	78.00	308
NEMA 4	"	910	120	1,030
NEMA 12	"	190	120	310
60a				
NEMA1	EA.	180	78.00	258
NEMA 3r	"	310	78.00	388
NEMA 4	"	890	120	1,010
NEMA 12	"	300	120	420
100a				
NEMA 1	EA.	280	120	400
NEMA 3r	"	440	120	560
NEMA 4	"	1,710	170	1,880
NEMA 12	"	420	170	590
200a, NEMA 1	"	430	170	600
600a, NEMA 12	"	2,260	840	3,100

16365.10 — Fuses

	UNIT	MAT.	INST.	TOTAL
Fuse, one-time, 250v				
30a	EA.	2.09	3.39	5.48
60a	"	3.52	3.39	6.91
100a	"	14.75	3.39	18.14
200a	"	35.75	3.39	39.14
400a	"	80.00	3.39	83.39
600a	"	140	3.39	143
600v				
30a	EA.	10.50	3.39	13.89
60a	"	16.75	3.39	20.14
100a	"	32.00	3.39	35.39
200a	"	85.00	3.39	88.39
400a	"	180	3.39	183
Fusetron, 600v				
200a	EA.	73.00	3.39	76.39
400a	"	150	3.39	153
Fuse, amp-trap, K1, 250v				
30a	EA.	6.76	3.39	10.15
60a	"	12.50	3.39	15.89
100a	"	27.75	3.39	31.14
200a	"	61.00	3.39	64.39
400a	"	140	3.39	143
600a	"	180	3.39	183
600v				
30a	EA.	21.25	3.39	24.64
60a	"	38.25	3.39	41.64
100a	"	77.00	3.39	80.39
200a	"	110	3.39	113
400a	"	240	3.39	243
K5, 250v				
30a	EA.	4.66	3.39	8.05
60a	"	8.49	3.39	11.88
100a	"	19.00	3.39	22.39
200a	"	42.00	3.39	45.39

Power Generation	UNIT	MAT.	INST.	TOTAL
16365.10 — **Fuses** *(Cont.)*				
400a	EA.	76.00	3.39	79.39
600a	"	120	3.39	123
600v				
30a	EA.	10.25	3.39	13.64
60a	"	17.75	3.39	21.14
100a	"	36.50	3.39	39.89
200a	"	73.00	3.39	76.39
400a	"	150	3.39	153
600a	"	210	3.39	213
J, 600v				
30a	EA.	18.25	3.39	21.64
60a	"	31.25	3.39	34.64
100a	"	70.00	3.39	73.39
200a	"	120	3.39	123
400a	"	220	3.39	223
L, 600v				
1200a	EA.	520	27.25	547
1600a	"	660	27.25	687
2000a	"	890	27.25	917
2500a	"	1,180	27.25	1,207
3000a	"	1,360	27.25	1,387
4000a	"	1,860	27.25	1,887
5000a	"	2,920	27.25	2,947
Fuse cl-ay 250v				
600a	EA.	460	20.00	480
1200a	"	460	20.00	480
1600a	"	570	20.00	590
2000a	"	740	20.00	760
600v				
1200a	EA.	550	20.00	570
1600a	"	640	20.00	660
2000a	"	770	20.00	790
16395.10 — **Grounding**				
Ground rods, copper clad, 1/2" x				
6'	EA.	10.75	45.25	56.00
8'	"	14.75	49.25	64.00
10'	"	18.50	68.00	86.50
5/8" x				
5'	EA.	13.25	41.75	55.00
6'	"	14.25	49.25	63.50
8'	"	18.50	68.00	86.50
10'	"	22.75	85.00	108
3/4" x				
8'	EA.	32.75	49.25	82.00
10'	"	35.75	54.00	89.75
Ground rod clamp				
5/8"	EA.	5.46	8.35	13.81
3/4"	"	7.73	8.35	16.08

Service And Distribution

16425.10 Switchboards

	UNIT	MAT.	INST.	TOTAL
Switchboard, 90" high, no main disconnect, 208/120v				
400a	EA.	2,630	540	3,170
600a	"	4,080	540	4,620
1000a	"	5,140	540	5,680
1200a	"	5,430	680	6,110
1600a	"	5,970	810	6,780
2000a	"	6,410	950	7,360
2500a	"	6,490	1,090	7,580
277/480v				
600a	EA.	4,690	550	5,240
800a	"	5,140	550	5,690
1600a	"	6,470	810	7,280
2000a	"	6,920	950	7,870
2500a	"	7,370	1,090	8,460
3000a	"	8,480	1,870	10,350
4000a	"	10,270	2,010	12,280
1500 kva	"	52,600	3,880	56,480

16470.10 Panelboards

	UNIT	MAT.	INST.	TOTAL
3 phase, 480/277v, main lugs only, 120a, 30 circuits	EA.	1,040	240	1,280
277/480v, 4 wire, flush surface				
225a, 30 circuits	EA.	1,710	270	1,980
400a, 30 circuits	"	2,310	340	2,650
600a, 42 circuits	"	4,450	410	4,860
208/120v, main circuit breaker, 3 phase, 4 wire				
100a				
12 circuits	EA.	900	350	1,250
20 circuits	"	1,120	430	1,550
30 circuits	"	1,650	480	2,130
225a				
30 circuits	EA.	1,410	530	1,940
42 circuits	"	2,550	650	3,200
400a				
30 circuits	EA.	3,480	1,010	4,490
42 circuits	"	4,170	1,090	5,260
600a, 42 circuits	"	8,110	1,230	9,340
120/208v, flush, 3 ph., 4 wire, main only				
100a				
12 circuits	EA.	640	350	990
20 circuits	"	880	430	1,310
30 circuits	"	1,310	480	1,790
225a				
30 circuits	EA.	1,330	530	1,860
42 circuits	"	1,680	650	2,330
400a				
30 circuits	EA.	2,550	1,010	3,560
42 circuits	"	3,720	1,090	4,810
600a, 42 circuits	"	5,800	1,230	7,030
Panelboard accessories				
Grounding bus	EA.	29.75	23.50	53.25
Handle lock device	"	14.50	9.86	24.36
Factory assembled panel				
1 pole space	EA.	18.50	23.50	42.00

Service And Distribution	UNIT	MAT.	INST.	TOTAL
16470.10 — **Panelboards** *(Cont.)*				
2 pole space	EA.	39.50	9.86	49.36
3 pole space	"	58.00	9.04	67.04
Panelboards 1 phase, 240/120v main circuit breaker				
Single phase, 3 wire, 120/240v flush				
100a, 20 circuits	EA.	830	240	1,070
225a, 30 circuits	"	1,660	270	1,930
240/120v, main lugs only				
100a				
8 circuits	EA.	450	200	650
12 circuits	"	480	200	680
20 circuits	"	510	200	710
225a				
24 circuits	EA.	540	240	780
30 circuits	"	620	260	880
42 circuits	"	680	260	940
16480.10 — **Motor Controls**				
Motor generator set, 3 phase, 480/277v, w/controls				
10kw	EA.	11,680	1,870	13,550
15kw	"	15,230	2,090	17,320
20kw	"	16,910	2,170	19,080
25kw	"	19,500	2,360	21,860
30kw	"	21,820	2,470	24,290
40kw	"	23,800	2,580	26,380
50kw	"	26,560	2,710	29,270
60kw	"	30,100	3,020	33,120
75kw	"	33,800	3,390	37,190
100kw	"	39,010	4,180	43,190
125kw	"	70,590	4,520	75,110
150kw	"	78,020	4,520	82,540
200kw	"	89,170	4,930	94,100
250kw	"	96,590	4,930	101,520
300kw	"	111,450	5,430	116,880
2 pole, 230 volt starter, w/NEMA-1				
1 hp, 9a, size 00	EA.	130	68.00	198
2 hp, 18a, size 0	"	140	68.00	208
3 hp, 27a, size 1	"	160	68.00	228
5 hp, 45a, size 1p	"	210	68.00	278
7-1/2 hp, 45a, size 2	"	560	68.00	628
15 hp, 90a, size 3	"	840	68.00	908
2 pole, w/NEMA-4 enclosure				
2 hp, 18a, size 1	EA.	470	110	580
5 hp, 45a, size 1p	"	600	110	710
7-1/2 hp, 45a, size 2	"	950	110	1,060
3 pole, 2 hp, 9a, 200-575v starter				
W/NEMA-1, size 00	EA.	290	90.00	380
W/NEMA-4 enclosure, size 00	"	470	120	590
5hp, 18a				
W/NEMA-1 enclosure, size 0	EA.	360	90.00	450
W/NEMA-4 enclosure, size 0	"	720	120	840
7.5-10hp, 27a				
7.5-10hp 27a, w/NEMA-1 enclosure, size 1	EA.	420	90.00	510
W/NEMA-4 enclosure size 1	"	780	120	900

Service And Distribution	UNIT	MAT.	INST.	TOTAL

16480.10 Motor Controls *(Cont.)*

	UNIT	MAT.	INST.	TOTAL
10-25hp, 45a				
W/NEMA-1 enclosure, size 2	EA.	820	90.00	910
W/NEMA-4 enclosure, size 2	"	1,540	120	1,660
25-50hp, 90a				
W/NEMA-1 enclosure, size 3	EA.	1,280	120	1,400
W/NEMA-4 enclosure, size 3	"	1,210	170	1,380
40-100hp, 135a				
W/NEMA-1 enclosure, size 4	EA.	2,070	170	2,240
W/NEMA-4 enclosure, size 4	"	3,190	240	3,430
75-200hp, 270a				
W/NEMA-1 enclosure, size 5	EA.	4,870	370	5,240
W/NEMA-4 enclosure, size 5	"	6,180	480	6,660
Magnetic starter accessories				
On-off-auto selector switch kit	EA.	33.50	21.75	55.25
With pilot light	"	63.00	23.50	86.50

16490.10 Switches

	UNIT	MAT.	INST.	TOTAL
Oil switches, medium voltage, bus components				
Switches, 277/120v, toggle device only	EA.	510	110	620
With oil 35kv, g&w gram 44, 4 way switch	"	16,740	540	17,280
Weatherproof enclosure				
3 way switch	EA.	20,680	680	21,360
4 way switch	"	19,690	740	20,430
Fused interrupter load, 35kv				21,660
20A				
1 pole	EA.	23,640	1,090	24,730
2 pole	"	25,610	1,150	26,760
3 way	"	27,580	1,150	28,730
4 way	"	29,550	1,230	30,780
30a, 1 pole	"	23,640	1,090	24,730
3 way	"	27,580	1,150	28,730
4 way	"	29,550	1,230	30,780
Weatherproof switch, including box & cover, 20a				
1 pole	EA.	25,600	1,090	26,690
2 pole	"	27,580	1,150	28,730
3 way	"	29,550	1,230	30,780
4 way	"	31,530	1,230	32,760
3 way, oil switch, 15kv enclosure	"	21,660	810	22,470
Pedestal for 35kv double breaker switch	"	860	340	1,200
Bus terminal connector, 2	"	750	170	920
2 to 3	"	910	170	1,080
Support connector, 3	"	480	110	590
Tee connector, 2 to 3	"	690	140	830
Flexible bus stud connector	"	590	120	710
End cap 3	"	640	90.00	730
Weldment connection, 3	"	260	68.00	328
Plate switch, 1 gang	"	0.64	3.39	4.03
Start stop stations, manual motor starters	"	64.00	49.25	113
Lockout switch	"	10.75	17.00	27.75
Forward-reverse switch	"	75.00	49.25	124
On-off switch	"	81.00	49.25	130
Open-close switch	"	75.00	49.25	124
Forward-reverse-stop switch	"	120	68.00	188

Service And Distribution

16490.10 — Switches *(Cont.)*

Service And Distribution	UNIT	MAT.	INST.	TOTAL
Standard 3 button switch any standard legend	EA.	86.00	68.00	154
Standard 3 button with lockout	"	86.00	68.00	154
Manual motor starters, tog, 115/230v				
Size 1 gp	EA.	140	68.00	208
Size 2	"	170	68.00	238
Button				
Size 0	EA.	100	68.00	168
Size 1	"	150	68.00	218
Size 2	"	180	68.00	248
3-phase				
Size 0	EA.	220	90.00	310
Size 1	"	270	90.00	360
Time & float switches	"	340	110	450
Astronomical time switch, 40a, 240v	"	530	68.00	598
Timer switch 0-5 minute, with box	"	25.75	34.00	59.75
Single pole/single throw time, 277v, NEMA-1	"	96.00	49.25	145
Single toggle switch, 20a, 120v, with pilot	"	23.00	17.00	40.00
3-way toggle	"	64.00	20.00	84.00
Photo electric switches				
1000 watt				
105-135v	EA.	33.50	49.25	82.75
208-277v	"	45.25	49.25	94.50
3000 watt, 105-130v				
Double throw	EA.	130	68.00	198
Single throw	"	120	68.00	188
Double pole/single throw, 210-250v	"	150	90.00	240
Dimmer switch and switch plate				
600w	EA.	30.75	20.75	51.50
1000w	"	45.00	23.50	68.50
Dimmer switch incandescent				
1500w	EA.	77.00	47.50	125
2000w	"	100	51.00	151
Fluorescent				
12 lamps	EA.	50.00	34.00	84.00
20 lamps	"	160	37.50	198
30 lamps	"	280	40.75	321
40 lamps	"	260	47.50	308
Specification grade toggle switches, 20a, 120-277v				
Single pole	EA.	3.57	13.50	17.07
Double pole	"	8.58	20.00	28.58
3 way	"	7.42	17.00	24.42
4 way	"	21.50	20.00	41.50
Switch plates, plastic ivory				
1 gang	EA.	0.34	5.42	5.76
2 gang	"	0.69	6.78	7.47
3 gang	"	1.02	8.10	9.12
4 gang	"	3.24	9.86	13.10
5 gang	"	3.68	10.75	14.43
6 gang	"	4.34	12.25	16.59
Stainless steel				
1 gang	EA.	3.16	5.42	8.58
2 gang	"	4.40	6.78	11.18
3 gang	"	6.75	8.35	15.10

Service And Distribution	UNIT	MAT.	INST.	TOTAL
16490.10 Switches *(Cont.)*				
4 gang	EA.	11.50	9.86	21.36
5 gang	"	13.50	10.75	24.25
6 gang	"	17.00	12.25	29.25
Brass				
1 gang	EA.	5.90	5.42	11.32
2 gang	"	12.75	6.78	19.53
3 gang	"	19.50	8.35	27.85
4 gang	"	22.50	9.86	32.36
5 gang	"	28.00	10.75	38.75
6 gang	"	33.75	12.25	46.00
16490.20 Transfer Switches				
Automatic transfer switch 600v, 3 pole				
30a	EA.	2,960	240	3,200
60a	"	3,580	240	3,820
100a	"	3,920	320	4,240
150a	"	5,230	410	5,640
225a	"	6,530	540	7,070
260a	"	7,130	540	7,670
400a	"	8,840	680	9,520
600a	"	12,860	1,020	13,880
800a	"	16,280	1,230	17,510
1000a	"	23,210	1,430	24,640
1200a	"	26,830	1,550	28,380
1600a	"	33,860	1,700	35,560
2000a	"	34,250	2,010	36,260
2600a	"	70,300	2,860	73,160
3000a	"	110,470	3,390	113,860
16490.80 Safety Switches				
Safety switch, 600v, 3 pole, heavy duty, NEMA-1				
30a	EA.	200	68.00	268
60a	"	260	78.00	338
100a	"	510	110	620
200a	"	790	170	960
400a	"	2,040	370	2,410
600a	"	3,620	540	4,160
800a	"	8,200	710	8,910
1200a	"	10,180	970	11,150

Lighting	UNIT	MAT.	INST.	TOTAL

16510.05 — Interior Lighting

	UNIT	MAT.	INST.	TOTAL
Recessed fluorescent fixtures, 2'x2'				
2 lamp	EA.	63.00	49.25	112
4 lamp	"	85.00	49.25	134
2 lamp w/flange	"	79.00	68.00	147
4 lamp w/flange	"	97.00	68.00	165
1'x4'				
2 lamp	EA.	64.00	45.25	109
3 lamp	"	88.00	45.25	133
2 lamp w/flange	"	79.00	49.25	128
3 lamp w/flange	"	110	49.25	159
2'x4'				
2 lamp	EA.	79.00	49.25	128
3 lamp	"	97.00	49.25	146
4 lamp	"	88.00	49.25	137
2 lamp w/flange	"	97.00	68.00	165
3 lamp w/flange	"	110	68.00	178
4 lamp w/flange	"	110	68.00	178
4'x4'				
4 lamp	EA.	320	68.00	388
6 lamp	"	380	68.00	448
8 lamp	"	400	68.00	468
4 lamp w/flange	"	390	100	490
6 lamp w/flange	"	490	100	590
8 lamp, w/flange	"	540	100	640
Surface mounted incandescent fixtures				
40w	EA.	75.00	45.25	120
75w	"	82.00	45.25	127
100w	"	97.00	45.25	142
150w	"	110	45.25	155
Pendant				
40w	EA.	78.00	54.00	132
75w	"	86.00	54.00	140
100w	"	98.00	54.00	152
150w	"	110	54.00	164
Recessed incandescent fixtures				
40w	EA.	130	100	230
75w	"	140	100	240
100w	"	150	100	250
150w	"	160	100	260
Exit lights, 120v				
Recessed	EA.	33.00	85.00	118
Back mount	"	55.00	49.25	104
Universal mount	"	33.00	49.25	82.25
Emergency battery units, 6v-120v, 50 unit	"	150	100	250
With 1 head	"	180	100	280
With 2 heads	"	200	100	300
Light track single circuit				
2'	EA.	32.00	34.00	66.00
4'	"	38.00	34.00	72.00
8'	"	52.00	68.00	120
12'	"	73.00	100	173
Fittings and accessories				
Dead end	EA.	15.25	9.86	25.11

Lighting	UNIT	MAT.	INST.	TOTAL
16510.05		**Interior Lighting** *(Cont.)*		
Starter kit	EA.	20.50	17.00	37.50
Conduit feed	"	19.75	9.86	29.61
Straight connector	"	17.50	9.86	27.36
Center feed	"	28.00	9.86	37.86
L-connector	"	19.75	9.86	29.61
T-connector	"	26.50	9.86	36.36
X-connector	"	32.00	13.50	45.50
Cord and plug	"	32.00	6.78	38.78
Rigid corner	"	42.25	9.86	52.11
Flex connector	"	33.00	9.86	42.86
2 way connector	"	92.00	13.50	106
Spacer clip	"	1.43	3.39	4.82
Grid box	"	7.97	9.86	17.83
T-bar clip	"	2.13	3.39	5.52
Utility hook	"	6.18	9.86	16.04
Fixtures, square				
R-20	EA.	39.25	9.86	49.11
R-30	"	61.00	9.86	70.86
40w flood	"	99.00	9.86	109
40w spot	"	99.00	9.86	109
100w flood	"	110	9.86	120
100w spot	"	88.00	9.86	97.86
Mini spot	"	37.25	9.86	47.11
Mini flood	"	86.00	9.86	95.86
Quartz, 500w	"	220	9.86	230
R-20 sphere	"	66.00	9.86	75.86
R-30 sphere	"	34.50	9.86	44.36
R-20 cylinder	"	46.25	9.86	56.11
R-30 cylinder	"	54.00	9.86	63.86
R-40 cylinder	"	54.00	9.86	63.86
R-30 wall wash	"	86.00	9.86	95.86
R-40 wall wash	"	87.00	9.86	96.86
Energy saving rapid start fluor. lamps				
F30 cw	EA.	8.14	6.78	14.92
F40 cw	"	8.14	6.78	14.92
F40 cwx	"	9.13	6.78	15.91
F30 ww	"	12.25	6.78	19.03
F40 ww	"	6.71	6.78	13.49
F40 wwx	"	9.13	6.78	15.91
Incandescent lamps				
200w	EA.	5.33	6.78	12.11
300w	"	5.83	6.78	12.61
500w	"	13.00	6.78	19.78
750w	"	31.75	9.86	41.61
1000w	"	34.75	13.50	48.25
1500w	"	51.00	13.50	64.50
Energy saving reflector floodlight lamps				
25w	EA.	9.35	6.78	16.13
30w	"	10.00	6.78	16.78
50w	"	10.00	6.78	16.78
75w	"	11.50	6.78	18.28
120w	"	11.50	6.78	18.28
150w	"	16.00	6.78	22.78

Lighting	UNIT	MAT.	INST.	TOTAL
16510.05 Interior Lighting *(Cont.)*				
200w	EA.	16.75	6.78	23.53
300w	"	22.50	6.78	29.28
500w	"	37.25	9.86	47.11
750w	"	48.50	9.86	58.36
Reflector spotlight				
75w	EA.	8.69	6.78	15.47
100w	"	10.50	6.78	17.28
125w	"	18.00	6.78	24.78
150w	"	19.00	6.78	25.78
250w	"	31.25	6.78	38.03
300w	"	49.75	6.78	56.53
400w	"	55.00	9.86	64.86
500w	"	69.00	9.86	78.86
1000w	"	150	13.50	164
Medium par flood lamps				
75w	EA.	8.58	6.78	15.36
100w	"	9.57	6.78	16.35
150w	"	10.25	6.78	17.03
200w	"	39.75	6.78	46.53
300w	"	44.00	6.78	50.78
500w	"	86.00	9.86	95.86
Medium par spot lamps				
75w	EA.	10.75	6.78	17.53
120w	"	11.00	6.78	17.78
150w	"	11.50	6.78	18.28
Ballast replacements rapid start fluor				
1f-40-120v	EA.	23.25	49.25	72.50
1f-40-277v	"	39.00	49.25	88.25
1f-96-120v	"	120	49.25	169
1f-96-277v	"	140	49.25	189
2f-40-120v	"	33.00	49.25	82.25
2f-40-277v	"	39.00	49.25	88.25
2f-96-120v	"	100	49.25	149
2f-96-277v	"	110	49.25	159
16510.10 Lighting Industrial				
Surface mounted fluorescent, wrap around lens				
1 lamp	EA.	86.00	54.00	140
2 lamps	"	140	60.00	200
4 lamps	"	130	68.00	198
Wall mounted fluorescent				
2-20w lamps	EA.	88.00	34.00	122
2-30w lamps	"	100	34.00	134
2-40w lamps	"	100	45.25	145
Indirect, with wood shielding, 2049w lamps				
4'	EA.	100	68.00	168
8'	"	130	110	240
Industrial fluorescent, 2 lamp				
4'	EA.	51.00	49.25	100
8'	"	100	90.00	190
Strip fluorescent				
4'				
1 lamp	EA.	43.25	45.25	88.50

Lighting	UNIT	MAT.	INST.	TOTAL
16510.10 **Lighting Industrial** *(Cont.)*				
2 lamps	EA.	53.00	45.25	98.25
8'				
1 lamp	EA.	48.25	49.25	97.50
2 lamps	"	88.00	60.00	148
Wire guard for strip fixture, 4' long	"	9.80	23.50	33.30
Strip fluorescent, 8' long, two 4' lamps	"	130	90.00	220
With four 4' lamps	"	160	110	270
Wet location fluorescent, plastic housing				
4' long				
1 lamp	EA.	140	68.00	208
2 lamps	"	150	90.00	240
8' long				
2 lamps	EA.	260	110	370
4 lamps	"	350	120	470
Parabolic troffer, 2'x2'				
With 2 "U" lamps	EA.	120	68.00	188
With 3 "U" lamps	"	140	78.00	218
2'x4'				
With 2 40w lamps	EA.	110	78.00	188
With 3 40w lamps	"	110	90.00	200
With 4 40w lamps	"	110	90.00	200
1'x4'				
With 1 T-12 lamp, 9 cell	EA.	110	49.25	159
With 2 T-12 lamps	"	130	60.00	190
With 1 T-12 lamp, 20 cell	"	130	49.25	179
With 2 T-12 lamps	"	140	60.00	200
Steel sided surface fluorescent, 2'x4'				
3 lamps	EA.	100	90.00	190
4 lamps	"	110	90.00	200
Outdoor sign fluor., 1 lamp, remote ballast				
4' long	EA.	3,010	410	3,420
6' long	"	3,620	540	4,160
Recess mounted, commercial, 2'x2', 13" high				
100w	EA.	1,000	270	1,270
250w	"	1,100	300	1,400
High pressure sodium, hi-bay open				
400w	EA.	430	120	550
1000w	"	750	160	910
Enclosed				
400w	EA.	700	160	860
1000w	"	980	200	1,180
Metal halide hi-bay, open				
400w	EA.	270	120	390
1000w	"	460	160	620
Enclosed				
400w	EA.	610	160	770
1000w	"	580	200	780
High pressure sodium, low bay, surface mounted				
100w	EA.	220	68.00	288
150w	"	240	78.00	318
250w	"	270	90.00	360
400w	"	340	110	450
Metal halide, low bay, pendant mounted				

Lighting	UNIT	MAT.	INST.	TOTAL
16510.10 — **Lighting Industrial** *(Cont.)*				
175w	EA.	350	90.00	440
250w	"	480	110	590
400w	"	520	150	670
Indirect luminare, square, metal halide, freestanding				
175w	EA.	440	68.00	508
250w	"	470	68.00	538
400w	"	490	68.00	558
High pressure sodium				
150w	EA.	790	68.00	858
250w	"	850	68.00	918
400w	"	930	68.00	998
Round, metal halide				
175w	EA.	910	68.00	978
250w	"	950	68.00	1,018
400w	"	990	68.00	1,058
High pressure sodium				
150w	EA.	870	68.00	938
250w	"	1,010	68.00	1,078
400w	"	1,060	68.00	1,128
Wall mounted, metal halide				
175w	EA.	400	170	570
250w	"	390	170	560
400w	"	430	220	650
High pressure sodium				
150w	EA.	370	170	540
250w	"	390	170	560
400w	"	400	220	620
Wall pack lithonia, high pressure sodium				
35w	EA.	57.00	60.00	117
55w	"	72.00	68.00	140
150w	"	180	110	290
250w	"	190	120	310
Low pressure sodium				
35w	EA.	310	120	430
55w	"	420	140	560
Wall pack hubbell, high pressure sodium				
35w	EA.	250	60.00	310
150w	"	320	110	430
250w	"	410	120	530
Compact fluorescent				
2-7w	EA.	150	68.00	218
2-13w	"	180	90.00	270
1-18w	"	210	90.00	300
Handball & racquet ball court, 2'x2', metal halide				
250w	EA.	550	170	720
400w	"	660	190	850
High pressure sodium				
250w	EA.	600	170	770
400w	"	660	190	850
Bollard light, 42" w/found., high pressure sodium				
70w	EA.	920	180	1,100
100w	"	950	180	1,130
150w	"	960	180	1,140

Lighting	UNIT	MAT.	INST.	TOTAL

16510.10 — Lighting Industrial *(Cont.)*

	UNIT	MAT.	INST.	TOTAL
Light fixture lamps				
Lamp				
20w med. bipin base, cool white, 24"	EA.	17.00	9.86	26.86
30w cool white, rapid start, 36"	"	19.00	9.86	28.86
40w cool white "U", 3'	"	20.50	9.86	30.36
40w cool white, rapid start, 48"	"	4.07	9.86	13.93
70w high pressure sodium, mogul base	"	59.00	13.50	72.50
75w slimline, 96"	"	11.25	13.50	24.75
100w				
Incandescent, 100a, inside frost	EA.	3.52	6.78	10.30
Mercury vapor, clear, mogul base	"	50.00	13.50	63.50
High pressure sodium, mogul base	"	94.00	13.50	108
150w				
Par 38 flood or spot, incandescent	EA.	10.50	6.78	17.28
High pressure sodium, 1/2 mogul base	"	63.00	13.50	76.50
175w				
Mercury vapor, clear, mogul base	EA.	32.00	13.50	45.50
Metal halide, clear, mogul base	"	27.25	13.50	40.75
High pressure sodium, mogul base	"	45.00	13.50	58.50
250w				
Mercury vapor, clear, mogul base	EA.	42.25	13.50	55.75
Metal halide, clear, mogul base	"	27.50	13.50	41.00
High pressure sodium, mogul base	"	97.00	13.50	111
400w				
Mercury vapor, clear, mogul base	EA.	46.00	13.50	59.50
Metal halide, clear, mogul base	"	28.00	13.50	41.50
High pressure sodium, mogul base	"	100	13.50	114
1000w				
Mercury vapor, clear, mogul base	EA.	140	17.00	157
High pressure sodium, mogul base	"	230	17.00	247

16510.30 — Exterior Lighting

	UNIT	MAT.	INST.	TOTAL
Exterior light fixtures				
Rectangle, high pressure sodium				
70w	EA.	300	170	470
100w	"	310	180	490
150w	"	330	180	510
250w	"	450	190	640
400w	"	500	240	740
Flood, rectangular, high pressure sodium				
70w	EA.	180	170	350
100w	"	210	180	390
150w	"	200	180	380
400w	"	250	240	490
1000w	"	410	300	710
Round				
400w	EA.	550	240	790
1000w	"	860	300	1,160
Round, metal halide				
400w	EA.	600	240	840
1000w	"	900	300	1,200
Light fixture arms, cobra head, 6', high press. sodium				
100w	EA.	330	140	470

Lighting	UNIT	MAT.	INST.	TOTAL
16510.30 **Exterior Lighting** *(Cont.)*				
150w	EA.	520	170	690
250w	"	540	170	710
400w	"	560	200	760
Flood, metal halide				
400w	EA.	540	240	780
1000w	"	740	300	1,040
1500w	"	930	410	1,340
Mercury vapor				
250w	EA.	360	190	550
400w	"	400	240	640
Incandescent				
300w	EA.	83.00	120	203
500w	"	150	140	290
1000w	"	160	220	380
16510.90 **Power Line Filters**				
Heavy duty power line filter, 240v				
100a	EA.	4,510	680	5,190
300a	"	14,970	1,090	16,060
600a	"	20,750	1,640	22,390
16610.30 **Uninterruptible Power**				
Uninterruptible power systems, (U.P.S.), 3kva	EA.	7,810	540	8,350
5 kva	"	8,760	750	9,510
7.5 kva	"	10,510	1,090	11,600
10 kva	"	13,140	1,490	14,630
15 kva	"	15,770	1,550	17,320
20 kva	"	21,900	1,630	23,530
25 kva	"	28,030	1,700	29,730
30 kva	"	28,910	1,760	30,670
35 kva	"	30,660	1,830	32,490
40 kva	"	33,290	1,900	35,190
45 kva	"	35,040	1,970	37,010
50 kva	"	37,670	2,030	39,700
62.5 kva	"	44,680	2,170	46,850
75 kva	"	51,680	2,370	54,050
100 kva	"	69,210	2,450	71,660
150 kva	"	105,130	3,390	108,520
200 kva	"	140,170	3,740	143,910
300 kva	"	210,250	5,070	215,320
400 kva	"	314,530	6,100	320,630
500 kva	"	393,170	7,440	400,610
16670.10 **Lightning Protection**				
Lightning protection				
Copper point, nickel plated, 12'				
1/2" dia.	EA.	44.00	68.00	112
5/8" dia.	"	49.50	68.00	118

Communications

16720.10   Fire Alarm Systems

	UNIT	MAT.	INST.	TOTAL
Master fire alarm box, pedestal mounted	EA.	7,110	1,090	8,200
Master fire alarm box	"	3,650	410	4,060
Box light	"	130	34.00	164
Ground assembly for box	"	100	45.25	145
Bracket for pole type box	"	130	49.25	179
Pull station				
Waterproof	EA.	65.00	34.00	99.00
Manual	"	48.50	27.25	75.75
Horn, waterproof	"	91.00	68.00	159
Interior alarm	"	61.00	49.25	110
Coded transmitter, automatic	"	930	140	1,070
Control panel, 8 zone	"	2,140	540	2,680
Battery charger and cabinet	"	720	140	860
Batteries, nickel cadmium or lead calcium	"	540	340	880
CO2 pressure switch connection	"	100	49.25	149
Annunciator panels				
Fire detection annunciator, remote type, 8 zone	EA.	360	120	480
12 zone	"	460	140	600
16 zone	"	580	170	750
Fire alarm systems				
Bell	EA.	110	41.75	152
Weatherproof bell	"	71.00	45.25	116
Horn	"	65.00	49.25	114
Siren	"	660	140	800
Chime	"	81.00	41.75	123
Audio/visual	"	120	49.25	169
Strobe light	"	110	49.25	159
Smoke detector	"	180	45.25	225
Heat detection	"	30.50	34.00	64.50
Thermal detector	"	28.50	34.00	62.50
Ionization detector	"	140	36.25	176
Duct detector	"	470	190	660
Test switch	"	81.00	34.00	115
Remote indicator	"	51.00	38.75	89.75
Door holder	"	180	49.25	229
Telephone jack	"	3.30	20.00	23.30
Fireman phone	"	430	68.00	498
Speaker	"	87.00	54.00	141
Remote fire alarm annunciator panel				
24 zone	EA.	2,230	450	2,680
48 zone	"	4,470	880	5,350
Control panel				
12 zone	EA.	1,520	200	1,720
16 zone	"	1,990	300	2,290
24 zone	"	3,050	450	3,500
48 zone	"	5,690	1,090	6,780
Power supply	"	350	100	450
Status command	"	9,640	340	9,980
Printer	"	3,080	100	3,180
Transponder	"	150	61.00	211
Transformer	"	220	45.25	265
Transceiver	"	310	49.25	359
Relays	"	120	34.00	154

Communications	UNIT	MAT.	INST.	TOTAL
16720.10 Fire Alarm Systems *(Cont.)*				
Flow switch	EA.	400	140	540
Tamper switch	"	240	200	440
End of line resistor	"	17.25	23.50	40.75
Printed ckt. card	"	150	34.00	184
Central processing unit	"	10,540	420	10,960
UPS backup to c.p.u.	"	19,290	610	19,900
Smoke detector, fixed temp. & rate of rise comb.	"	310	110	420
16720.50 Security Systems				
Sensors				
Balanced magnetic door switch, surface mounted	EA.	150	34.00	184
With remote test	"	200	68.00	268
Flush mounted	"	140	130	270
Mounted bracket	"	11.00	23.50	34.50
Mounted bracket spacer	"	9.84	23.50	33.34
Photoelectric sensor, for fence				
6 beam	EA.	16,060	190	16,250
9 beam	"	19,630	290	19,920
Photoelectric sensor, 12 volt dc				
500' range	EA.	470	110	580
800' range	"	530	140	670
Capacitance wire grid kit				
Surface	EA.	130	68.00	198
Duct	"	98.00	110	208
Tube grid kit	"	160	34.00	194
Vibration sensor, 30 max per zone	"	200	34.00	234
Audio sensor, 30 max per zone	"	210	34.00	244
Inertia sensor				
Outdoor	EA.	150	49.25	199
Indoor	"	98.00	34.00	132
Ultrasonic transmitter, 20 max per zone				
Omni-directional	EA.	110	110	220
Directional	"	120	90.00	210
Transceiver				
Omni-directional	EA.	120	68.00	188
Directional	"	130	68.00	198
Passive infra-red sensor, 20 max per zone	"	850	110	960
Access/secure control unit, balanced magnetic switch	"	530	110	640
Photoelectric sensor	"	870	110	980
Photoelectric fence sensor	"	890	110	1,000
Capacitance sensor	"	1,030	120	1,150
Audio and vibration sensor	"	910	110	1,020
Inertia sensor	"	1,220	110	1,330
Ultrasonic sensor	"	1,400	120	1,520
Infra-red sensor	"	870	140	1,010
Monitor panel, with access/secure tone, standard	"	590	120	710
High security	"	870	140	1,010
Emergency power indicator	"	350	34.00	384
Monitor rack with 115v power supply				
1 zone	EA.	500	68.00	568
10 zone	"	2,520	170	2,690
Monitor cabinet, wall mounted				
1 zone	EA.	760	68.00	828

Communications	UNIT	MAT.	INST.	TOTAL

16720.50 — Security Systems *(Cont.)*

	UNIT	MAT.	INST.	TOTAL
5 zone	EA.	2,740	110	2,850
10 zone	"	1,250	120	1,370
20 zone	"	3,820	140	3,960
Floor mounted, 50 zone	"	3,970	270	4,240
Security system accessories				
Tamper assembly for monitor cabinet	EA.	98.00	30.25	128
Monitor panel blank	"	13.25	23.50	36.75
Audible alarm	"	110	34.00	144
Audible alarm control	"	460	23.50	484
Termination screw, terminal cabinet				
25 pair	EA.	330	110	440
50 pair	"	530	170	700
150 pair	"	860	340	1,200
Universal termination, cabinets & panel				
Remote test	EA.	77.00	120	197
No remote test	"	55.00	49.25	104
High security line supervision termination	"	390	68.00	458
Door cord for capacitance sensor, 12"	"	13.25	34.00	47.25
Insulation block kit for capacitance sensor	"	62.00	23.50	85.50
Termination block for capacitance sensor	"	13.25	23.50	36.75
Guard alert display	"	1,400	41.75	1,442
Uninterrupted power supply	"	1,420	540	1,960
Plug-in 40kva transformer				
12 volt	EA.	56.00	23.50	79.50
18 volt	"	37.25	23.50	60.75
24 volt	"	26.25	23.50	49.75
Test relay	"	85.00	23.50	109
Coaxial cable, 50 ohm	L.F.	0.38	0.40	0.78
Door openers	EA.	98.00	34.00	132
Push buttons				
Standard	EA.	22.00	23.50	45.50
Weatherproof	"	33.00	30.25	63.25
Bells	"	84.00	49.25	133
Horns				
Standard	EA.	87.00	68.00	155
Weatherproof	"	160	85.00	245
Chimes	"	130	45.25	175
Flasher	"	92.00	41.75	134
Motion detectors	"	390	100	490
Intercom units	"	87.00	49.25	136
Remote annunciator	"	3,950	340	4,290

16740.10 — Telephone Systems

	UNIT	MAT.	INST.	TOTAL
Communication cable				
25 pair	L.F.	0.93	1.75	2.68
100 pair	"	4.45	1.93	6.38
150 pair	"	6.76	2.26	9.02
200 pair	"	9.13	2.71	11.84
300 pair	"	11.50	2.85	14.35
400 pair	"	15.75	3.01	18.76
Cable tap in manhole or junction box				
25 pair cable	EA.	6.49	260	266
50 pair cable	"	13.25	510	523

Communications	UNIT	MAT.	INST.	TOTAL
16740.10 **Telephone Systems** *(Cont.)*				
75 pair cable	EA.	19.75	760	780
100 pair cable	"	26.25	1,020	1,046
150 pair cable	"	39.25	1,510	1,549
200 pair cable	"	52.00	2,010	2,062
300 pair cable	"	79.00	3,020	3,099
400 pair cable	"	110	4,180	4,290
Cable terminations, manhole or junction box				
25 pair cable	EA.	6.49	250	256
50 pair cable	"	13.25	510	523
100 pair cable	"	26.25	1,020	1,046
150 pair cable	"	39.25	1,510	1,549
200 pair cable	"	52.00	2,010	2,062
300 pair cable	"	79.00	3,020	3,099
400 pair cable	"	81.00	4,180	4,261
16780.10 **Antennas And Towers**				
Guy cable, alumaweld				
1x3, 7/32"	L.F.	0.45	3.39	3.84
1x3, 1/4"	"	0.52	3.39	3.91
1x3, 25/64"	"	0.74	4.02	4.76
1x19, 1/2"	"	1.87	4.72	6.59
1x7, 35/64"	"	2.14	5.42	7.56
1x19, 13/16"	"	2.31	6.78	9.09
Preformed alumaweld end grip				
1/4" cable	EA.	3.41	6.78	10.19
3/8" cable	"	4.51	6.78	11.29
1/2" cable	"	5.66	9.86	15.52
9/16" cable	"	7.04	13.50	20.54
5/8" cable	"	8.63	17.00	25.63
Fiberglass guy rod, white epoxy coated				
1/4" dia.	L.F.	2.31	9.86	12.17
3/8" dia	"	3.46	9.86	13.32
1/2" dia	"	4.62	13.50	18.12
5/8" dia	"	5.77	17.00	22.77
Preformed glass grip end grip, guy rod				
1/4" dia.	EA.	12.75	9.86	22.61
3/8" dia.	"	14.75	13.50	28.25
1/2" dia.	"	18.00	17.00	35.00
5/8" dia.	"	20.25	17.00	37.25
Spelter socket end grip, 1/4" dia. guy rod				
Standard strength	EA.	37.75	34.00	71.75
High performance	"	46.50	34.00	80.50
3/8" dia. guy rod				
Standard strength	EA.	36.75	23.50	60.25
High performance	"	46.50	34.00	80.50
Timber pole, Douglas Fir				
80-85 ft	EA.	3,530	1,320	4,850
90-95 ft	"	4,420	1,510	5,930
Southern yellow pine				
35-45 ft	EA.	2,110	740	2,850
50-55 ft	"	2,950	950	3,900

Communications	UNIT	MAT.	INST.	TOTAL
16780.50 **Television Systems**				
TV outlet, self terminating, w/cover plate	EA.	5.97	20.75	26.72
Thru splitter	"	13.00	110	123
End of line	"	10.75	90.00	101
In line splitter multitap				
4 way	EA.	21.75	120	142
2 way	"	16.25	120	136
Equipment cabinet	"	54.00	110	164
Antenna				
Broad band uhf	EA.	110	240	350
Lightning arrester	"	33.00	49.25	82.25
TV cable	L.F.	0.49	0.33	0.82
Coaxial cable rg	"	0.33	0.33	0.66
Cable drill, with replacement tip	EA.	5.44	34.00	39.44
Cable blocks for in-line taps	"	10.75	49.25	60.00
In-line taps ptu-series 36 tv system	"	13.00	78.00	91.00
16850.10 **Electric Heating**				
Baseboard heater				
2', 375w	EA.	41.75	68.00	110
3', 500w	"	49.50	68.00	118
4', 750w	"	55.00	78.00	133
5', 935w	"	78.00	90.00	168
6', 1125w	"	92.00	110	202
7', 1310w	"	100	120	220
8', 1500w	"	120	140	260
9', 1680w	"	130	150	280
10', 1875w	"	180	160	340
Unit heater, wall mounted				
1500w	EA.	220	110	330
2500w	"	240	120	360
4000w	"	320	160	480
Thermostat				
Integral	EA.	37.50	34.00	71.50
Line voltage	"	38.50	34.00	72.50
Electric heater connection	"	1.65	17.00	18.65

BNi Building News

Man-Hour Tables

The man-hour productivities used to develop the labor costs are listed in the following section of this book. These productivities represent typical installation labor for thousands of construction items. The data takes into account all activities involved in normal construction under commonly experienced working conditions. As with the Costbook pages, these items are listed according to the CSI MASTERFORMAT. In order to best use the information in this book, please review this sample page and read the "Features in this Book" section.

CSI MASTERFORMAT Division

CSI Broadscope Category

CSI Mediumscope Category (First 5 Digits)

Detailed Descriptions

Complete descriptions of items may include information listed above a particular line. Review of the whole category is recommended for a complete description.

Unit of Measurement

Each item is defined in terms of the common estimating unit. Quantities listed are defined as man-hour per unit.

Man-Hours

Man-hour quantities represent typical installation times and take into account all activities involved in normal construction under commonly experienced working conditions.

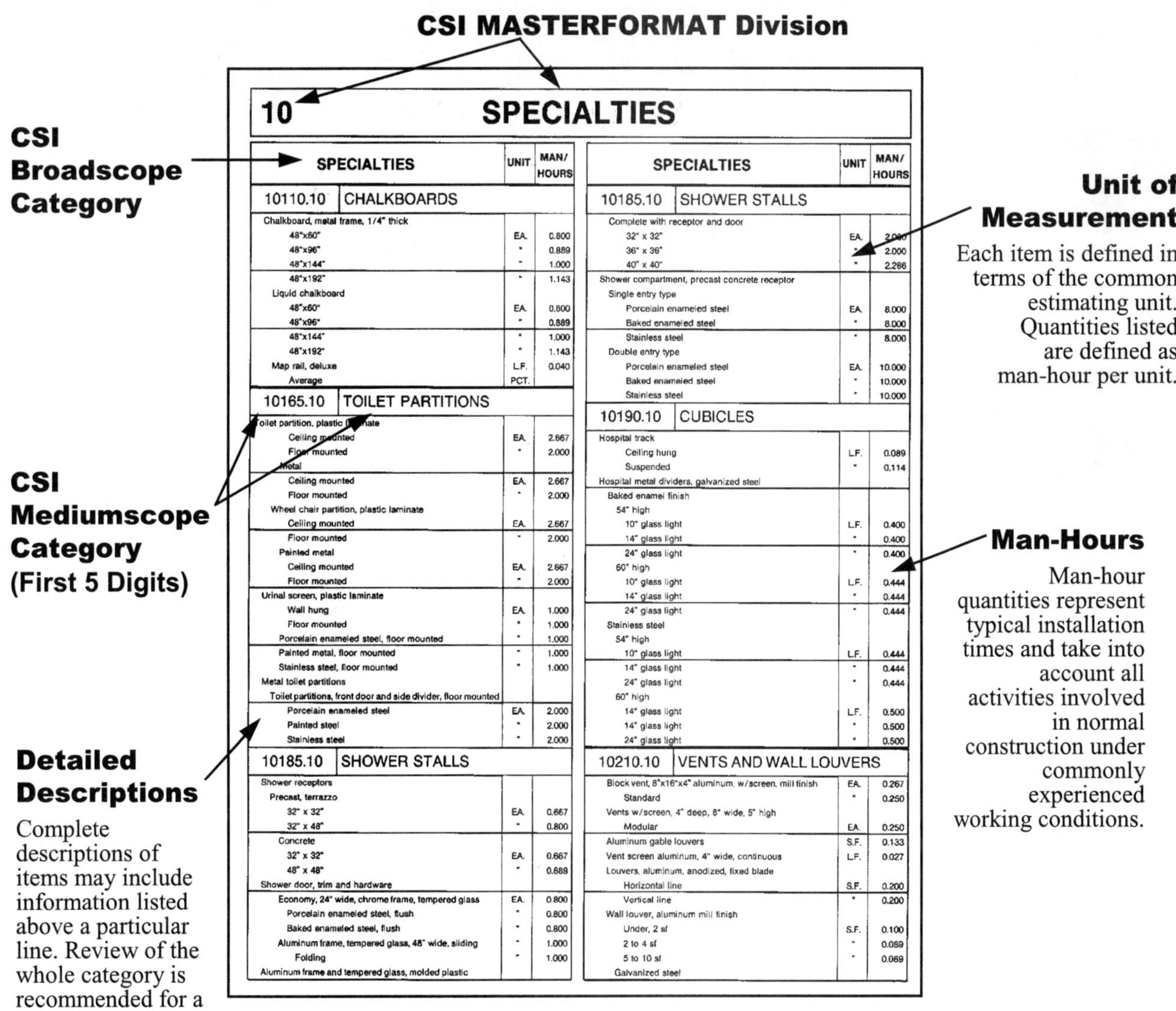

10 SPECIALTIES

SPECIALTIES		UNIT	MAN/HOURS
10110.10	**CHALKBOARDS**		
Chalkboard, metal frame, 1/4" thick			
48"x60"		EA.	0.800
48"x96"		"	0.889
48"x144"		"	1.000
48"x192"		"	1.143
Liquid chalkboard			
48"x60"		EA.	0.800
48"x96"		"	0.889
48"x144"		"	1.000
48"x192"		"	1.143
Map rail, deluxe		L.F.	0.040
Average		PCT.	
10165.10	**TOILET PARTITIONS**		
Toilet partition, plastic laminate			
Ceiling mounted		EA.	2.667
Floor mounted		"	2.000
Metal			
Ceiling mounted		EA.	2.667
Floor mounted		"	2.000
Wheel chair partition, plastic laminate			
Ceiling mounted		EA.	2.667
Floor mounted		"	2.000
Painted metal			
Ceiling mounted		EA.	2.667
Floor mounted		"	2.000
Urinal screen, plastic laminate			
Wall hung		EA.	1.000
Floor mounted		"	1.000
Porcelain enameled steel, floor mounted		"	1.000
Painted metal, floor mounted		"	1.000
Stainless steel, floor mounted		"	1.000
Metal toilet partitions			
Toilet partitions, front door and side divider, floor mounted			
Porcelain enameled steel		EA.	2.000
Painted steel		"	2.000
Stainless steel		"	2.000
10185.10	**SHOWER STALLS**		
Shower receptors			
Precast, terrazzo			
32" x 32"		EA.	0.667
32" x 48"		"	0.800
Concrete			
32" x 32"		EA.	0.667
48" x 48"		"	0.889
Shower door, trim and hardware			
Economy, 24" wide, chrome frame, tempered glass		EA.	0.800
Porcelain enameled steel, flush		"	0.800
Baked enameled steel, flush		"	0.800
Aluminum frame, tempered glass, 48" wide, sliding		"	1.000
Folding		"	1.000
Aluminum frame and tempered glass, molded plastic			

SPECIALTIES		UNIT	MAN/HOURS
10185.10	**SHOWER STALLS**		
Complete with receptor and door			
32" x 32"		EA.	2.000
36" x 36"		"	2.000
40" x 40"		"	2.286
Shower compartment, precast concrete receptor			
Single entry type			
Porcelain enameled steel		EA.	8.000
Baked enameled steel		"	8.000
Stainless steel		"	8.000
Double entry type			
Porcelain enameled steel		EA.	10.000
Baked enameled steel		"	10.000
Stainless steel		"	10.000
10190.10	**CUBICLES**		
Hospital track			
Ceiling hung		L.F.	0.089
Suspended		"	0.114
Hospital metal dividers, galvanized steel			
Baked enamel finish			
54" high			
10" glass light		L.F.	0.400
14" glass light		"	0.400
24" glass light		"	0.400
60" high			
10" glass light		L.F.	0.444
14" glass light		"	0.444
24" glass light		"	0.444
Stainless steel			
54" high			
10" glass light		L.F.	0.444
14" glass light		"	0.444
24" glass light		"	0.444
60" high			
14" glass light		L.F.	0.500
14" glass light		"	0.500
24" glass light		"	0.500
10210.10	**VENTS AND WALL LOUVERS**		
Block vent, 8"x16"x4" aluminum, w/screen, mill finish		EA.	0.267
Standard		"	0.250
Vents w/screen, 4" deep, 8" wide, 5" high			
Modular		EA.	0.250
Aluminum gable louvers		S.F.	0.133
Vent screen aluminum, 4" wide, continuous		L.F.	0.027
Louvers, aluminum, anodized, fixed blade			
Horizontal line		S.F.	0.200
Vertical line		"	0.200
Wall louver, aluminum mill finish			
Under, 2 sf		S.F.	0.100
2 to 4 sf		"	0.089
5 to 10 sf		"	0.089
Galvanized steel			

Soil Tests	UNIT	MAN/ HOURS
02010.10 Soil Boring		
Borings, uncased, stable earth		
2-1/2" dia.		
Minimum	L.F.	0.200
Average	"	0.300
Maximum	"	0.480
4" dia.		
Minimum	L.F.	0.218
Average	"	0.343
Maximum	"	0.600
Cased, including samples		
2-1/2" dia.		
Minimum	L.F.	0.240
Average	"	0.400
Maximum	"	0.800
4" dia.		
Minumum	L.F.	0.480
Average	"	0.686
Maximum	"	0.960
Drilling in rock		
No sampling		
Minimum	L.F.	0.436
Average	"	0.632
Maximum	"	0.857
With casing and sampling		
Minimum	L.F.	0.600
Average	"	0.800
Maximum	"	1.200
Test pits		
Light soil		
Minimum	EA.	3.000
Average	"	4.000
Maximum	"	8.000
Heavy soil		
Minimum	EA.	4.800
Average	"	6.000
Maximum	"	12.000

Demolition	UNIT	MAN/ HOURS
02060.10 Building Demolition		
Building, complete with disposal		
Wood frame	C.F.	0.003
Concrete	"	0.004
Steel frame	"	0.005
Partition removal		
Concrete block partitions		

Demolition	UNIT	MAN/ HOURS
02060.10 Building Demolition (Cont.)		
4" thick	S.F.	0.040
8" thick	"	0.053
12" thick	"	0.073
Brick masonry partitions		
4" thick	S.F.	0.040
8" thick	"	0.050
12" thick	"	0.067
16" thick	"	0.100
Cast in place concrete partitions		
Unreinforced		
6" thick	S.F.	0.160
8" thick	"	0.171
10" thick	"	0.200
12" thick	"	0.240
Reinforced		
6" thick	S.F.	0.185
8" thick	"	0.240
10" thick	"	0.267
12" thick	"	0.320
Terra cotta		
To 6" thick	S.F.	0.040
Stud partitions		
Metal or wood, with drywall both sides	S.F.	0.040
Metal studs, both sides, lath and plaster	"	0.053
Door and frame removal		
Hollow metal in masonry wall		
Single		
2'6"x6'8"	EA.	1.000
3'x7'	"	1.333
Double		
3'x7'	EA.	1.600
4'x8'	"	1.600
Wood in framed wall		
Single		
2'6"x6'8"	EA.	0.571
3'x6'8"	"	0.667
Double		
2'6"x6'8"	EA.	0.800
3'x6'8"	"	0.889
Remove for re-use		
Hollow metal	EA.	2.000
Wood	"	1.333
Floor removal		
Brick flooring	S.F.	0.032
Ceramic or quarry tile	"	0.018
Terrazzo	"	0.036
Heavy wood	"	0.021
Residential wood	"	0.023
Resilient tile or linoleum	"	0.008
Ceiling removal		
Acoustical tile ceiling		
Adhesive fastened	S.F.	0.008
Furred and glued	"	0.007

Demolition

02060.10	Building Demolition *(Cont.)*	UNIT	MAN/HOURS
Suspended grid		S.F.	0.005
Drywall ceiling			
Furred and nailed		S.F.	0.009
Nailed to framing		"	0.008
Plastered ceiling			
Furred on framing		S.F.	0.020
Suspended system		"	0.027
Roofing removal			
Steel frame			
Corrugated metal roofing		S.F.	0.016
Built-up roof on metal deck		"	0.027
Wood frame			
Built up roof on wood deck		S.F.	0.025
Roof shingles		"	0.013
Roof tiles		"	0.027
Concrete frame		C.F.	0.053
Concrete plank		S.F.	0.040
Built-up roof on concrete		"	0.023
Cut-outs			
Concrete, elevated slabs, mesh reinforcing			
Under 5 cf		C.F.	0.800
Over 5 cf		"	0.667
Bar reinforcing			
Under 5 cf		C.F.	1.333
Over 5 cf		"	1.000
Window removal			
Metal windows, trim included			
2'x3'		EA.	0.800
2'x4'		"	0.889
2'x6'		"	1.000
3'x4'		"	1.000
3'x6'		"	1.143
3'x8'		"	1.333
4'x4'		"	1.333
4'x6'		"	1.600
4'x8'		"	2.000
Wood windows, trim included			
2'x3'		EA.	0.444
2'x4'		"	0.471
2'x6'		"	0.500
3'x4'		"	0.533
3'x6'		"	0.571
3'x8'		"	0.615
6'x4'		"	0.667
6'x6'		"	0.727
6'x8'		"	0.800
Walls, concrete, bar reinforcing			
Small jobs		C.F.	0.533
Large jobs		"	0.444
Brick walls, not including toothing			
4" thick		S.F.	0.040
8" thick		"	0.050
12" thick		"	0.067

Demolition

02060.10	Building Demolition *(Cont.)*	UNIT	MAN/HOURS
16" thick		S.F.	0.100
Concrete block walls, not including toothing			
4" thick		S.F.	0.044
6" thick		"	0.047
8" thick		"	0.050
10" thick		"	0.057
12" thick		"	0.067
Rubbish handling			
Load in dumpster or truck			
Minimum		C.F.	0.018
Maximum		"	0.027
For use of elevators, add			
Minimum		C.F.	0.004
Maximum		"	0.008
Rubbish hauling			
Hand loaded on trucks, 2 mile trip		C.Y.	0.320
Machine loaded on trucks, 2 mile trip		"	0.240

Highway Demolition

02065.10	Pavement Demolition	UNIT	MAN/HOURS
Bituminous pavement, up to 3" thick			
On streets		S.Y.	0.096
On pipe trench		"	0.120
Concrete pavement, 6" thick			
No reinforcement		S.Y.	0.160
With wire mesh		"	0.240
With rebars		"	0.300
9" thick			
No reinforcement		S.Y.	0.200
With wire mesh		"	0.300
With rebars		"	0.400
12" thick			
No reinforcement		S.Y.	0.240
With wire mesh		"	0.343
With rebars		"	0.480
Sidewalk, 4" thick, with disposal		"	0.080
Removal of pavement markings by waterblasting		S.F.	0.004

02065.15	Saw Cutting Pavement		
Pavement, bituminous			
2" thick		L.F.	0.016
3" thick		"	0.020
4" thick		"	0.025
5" thick		"	0.027
6" thick		"	0.029

Highway Demolition

Highway Demolition	UNIT	MAN/HOURS
02065.15 **Saw Cutting Pavement** *(Cont.)*		
Concrete pavement, with wire mesh		
4" thick	L.F.	0.031
5" thick	"	0.033
6" thick	"	0.036
8" thick	"	0.040
10" thick	"	0.044
Plain concrete, unreinforced		
4" thick	L.F.	0.027
5" thick	"	0.031
6" thick	"	0.033
8" thick	"	0.036
10" thick	"	0.040
02065.80 **Curb & Gutter**		
Removal, plain concrete curb	L.F.	0.060
Plain concrete curb and 2' gutter	"	0.083
Curb removal		
Concrete, unreinforced		
Minimum	L.F.	0.048
Average	"	0.060
Maximum	"	0.075
Reinforced		
Minimum	L.F.	0.077
Average	"	0.086
Maximum	"	0.096
Combination curb and 2' gutter		
Unreinforced		
Minimum	L.F.	0.063
Average	"	0.083
Maximum	"	0.120
Reinforced		
Minimum	L.F.	0.100
Average	"	0.133
Maximum	"	0.240
Granite curb		
Minimum	L.F.	0.069
Average	"	0.080
Maximum	"	0.092
Asphalt curb		
Minimum	L.F.	0.040
Average	"	0.048
Maximum	"	0.057
02065.85 **Guardrails**		
Remove standard guardrail		
Steel	L.F.	0.080
Wood	"	0.062

Highway Demolition

Highway Demolition	UNIT	MAN/HOURS
02075.80 **Core Drilling**		
Concrete		
6" thick		
3" dia.	EA.	0.571
4" dia.	"	0.667
6" dia.	"	0.800
8" dia.	"	1.333
8" thick		
3" dia.	EA.	0.800
4" dia.	"	1.000
6" dia.	"	1.143
8" dia.	"	1.600
10" thick		
3" dia.	EA.	1.000
4" dia.	"	1.143
6" dia.	"	1.333
8" dia.	"	2.000
12" thick		
3" dia.	EA.	1.333
4" dia.	"	1.600
6" dia.	"	2.000
8" dia.	"	2.667

Hazardous Waste

Hazardous Waste	UNIT	MAN/HOURS
02080.10 **Asbestos Removal**		
Enclosure using wood studs & poly, install & remove	S.F.	0.020
02080.12 **Duct Insulation Removal**		
Remove duct insulation, duct size		
6" x 12"	L.F.	0.044
x 18"	"	0.062
x 24"	"	0.089
8" x 12"	"	0.067
x 18"	"	0.073
x 24"	"	0.100
12" x 12"	"	0.067
x 18"	"	0.089
x 24"	"	0.114
02080.15 **Pipe Insulation Removal**		
Removal, asbestos insulation		
2" thick, pipe		
1" to 3" dia.	L.F.	0.067
4" to 6" dia.	"	0.076

Hazardous Waste

	UNIT	MAN/HOURS

02080.15 — Pipe Insulation Removal *(Cont.)*

	UNIT	MAN/HOURS
3" thick		
7" to 8" dia.	L.F.	0.080
9" to 10" dia.	"	0.084
11" to 12" dia.	"	0.089
13" to 14" dia.	"	0.094
15" to 18" dia.	"	0.100

Site Demolition

	UNIT	MAN/HOURS

02105.10 — Catch Basins/manholes

	UNIT	MAN/HOURS
Abandon catch basin or manhole (fill with sand)	EA.	4.800
Remove and reset frame and cover	"	2.400
Remove catch basin, to 10' deep		
Masonry	EA.	6.000
Concrete	"	8.000

02105.20 — Fences

	UNIT	MAN/HOURS
Remove fencing		
Chain link, 8' high		
For disposal	L.F.	0.040
For reuse	"	0.100
Wood		
4' high	S.F.	0.027
6' high	"	0.032
8' high	"	0.040
Masonry		
8" thick		
4' high	S.F.	0.080
6' high	"	0.100
8' high	"	0.114
12" thick		
4' high	S.F.	0.133
6' high	"	0.160
8' high	"	0.200
12' high	"	0.267

02105.42 — Drainage Piping

	UNIT	MAN/HOURS
Remove drainage pipe, not including excavation		
12" dia.	L.F.	0.100
18" dia.	"	0.126
24" dia.	"	0.160
36" dia.	"	0.200

Site Demolition

	UNIT	MAN/HOURS

02105.43 — Gas Piping

	UNIT	MAN/HOURS
Remove welded steel pipe, not including excavation		
4" dia.	L.F.	0.150
5" dia.	"	0.240
6" dia.	"	0.300
8" dia.	"	0.480
10" dia.	"	0.600

02105.45 — Sanitary Piping

	UNIT	MAN/HOURS
Remove sewer pipe, not including excavation		
4" dia.	L.F.	0.096
6" dia.	"	0.109
8" dia.	"	0.120
10" dia.	"	0.126
12" dia.	"	0.133
15" dia.	"	0.141
18" dia.	"	0.160
24" dia.	"	0.200
30" dia.	"	0.240
36" dia.	"	0.300

02105.48 — Water Piping

	UNIT	MAN/HOURS
Remove water pipe, not including excavation		
4" dia.	L.F.	0.109
6" dia.	"	0.114
8" dia.	"	0.126
10" dia.	"	0.133
12" dia.	"	0.141
14" dia.	"	0.150
16" dia.	"	0.160
18" dia.	"	0.171
20" dia.	"	0.185
Remove valves		
6"	EA.	1.200
10"	"	1.333
14"	"	1.500
18"	"	2.000

02105.60 — Underground Tanks

	UNIT	MAN/HOURS
Remove underground storage tank, and backfill		
50 to 250 gals	EA.	8.000
600 gals	"	8.000
1000 gals	"	12.000
4000 gals	"	19.200
5000 gals	"	19.200
10,000 gals	"	32.000
12,000 gals	"	40.000
15,000 gals	"	48.000
20,000 gals	"	60.000

Site Demolition	UNIT	MAN/HOURS
02105.66 Septic Tanks		
Remove septic tank		
1000 gals	EA.	2.000
2000 gals	"	2.400
5000 gals	"	3.000
15,000 gals	"	24.000
25,000 gals	"	32.000
40,000 gals	"	48.000
02105.80 Walls, Exterior		
Concrete wall		
Light reinforcing		
6" thick	S.F.	0.120
8" thick	"	0.126
10" thick	"	0.133
12" thick	"	0.150
Medium reinforcing		
6" thick	S.F.	0.126
8" thick	"	0.133
10" thick	"	0.150
12" thick	"	0.171
Heavy reinforcing		
6" thick	S.F.	0.141
8" thick	"	0.150
10" thick	"	0.171
12" thick	"	0.200
Masonry		
No reinforcing		
8" thick	S.F.	0.053
12" thick	"	0.060
16" thick	"	0.069
Horizontal reinforcing		
8" thick	S.F.	0.060
12" thick	"	0.065
16" thick	"	0.077
Vertical reinforcing		
8" thick	S.F.	0.077
12" thick	"	0.089
16" thick	"	0.109
Remove concrete headwall		
15" pipe	EA.	1.714
24" pipe	"	2.182
02162.10 Trench Sheeting		
Closed timber, including pull and salvage, excavation		
8' deep	S.F.	0.064
20' deep	"	0.098

Earthwork	UNIT	MAN/HOURS
02210.10 Hauling Material		
Haul material by 10 cy dump truck, round trip distance		
1 mile	C.Y.	0.044
2 mile	"	0.053
5 mile	"	0.073
10 mile	"	0.080
20 mile	"	0.089
30 mile	"	0.107
Site grading, cut & fill, sandy clay, 200' haul, 75 hp dozer	"	0.032
Spread topsoil by equipment on site	"	0.036
Site grading (cut and fill to 6") less than 1 acre		
75 hp dozer	C.Y.	0.053
1.5 cy backhoe/loader	"	0.080
02210.30 Bulk Excavation		
Excavation, by small dozer		
Small areas	C.Y.	0.027
Trim banks	"	0.040
Hydraulic excavator		
1 cy capacity		
Light material	C.Y.	0.040
Medium material	"	0.048
Wet material	"	0.060
Blasted rock	"	0.069
Wheel mounted front-end loader		
7/8 cy capacity		
Light material	C.Y.	0.020
Medium material	"	0.023
Wet material	"	0.027
Blasted rock	"	0.032
Track mounted front-end loader		
1-1/2 cy capacity		
Light material	C.Y.	0.013
Medium material	"	0.015
Wet material	"	0.016
Blasted rock	"	0.018
02220.10 Borrow		
Borrow fill, F.O.B. at pit		
Sand, haul to site, round trip		
10 mile	C.Y.	0.080
20 mile	"	0.133
30 mile	"	0.200
Place borrow fill and compact		
Less than 1 in 4 slope	C.Y.	0.040
Greater than 1 in 4 slope	"	0.053
02220.40 Building Excavation		
Structural excavation, unclassified earth		
3/8 cy backhoe	C.Y.	0.107
3/4 cy backhoe	"	0.080
1 cy backhoe	"	0.067
Foundation backfill and compaction by machine	"	0.160

Earthwork	UNIT	MAN/HOURS
02220.50 **Utility Excavation**		
Trencher, sandy clay, 8" wide trench		
18" deep	L.F.	0.018
24" deep	"	0.020
36" deep	"	0.023
Trench backfill, 95% compaction		
Tamp by hand	C.Y.	0.500
Vibratory compaction	"	0.400
Trench backfilling, with borrow sand, place & compact	"	0.400
02220.60 **Trenching**		
Trenching and continuous footing excavation		
By gradall		
1 cy capacity		
Light soil	C.Y.	0.023
Medium soil	"	0.025
Heavy/wet soil	"	0.027
Loose rock	"	0.029
Blasted rock	"	0.031
By hydraulic excavator		
1/2 cy capacity		
Light soil	C.Y.	0.027
Medium soil	"	0.029
Heavy/wet soil	"	0.032
Loose rock	"	0.036
Blasted rock	"	0.040
Hand excavation		
Bulk, wheeled 100'		
Normal soil	C.Y.	0.889
Sand or gravel	"	0.800
Medium clay	"	1.143
Heavy clay	"	1.600
Loose rock	"	2.000
Trenches, up to 2' deep		
Normal soil	C.Y.	1.000
Sand or gravel	"	0.889
Medium clay	"	1.333
Heavy clay	"	2.000
Loose rock	"	2.667
Trenches, to 6' deep		
Normal soil	C.Y.	1.143
Sand or gravel	"	1.000
Medium clay	"	1.600
Heavy clay	"	2.667
Loose rock	"	4.000
Backfill trenches		
With compaction		
By hand	C.Y.	0.667
By 60 hp tracked dozer	"	0.020
By small front-end loader	"	0.023
Spread dumped fill or gravel, no compaction		
6" layers	S.Y.	0.013

Earthwork	UNIT	MAN/HOURS
02220.60 **Trenching** *(Cont.)*		
12" layers	S.Y.	0.016
Compaction in 6" layers		
By hand with air tamper	S.Y.	0.016
Backfill trenches, sand bedding, no compaction		
By hand	C.Y.	0.667
By small front-end loader	"	0.023
02220.90 **Hand Excavation**		
Excavation		
To 2' deep		
Normal soil	C.Y.	0.889
Sand and gravel	"	0.800
Medium clay	"	1.000
Heavy clay	"	1.143
Loose rock	"	1.333
To 6' deep		
Normal soil	C.Y.	1.143
Sand and gravel	"	1.000
Medium clay	"	1.333
Heavy clay	"	1.600
Loose rock	"	2.000
Backfilling foundation without compaction, 6" lifts	"	0.500
Compaction of backfill around structures or in trench		
By hand with air tamper	C.Y.	0.571
By hand with vibrating plate tamper	"	0.533
1 ton roller	"	0.400
Miscellaneous hand labor		
Trim slopes, sides of excavation	S.F.	0.001
Trim bottom of excavation	"	0.002
Excavation around obstructions and services	C.Y.	2.667
02240.05 **Soil Stabilization**		
Straw bale secured with rebar	L.F.	0.027
Filter barrier, 18" high filter fabric	"	0.080
Sediment fence, 36" fabric with 6" mesh	"	0.100
02280.20 **Soil Treatment**		
Soil treatment, termite control pretreatment		
Under slabs	S.F.	0.004
By walls	"	0.005
02290.30 **Weed Control**		
Weed control, bromicil, 15 lb./acre, wettable powder	ACRE	4.000
Vegetation control, by application of plant killer	S.Y.	0.003
Weed killer, lawns and fields	"	0.002

Piles And Caissons	UNIT	MAN/ HOURS
02360.50 Prestressed Piling		
Prestressed concrete piling, less than 60' long		
10" sq.	L.F.	0.040
12" sq.	"	0.042
14" sq.	"	0.043
16" sq.	"	0.044
18" sq.	"	0.047
20" sq.	"	0.048
24" sq.	"	0.049
02360.60 Steel Piles		
H-section piles		
8x8		
36 lb/ft		
30' long	L.F.	0.080
40' long	"	0.064
50' long	"	0.053
10x10		
42 lb/ft		
30' long	L.F.	0.080
40' long	"	0.064
50' long	"	0.053
57 lb/ft		
30' long	L.F.	0.080
40' long	"	0.064
50' long	"	0.053
Splice		
8"	EA.	1.333
10"	"	1.600
Driving cap		
8"	EA.	0.800
10"	"	1.000
Standard point		
8"	EA.	0.800
10"	"	1.000
Heavy duty point		
8"	EA.	0.889
10"	"	1.143
02360.65 Steel Pipe Piles		
Concrete filled, 3000# concrete, up to 40'		
8" dia.	L.F.	0.069
10" dia.	"	0.071
12" dia.	"	0.074
Pipe piles, non-filled		
8" dia.	L.F.	0.053
10" dia.	"	0.055
12" dia.	"	0.056
Splice		
8" dia.	EA.	1.600
10" dia.	"	1.600
12" dia.	"	2.000
Standard point		
8" dia.	EA.	1.600

Piles And Caissons	UNIT	MAN/ HOURS
02360.65 Steel Pipe Piles (Cont.)		
10" dia.	EA.	1.600
12" dia.	"	2.000
Heavy duty point		
8" dia.	EA.	2.000
10" dia.	"	2.000
12" dia.	"	2.667
02360.70 Steel Sheet Piling		
Steel sheet piling,12" wide		
20' long	S.F.	0.096
35' long	"	0.069
02360.80 Wood And Timber Piles		
Treated wood piles, 12" butt, 8" tip		
25' long	L.F.	0.096
30' long	"	0.080
35' long	"	0.069
40' long	"	0.060
02380.10 Caissons		
Caisson, including 3000# concrete, in stable ground		
18" dia.	L.F.	0.192
24" dia.	"	0.200
Wet ground, casing required but pulled		
18" dia.	L.F.	0.240
24" dia.	"	0.267
Soft rock		
18" dia.	L.F.	0.686
24" dia.	"	1.200

Paving And Surfacing	UNIT	MAN/ HOURS
02510.20 Asphalt Surfaces		
Asphalt wearing surface, for flexible pavement		
1" thick	S.Y.	0.016
1-1/2" thick	"	0.019
2" thick	"	0.024
3" thick	"	0.032
Binder course		
1-1/2" thick	S.Y.	0.018
2" thick	"	0.022
3" thick	"	0.029

Paving And Surfacing

02510.20 — Asphalt Surfaces *(Cont.)*

	UNIT	MAN/HOURS
4" thick	S.Y.	0.032
5" thick	"	0.036
6" thick	"	0.040
Bituminous sidewalk, no base		
2" thick	S.Y.	0.028
3" thick	"	0.030

02520.10 — Concrete Paving

	UNIT	MAN/HOURS
Concrete paving, reinforced, 5000 psi concrete		
6" thick	S.Y.	0.150
7" thick	"	0.160
8" thick	"	0.171
Concrete paving, for pipe trench, reinforced		
7" thick	S.Y.	0.240
8" thick	"	0.267

02545.10 — Asphalt Repair

	UNIT	MAN/HOURS
Coal tar emulsion seal coat, rubber add., fuel resistant	S.Y.	0.011
Bituminous surface treatment, single	"	0.008
Double	"	0.001
Bituminous prime coat	"	0.001
Tack coat	"	0.001
Crack sealing, concrete paving	L.F.	0.005
Bituminous paving for pipe trench, 4" thick	S.Y.	0.160
Polypropylene, nonwoven paving fabric	"	0.004
Rubberized asphalt	"	0.073
Asphalt slurry seal	"	0.047

02580.10 — Pavement Markings

	UNIT	MAN/HOURS
Pavement line marking, paint		
4" wide	L.F.	0.002
6" wide	"	0.004
8" wide	"	0.007
Reflective paint, 4" wide	"	0.007
Preformed tape, 4" wide		
Inlaid reflective	L.F.	0.001
Reflective paint	"	0.002
Thermoplastic		
White	L.F.	0.004
Yellow	"	0.004
12" wide, thermoplastic, white	"	0.011
Directional arrows, reflective preformed tape	EA.	0.800
Messages, reflective preformed tape (per letter)	"	0.400
Handicap symbol, preformed tape	"	0.800
Parking stall painting	"	0.160

Site Improvements

02810.40 — Lawn Irrigation

	UNIT	MAN/HOURS
Pipe		
Schedule 40, PVC		
1/2"	L.F.	0.042
3/4"	"	0.044
1"	"	0.046
Fittings		
Tee		
1/2"	EA.	0.133
3/4"	"	0.133
1"	"	0.133
El		
1/2"	EA.	0.123
3/4"	"	0.133
1"	"	0.133
Coupling		
1/2"	EA.	0.123
3/4"	"	0.133
1"	"	0.133
45 El		
1/2"	EA.	0.123
3/4"	"	0.133
1"	"	0.133
Riser, 1/2" diameter		
2" (close)	EA.	0.200
6"	"	0.229
3/4" diameter		
2" (close)	EA.	0.200
6"	"	0.229
1" diameter		
2" (close)	EA.	0.200
6"	"	0.229
Valve Box		
Concrete, Square		
12" x 22"	EA.	1.000
Round		
12"	EA.	0.800
Plastic		
Square		
12"	EA.	1.000
Round		
6"	EA.	1.000
Sprinkler, Pop-Up		
Spray		
2" high	EA.	1.333
6" high	"	1.600
Rotor		
4" high	EA.	1.333
Impact		
Brass	EA.	1.333
Plastic	"	1.600
Shrub Head		
Spray	EA.	1.333
Rotor	"	1.600

Site Improvements	UNIT	MAN/HOURS
02810.40 **Lawn Irrigation** *(Cont.)*		
Time Clocks		
Minimum	EA.	2.000
Average	"	2.667
Maximum	"	4.000
Valves		
Anti-siphon		
Plastic		
3/4"	EA.	1.333
1"	"	1.333
Ball Valve		
Plastic		
1/2"	EA.	1.333
3/4"	"	1.333
1"	"	1.333
Brass		
1/2"	EA.	1.333
3/4"	"	1.333
1"	"	1.333
Gate valves, Brass		
1/2"	EA.	1.333
3/4"	"	1.333
1"	"	1.333
Vacuum Breakers		
Brass		
3/4"	EA.	2.667
1"	"	2.667
Plastic		
3/4"	EA.	2.667
1"	"	2.667
Backflow Preventors, Brass		
3/4"	EA.	26.667
1"	"	26.667
Pressure Regulators, Brass		
3/4"	EA.	0.800
1"	"	0.800
02830.10 **Chain Link Fence**		
Chain link fence, 9 ga., galvanized, with posts 10' o.c.		
4' high	L.F.	0.057
5' high	"	0.073
6' high	"	0.100
7' high	"	0.123
8' high	"	0.160
For barbed wire with hangers, add		
3 strand	L.F.	0.040
6 strand	"	0.067
Corner or gate post, 3" post		
4' high	EA.	0.267
5' high	"	0.296
6' high	"	0.348
7' high	"	0.400
8' high	"	0.444
4" post		

Site Improvements	UNIT	MAN/HOURS
02830.10 **Chain Link Fence** *(Cont.)*		
4' high	EA.	0.296
5' high	"	0.348
6' high	"	0.400
7' high	"	0.444
8' high	"	0.500
Gate with gate posts, galvanized, 3' wide		
4' high	EA.	2.000
5' high	"	2.667
6' high	"	2.667
7' high	"	4.000
8' high	"	4.000
Fabric, galvanized chain link, 2" mesh, 9 ga.		
4' high	L.F.	0.027
5' high	"	0.032
6' high	"	0.040
8' high	"	0.053
Line post, no rail fitting, galvanized, 2-1/2" dia.		
4' high	EA.	0.229
5' high	"	0.250
6' high	"	0.267
7' high	"	0.320
8' high	"	0.400
1-7/8" H beam		
4' high	EA.	0.229
5' high	"	0.250
6' high	"	0.267
7' high	"	0.320
8' high	"	0.400
2-1/4" H beam		
4' high	EA.	0.229
5' high	"	0.250
6' high	"	0.267
7' high	"	0.320
8' high	"	0.400
Vinyl coated, 9 ga., with posts 10' o.c.		
4' high	L.F.	0.057
5' high	"	0.073
6' high	"	0.100
7' high	"	0.123
8' high	"	0.160
For barbed wire w/hangers, add		
3 strand	L.F.	0.040
6 Strand	"	0.067
Corner, or gate post, 4' high		
3" dia.	EA.	0.267
4" dia.	"	0.267
6" dia.	"	0.320
Gate, with posts, 3' wide		
4' high	EA.	2.000
5' high	"	2.667
6' high	"	2.667
7' high	"	4.000
8' high	"	4.000

Site Improvements	UNIT	MAN/ HOURS
02830.10 **Chain Link Fence** *(Cont.)*		
Line post, no rail fitting, 2-1/2" dia.		
4' high	EA.	0.229
5' high	"	0.250
6' high	"	0.267
7' high	"	0.320
8' high	"	0.400
Corner post, no top rail fitting, 4" dia.		
4' high	EA.	0.267
5' high	"	0.296
6' high	"	0.348
7' high	"	0.400
8' high	"	0.444
Fabric, vinyl, chain link, 2" mesh, 9 ga.		
4' high	L.F.	0.027
5' high	"	0.032
6' high	"	0.040
8' high	"	0.053
Swing gates, galvanized, 4' high		
Single gate		
3' wide	EA.	2.000
4' wide	"	2.000
Double gate		
10' wide	EA.	3.200
12' wide	"	3.200
14' wide	"	3.200
16' wide	"	3.200
18' wide	"	4.571
20' wide	"	4.571
22' wide	"	4.571
24' wide	"	5.333
26' wide	"	5.333
28' wide	"	6.400
30' wide	"	6.400
5' high		
Single gate		
3' wide	EA.	2.667
4' wide	"	2.667
Double gate		
10' wide	EA.	4.000
12' wide	"	4.000
14' wide	"	4.000
16' wide	"	4.000
18' wide	"	4.571
20' wide	"	4.571
22' wide	"	4.571
24' wide	"	5.333
26' wide	"	5.333
28' wide	"	6.400
30' wide	"	6.400
6' high		
Single gate		
3' wide	EA.	2.667
4' wide	"	2.667

Site Improvements	UNIT	MAN/ HOURS
02830.10 **Chain Link Fence** *(Cont.)*		
Double gate		
10' wide	EA.	4.000
12' wide	"	4.000
14' wide	"	4.000
16' wide	"	4.000
18' wide	"	4.571
20' wide	"	4.571
22' wide	"	4.571
24' wide	"	5.333
26' wide	"	5.333
28' wide	"	6.400
30' wide	"	6.400
7' high		
Single gate		
3' wide	EA.	4.000
4' wide	"	4.000
Double gate		
10' wide	EA.	5.333
12' wide	"	5.333
14' wide	"	5.333
16' wide	"	5.333
18' wide	"	6.400
20' wide	"	6.400
22' wide	"	6.400
24' wide	"	8.000
26' wide	"	8.000
28' wide	"	10.000
30' wide	"	10.000
8' high		
Single gate		
3' wide	EA.	4.000
4' wide	"	4.000
Double gate		
10' wide	EA.	5.333
12' wide	"	5.333
14' wide	"	5.333
16' wide	"	5.333
18' wide	"	6.400
20' wide	"	6.400
22' wide	"	6.400
24' wide	"	8.000
26' wide	"	8.000
28' wide	"	10.000
30' wide	"	10.000
Vinyl coated swing gates, 4' high		
Single gate		
3' wide	EA.	2.000
4' wide	"	2.000
Double gate		
10' wide	EA.	3.200
12' wide	"	3.200
14' wide	"	3.200
16' wide	"	3.200

Site Improvements	UNIT	MAN/HOURS

02830.10 — **Chain Link Fence** *(Cont.)*

Site Improvements	UNIT	MAN/HOURS
18' wide	EA.	4.571
20' wide	"	4.571
22' wide	"	4.571
24' wide	"	5.333
26' wide	"	5.333
28' wide	"	6.400
30' wide	"	6.400
5' high		
Single gate		
3' wide	EA.	2.667
4' wide	"	2.667
Double gate		
10' wide	EA.	4.000
12' wide	"	4.000
14' wide	"	4.000
16' wide	"	4.000
18' wide	"	4.571
20' wide	"	4.571
22' wide	"	4.571
24' wide	"	5.333
26' wide	"	5.333
28' wide	"	6.400
30' wide	"	6.400
6' high		
Single gate		
3' wide	EA.	2.667
4' wide	"	2.667
Double gate		
10' wide	EA.	4.000
12' wide	"	4.000
14' wide	"	4.000
16' wide	"	4.000
18' wide	"	4.571
20' wide	"	4.571
22' wide	"	4.571
24' wide	"	5.333
26' wide	"	5.333
28' wide	"	6.400
30' wide	"	6.400
7' high		
Single gate		
3' wide	EA.	4.000
4' wide	"	4.000
Double gate		
10' wide	EA.	5.333
12' wide	"	5.333
14' wide	"	5.333
16' wide	"	5.333
18' wide	"	6.400
20' wide	"	6.400
22' wide	"	6.400
24' wide	"	8.000
26' wide	"	8.000

02830.10 — **Chain Link Fence** *(Cont.)*

Site Improvements	UNIT	MAN/HOURS
28' wide	EA.	10.000
30' wide	"	10.000
8' high		
Single gate		
3' wide	EA.	4.000
4' wide	"	4.000
Double gate		
10' wide	EA.	5.333
12' wide	"	5.333
14' wide	"	5.333
16' wide	"	5.333
18' wide	"	6.400
20' wide	"	6.400
22' wide	"	6.400
24' wide	"	8.000
28' wide	"	8.000
30' wide	"	10.000
Drilling fence post holes		
In soil		
By hand	EA.	0.400
By machine auger	"	0.200
In rock		
By jackhammer	EA.	2.667
By rock drill	"	0.800
Aluminum privacy slats, installed vertically	S.F.	0.020
Post hole, dig by hand	EA.	0.533
Set fence post in concrete	"	0.400

02840.30 — **Guardrails**

	UNIT	MAN/HOURS
Pipe bollard, steel pipe, concrete filled, painted		
6" dia.	EA.	0.667
8" dia.	"	1.000
12" dia.	"	2.667
Corrugated steel, guardrail, galvanized	L.F.	0.040
End section, wrap around or flared	EA.	0.800
Timber guardrail, 4" x 8"	L.F.	0.030
Guard rail, 3 cables, 3/4" dia.		
Steel posts	L.F.	0.120
Wood posts	"	0.096
Steel box beam		
6" x 6"	L.F.	0.133
6" x 8"	"	0.150
Concrete posts	EA.	0.400
Barrel type impact barrier	"	0.800
Light shield, 6' high	L.F.	0.160

02840.40 — **Parking Barriers**

	UNIT	MAN/HOURS
Timber, treated, 4' long		
4" x 4"	EA.	0.667
6" x 6"	"	0.800
Precast concrete, 6' long, with dowels		
12" x 6"	EA.	0.400
12" x 8"	"	0.444

Site Improvements	UNIT	MAN/HOURS
02870.10 — **Prefabricated Planters**		
Concrete precast, circular		
24" dia., 18" high	EA.	0.800
42" dia., 30" high	"	1.000
Fiberglass, circular		
36" dia., 27" high	EA.	0.400
60" dia., 39" high	"	0.444
Tapered, circular		
24" dia., 36" high	EA.	0.364
40" dia., 36" high	"	0.400
Square		
2' by 2', 17" high	EA.	0.364
4' by 4', 39" high	"	0.444
Rectangular		
4' by 1', 18" high	EA.	0.400

03110.05 — Beam Formwork

Formwork	UNIT	MAN/HOURS
Beam forms, job built		
Beam bottoms		
1 use	S.F.	0.133
3 uses	"	0.123
5 uses	"	0.114
Beam sides		
1 use	S.F.	0.089
3 uses	"	0.080
5 uses	"	0.073

03110.15 — Column Formwork

Formwork	UNIT	MAN/HOURS
Column, square forms, job built		
8" x 8" columns		
1 use	S.F.	0.160
3 uses	"	0.148
5 uses	"	0.138
12" x 12" columns		
1 use	S.F.	0.145
3 uses	"	0.136
5 uses	"	0.127
16" x 16" columns		
1 use	S.F.	0.133
3 uses	"	0.125
5 uses	"	0.118
3 uses	"	0.116
3 uses	"	0.108
Round fiber forms, 1 use		
10" dia.	L.F.	0.160
12" dia.	"	0.163
18" dia.	"	0.190
24" dia.	"	0.205
36" dia.	"	0.242

03110.18 — Curb Formwork

Formwork	UNIT	MAN/HOURS
Curb forms		
Straight, 6" high		
1 use	L.F.	0.080
3 uses	"	0.073
5 uses	"	0.067
Curved, 6" high		
1 use	L.F.	0.100
3 uses	"	0.089
5 uses	"	0.082

03110.20 — Elevated Slab Formwork

Formwork	UNIT	MAN/HOURS
Elevated slab formwork		
Slab, with drop panels		
1 use	S.F.	0.064
3 uses	"	0.059
5 uses	"	0.055
Floor slab, hung from steel beams		
1 use	S.F.	0.062
3 uses	"	0.057

03110.20 — Elevated Slab Formwork *(Cont.)*

Formwork	UNIT	MAN/HOURS
5 uses	S.F.	0.053
Floor slab, with pans or domes		
1 use	S.F.	0.073
3 uses	"	0.067
5 uses	"	0.062
Equipment curbs, 12" high		
1 use	L.F.	0.080
3 uses	"	0.073
5 uses	"	0.067

03110.25 — Equipment Pad Formwork

Formwork	UNIT	MAN/HOURS
Equipment pad, job built		
1 use	S.F.	0.100
2 uses	"	0.094
3 uses	"	0.089
4 uses	"	0.084
5 uses	"	0.080

03110.35 — Footing Formwork

Formwork	UNIT	MAN/HOURS
Wall footings, job built, continuous		
1 use	S.F.	0.080
3 uses	"	0.073
5 uses	"	0.067
Column footings, spread		
1 use	S.F.	0.100
3 uses	"	0.089
5 uses	"	0.080

03110.50 — Grade Beam Formwork

Formwork	UNIT	MAN/HOURS
Grade beams, job built		
1 use	S.F.	0.080
3 uses	"	0.073
5 uses	"	0.067

03110.53 — Pile Cap Formwork

Formwork	UNIT	MAN/HOURS
Pile cap forms, job built		
Square		
1 use	S.F.	0.100
3 uses	"	0.089
5 uses	"	0.080
Triangular		
1 use	S.F.	0.114
3 uses	"	0.100
5 uses	"	0.089

03110.55 — Slab/mat Formwork

Formwork	UNIT	MAN/HOURS
Mat foundations, job built		
1 use	S.F.	0.100
3 uses	"	0.089
5 uses	"	0.080
Edge forms		
6" high		

Formwork	UNIT	MAN/HOURS
03110.55 Slab/mat Formwork *(Cont.)*		
1 use	L.F.	0.073
3 uses	"	0.067
5 uses	"	0.062
12" high		
1 use	L.F.	0.080
3 uses	"	0.073
5 uses	"	0.067
Formwork for openings		
1 use	S.F.	0.160
3 uses	"	0.133
5 uses	"	0.114
03110.60 Stair Formwork		
Stairway forms, job built		
1 use	S.F.	0.160
3 uses	"	0.133
5 uses	"	0.114
Stairs, elevated		
1 use	S.F.	0.160
3 uses	"	0.114
5 uses	"	0.100
03110.65 Wall Formwork		
Wall forms, exterior, job built		
Up to 8' high wall		
1 use	S.F.	0.080
3 uses	"	0.073
5 uses	"	0.067
Over 8' high wall		
1 use	S.F.	0.100
3 uses	"	0.089
5 uses	"	0.080
Column pier and pilaster		
1 use	S.F.	0.160
3 uses	"	0.133
5 uses	"	0.114
Interior wall forms		
Up to 8' high		
1 use	S.F.	0.073
3 uses	"	0.067
5 uses	"	0.062
Over 8' high		
1 use	S.F.	0.089
3 uses	"	0.080
5 uses	"	0.073
PVC form liner, per side, smooth finish		
1 use	S.F.	0.067
3 uses	"	0.062
5 uses	"	0.053

Formwork	UNIT	MAN/HOURS
03110.90 Miscellaneous Formwork		
Keyway forms (5 uses)		
2 x 4	L.F.	0.040
2 x 6	"	0.044
Bulkheads		
Walls, with keyways		
2 piece	L.F.	0.073
3 piece	"	0.080
Elevated slab, with keyway		
2 piece	L.F.	0.067
3 piece	"	0.073
Ground slab, with keyway		
2 piece	L.F.	0.057
3 piece	"	0.062
Chamfer strips		
Wood		
1/2" wide	L.F.	0.018
3/4" wide	"	0.018
1" wide	"	0.018
PVC		
1/2" wide	L.F.	0.018
3/4" wide	"	0.018
1" wide	"	0.018
Radius		
1"	L.F.	0.019
1-1/2"	"	0.019
Reglets		
Galvanized steel, 24 ga.	L.F.	0.032
Metal formwork		
Straight edge forms		
4" high	L.F.	0.050
6" high	"	0.053
8" high	"	0.057
12" high	"	0.062
16" high	"	0.067
Curb form, S-shape		
12" x		
1'-6"	L.F.	0.114
2'	"	0.107
2'-6"	"	0.100
3'	"	0.089

Reinforcement	UNIT	MAN/HOURS
03210.05 **Beam Reinforcing**		
Beam-girders		
#3 - #4	TON	20.000
#5 - #6	"	16.000
#7 - #8	"	13.333
Galvanized		
#3 - #4	TON	20.000
#5 - #6	"	16.000
#7 - #8	"	13.333
Bond Beams		
#3 - #4	TON	26.667
#5 - #6	"	20.000
#7 - #8	"	17.778
Galvanized		
#3 - #4	TON	26.667
#5 - #6	"	20.000
#7 - #8	"	17.778
03210.15 **Column Reinforcing**		
Columns		
#3 - #4	TON	22.857
#5 - #6	"	17.778
#7 - #8	"	16.000
Galvanized		
#3 - #4	TON	22.857
#5 - #6	"	17.778
#7 - #8	"	16.000
03210.20 **Elevated Slab Reinforcing**		
Elevated slab		
#3 - #4	TON	10.000
#5 - #6	"	8.889
#7 - #8	"	8.000
Galvanized		
#3 - #4	TON	10.000
#5 - #6	"	8.889
03210.25 **Equip. Pad Reinforcing**		
Equipment pad		
#3 - #4	TON	16.000
#5 - #6	"	14.545
#7 - #8	"	13.333
#9 - #10	"	12.308
#11 - #12	"	11.429
03210.35 **Footing Reinforcing**		
Footings		
Grade 50		
#3 - #4	TON	13.333
#5 - #6	"	11.429
#7 - #8	"	10.000
#9 - #10	"	8.889
Grade 60		
#3 - #4	TON	13.333

Reinforcement	UNIT	MAN/HOURS
03210.35 **Footing Reinforcing** (Cont.)		
#5 - #6	TON	11.429
#7 - #8	"	10.000
Straight dowels, 24" long		
1" dia. (#8)	EA.	0.080
3/4" dia. (#6)	"	0.080
5/8" dia. (#5)	"	0.067
1/2" dia. (#4)	"	0.057
03210.45 **Foundation Reinforcing**		
Foundations		
#3 - #4	TON	13.333
#5 - #6	"	11.429
#7 - #8	"	10.000
Galvanized		
#3 - #4	TON	13.333
#5 - #6	"	11.429
#7 - #8	"	10.000
03210.50 **Grade Beam Reinforcing**		
Grade beams		
#3 - #4	TON	12.308
#5 - #6	"	10.667
#7 - #8	"	9.412
Galvanized		
#3 - #4	TON	12.308
#5 - #6	"	10.667
#7 - #8	"	9.412
03210.53 **Pile Cap Reinforcing**		
Pile caps		
#3 - #4	TON	20.000
#5 - #6	"	17.778
#7 - #8	"	16.000
Galvanized		
#3 - #4	TON	20.000
#5 - #6	"	17.778
#7 - #8	"	16.000
03210.55 **Slab/mat Reinforcing**		
Bars, slabs		
#3 - #4	TON	13.333
#5 - #6	"	11.429
#7 - #8	"	10.000
Galvanized		
#3 - #4	TON	13.333
#5 - #6	"	11.429
#7 - #8	"	10.000
Wire mesh, slabs		
Galvanized		
4x4		
W1.4xW1.4	S.F.	0.005
W2.0xW2.0	"	0.006
W2.9xW2.9	"	0.006

Reinforcement	UNIT	MAN/HOURS
03210.55 **Slab/mat Reinforcing** *(Cont.)*		
W4.0xW4.0	S.F.	0.007
6x6		
W1.4xW1.4	S.F.	0.004
W2.0xW2.0	"	0.004
W2.9xW2.9	"	0.005
W4.0xW4.0	"	0.005
Standard		
2x2		
W.9xW.9	S.F.	0.005
4x4		
W1.4xW1.4	S.F.	0.005
W4.0xW4.0	"	0.007
6x6		
W1.4xW1.4	S.F.	0.004
W4.0xW4.0	"	0.005
03210.60 **Stair Reinforcing**		
Stairs		
#3 - #4	TON	16.000
#5 - #6	"	13.333
Galvanized		
#3 - #4	TON	16.000
#5 - #6	"	13.333
03210.65 **Wall Reinforcing**		
Walls		
#3 - #4	TON	11.429
#5 - #6	"	10.000
#7 - #8	"	8.889
Galvanized		
#3 - #4	TON	11.429
#5 - #6	"	10.000
#7 - #8	"	8.889
Masonry wall (horizontal)		
#3 - #4	TON	32.000
#5 - #6	"	26.667
Galvanized		
#3 - #4	TON	32.000
#5 - #6	"	26.667
Masonry wall (vertical)		
#3 - #4	TON	40.000
#5 - #6	"	32.000
Galvanized		
#3 - #4	TON	40.000
#5 - #6	"	32.000

Accessories	UNIT	MAN/HOURS
03250.40 **Concrete Accessories**		
Expansion joint, poured		
Asphalt		
1/2" x 1"	L.F.	0.016
1" x 2"	"	0.017
Liquid neoprene, cold applied		
1/2" x 1"	L.F.	0.016
1" x 2"	"	0.018
Polyurethane, 2 parts		
1/2" x 1"	L.F.	0.027
1" x 2"	"	0.029
Rubberized asphalt, cold		
1/2" x 1"	L.F.	0.016
1" x 2"	"	0.017
Hot, fuel resistant		
1/2" x 1"	L.F.	0.016
1" x 2"	"	0.017
Expansion joint, premolded, in slabs		
Asphalt		
1/2" x 6"	L.F.	0.020
1" x 12"	"	0.027
Cork		
1/2" x 6"	L.F.	0.020
1" x 12"	"	0.027
Neoprene sponge		
1/2" x 6"	L.F.	0.020
1" x 12"	"	0.027
Polyethylene foam		
1/2" x 6"	L.F.	0.020
1" x 12"	"	0.027
Polyurethane foam		
1/2" x 6"	L.F.	0.020
1" x 12"	"	0.027
Polyvinyl chloride foam		
1/2" x 6"	L.F.	0.020
1" x 12"	"	0.027
Rubber, gray sponge		
1/2" x 6"	L.F.	0.020
1" x 12"	"	0.027
Asphalt felt control joints or bond breaker, screed joints		
4" slab	L.F.	0.016
6" slab	"	0.018
8" slab	"	0.020
10" slab	"	0.023
Keyed cold expansion and control joints, 24 ga.		
4" slab	L.F.	0.050
5" slab	"	0.050
6" slab	"	0.053
8" slab	"	0.057
10" slab	"	0.062
Waterstops		
Polyvinyl chloride		
Ribbed		
3/16" thick x		

Accessories

03250.40 — Concrete Accessories *(Cont.)*

	UNIT	MAN/HOURS
4" wide	L.F.	0.040
6" wide	"	0.044
1/2" thick x		
9" wide	L.F.	0.050
Ribbed with center bulb		
3/16" thick x 9" wide	L.F.	0.050
3/8" thick x 9" wide	"	0.050
Dumbbell type, 3/8" thick x 6" wide	"	0.044
Plain, 3/8" thick x 9" wide	"	0.050
Center bulb, 3/8" thick x 9" wide	"	0.050
Rubber		
Vapor barrier		
4 mil polyethylene	S.F.	0.003
6 mil polyethylene	"	0.003
Gravel porous fill, under floor slabs, 3/4" stone	C.Y.	1.333

Cast-in-place Concrete

03350.10 — Concrete Finishes

	UNIT	MAN/HOURS
Floor finishes		
Broom	S.F.	0.011
Screed	"	0.010
Darby	"	0.010
Steel float	"	0.013
Wall finishes		
Burlap rub, with cement paste	S.F.	0.013
Break ties and patch holes	"	0.016
Carborundum		
Dry rub	S.F.	0.027
Wet rub	"	0.040
Floor hardeners		
Metallic		
Light service	S.F.	0.010
Heavy service	"	0.013
Non-metallic		
Light service	S.F.	0.010
Heavy service	"	0.013

03360.10 — Pneumatic Concrete

	UNIT	MAN/HOURS
Pneumatic applied concrete (gunite)		
2" thick	S.F.	0.030
3" thick	"	0.040
4" thick	"	0.048
Finish surface		
Minimum	S.F.	0.040
Maximum	"	0.080

Cast-in-place Concrete

03370.10 — Curing Concrete

	UNIT	MAN/HOURS
Sprayed membrane		
Slabs	S.F.	0.002
Walls	"	0.002
Curing paper		
Slabs	S.F.	0.002
Walls	"	0.002
Burlap		
7.5 oz.	S.F.	0.003
12 oz.	"	0.003

Placing Concrete

03380.05 — Beam Concrete

	UNIT	MAN/HOURS
Beams and girders		
2500# or 3000# concrete		
By pump	C.Y.	0.873
By hand buggy	"	0.800
5000# concrete		
By pump	C.Y.	0.873
By hand buggy	"	0.800
Bond beam, 3000# concrete		
By pump		
8" high		
4" wide	L.F.	0.019
6" wide	"	0.022
8" wide	"	0.024
10" wide	"	0.027
12" wide	"	0.030

03380.15 — Column Concrete

	UNIT	MAN/HOURS
Columns		
2500# or 3000# concrete		
By pump	C.Y.	0.800
5000# concrete		
By pump	C.Y.	0.800

03380.25 — Equipment Pad Concrete

	UNIT	MAN/HOURS
Equipment pad		
2500# or 3000# concrete		
By chute	C.Y.	0.267
By pump	"	0.686

Placing Concrete

Placing Concrete	UNIT	MAN/HOURS
03380.25 — Equipment Pad Concrete (Cont.)		
3500# or 4000# concrete		
By chute	C.Y.	0.267
By pump	"	0.686
5000# concrete		
By chute	C.Y.	0.267
By pump	"	0.686
03380.35 — Footing Concrete		
Continuous footing		
2500# or 3000# concrete		
By chute	C.Y.	0.267
By pump	"	0.600
5000# concrete		
By chute	C.Y.	0.267
By pump	"	0.600
Spread footing		
2500# or 3000# concrete		
Under 5 cy		
By chute	C.Y.	0.267
By pump	"	0.640
5000# concrete		
Under 5 c.y.		
By chute	C.Y.	0.267
By pump	"	0.640
03380.50 — Grade Beam Concrete		
Grade beam		
2500# or 3000# concrete		
By chute	C.Y.	0.267
By pump	"	0.600
By hand buggy	"	0.800
5000# concrete		
By chute	C.Y.	0.267
By pump	"	0.600
By hand buggy	"	0.800
03380.53 — Pile Cap Concrete		
Pile cap		
2500# or 3000 concrete		
By chute	C.Y.	0.267
By pump	"	0.686
By hand buggy	"	0.800
5000# concrete		
By chute	C.Y.	0.267
By pump	"	0.686
By hand buggy	"	0.800
03380.55 — Slab/mat Concrete		
Slab on grade		
2500# or 3000# concrete		
By chute	C.Y.	0.200

Placing Concrete	UNIT	MAN/HOURS
03380.55 — Slab/mat Concrete (Cont.)		
By pump	C.Y.	0.343
By hand buggy	"	0.533
5000# concrete		
By chute	C.Y.	0.200
By pump	"	0.343
By hand buggy	"	0.533
03380.58 — Sidewalks		
Walks, cast in place with wire mesh, base not incl.		
4" thick	S.F.	0.027
5" thick	"	0.032
6" thick	"	0.040
03380.60 — Stair Concrete		
Stairs		
2500# or 3000# concrete		
By chute	C.Y.	0.267
By pump	"	0.686
By hand buggy	"	0.800
3500# or 4000# concrete		
By chute	C.Y.	0.267
By pump	"	0.686
By hand buggy	"	0.800
5000# concrete		
By chute	C.Y.	0.267
By pump	"	0.686
By hand buggy	"	0.800
03380.65 — Wall Concrete		
Walls		
2500# or 3000# concrete		
To 4'		
By chute	C.Y.	0.229
By pump	"	0.738
To 8'		
By pump	C.Y.	0.800
3500# or 4000# concrete		
To 4'		
By chute	C.Y.	0.229
By pump	"	0.738
To 8'		
By pump	C.Y.	0.800
Filled block (CMU)		
3000# concrete, by pump		
4" wide	S.F.	0.034
6" wide	"	0.040
8" wide	"	0.048
10" wide	"	0.056
12" wide	"	0.069
Pilasters, 3000# concrete	C.F.	0.960
Wall cavity, 2" thick, 3000# concrete	S.F.	0.032

Cementitous Toppings

Cementitous Toppings	UNIT	MAN/ HOURS
03550.10 Concrete Toppings		
Gypsum fill		
2" thick	S.F.	0.005
2-1/2" thick	"	0.005
3" thick	"	0.005
3-1/2" thick	"	0.005
4" thick	"	0.006
Formboard		
Mineral fiber board		
1" thick	S.F.	0.020
1-1/2" thick	"	0.023
Cement fiber board		
1" thick	S.F.	0.027
1-1/2" thick	"	0.031
Glass fiber board		
1" thick	S.F.	0.020
1-1/2" thick	"	0.023
Poured deck		
Vermiculite or perlite		
1 to 4 mix	C.Y.	0.800
1 to 6 mix	"	0.738
Vermiculite or perlite		
2" thick		
1 to 4 mix	S.F.	0.005
1 to 6 mix	"	0.005
3" thick		
1 to 4 mix	S.F.	0.007
1 to 6 mix	"	0.007
Concrete plank, lightweight		
2" thick	S.F.	0.024
2-1/2" thick	"	0.024
3-1/2" thick	"	0.027
4" thick	"	0.027
Channel slab, lightweight, straight		
2-3/4" thick	S.F.	0.024
3-1/2" thick	"	0.024
3-3/4" thick	"	0.024
4-3/4" thick	"	0.027
Gypsum plank		
2" thick	S.F.	0.024
3" thick	"	0.024
Cement fiber, T and G planks		
1" thick	S.F.	0.022
1-1/2" thick	"	0.022
2" thick	"	0.024
2-1/2" thick	"	0.024
3" thick	"	0.024
3-1/2" thick	"	0.027
4" thick	"	0.027

Concrete Restoration

Concrete Restoration	UNIT	MAN/ HOURS
03730.10 Concrete Repair		
Epoxy grout floor patch, 1/4" thick	S.F.	0.080
Epoxy gel grout	"	0.800
Injection valve, 1 way, threaded plastic	EA.	0.160
Grout crack seal, 2 component	C.F.	0.800
Grout, non shrink	"	0.800
Concrete, epoxy modified		
Sand mix	C.F.	0.320
Gravel mix	"	0.296
Concrete repair		
Soffit repair		
16" wide	L.F.	0.160
18" wide	"	0.167
24" wide	"	0.178
30" wide	"	0.190
32" wide	"	0.200
Edge repair		
2" spall	L.F.	0.200
3" spall	"	0.211
4" spall	"	0.216
6" spall	"	0.222
8" spall	"	0.235
9" spall	"	0.267
Crack repair, 1/8" crack	"	0.080
Reinforcing steel repair		
1 bar, 4 ft		
#4 bar	L.F.	0.100
#5 bar	"	0.100
#6 bar	"	0.107
#8 bar	"	0.107
#9 bar	"	0.114
#11 bar	"	0.114
Pile repairs		
Polyethylene wrap		
30 mil thick		
60" wide	S.F.	0.267
72" wide	"	0.320
60 mil thick		
60" wide	S.F.	0.267
80" wide	"	0.364
Pile spall, average repair 3'		
18" x 18"	EA.	0.667
20" x 20"	"	0.800

Mortar And Grout	UNIT	MAN/HOURS
04100.10 Masonry Grout		
Grout, non shrink, non-metallic, trowelable	C.F.	0.016
Grout door frame, hollow metal		
Single	EA.	0.600
Double	"	0.632
Grout-filled concrete block (CMU)		
4" wide	S.F.	0.020
6" wide	"	0.022
8" wide	"	0.024
12" wide	"	0.025
Grout-filled individual CMU cells		
4" wide	L.F.	0.012
6" wide	"	0.012
8" wide	"	0.012
10" wide	"	0.014
12" wide	"	0.014
Bond beams or lintels, 8" deep		
6" thick	L.F.	0.022
8" thick	"	0.024
10" thick	"	0.027
12" thick	"	0.030
Cavity walls		
2" thick	S.F.	0.032
3" thick	"	0.032
4" thick	"	0.034
6" thick	"	0.040
04150.10 Masonry Accessories		
Foundation vents	EA.	0.320
Bar reinforcing		
Horizontal		
#3 - #4	Lb.	0.032
#5 - #6	"	0.027
Vertical		
#3 - #4	Lb.	0.040
#5 - #6	"	0.032
Horizontal joint reinforcing		
Truss type		
4" wide, 6" wall	L.F.	0.003
6" wide, 8" wall	"	0.003
8" wide, 10" wall	"	0.003
10" wide, 12" wall	"	0.004
12" wide, 14" wall	"	0.004
Ladder type		
4" wide, 6" wall	L.F.	0.003
6" wide, 8" wall	"	0.003
8" wide, 10" wall	"	0.003
10" wide, 12" wall	"	0.003
Rectangular wall ties		
3/16" dia., galvanized		
2" x 6"	EA.	0.013
2" x 8"	"	0.013
2" x 10"	"	0.013
2" x 12"	"	0.013

Mortar And Grout	UNIT	MAN/HOURS
04150.10 Masonry Accessories *(Cont.)*		
4" x 6"	EA.	0.016
4" x 8"	"	0.016
4" x 10"	"	0.016
4" x 12"	"	0.016
1/4" dia., galvanized		
2" x 6"	EA.	0.013
2" x 8"	"	0.013
2" x 10"	"	0.013
2" x 12"	"	0.013
4" x 6"	"	0.016
4" x 8"	"	0.016
4" x 10"	"	0.016
4" x 12"	"	0.016
"Z" type wall ties, galvanized		
6" long		
1/8" dia.	EA.	0.013
3/16" dia.	"	0.013
1/4" dia.	"	0.013
8" long		
1/8" dia.	EA.	0.013
3/16" dia.	"	0.013
1/4" dia.	"	0.013
10" long		
1/8" dia.	EA.	0.013
3/16" dia.	"	0.013
1/4" dia.	"	0.013
Dovetail anchor slots		
Galvanized steel, filled		
24 ga.	L.F.	0.020
20 ga.	"	0.020
16 oz. copper, foam filled	"	0.020
Dovetail anchors		
16 ga.		
3-1/2" long	EA.	0.013
5-1/2" long	"	0.013
12 ga.		
3-1/2" long	EA.	0.013
5-1/2" long	"	0.013
Dovetail, triangular galvanized ties, 12 ga.		
3" x 3"	EA.	0.013
5" x 5"	"	0.013
7" x 7"	"	0.013
7" x 9"	"	0.013
Brick anchors		
Corrugated, 3-1/2" long		
16 ga.	EA.	0.013
12 ga.	"	0.013
Non-corrugated, 3-1/2" long		
16 ga.	EA.	0.013
12 ga.	"	0.013
Cavity wall anchors, corrugated, galvanized		
5" long		
16 ga.	EA.	0.013

Mortar And Grout

04150.10   Masonry Accessories *(Cont.)*

	UNIT	MAN/HOURS
12 ga.	EA.	0.013
7" long		
28 ga.	EA.	0.013
24 ga.	"	0.013
22 ga.	"	0.013
16 ga.	"	0.013
Mesh ties, 16 ga., 3" wide		
8" long	EA.	0.013
12" long	"	0.013
20" long	"	0.013
24" long	"	0.013

04150.20   Masonry Control Joints

	UNIT	MAN/HOURS
Control joint, cross shaped PVC	L.F.	0.020
Closed cell joint filler		
1/2"	L.F.	0.020
3/4"	"	0.020
Rubber, for		
4" wall	L.F.	0.020
6" wall	"	0.021
8" wall	"	0.022
PVC, for		
4" wall	L.F.	0.020
6" wall	"	0.021
8" wall	"	0.022

04150.50   Masonry Flashing

	UNIT	MAN/HOURS
Through-wall flashing		
5 oz. coated copper	S.F.	0.067
0.030" elastomeric	"	0.053

Unit Masonry

04210.10   Brick Masonry

	UNIT	MAN/HOURS
Standard size brick, running bond		
Face brick, red (6.4/sf)		
Veneer	S.F.	0.133
Cavity wall	"	0.114
9" solid wall	"	0.229
Common brick (6.4/sf)		
Select common for veneers	S.F.	0.133
Back-up		
4" thick	S.F.	0.100
8" thick	"	0.160
Firewall		
12" thick	S.F.	0.267

Unit Masonry

04210.10   Brick Masonry *(Cont.)*

	UNIT	MAN/HOURS
16" thick	S.F.	0.364
Glazed brick (7.4/sf)		
Veneer	S.F.	0.145
Buff or gray face brick (6.4/sf)		
Veneer	S.F.	0.133
Cavity wall	"	0.114
Jumbo or oversize brick (3/sf)		
4" veneer	S.F.	0.080
4" back-up	"	0.067
8" back-up	"	0.114
12" firewall	"	0.200
16" firewall	"	0.267
Norman brick, red face, (4.5/sf)		
4" veneer	S.F.	0.100
Cavity wall	"	0.089
Chimney, standard brick, including flue		
16" x 16"	L.F.	0.800
16" x 20"	"	0.800
16" x 24"	"	0.800
20" x 20"	"	1.000
20" x 24"	"	1.000
20" x 32"	"	1.143
Window sill, face brick on edge	"	0.200

04210.20   Structural Tile

	UNIT	MAN/HOURS
Structural glazed tile		
6T series, 5-1/2" x 12"		
Glazed on one side		
2" thick	S.F.	0.080
4" thick	"	0.080
6" thick	"	0.089
8" thick	"	0.100
Glazed on two sides		
4" thick	S.F.	0.100
6" thick	"	0.114
Special shapes		
Group 1	S.F.	0.160
Group 2	"	0.160
Group 3	"	0.160
Group 4	"	0.160
Group 5	"	0.160
Fire rated		
4" thick, 1 hr rating	S.F.	0.080
6" thick, 2 hr rating	"	0.089
8W series, 8" x 16"		
Glazed on one side		
2" thick	S.F.	0.053
4" thick	"	0.053
6" thick	"	0.062
8" thick	"	0.062
Glazed on two sides		
4" thick	S.F.	0.067
6" thick	"	0.080

Unit Masonry

04210.20 — Structural Tile (Cont.)

Description	UNIT	MAN/HOURS
8" thick	S.F.	0.080
Special shapes		
Group 1	S.F.	0.114
Group 2	"	0.114
Group 3	"	0.114
Group 4	"	0.114
Group 5	"	0.114
Fire rated		
4" thick, 1 hr rating	S.F.	0.114
6" thick, 2 hr rating	"	0.114

04210.60 — Pavers, Masonry

Description	UNIT	MAN/HOURS
Brick walk laid on sand, sand joints		
Laid flat, (4.5 per sf)	S.F.	0.089
Laid on edge, (7.2 per sf)	"	0.133
Precast concrete patio blocks		
2" thick		
Natural	S.F.	0.027
Colors	"	0.027
Exposed aggregates, local aggregate		
Natural	S.F.	0.027
Colors	"	0.027
Granite or limestone aggregate	"	0.027
White tumblestone aggregate	"	0.027
Stone pavers, set in mortar		
Bluestone		
1" thick		
Irregular	S.F.	0.200
Snapped rectangular	"	0.160
1-1/2" thick, random rectangular	"	0.200
2" thick, random rectangular	"	0.229
Slate		
Natural cleft		
Irregular, 3/4" thick	S.F.	0.229
Random rectangular		
1-1/4" thick	S.F.	0.200
1-1/2" thick	"	0.222
Granite blocks		
3" thick, 3" to 6" wide		
4" to 12" long	S.F.	0.267
6" to 15" long	"	0.229
Crushed stone, white marble, 3" thick	"	0.016

04220.10 — Concrete Masonry Units

Description	UNIT	MAN/HOURS
Hollow, load bearing		
4"	S.F.	0.059
6"	"	0.062
8"	"	0.067
10"	"	0.073
12"	"	0.080
Solid, load bearing		
4"	S.F.	0.059
6"	"	0.062

04220.10 — Concrete Masonry Units (Cont.)

Description	UNIT	MAN/HOURS
8"	S.F.	0.067
10"	"	0.073
12"	"	0.080
Back-up block, 8" x 16"		
2"	S.F.	0.046
4"	"	0.047
6"	"	0.050
8"	"	0.053
10"	"	0.057
12"	"	0.062
Foundation wall, 8" x 16"		
6"	S.F.	0.057
8"	"	0.062
10"	"	0.067
12"	"	0.073
Solid		
6"	S.F.	0.062
8"	"	0.067
10"	"	0.073
12"	"	0.080
Exterior, styrofoam inserts, standard weight, 8" x 16"		
6"	S.F.	0.062
8"	"	0.067
10"	"	0.073
12"	"	0.080
Lightweight		
6"	S.F.	0.062
8"	"	0.067
10"	"	0.073
12"	"	0.080
Acoustical slotted block		
4"	S.F.	0.073
6"	"	0.073
8"	"	0.080
Filled cavities		
4"	S.F.	0.089
6"	"	0.094
8"	"	0.100
Hollow, split face		
4"	S.F.	0.059
6"	"	0.062
8"	"	0.067
10"	"	0.073
12"	"	0.080
Split rib profile		
4"	S.F.	0.073
6"	"	0.073
8"	"	0.080
10"	"	0.080
12"	"	0.080
High strength block, 3500 psi		
2"	S.F.	0.059
4"	"	0.062

Unit Masonry

04220.10 — Concrete Masonry Units (Cont.)

	UNIT	MAN/HOURS
6"	S.F.	0.062
8"	"	0.067
10"	"	0.073
12"	"	0.080
Solar screen concrete block		
4" thick		
6" x 6"	S.F.	0.178
8" x 8"	"	0.160
12" x 12"	"	0.123
8" thick		
8" x 16"	S.F.	0.114
Glazed block		
Cove base, glazed 1 side, 2"	L.F.	0.089
4"	"	0.089
6"	"	0.100
8"	"	0.100
Single face		
2"	S.F.	0.067
4"	"	0.067
6"	"	0.073
8"	"	0.080
10"	"	0.089
12"	"	0.094
Double face		
4"	S.F.	0.084
6"	"	0.089
8"	"	0.100
Corner or bullnose		
2"	EA.	0.100
4"	"	0.114
6"	"	0.114
8"	"	0.133
10"	"	0.145
12"	"	0.160
Gypsum unit masonry		
Partition blocks (12"x30")		
Solid		
2"	S.F.	0.032
Hollow		
3"	S.F.	0.032
4"	"	0.033
6"	"	0.036
Vertical reinforcing		
4' o.c., add 5% to labor		
2'8" o.c., add 15% to labor		
Interior partitions, add 10% to labor		

04220.90 — Bond Beams & Lintels

	UNIT	MAN/HOURS
Bond beam, no grout or reinforcement		
8" x 16" x		
4" thick	L.F.	0.062
6" thick	"	0.064
8" thick	"	0.067

04220.90 — Bond Beams & Lintels (Cont.)

	UNIT	MAN/HOURS
10" thick	L.F.	0.070
12" thick	"	0.073
Beam lintel, no grout or reinforcement		
8" x 16" x		
10" thick	L.F.	0.080
12" thick	"	0.089
Precast masonry lintel		
6 lf, 8" high x		
4" thick	L.F.	0.133
6" thick	"	0.133
8" thick	"	0.145
10" thick	"	0.145
10 lf, 8" high x		
4" thick	L.F.	0.080
6" thick	"	0.080
8" thick	"	0.089
10" thick	"	0.089
Steel angles and plates		
Minimum	Lb.	0.011
Maximum	"	0.020
Various size angle lintels		
1/4" stock		
3" x 3"	L.F.	0.050
3" x 3-1/2"	"	0.050
3/8" stock		
3" x 4"	L.F.	0.050
3-1/2" x 4"	"	0.050
4" x 4"	"	0.050
5" x 3-1/2"	"	0.050
6" x 3-1/2"	"	0.050
1/2" stock		
6" x 4"	L.F.	0.050

04240.10 — Clay Tile

	UNIT	MAN/HOURS
Hollow clay tile, for back-up, 12" x 12"		
Scored face		
Load bearing		
4" thick	S.F.	0.057
6" thick	"	0.059
8" thick	"	0.062
10" thick	"	0.064
12" thick	"	0.067
Non-load bearing		
3" thick	S.F.	0.055
4" thick	"	0.057
6" thick	"	0.059
8" thick	"	0.062
12" thick	"	0.067
Partition, 12" x 12"		
In walls		
3" thick	S.F.	0.067
4" thick	"	0.067
6" thick	"	0.070

Unit Masonry

04240.10 — Clay Tile (Cont.)

	UNIT	MAN/HOURS
8" thick	S.F.	0.073
10" thick	"	0.076
12" thick	"	0.080
Clay tile floors		
4" thick	S.F.	0.044
6" thick	"	0.047
8" thick	"	0.050
10" thick	"	0.053
12" thick	"	0.057
Terra cotta		
Coping, 10" or 12" wide, 3" thick	L.F.	0.160

04270.10 — Glass Block

	UNIT	MAN/HOURS
Glass block, 4" thick		
6" x 6"	S.F.	0.267
8" x 8"	"	0.200
12" x 12"	"	0.160
Replacement glass blocks, 4" x 8" x 8"		
Minimum	S.F.	0.800
Maximum	"	1.600

04295.10 — Parging/masonry Plaster

	UNIT	MAN/HOURS
Parging		
1/2" thick	S.F.	0.053
3/4" thick	"	0.067
1" thick	"	0.080

Stone

04400.10 — Stone

	UNIT	MAN/HOURS
Rubble stone		
Walls set in mortar		
8" thick	S.F.	0.200
12" thick	"	0.320
18" thick	"	0.400
24" thick	"	0.533
Dry set wall		
8" thick	S.F.	0.133
12" thick	"	0.200
18" thick	"	0.267
24" thick	"	0.320
Cut stone		
Imported marble		
Facing panels		
3/4" thick	S.F.	0.320
1-1/2" thick	"	0.364

Stone

04400.10 — Stone (Cont.)

	UNIT	MAN/HOURS
2-1/4" thick	S.F.	0.444
Base		
1" thick		
4" high	L.F.	0.400
6" high	"	0.400
Columns, solid		
Plain faced	C.F.	5.333
Fluted	"	5.333
Flooring, travertine, minimum	S.F.	0.123
Average	"	0.160
Maximum	"	0.178
Domestic marble		
Facing panels		
7/8" thick	S.F.	0.320
1-1/2" thick	"	0.364
2-1/4" thick	"	0.444
Stairs		
12" treads	L.F.	0.400
6" risers	"	0.267
Thresholds, 7/8" thick, 3' long, 4" to 6" wide		
Plain	EA.	0.667
Beveled	"	0.667
Window sill		
6" wide, 2" thick	L.F.	0.320
Stools		
5" wide, 7/8" thick	L.F.	0.320
Limestone panels up to 12' x 5', smooth finish		
2" thick	S.F.	0.096
3" thick	"	0.096
4" thick	"	0.096
Miscellaneous limestone items		
Steps, 14" wide, 6" deep	L.F.	0.533
Coping, smooth finish	C.F.	0.267
Sills, lintels, jambs, smooth finish	"	0.320
Granite veneer facing panels, polished		
7/8" thick		
Black	S.F.	0.320
Gray	"	0.320
Base		
4" high	L.F.	0.160
6" high	"	0.178
Curbing, straight, 6" x 16"	"	0.400
Radius curbs, radius over 5'	"	0.533
Ashlar veneer		
4" thick, random	S.F.	0.320
Pavers, 4" x 4" split		
Gray	S.F.	0.160
Pink	"	0.160
Black	"	0.160
Slate, panels		
1" thick	S.F.	0.320
2" thick	"	0.364
Sills or stools		

Stone	UNIT	MAN/ HOURS
04400.10 — **Stone** *(Cont.)*		
1" thick		
6" wide	L.F.	0.320
10" wide	"	0.348
2" thick		
6" wide	L.F.	0.364
10" wide	"	0.400

Masonry Restoration	UNIT	MAN/ HOURS
04520.10 — **Restoration And Cleaning**		
Masonry cleaning		
Washing brick		
Smooth surface	S.F.	0.013
Rough surface	"	0.018
Steam clean masonry		
Smooth face		
Minimum	S.F.	0.010
Maximum	"	0.015
Rough face		
Minimum	S.F.	0.013
Maximum	"	0.020
Sandblast masonry		
Minimum	S.F.	0.016
Maximum	"	0.027
Pointing masonry		
Brick	S.F.	0.032
Concrete block	"	0.023
Cut and repoint		
Brick		
Minimum	S.F.	0.040
Maximum	"	0.080
Stone work	L.F.	0.062
Cut and recaulk		
Oil base caulks	L.F.	0.053
Butyl caulks	"	0.053
Polysulfides and acrylics	"	0.053
Silicones	"	0.053
Cement and sand grout on walls, to 1/8" thick		
Minimum	S.F.	0.032
Maximum	"	0.040
Brick removal and replacement		
Minimum	EA.	0.100
Average	"	0.133
Maximum	"	0.400

Masonry Restoration	UNIT	MAN/ HOURS
04550.10 — **Refractories**		
Flue liners		
Rectangular		
8" x 12"	L.F.	0.133
12" x 12"	"	0.145
12" x 18"	"	0.160
16" x 16"	"	0.178
18" x 18"	"	0.190
20" x 20"	"	0.200
24" x 24"	"	0.229
Round		
18" dia.	L.F.	0.190
24" dia.	"	0.229

Metal Fastening	UNIT	MAN/ HOURS
05050.10 **Structural Welding**		
Welding		
Single pass		
1/8"	L.F.	0.040
3/16"	"	0.053
1/4"	"	0.067
Miscellaneous steel shapes		
Plain	Lb.	0.002
Galvanized	"	0.003
Plates		
Plain	Lb.	0.002
Galvanized	"	0.003
05050.95 **Metal Lintels**		
Lintels, steel		
Plain	Lb.	0.020
Galvanized	"	0.020
05300.10 **Metal Decking**		
Roof, 1-1/2" deep, non-composite		
16 ga.		
Primed	S.F.	0.008
Galvanized	"	0.008
Open type decking, galvanized		
1-1/2" deep		
18 ga.	S.F.	0.008
20 ga.	"	0.008
3" deep		
16 ga.	S.F.	0.009
18 ga.	"	0.009
20 ga.	"	0.009
Cellular type		
1-1/2" deep, galvanized		
18-18 ga.	S.F.	0.010
22-18 ga.	"	0.010
3" deep, galvanized		
16-16 ga.	S.F.	0.011
18-16 ga.	"	0.011
18-18 ga.	"	0.011
20-18 ga.	"	0.011
Composite deck, non-cellular, galvanized		
1-1/2" deep		
18 ga.	S.F.	0.009
20 ga.	"	0.009
3" deep		
18 ga.	S.F.	0.009
20 ga.	"	0.009

Cold Formed Framing	UNIT	MAN/ HOURS
05410.10 **Metal Framing**		
Furring channel, galvanized		
Beams and columns, 3/4"		
12" o.c.	S.F.	0.080
16" o.c.	"	0.073
Walls, 3/4"		
12" o.c.	S.F.	0.040
16" o.c.	"	0.033
24" o.c.	"	0.027
1-1/2"		
12" o.c.	S.F.	0.040
16" o.c.	"	0.033
24" o.c.	"	0.027
Stud, load bearing		
16" o.c.		
16 ga.		
2-1/2"	S.F.	0.036
3-5/8"	"	0.036
4"	"	0.036
6"	"	0.040
18 ga.		
2-1/2"	S.F.	0.036
3-5/8"	"	0.036
4"	"	0.036
6"	"	0.040
8"	"	0.040
20 ga.		
2-1/2"	S.F.	0.036
3-5/8"	"	0.036
4"	"	0.036
6"	"	0.040
8"	"	0.040
24" o.c.		
16 ga.		
2-1/2"	S.F.	0.031
3-5/8"	"	0.031
4"	"	0.031
6"	"	0.033
8"	"	0.033
18 ga.		
2-1/2"	S.F.	0.031
3-5/8"	"	0.031
4"	"	0.031
6"	"	0.033
8"	"	0.033
20 ga.		
2-1/2"	S.F.	0.031
3-5/8"	"	0.031
4"	"	0.031
6"	"	0.033
8"	"	0.033

Metal Fabrications

Metal Fabrications	UNIT	MAN/HOURS
05510.10 — Stairs		
Stock unit, steel, complete, per riser		
Tread		
3'-6" wide	EA.	1.000
4' wide	"	1.143
5' wide	"	1.333
Metal pan stair, cement filled, per riser		
3'-6" wide	EA.	0.800
4' wide	"	0.889
5' wide	"	1.000
Landing, steel pan	S.F.	0.200
Cast iron tread, steel stringers, stock units, per riser		
Tread		
3'-6" wide	EA.	1.000
4' wide	"	1.143
5' wide	"	1.333
Stair treads, abrasive, 12" x 3'-6"		
Cast iron		
3/8"	EA.	0.400
1/2"	"	0.400
Cast aluminum		
5/16"	EA.	0.400
3/8"	"	0.400
1/2"	"	0.400
05515.10 — Ladders		
Ladder, 18" wide		
With cage	L.F.	0.533
Without cage	"	0.400
05520.10 — Railings		
Railing, pipe		
1-1/4" diameter, welded steel		
2-rail		
Primed	L.F.	0.160
Galvanized	"	0.160
3-rail		
Primed	L.F.	0.200
Galvanized	"	0.200
Wall mounted, single rail, welded steel		
Primed	L.F.	0.123
Galvanized	"	0.123
1-1/2" diameter, welded steel		
2-rail		
Primed	L.F.	0.160
Galvanized	"	0.160
3-rail		
Primed	L.F.	0.200
Galvanized	"	0.200
Wall mounted, single rail, welded steel		
Primed	L.F.	0.123
Galvanized	"	0.123
2" diameter, welded steel		
2-rail		

Metal Fabrications	UNIT	MAN/HOURS
05520.10 — Railings *(Cont.)*		
Primed	L.F.	0.178
Galvanized	"	0.178
3-rail		
Primed	L.F.	0.229
Galvanized	"	0.229
Wall mounted, single rail, welded steel		
Primed	L.F.	0.133
Galvanized	"	0.133
05530.10 — Metal Grating		
Floor plate, checkered, steel		
1/4"		
Primed	S.F.	0.011
Galvanized	"	0.011
3/8"		
Primed	S.F.	0.012
Galvanized	"	0.012
Aluminum grating, pressure-locked bearing bars		
3/4" x 1/8"	S.F.	0.020
1" x 1/8"	"	0.020
1-1/4" x 1/8"	"	0.020
1-1/4" x 3/16"	"	0.020
1-1/2" x 1/8"	"	0.020
1-3/4" x 3/16"	"	0.020
Miscellaneous expenses		
Cutting		
Minimum	L.F.	0.053
Maximum	"	0.080
Banding		
Minimum	L.F.	0.133
Maximum	"	0.160
Toe plates		
Minimum	L.F.	0.160
Maximum	"	0.200
Steel grating, primed		
3/4" x 1/8"	S.F.	0.027
1" x 1/8"	"	0.027
1-1/4" x 1/8"	"	0.027
1-1/4" x 3/16"	"	0.027
1-1/2" x 1/8"	"	0.027
1-3/4" x 3/16"	"	0.027
Galvanized		
3/4" x 1/8"	S.F.	0.027
1" x 1/8"	"	0.027
1-1/4" x 1/8"	"	0.027
1-1/4" x 3/16"	"	0.027
1-1/2" x 1/8"	"	0.027
1-3/4" x 3/16"	"	0.027
Miscellaneous expenses		
Cutting		
Minimum	L.F.	0.057
Maximum	"	0.089

Metal Fabrications	UNIT	MAN/HOURS
05530.10 — Metal Grating *(Cont.)*		
Banding		
Minimum	L.F.	0.145
Maximum	"	0.178
Toe plates		
Minimum	L.F.	0.178
Maximum	"	0.229
05540.10 — Castings		
Miscellaneous castings		
Light sections	Lb.	0.016
Heavy sections	"	0.011

Misc. Fabrications	UNIT	MAN/HOURS
05580.10 — Metal Specialties		
Kick plate		
4" high x 1/4" thick		
Primed	L.F.	0.160
Galvanized	"	0.160
6" high x 1/4" thick		
Primed	L.F.	0.178
Galvanized	"	0.178
05700.10 — Ornamental Metal		
Railings, vertical square bars, 6" o.c., with shaped top rails		
Steel	L.F.	0.400
Aluminum	"	0.400
Bronze	"	0.533
Stainless steel	"	0.533
Laminated metal or wood handrails with metal supports		
2-1/2" round or oval shape	L.F.	0.400
Grilles and louvers		
Fixed type louvers		
4 through 10 sf	S.F.	0.133
Over 10 sf	"	0.100
Movable type louvers		
4 through 10 sf	S.F.	0.133
Over 10 sf	"	0.100
Aluminum louvers		
Residential use, fixed type, with screen		
8" x 8"	EA.	0.400
12" x 12"	"	0.400
12" x 18"	"	0.400
14" x 24"	"	0.400
18" x 24"	"	0.400
30" x 24"	"	0.444

Misc. Fabrications	UNIT	MAN/HOURS
05800.10 — Expansion Control		
Expansion joints with covers, floor assembly type		
With 1" space		
Aluminum	L.F.	0.133
Bronze	"	0.133
Stainless steel	"	0.133
Aluminum	"	0.133
Ceiling and wall assembly type		
With 1" space		
Aluminum	L.F.	0.160
Bronze	"	0.160
Stainless steel	"	0.160
Exterior roof and wall, aluminum		
Roof to roof		
With 1" space	L.F.	0.133
With 2" space	"	0.133
Roof to wall		
With 1" space	L.F.	0.145
With 2" space	"	0.145
Flat wall to wall		
With 1" space	L.F.	0.133
With 2" space	"	0.133
Corner to flat wall		
With 1" space	L.F.	0.160
With 2" in space	"	0.160

Fasteners And Adhesives	UNIT	MAN/HOURS
06050.10 **Accessories**		
Column/post base, cast aluminum		
4" x 4"	EA.	0.200
6" x 6"	"	0.200
Bridging, metal, per pair		
12" o.c.	EA.	0.080
16" o.c.	"	0.073
Anchors		
Bolts, threaded two ends, with nuts and washers		
1/2" dia.		
4" long	EA.	0.050
7-1/2" long	"	0.050
3/4" dia.		
7-1/2" long	EA.	0.050
15" long	"	0.050
Framing anchors		
10 gauge	EA.	0.067
Bolts, carriage		
1/4 x 4	EA.	0.080
5/16 x 6	"	0.084
3/8 x 6	"	0.084
1/2 x 6	"	0.084
Joist and beam hangers		
18 ga.		
2 x 4	EA.	0.080
2 x 6	"	0.080
2 x 8	"	0.080
2 x 10	"	0.089
2 x 12	"	0.100
16 ga.		
3 x 6	EA.	0.089
3 x 8	"	0.089
3 x 10	"	0.094
3 x 12	"	0.107
3 x 14	"	0.114
4 x 6	"	0.089
4 x 8	"	0.089
4 x 10	"	0.094
4 x 12	"	0.107
4 x 14	"	0.114
Rafter anchors, 18 ga., 1-1/2" wide		
5-1/4" long	EA.	0.067
10-3/4" long	"	0.067
Shear plates		
2-5/8" dia.	EA.	0.062
4" dia.	"	0.067
Sill anchors		
Embedded in concrete	EA.	0.080
Split rings		
2-1/2" dia.	EA.	0.089
4" dia.	"	0.100
Strap ties, 14 ga., 1-3/8" wide		
12" long	EA.	0.067

Fasteners And Adhesives	UNIT	MAN/HOURS
06050.10 **Accessories** (Cont.)		
18" long	EA.	0.073
24" long	"	0.080
36" long	"	0.089
Toothed rings		
2-5/8" dia.	EA.	0.133
4" dia.	"	0.160

Rough Carpentry	UNIT	MAN/HOURS
06110.10 **Blocking**		
Steel construction		
Walls		
2x4	L.F.	0.053
2x6	"	0.062
2x8	"	0.067
2x10	"	0.073
2x12	"	0.080
Ceilings		
2x4	L.F.	0.062
2x6	"	0.073
2x8	"	0.080
2x10	"	0.089
2x12	"	0.100
Wood construction		
Walls		
2x4	L.F.	0.044
2x6	"	0.050
2x8	"	0.053
2x10	"	0.057
2x12	"	0.062
Ceilings		
2x4	L.F.	0.050
2x6	"	0.057
2x8	"	0.062
2x10	"	0.067
2x12	"	0.073
06110.20 **Ceiling Framing**		
Ceiling joists		
12" o.c.		
2x4	S.F.	0.019
2x6	"	0.020
2x8	"	0.021
2x10	"	0.022
2x12	"	0.024
16" o.c.		

Rough Carpentry	UNIT	MAN/HOURS
06110.20 **Ceiling Framing** *(Cont.)*		
2x4	S.F.	0.015
2x6	"	0.016
2x8	"	0.017
2x10	"	0.017
2x12	"	0.018
24" o.c.		
2x4	S.F.	0.013
2x6	"	0.013
2x8	"	0.014
2x10	"	0.015
2x12	"	0.016
Headers and nailers		
2x4	L.F.	0.026
2x6	"	0.027
2x8	"	0.029
2x10	"	0.031
2x12	"	0.033
Sister joists for ceilings		
2x4	L.F.	0.057
2x6	"	0.067
2x8	"	0.080
2x10	"	0.100
2x12	"	0.133
06110.30 **Floor Framing**		
Floor joists		
12" o.c.		
2x6	S.F.	0.016
2x8	"	0.016
2x10	"	0.017
2x12	"	0.017
2x14	"	0.017
3x6	"	0.017
3x8	"	0.017
3x10	"	0.018
3x12	"	0.019
3x14	"	0.020
4x6	"	0.017
4x8	"	0.017
4x10	"	0.018
4x12	"	0.019
4x14	"	0.020
16" o.c.		
2x6	S.F.	0.013
2x8	"	0.014
2x10	"	0.014
2x12	"	0.014
2x14	"	0.015
3x6	"	0.014
3x8	"	0.014
3x10	"	0.015
3x12	"	0.015
3x14	"	0.016

Rough Carpentry	UNIT	MAN/HOURS
06110.30 **Floor Framing** *(Cont.)*		
4x6	S.F.	0.014
4x8	"	0.014
4x10	"	0.015
4x12	"	0.015
4x14	"	0.016
Sister joists for floors		
2x4	L.F.	0.050
2x6	"	0.057
2x8	"	0.067
2x10	"	0.080
2x12	"	0.100
3x6	"	0.080
3x8	"	0.089
3x10	"	0.100
3x12	"	0.114
4x6	"	0.080
4x8	"	0.089
4x10	"	0.100
4x12	"	0.114
06110.40 **Furring**		
Furring, wood strips		
Walls		
On masonry or concrete walls		
1x2 furring		
12" o.c.	S.F.	0.025
16" o.c.	"	0.023
24" o.c.	"	0.021
1x3 furring		
12" o.c.	S.F.	0.025
16" o.c.	"	0.023
24" o.c.	"	0.021
On wood walls		
1x2 furring		
12" o.c.	S.F.	0.018
16" o.c.	"	0.016
24" o.c.	"	0.015
1x3 furring		
12" o.c.	S.F.	0.018
16" o.c.	"	0.016
24" o.c.	"	0.015
Ceilings		
On masonry or concrete ceilings		
1x2 furring		
12" o.c.	S.F.	0.044
16" o.c.	"	0.040
24" o.c.	"	0.036
1x3 furring		
12" o.c.	S.F.	0.044
16" o.c.	"	0.040
24" o.c.	"	0.036
On wood ceilings		
1x2 furring		

Rough Carpentry	UNIT	MAN/HOURS
06110.40 **Furring** *(Cont.)*		
12" o.c.	S.F.	0.030
16" o.c.	"	0.027
24" o.c.	"	0.024
1x3		
12" o.c.	S.F.	0.030
16" o.c.	"	0.027
24" o.c.	"	0.024
06110.50 **Roof Framing**		
Roof framing		
Rafters, gable end		
0-2 pitch (flat to 2-in-12)		
12" o.c.		
2x4	S.F.	0.017
2x6	"	0.017
2x8	"	0.018
2x10	"	0.019
2x12	"	0.020
16" o.c.		
2x6	S.F.	0.014
2x8	"	0.015
2x10	"	0.015
2x12	"	0.016
24" o.c.		
2x6	S.F.	0.012
2x8	"	0.013
2x10	"	0.013
2x12	"	0.013
4-6 pitch (4-in-12 to 6-in-12)		
12" o.c.		
2x4	S.F.	0.017
2x6	"	0.018
2x8	"	0.019
2x10	"	0.020
2x12	"	0.021
16" o.c.		
2x6	S.F.	0.015
2x8	"	0.015
2x10	"	0.016
2x12	"	0.017
24" o.c.		
2x6	S.F.	0.013
2x8	"	0.013
2x10	"	0.014
2x12	"	0.015
8-12 pitch (8-in-12 to 12-in-12)		
12" o.c.		
2x4	S.F.	0.018
2x6	"	0.019
2x8	"	0.020
2x10	"	0.021
2x12	"	0.022
16" o.c.		

Rough Carpentry	UNIT	MAN/HOURS
06110.50 **Roof Framing** *(Cont.)*		
2x6	S.F.	0.015
2x8	"	0.016
2x10	"	0.017
2x12	"	0.017
24" o.c.		
2x6	S.F.	0.013
2x8	"	0.013
2x10	"	0.014
2x12	"	0.014
Ridge boards		
2x6	L.F.	0.040
2x8	"	0.044
2x10	"	0.050
2x12	"	0.057
Hip rafters		
2x6	L.F.	0.029
2x8	"	0.030
2x10	"	0.031
2x12	"	0.032
Jack rafters		
4-6 pitch (4-in-12 to 6-in-12)		
16" o.c.		
2x6	S.F.	0.024
2x8	"	0.024
2x10	"	0.026
2x12	"	0.027
24" o.c.		
2x6	S.F.	0.018
2x8	"	0.019
2x10	"	0.020
2x12	"	0.020
8-12 pitch (8-in-12 to 12-in-12)		
16" o.c.		
2x6	S.F.	0.025
2x8	"	0.026
2x10	"	0.027
2x12	"	0.028
24" o.c.		
2x6	S.F.	0.019
2x8	"	0.020
2x10	"	0.020
2x12	"	0.021
Sister rafters		
2x4	L.F.	0.057
2x6	"	0.067
2x8	"	0.080
2x10	"	0.100
2x12	"	0.133
Fascia boards		
2x4	L.F.	0.040
2x6	"	0.040
2x8	"	0.044
2x10	"	0.044

Rough Carpentry

06110.50 — Roof Framing *(Cont.)*

	UNIT	MAN/HOURS
2x12	L.F.	0.050
Cant strips		
Fiber		
3x3	L.F.	0.023
4x4	"	0.024
Wood		
3x3	L.F.	0.024

06110.60 — Sleepers

	UNIT	MAN/HOURS
Sleepers, over concrete		
12" o.c.		
1x2	S.F.	0.018
1x3	"	0.019
2x4	"	0.022
2x6	"	0.024
16" o.c.		
1x2	S.F.	0.016
1x3	"	0.016
2x4	"	0.019
2x6	"	0.020

06110.65 — Soffits

	UNIT	MAN/HOURS
Soffit framing		
2x3	L.F.	0.057
2x4	"	0.062
2x6	"	0.067
2x8	"	0.073

06110.70 — Wall Framing

	UNIT	MAN/HOURS
Framing wall, studs		
12" o.c.		
2x3	S.F.	0.015
2x4	"	0.015
2x6	"	0.016
2x8	"	0.017
16" o.c.		
2x3	S.F.	0.013
2x4	"	0.013
2x6	"	0.013
2x8	"	0.014
24" o.c.		
2x3	S.F.	0.011
2x4	"	0.011
2x6	"	0.011
2x8	"	0.012
Plates, top or bottom		
2x3	L.F.	0.024
2x4	"	0.025
2x6	"	0.027
2x8	"	0.029
Headers, door or window		
2x6		
Single		

Rough Carpentry

06110.70 — Wall Framing *(Cont.)*

	UNIT	MAN/HOURS
3' long	EA.	0.400
6' long	"	0.500
Double		
3' long	EA.	0.444
6' long	"	0.571
2x8		
Single		
4' long	EA.	0.500
8' long	"	0.615
Double		
4' long	EA.	0.571
8' long	"	0.727
2x10		
Single		
5' long	EA.	0.615
10' long	"	0.800
Double		
5' long	EA.	0.667
10' long	"	0.800
2x12		
Single		
6' long	EA.	0.615
12' long	"	0.800
Double		
6' long	EA.	0.727
12' long	"	0.889

06115.10 — Floor Sheathing

	UNIT	MAN/HOURS
Sub-flooring, plywood, CDX		
1/2" thick	S.F.	0.010
5/8" thick	"	0.011
3/4" thick	"	0.013
Structural plywood		
1/2" thick	S.F.	0.010
5/8" thick	"	0.011
3/4" thick	"	0.012
Board type subflooring		
1x6		
Minimum	S.F.	0.018
Maximum	"	0.020
1x8		
Minimum	S.F.	0.017
Maximum	"	0.019
1x10		
Minimum	S.F.	0.016
Maximum	"	0.018
Underlayment		
Hardboard, 1/4" tempered	S.F.	0.010
Plywood, CDX		
3/8" thick	S.F.	0.010
1/2" thick	"	0.011
5/8" thick	"	0.011
3/4" thick	"	0.012

Rough Carpentry	UNIT	MAN/HOURS
06115.20 — Roof Sheathing		
Sheathing		
Plywood, CDX		
3/8" thick	S.F.	0.010
1/2" thick	"	0.011
5/8" thick	"	0.011
3/4" thick	"	0.012
Structural plywood		
3/8" thick	S.F.	0.010
1/2" thick	"	0.011
5/8" thick	"	0.011
3/4" thick	"	0.012
06115.30 — Wall Sheathing		
Sheathing		
Plywood, CDX		
3/8" thick	S.F.	0.012
1/2" thick	"	0.012
5/8" thick	"	0.013
3/4" thick	"	0.015
Waferboard		
3/8" thick	S.F.	0.012
1/2" thick	"	0.012
5/8" thick	"	0.013
3/4" thick	"	0.015
Structural plywood		
3/8" thick	S.F.	0.012
1/2" thick	"	0.012
5/8" thick	"	0.013
3/4" thick	"	0.015
Gypsum, 1/2" thick	"	0.012
Asphalt impregnated fiberboard, 1/2" thick	"	0.012
06125.10 — Wood Decking		
Decking, T&G solid		
Cedar		
3" thick	S.F.	0.020
4" thick	"	0.021
Fir		
3" thick	S.F.	0.020
4" thick	"	0.021
Southern yellow pine		
3" thick	S.F.	0.023
4" thick	"	0.025
White pine		
3" thick	S.F.	0.020
4" thick	"	0.021
06130.10 — Heavy Timber		
Mill framed structures		
Beams to 20' long		
Douglas fir		
6x8	L.F.	0.080
6x10	"	0.083

Rough Carpentry	UNIT	MAN/HOURS
06130.10 — Heavy Timber *(Cont.)*		
6x12	L.F.	0.089
6x14	"	0.092
6x16	"	0.096
8x10	"	0.083
8x12	"	0.089
8x14	"	0.092
8x16	"	0.096
Southern yellow pine		
6x8	L.F.	0.080
6x10	"	0.083
6x12	"	0.089
6x14	"	0.092
6x16	"	0.096
8x10	"	0.083
8x12	"	0.089
8x14	"	0.092
8x16	"	0.096
Columns to 12' high		
Douglas fir		
6x6	L.F.	0.120
8x8	"	0.120
10x10	"	0.133
12x12	"	0.133
Southern yellow pine		
6x6	L.F.	0.120
8x8	"	0.120
10x10	"	0.133
12x12	"	0.133
Posts, treated		
4x4	L.F.	0.032
6x6	"	0.040
06190.20 — Wood Trusses		
Truss, fink, 2x4 members		
3-in-12 slope		
24' span	EA.	0.686
26' span	"	0.686
28' span	"	0.727
30' span	"	0.727
34' span	"	0.774
38' span	"	0.774
5-in-12 slope		
24' span	EA.	0.706
28' span	"	0.727
30' span	"	0.750
32' span	"	0.750
40' span	"	0.800
Gable, 2x4 members		
5-in-12 slope		
24' span	EA.	0.706
26' span	"	0.706
28' span	"	0.727
30' span	"	0.750

Rough Carpentry

Rough Carpentry	UNIT	MAN/HOURS
06190.20 **Wood Trusses** *(Cont.)*		
32' span	EA.	0.750
36' span	"	0.774
40' span	"	0.800
King post type, 2x4 members		
4-in-12 slope		
16' span	EA.	0.649
18' span	"	0.667
24' span	"	0.706
26' span	"	0.706
30' span	"	0.750
34' span	"	0.750
38' span	"	0.774
42' span	"	0.828

Finish Carpentry

Finish Carpentry	UNIT	MAN/HOURS
06200.10 **Finish Carpentry**		
Mouldings and trim		
Apron, flat		
9/16 x 2	L.F.	0.040
9/16 x 3-1/2	"	0.042
Base		
Colonial		
7/16 x 2-1/4	L.F.	0.040
7/16 x 3	"	0.040
7/16 x 3-1/4	"	0.040
9/16 x 3	"	0.042
9/16 x 3-1/4	"	0.042
11/16 x 2-1/4	"	0.044
Ranch		
7/16 x 2-1/4	L.F.	0.040
7/16 x 3-1/4	"	0.040
9/16 x 2-1/4	"	0.042
9/16 x 3	"	0.042
9/16 x 3-1/4	"	0.042
Casing		
11/16 x 2-1/2	L.F.	0.036
11/16 x 3-1/2	"	0.038
Chair rail		
9/16 x 2-1/2	L.F.	0.040
9/16 x 3-1/2	"	0.040
Closet pole		
1-1/8" dia.	L.F.	0.053
1-5/8" dia.	"	0.053
Cove		
9/16 x 1-3/4	L.F.	0.040

Finish Carpentry	UNIT	MAN/HOURS
06200.10 **Finish Carpentry** *(Cont.)*		
11/16 x 2-3/4	L.F.	0.040
Crown		
9/16 x 1-5/8	L.F.	0.053
9/16 x 2-5/8	"	0.062
11/16 x 3-5/8	"	0.067
11/16 x 4-1/4	"	0.073
11/16 x 5-1/4	"	0.080
Drip cap		
1-1/16 x 1-5/8	L.F.	0.040
Glass bead		
3/8 x 3/8	L.F.	0.050
1/2 x 9/16	"	0.050
5/8 x 5/8	"	0.050
3/4 x 3/4	"	0.050
Half round		
1/2	L.F.	0.032
5/8	"	0.032
3/4	"	0.032
Lattice		
1/4 x 7/8	L.F.	0.032
1/4 x 1-1/8	"	0.032
1/4 x 1-3/8	"	0.032
1/4 x 1-3/4	"	0.032
1/4 x 2	"	0.032
Ogee molding		
5/8 x 3/4	L.F.	0.040
11/16 x 1-1/8	"	0.040
11/16 x 1-3/8	"	0.040
Parting bead		
3/8 x 7/8	L.F.	0.050
Quarter round		
1/4 x 1/4	L.F.	0.032
3/8 x 3/8	"	0.032
1/2 x 1/2	"	0.032
11/16 x 11/16	"	0.035
3/4 x 3/4	"	0.035
1-1/16 x 1-1/16	"	0.036
Railings, balusters		
1-1/8 x 1-1/8	L.F.	0.080
1-1/2 x 1-1/2	"	0.073
Screen moldings		
1/4 x 3/4	L.F.	0.067
5/8 x 5/16	"	0.067
Shoe		
7/16 x 11/16	L.F.	0.032
Sash beads		
1/2 x 3/4	L.F.	0.067
1/2 x 7/8	"	0.067
1/2 x 1-1/8	"	0.073
5/8 x 7/8	"	0.073
Stop		
5/8 x 1-5/8		
Colonial	L.F.	0.050

Finish Carpentry	UNIT	MAN/HOURS
06200.10 **Finish Carpentry** (Cont.)		
Ranch	L.F.	0.050
Stools		
11/16 x 2-1/4	L.F.	0.089
11/16 x 2-1/2	"	0.089
11/16 x 5-1/4	"	0.100
Exterior trim, casing, select pine, 1x3	"	0.040
Douglas fir		
1x3	L.F.	0.040
1x4	"	0.040
1x6	"	0.044
1x8	"	0.050
Cornices, white pine, #2 or better		
1x2	L.F.	0.040
1x4	"	0.040
1x6	"	0.044
1x8	"	0.047
1x10	"	0.050
1x12	"	0.053
Shelving, pine		
1x8	L.F.	0.062
1x10	"	0.064
1x12	"	0.067
Plywood shelf, 3/4", with edge band, 12" wide	"	0.080
Adjustable shelf, and rod, 12" wide		
3' to 4' long	EA.	0.200
5' to 8' long	"	0.267
Prefinished wood shelves with brackets and supports		
8" wide		
3' long	EA.	0.200
4' long	"	0.200
6' long	"	0.200
10" wide		
3' long	EA.	0.200
4' long	"	0.200
6' long	"	0.200
06220.10 **Millwork**		
Countertop, laminated plastic		
25" x 7/8" thick		
Minimum	L.F.	0.200
Average	"	0.267
Maximum	"	0.320
25" x 1-1/4" thick		
Minimum	L.F.	0.267
Average	"	0.320
Maximum	"	0.400
Add for cutouts	EA.	0.500
Backsplash, 4" high, 7/8" thick	L.F.	0.160
Plywood, sanded, A-C		
1/4" thick	S.F.	0.027
3/8" thick	"	0.029
1/2" thick	"	0.031
A-D		

Finish Carpentry	UNIT	MAN/HOURS
06220.10 **Millwork** (Cont.)		
1/4" thick	S.F.	0.027
3/8" thick	"	0.029
1/2" thick	"	0.031
Base cabinets, 34-1/2" high, 24" deep, hardwood, no tops		
Minimum	L.F.	0.320
Average	"	0.400
Maximum	"	0.533
Wall cabinets		
Minimum	L.F.	0.267
Average	"	0.320
Maximum	"	0.400
Oil borne		
Water borne		

Architectural Woodwork	UNIT	MAN/HOURS
06420.10 **Panel Work**		
Hardboard, tempered, 1/4" thick		
Natural faced	S.F.	0.020
Plastic faced	"	0.023
Pegboard, natural	"	0.020
Plastic faced	"	0.023
Untempered, 1/4" thick		
Natural faced	S.F.	0.020
Plastic faced	"	0.023
Pegboard, natural	"	0.020
Plastic faced	"	0.023
Plywood unfinished, 1/4" thick		
Birch		
Natural	S.F.	0.027
Select	"	0.027
Knotty pine	"	0.027
Cedar (closet lining)		
Standard boards T&G	S.F.	0.027
Particle board	"	0.027
Plywood, prefinished, 1/4" thick, premium grade		
Birch veneer	S.F.	0.032
Cherry veneer	"	0.032
Chestnut veneer	"	0.032
Lauan veneer	"	0.032
Mahogany veneer	"	0.032
Oak veneer (red)	"	0.032
Pecan veneer	"	0.032
Rosewood veneer	"	0.032
Teak veneer	"	0.032
Walnut veneer	"	0.032

Architectural Woodwork	UNIT	MAN/ HOURS
06430.10 **Stairwork**		
Risers, 1x8, 42" wide		
White oak	EA.	0.400
Pine	"	0.400
Treads, 1-1/16" x 9-1/2" x 42"		
White oak	EA.	0.500
06440.10 **Columns**		
Column, hollow, round wood		
12" diameter		
10' high	EA.	0.800
12' high	"	0.857
14' high	"	0.960
16' high	"	1.200
24" diameter		
16' high	EA.	1.200
18' high	"	1.263
20' high	"	1.263
22' high	"	1.333
24' high	"	1.333

Moisture Protection	UNIT	MAN/HOURS
07100.10 **Waterproofing**		
Membrane waterproofing, elastomeric		
Butyl		
1/32" thick	S.F.	0.032
1/16" thick	"	0.033
Butyl with nylon		
1/32" thick	S.F.	0.032
1/16" thick	"	0.033
Neoprene		
1/32" thick	S.F.	0.032
1/16" thick	"	0.033
Neoprene with nylon		
1/32" thick	S.F.	0.032
1/16" thick	"	0.033
Bituminous membrane waterproofing, asphalt felt, 15 lb.		
One ply	S.F.	0.020
Two ply	"	0.024
Three ply	"	0.029
Four ply	"	0.033
Five ply	"	0.042
Modified asphalt membrane waterproofing, fibrous asphalt		
One ply	S.F.	0.033
Two ply	"	0.040
Three ply	"	0.044
Four ply	"	0.053
Five ply	"	0.064
Asphalt coated protective board		
1/8" thick	S.F.	0.020
1/4" thick	"	0.020
3/8" thick	"	0.020
1/2" thick	"	0.021
Cement protective board		
3/8" thick	S.F.	0.027
1/2" thick	"	0.027
Fluid applied, neoprene		
50 mil	S.F.	0.027
90 mil	"	0.027
Tab extended polyurethane		
.050" thick	S.F.	0.020
Fluid applied rubber based polyurethane		
6 mil	S.F.	0.025
15 mil	"	0.020
Bentonite waterproofing, panels		
3/16" thick	S.F.	0.020
1/4" thick	"	0.020
5/8" thick	"	0.021
Granular admixtures, trowel on, 3/8" thick	"	0.020
Metallic oxide waterproofing, iron compound, troweled		
5/8" thick	S.F.	0.020
3/4" thick	"	0.023

Moisture Protection	UNIT	MAN/HOURS
07150.10 **Dampproofing**		
Silicone dampproofing, sprayed on		
Concrete surface		
1 coat	S.F.	0.004
2 coats	"	0.006
Concrete block		
1 coat	S.F.	0.005
2 coats	"	0.007
Brick		
1 coat	S.F.	0.006
2 coats	"	0.008
07160.10 **Bituminous Dampproofing**		
Building paper, asphalt felt		
15 lb	S.F.	0.032
30 lb	"	0.033
Asphalt dampproofing, troweled, cold, primer plus		
1 coat	S.F.	0.027
2 coats	"	0.040
3 coats	"	0.050
Fibrous asphalt dampproofing, hot troweled, primer plus		
1 coat	S.F.	0.032
2 coats	"	0.044
3 coats	"	0.057
Asphaltic paint dampproofing, per coat		
Brush on	S.F.	0.011
Spray on	"	0.009
07190.10 **Vapor Barriers**		
Vapor barrier, polyethylene		
2 mil	S.F.	0.004
6 mil	"	0.004
8 mil	"	0.004
10 mil	"	0.004

Insulation	UNIT	MAN/HOURS
07210.10 **Batt Insulation**		
Ceiling, fiberglass, unfaced		
3-1/2" thick, R11	S.F.	0.009
6" thick, R19	"	0.011
9" thick, R30	"	0.012
Suspended ceiling, unfaced		
3-1/2" thick, R11	S.F.	0.009
6" thick, R19	"	0.010
9" thick, R30	"	0.011
Crawl space, unfaced		

Insulation	UNIT	MAN/HOURS
07210.10 — **Batt Insulation** *(Cont.)*		
3-1/2" thick, R11	S.F.	0.012
6" thick, R19	"	0.013
9" thick, R30	"	0.015
Wall, fiberglass		
Paper backed		
2" thick, R7	S.F.	0.008
3" thick, R8	"	0.009
4" thick, R11	"	0.009
6" thick, R19	"	0.010
Foil backed, 1 side		
2" thick, R7	S.F.	0.008
3" thick, R11	"	0.009
4" thick, R14	"	0.009
6" thick, R21	"	0.010
Foil backed, 2 sides		
2" thick, R7	S.F.	0.009
3" thick, R11	"	0.010
4" thick, R14	"	0.011
6" thick, R21	"	0.011
Unfaced		
2" thick, R7	S.F.	0.008
3" thick, R9	"	0.009
4" thick, R11	"	0.009
6" thick, R19	"	0.010
Mineral wool batts		
Paper backed		
2" thick, R6	S.F.	0.008
4" thick, R12	"	0.009
6" thick, R19	"	0.010
Fasteners, self adhering, attached to ceiling deck		
2-1/2" long	EA.	0.013
4-1/2" long	"	0.015
Capped, self-locking washers for fastening insulation	"	0.008
07210.20 — **Board Insulation**		
Insulation, rigid		
Fiberglass, roof		
0.75" thick, R2.78	S.F.	0.007
1.06" thick, R4.17	"	0.008
1.31" thick, R5.26	"	0.008
1.63" thick, R6.67	"	0.008
2.25" thick, R8.33	"	0.009
Composite board, roof		
1-1/2" thick, R6.67	S.F.	0.008
1-5/8" thick, R7.69	"	0.008
2" thick, R10.0	"	0.009
2-1/4" thick, R12.50	"	0.009
2-1/2" thick, R14.29	"	0.010
2-3/4" thick, R16.67	"	0.011
3-1/4" thick, R20.00	"	0.011
Perlite board, roof		
1.00" thick, R2.78	S.F.	0.007
1.50" thick, R4.17	"	0.007

Insulation	UNIT	MAN/HOURS
07210.20 — **Board Insulation** *(Cont.)*		
2.00" thick, R5.92	S.F.	0.007
2.50" thick, R6.67	"	0.008
3.00" thick, R8.33	"	0.008
4.00" thick, R10.00	"	0.008
5.25" thick, R14.29	"	0.009
Rigid urethane		
Roof		
1" thick, R6.67	S.F.	0.007
1.20" thick, R8.33	"	0.007
1.50" thick, R11.11	"	0.007
2" thick, R14.29	"	0.007
2.25" thick, R16.67	"	0.008
Wall		
1" thick, R6.67	S.F.	0.008
1.5" thick, R11.11	"	0.009
2" thick, R14.29	"	0.009
Polystyrene		
Roof		
1.0" thick, R4.17	S.F.	0.007
1.5" thick, R6.26	"	0.007
2.0" thick, R8.33	"	0.007
Wall		
1.0" thick, R4.17	S.F.	0.008
1.5" thick, R6.26	"	0.009
2.0" thick, R8.33	"	0.009
Rigid board insulation, deck		
Mineral fiberboard		
1" thick, R3.0	S.F.	0.007
2" thick, R5.26	"	0.007
Fiberglass		
1" thick, R4.3	S.F.	0.007
2" thick, R8.5	"	0.007
Polystyrene		
1" thick, R5.4	S.F.	0.007
2" thick, R10.8	"	0.007
Urethane		
.75" thick, R5.4	S.F.	0.007
1" thick, R6.4	"	0.007
1.5" thick, R10.7	"	0.007
2" thick, R14.3	"	0.007
Foamglass		
1" thick, R1.8	S.F.	0.007
2" thick, R5.26	"	0.007
Wood fiber		
1" thick, R3.85	S.F.	0.007
2" thick, R7.7	"	0.007
Particle board		
3/4" thick, R2.08	S.F.	0.007
1" thick, R2.77	"	0.007
2" thick, R5.50	"	0.007

Insulation	UNIT	MAN/ HOURS
07210.60 — Loose Fill Insulation		
Blown-in type		
Fiberglass		
5" thick, R11	S.F.	0.007
6" thick, R13	"	0.008
9" thick, R19	"	0.011
Rockwool, attic application		
6" thick, R13	S.F.	0.008
8" thick, R19	"	0.010
10" thick, R22	"	0.012
12" thick, R26	"	0.013
15" thick, R30	"	0.016
Poured type		
Fiberglass		
1" thick, R4	S.F.	0.005
2" thick, R8	"	0.006
3" thick, R12	"	0.007
4" thick, R16	"	0.008
Mineral wool		
1" thick, R3	S.F.	0.005
2" thick, R6	"	0.006
3" thick, R9	"	0.007
4" thick, R12	"	0.008
Vermiculite or perlite		
2" thick, R4.8	S.F.	0.006
3" thick, R7.2	"	0.007
4" thick, R9.6	"	0.008
Masonry, poured vermiculite or perlite		
4" block	S.F.	0.004
6" block	"	0.005
8" block	"	0.006
10" block	"	0.006
12" block	"	0.007
07210.70 — Sprayed Insulation		
Foam, sprayed on		
Polystyrene		
1" thick, R4	S.F.	0.008
2" thick, R8	"	0.011
Urethane		
1" thick, R4	S.F.	0.008
2" thick, R8	"	0.011
07250.10 — Fireproofing		
Sprayed on		
1" thick		
On beams	S.F.	0.018
On columns	"	0.016
On decks		
Flat surface	S.F.	0.008
Fluted surface	"	0.010

Insulation	UNIT	MAN/ HOURS
07250.10 — Fireproofing (Cont.)		
1-1/2" thick		
On beams	S.F.	0.023
On columns	"	0.020
On decks		
Flat surface	S.F.	0.010
Fluted surface	"	0.013

Shingles And Tiles	UNIT	MAN/ HOURS
07310.10 — Asphalt Shingles		
Standard asphalt shingles, strip shingles		
210 lb/square	SQ.	0.800
235 lb/square	"	0.889
240 lb/square	"	1.000
260 lb/square	"	1.143
300 lb/square	"	1.333
385 lb/square	"	1.600
Roll roofing, mineral surface		
90 lb	SQ.	0.571
110 lb	"	0.667
140 lb	"	0.800
07310.50 — Metal Shingles		
Aluminum, .020" thick		
Plain	SQ.	1.600
Colors	"	1.600
Steel, galvanized		
26 ga.		
Plain	SQ.	1.600
Colors	"	1.600
24 ga.		
Plain	SQ.	1.600
Colors	"	1.600
Porcelain enamel, 22 ga.		
Minimum	SQ.	2.000
Average	"	2.000
Maximum	"	2.000
07310.60 — Slate Shingles		
Slate shingles		
Pennsylvania		
Ribbon	SQ.	4.000
Clear	"	4.000
Vermont		
Black	SQ.	4.000

Shingles And Tiles	UNIT	MAN/HOURS
07310.60 Slate Shingles (Cont.)		
Gray	SQ.	4.000
Green	"	4.000
Red	"	4.000
Replacement shingles		
Small jobs	EA.	0.267
Large jobs	S.F.	0.133
07310.70 Wood Shingles		
Wood shingles, on roofs		
White cedar, #1 shingles		
4" exposure	SQ.	2.667
5" exposure	"	2.000
#2 shingles		
4" exposure	SQ.	2.667
5" exposure	"	2.000
Resquared and rebutted		
4" exposure	SQ.	2.667
5" exposure	"	2.000
On walls		
White cedar, #1 shingles		
4" exposure	SQ.	4.000
5" exposure	"	3.200
6" exposure	"	2.667
#2 shingles		
4" exposure	SQ.	4.000
5" exposure	"	3.200
6" exposure	"	2.667
07310.80 Wood Shakes		
Shakes, hand split, 24" red cedar, on roofs		
5" exposure	SQ.	4.000
7" exposure	"	3.200
9" exposure	"	2.667
On walls		
6" exposure	SQ.	4.000
8" exposure	"	3.200
10" exposure	"	2.667

Roofing And Siding	UNIT	MAN/HOURS
07410.10 Manufactured Roofs		
Aluminum roof panels, for steel framing		
Corrugated		
Unpainted finish		
.024"	S.F.	0.020
.030"	"	0.020

Roofing And Siding	UNIT	MAN/HOURS
07410.10 Manufactured Roofs (Cont.)		
Painted finish		
.024"	S.F.	0.020
.030"	"	0.020
V-beam		
Unpainted finish		
.032"	S.F.	0.020
.040"	"	0.020
.050"	"	0.020
Painted finish		
.032"	S.F.	0.020
.040"	"	0.020
.050"	"	0.020
Steel roof panels, for structural steel framing		
Corrugated, painted		
18 ga.	S.F.	0.020
20 ga.	"	0.020
22 ga.	"	0.020
Box rib, painted		
18 ga.	S.F.	0.021
20 ga.	"	0.021
22 ga.	"	0.021
4" rib, painted		
18 ga.	S.F.	0.022
20 ga.	"	0.022
22 ga.	"	0.022
Standing seam roof		
2" high seam, painted		
22 ga.	S.F.	0.032
24 ga.	"	0.032
26 ga.	"	0.032
07410.30 Manufactured Walls		
Sandwich panels with 1-1/2" fiberglass insulation		
Galvanized 18 ga. steel interior panels		
Exterior panels		
16 ga. aluminum	S.F.	0.107
18 ga. galvanized steel	"	0.107
20 ga. painted steel	"	0.107
20 ga. stainless steel	"	0.107
Metal liner panels, 1-3/8" thick, 24" wide		
Galvanized		
22 ga.	S.F.	0.027
20 ga.	"	0.027
18 ga.	"	0.027
Primed		
22 ga.	S.F.	0.027
20 ga.	"	0.027
18 ga.	"	0.027

Roofing And Siding	UNIT	MAN/ HOURS
07440.10 — Aggregate Coated Panels		
Dryvit type system		
1" thick	S.F.	0.027
1-1/2" thick	"	0.029
2" thick	"	0.033
07460.10 — Metal Siding Panels		
Aluminum siding panels		
Corrugated		
Plain finish		
.024"	S.F.	0.032
.032"	"	0.032
Painted finish		
.024"	S.F.	0.032
.032"	"	0.032
V. beam		
Plain finish		
.032"	S.F.	0.032
.040"	"	0.032
.050"	"	0.032
Painted finish		
.032"	S.F.	0.032
.040"	"	0.032
.050"	"	0.032
4" rib		
Plain finish		
.032"	S.F.	0.036
.040"	"	0.036
.050"	"	0.036
Painted finish		
.032"	S.F.	0.036
.040"	"	0.036
.050"	"	0.036
Steel siding panels		
Corrugated		
22 ga.	S.F.	0.053
24 ga.	"	0.053
26 ga.	"	0.053
Box rib		
20 ga.	S.F.	0.053
22 ga.	"	0.053
24 ga.	"	0.053
26 ga.	"	0.053
07460.50 — Plastic Siding		
Horizontal vinyl siding, solid		
8" wide		
Standard	S.F.	0.031
Insulated	"	0.031
10" wide		
Standard	S.F.	0.029
Insulated	"	0.029
Vinyl moldings for doors and windows	L.F.	0.032

Roofing And Siding	UNIT	MAN/ HOURS
07460.60 — Plywood Siding		
Rough sawn cedar, 3/8" thick	S.F.	0.027
Fir, 3/8" thick	"	0.027
Texture 1-11, 5/8" thick		
Cedar	S.F.	0.029
Fir	"	0.029
Redwood	"	0.029
Southern Yellow Pine	"	0.029
07460.70 — Steel Siding		
Ribbed, sheets, galvanized		
22 ga.	S.F.	0.032
24 ga.	"	0.032
26 ga.	"	0.032
28 ga.	"	0.032
Primed		
24 ga.	S.F.	0.032
26 ga.	"	0.032
28 ga.	"	0.032
07460.80 — Wood Siding		
Beveled siding, cedar		
A grade		
1/2 x 6	S.F.	0.040
1/2 x 8	"	0.032
3/4 x 10	"	0.027
Clear		
1/2 x 6	S.F.	0.040
1/2 x 8	"	0.032
3/4 x 10	"	0.027
B grade		
1/2 x 6	S.F.	0.040
1/2 x 8	"	0.320
3/4 x 10	"	0.027
Board and batten		
Cedar		
1x6	S.F.	0.040
1x8	"	0.032
1x10	"	0.029
1x12	"	0.026
Pine		
1x6	S.F.	0.040
1x8	"	0.032
1x10	"	0.029
1x12	"	0.026
Redwood		
1x6	S.F.	0.040
1x8	"	0.032
1x10	"	0.029
1x12	"	0.026
Tongue and groove		
Cedar		
1x4	S.F.	0.044
1x6	"	0.042

Roofing And Siding

07460.80 — Wood Siding (Cont.)

	UNIT	MAN/HOURS
1x8	S.F.	0.040
1x10	"	0.038
Pine		
1x4	S.F.	0.044
1x6	"	0.042
1x8	"	0.040
1x10	"	0.038
Redwood		
1x4	S.F.	0.044
1x6	"	0.042
1x8	"	0.040
1x10	"	0.038

Membrane Roofing

07510.10 — Built-up Asphalt Roofing

	UNIT	MAN/HOURS
Built-up roofing, asphalt felt, including gravel		
2 ply	SQ.	2.000
3 ply	"	2.667
4 ply	"	3.200
Walkway, for built-up roofs		
3' x 3' x		
1/2" thick	S.F.	0.027
3/4" thick	"	0.027
1" thick	"	0.027
Cant strip, 4" x 4"		
Treated wood	L.F.	0.023
Foamglass	"	0.020
Mineral fiber	"	0.020
New gravel for built-up roofing, 400 lb/sq	SQ.	1.600
Roof gravel (ballast)	C.Y.	4.000
Aluminum coating, top surfacing, for built-up roofing	SQ.	1.333
Remove 4-ply built-up roof (includes gravel)	"	4.000
Remove & replace gravel, includes flood coat	"	2.667

07530.10 — Single-ply Roofing

	UNIT	MAN/HOURS
Elastic sheet roofing		
Neoprene, 1/16" thick	S.F.	0.010
EPDM rubber		
45 mil	S.F.	0.010
60 mil	"	0.010
PVC		
45 mil	S.F.	0.010
60 mil	"	0.010
Flashing		
Pipe flashing, 90 mil thick		

Membrane Roofing

07530.10 — Single-ply Roofing (Cont.)

	UNIT	MAN/HOURS
1" pipe	EA.	0.200
2" pipe	"	0.200
3" pipe	"	0.211
4" pipe	"	0.211
5" pipe	"	0.222
6" pipe	"	0.222
8" pipe	"	0.235
10" pipe	"	0.267
12" pipe	"	0.267
Neoprene flashing, 60 mil thick strip		
6" wide	L.F.	0.067
12" wide	"	0.100
18" wide	"	0.133
24" wide	"	0.200
Adhesives		
Mastic sealer, applied at joints only		
1/4" bead	L.F.	0.004
Fluid applied roofing		
Urethane, 2 components, elastomeric top membrane		
1" thick	S.F.	0.013
Vinyl liquid roofing, 2 coats, 2 mils per coat	"	0.011
Silicone roofing, 2 coats sprayed, 16 mil per coat	"	0.013
Inverted roof system		
Insulated membrane with coarse gravel ballast		
3 ply with 2" polystyrene	S.F.	0.013
Ballast, 3/4" through 1-1/2" dia. river gravel, 100lb/sf	"	0.800
Walkway for membrane roofs, 1/2" thick	"	0.027

Flashing And Sheet Metal

07610.10 — Metal Roofing

	UNIT	MAN/HOURS
Sheet metal roofing, copper, 16 oz, batten seam	SQ.	5.333
Standing seam	"	5.000
Aluminum roofing, natural finish		
Corrugated, on steel frame		
.0175" thick	SQ.	2.286
.0215" thick	"	2.286
.024" thick	"	2.286
.032" thick	"	2.286
V-beam, on steel frame		
.032" thick	SQ.	2.286
.040" thick	"	2.286
.050" thick	"	2.286
Ridge cap		
.019" thick	L.F.	0.027
Corrugated galvanized steel roofing, on steel frame		

Flashing And Sheet Metal	UNIT	MAN/ HOURS
07610.10 — **Metal Roofing** *(Cont.)*		
28 ga.	SQ.	2.286
26 ga.	"	2.286
24 ga.	"	2.286
22 ga.	"	2.286
26 ga., factory insulated with 1" polystyrene	"	3.200
Ridge roll		
10" wide	L.F.	0.027
20" wide	"	0.032
07620.10 — **Flashing And Trim**		
Counter flashing		
Aluminum, .032"	S.F.	0.080
Stainless steel, .015"	"	0.080
Copper		
16 oz.	S.F.	0.080
20 oz.	"	0.080
24 oz.	"	0.080
32 oz.	"	0.080
Valley flashing		
Aluminum, .032"	S.F.	0.050
Stainless steel, .015	"	0.050
Copper		
16 oz.	S.F.	0.050
20 oz.	"	0.067
24 oz.	"	0.050
32 oz.	"	0.050
Base flashing		
Aluminum, .040"	S.F.	0.067
Stainless steel, .018"	"	0.067
Copper		
16 oz.	S.F.	0.067
20 oz.	"	0.050
24 oz.	"	0.067
32 oz.	"	0.067
Waterstop, "T" section, 22 ga.		
1-1/2" x 3"	L.F.	0.040
2" x 2"	"	0.040
4" x 3"	"	0.040
6" x 4"	"	0.040
8" x 4"	"	0.040
Scupper outlets		
10" x 10" x 4"	EA.	0.200
22" x 4" x 4"	"	0.200
8" x 8" x 5"	"	0.200
Flashing and trim, aluminum		
.019" thick	S.F.	0.057
.032" thick	"	0.057
.040" thick	"	0.062
Neoprene sheet flashing, .060" thick	"	0.050
Copper, paper backed		
2 oz.	S.F.	0.080
Drainage boots, roof, cast iron		
2 x 3	L.F.	0.100

Flashing And Sheet Metal	UNIT	MAN/ HOURS
07620.10 — **Flashing And Trim** *(Cont.)*		
3 x 4	L.F.	0.100
4 x 5	"	0.107
4 x 6	"	0.107
5 x 7	"	0.114
Pitch pocket, copper, 16 oz.		
4 x 4	EA.	0.200
6 x 6	"	0.200
8 x 8	"	0.200
8 x 10	"	0.200
8 x 12	"	0.200
Reglets, copper 10 oz.	L.F.	0.053
Stainless steel, .020"	"	0.053
Gravel stop		
Aluminum, .032"		
4"	L.F.	0.027
6"	"	0.027
8"	"	0.031
10"	"	0.031
Copper, 16 oz.		
4"	L.F.	0.027
6"	"	0.027
8"	"	0.031
10"	"	0.031
07620.20 — **Gutters And Downspouts**		
Copper gutter and downspout		
Downspouts, 16 oz. copper		
Round		
3" dia.	L.F.	0.053
4" dia.	"	0.053
Rectangular, corrugated		
2" x 3"	L.F.	0.050
3" x 4"	"	0.050
Rectangular, flat surface		
2" x 3"	L.F.	0.053
3" x 4"	"	0.053
Lead-coated copper downspouts		
Round		
3" dia.	L.F.	0.050
4" dia.	"	0.057
Rectangular, corrugated		
2" x 3"	L.F.	0.053
3" x 4"	"	0.053
Rectangular, plain		
2" x 3"	L.F.	0.053
3" x 4"	"	0.053
Gutters, 16 oz. copper		
Half round		
4" wide	L.F.	0.080
5" wide	"	0.089
Type K		
4" wide	L.F.	0.080
5" wide	"	0.089

Flashing And Sheet Metal	UNIT	MAN/HOURS
07620.20 **Gutters And Downspouts** *(Cont.)*		
Lead-coated copper gutters		
Half round		
4" wide	L.F.	0.080
6" wide	"	0.089
Type K		
4" wide	L.F.	0.080
5" wide	"	0.089
Aluminum gutter and downspout		
Downspouts		
2" x 3"	L.F.	0.053
3" x 4"	"	0.057
4" x 5"	"	0.062
Round		
3" dia.	L.F.	0.053
4" dia.	"	0.057
Gutters, stock units		
4" wide	L.F.	0.084
5" wide	"	0.089
Galvanized steel gutter and downspout		
Downspouts, round corrugated		
3" dia.	L.F.	0.053
4" dia.	"	0.053
5" dia.	"	0.057
6" dia.	"	0.057
Rectangular		
2" x 3"	L.F.	0.053
3" x 4"	"	0.050
4" x 4"	"	0.050
Gutters, stock units		
5" wide		
Plain	L.F.	0.089
Painted	"	0.089
6" wide		
Plain	L.F.	0.094
Painted	"	0.094

Roofing Specialties	UNIT	MAN/HOURS
07700.10 **Manufactured Specialties**		
Moisture relief vent		
Aluminum	EA.	0.114
Copper	"	0.114
Expansion joint		
Aluminum		
Opening to 2.5"	L.F.	0.057
Opening to 3.5"	"	0.062

Roofing Specialties	UNIT	MAN/HOURS
07700.10 **Manufactured Specialties** *(Cont.)*		
Copper, 16 oz.		
Opening to 2.5"	L.F.	0.057
Opening to 3.5"	"	0.062
Butyl or neoprene		
4" wide		
16 oz. copper bellows	L.F.	0.067
28 ga. stainless steel bellows	"	0.067
6" wide		
Copper bellows	L.F.	0.073
Stainless steel		
Opening to 2.5"	L.F.	0.057
Opening to 3.5"	"	0.062
Smoke vent, 48" x 48"		
Aluminum	EA.	2.000
Galvanized steel	"	2.000
Heat/smoke vent, 48" x 96"		
Aluminum	EA.	2.667
Galvanized steel	"	2.667
Ridge vent strips		
Mill finish	L.F.	0.053
Connectors	EA.	0.200
End cap	"	0.229
Soffit vents		
Mill finish		
2-1/2" wide	L.F.	0.032
3" wide	"	0.032
6" wide	"	0.032
Roof hatches		
Steel, plain, primed		
2'6" x 3'0"	EA.	2.000
2'6" x 4'6"	"	2.667
2'6" x 8'0"	"	4.000
Galvanized steel		
2'6" x 3'0"	EA.	2.000
2'6" x 4'6"	"	2.667
2'6" x 8'0"	"	4.000
Aluminum		
2'6" x 3'0"	EA.	2.000
2'6" x 4'6"	"	2.667
2'6" x 8'0"	"	4.000
Ceiling access doors		
Swing up model, metal frame		
Steel door		
2'6" x 2'6"	EA.	0.800
2'6" x 3'0"	"	0.800
Aluminum door		
2'6" x 2'6"	EA.	0.800
2'6" x 3'0"	"	0.800
Swing down model, metal frame		
Steel door		
2'6" x 2'6"	EA.	0.800
2'6" x 3'0"	"	0.800
Aluminum door		

Roofing Specialties	UNIT	MAN/HOURS
07700.10 Manufactured Specialties *(Cont.)*		
2'6" x 2'6"	EA.	0.800
2'6" x 3'0"	"	0.800
Gravity ventilators, with curb, base, damper and screen		
Stationary siphon		
6" dia.	EA.	0.533
12" dia.	"	0.533
24" dia.	"	0.800
36" dia.	"	0.800
Wind driven spinner		
6" dia.	EA.	0.533
12" dia.	"	0.533
24" dia.	"	0.800
36" dia.	"	0.800
Stationary mushroom		
16" dia.	EA.	0.800
30" dia.	"	1.000
36" dia.	"	1.333
42" dia.	"	1.600

Skylights	UNIT	MAN/HOURS
07810.10 Plastic Skylights		
Single thickness, not including mounting curb		
2' x 4'	EA.	1.000
4' x 4'	"	1.333
5' x 5'	"	2.000
6' x 8'	"	2.667
Double thickness, not including mounting curb		
2' x 4'	EA.	1.000
4' x 4'	"	1.333
5' x 5'	"	2.000
6' x 8'	"	2.667
Metal framed skylights		
Translucent panels, 2-1/2" thick	S.F.	0.080
Continuous vaults, 8' wide		
Single glazed	S.F.	0.100
Double glazed	"	0.114

Joint Sealers	UNIT	MAN/HOURS
07920.10 Caulking		
Caulk exterior, two component		
1/4 x 1/2	L.F.	0.040
3/8 x 1/2	"	0.044
1/2 x 1/2	"	0.050
Caulk interior, single component		
1/4 x 1/2	L.F.	0.038
3/8 x 1/2	"	0.042
1/2 x 1/2	"	0.047
Butyl rubber fillers		
1/4" x 1/4"	L.F.	0.016
1/2" x 1/2"	"	0.027
1/2" x 3/4"	"	0.032
3/4" x 3/4"	"	0.032
1" x 1"	"	0.036
Seals, "O" ring type cord		
1/4" dia.	L.F.	0.020
1/2" dia.	"	0.021
1" dia.	"	0.022
1-1/4" dia.	"	0.024
1-1/2" dia.	"	0.025
1-3/4" dia.	"	0.026
2" dia.	"	0.027
Polyvinyl chloride, closed cell		
1/4" x 2"	L.F.	0.029
1/4" x 6"	"	0.036
Silicon foam penetration seal		
1/4" x 1/2"	L.F.	0.010
1/2" x 1/2"	"	0.013
1/2" x 3/4"	"	0.016
3/4" x 3/4"	"	0.020
1/8" x 1"	"	0.010
1/8" x 3"	"	0.016
1/4" x 3"	"	0.020
1/4" x 6"	"	0.027
1/2" x 6"	"	0.062
1/2" x 9"	"	0.100
1/2" x 12"	"	0.145
Oil base sealants and caulking		
1/4" x 1/4"	L.F.	0.020
1/4" x 3/8"	"	0.021
1/4" x 1/2"	"	0.022
3/8" x 3/8"	"	0.023
3/8" x 1/2"	"	0.024
3/8" x 5/8"	"	0.026
3/8" x 3/4"	"	0.028
1/2" x 1/2"	"	0.031
1/2" x 5/8"	"	0.035
1/2" x 3/4"	"	0.040
1/2" x 7/8"	"	0.041
1/2" x 1"	"	0.042
3/4" x 3/4"	"	0.043
1" x 1"	"	0.044
Polyurethane compounds		

Joint Sealers		UNIT	MAN/HOURS
07920.10	**Caulking** *(Cont.)*		
1/4" x 1/4"		L.F.	0.020
1/4" x 3/8"		"	0.021
1/4" x 1/2"		"	0.022
3/8" x 3/8"		"	0.023
3/8" x 1/2"		"	0.024
3/8" x 5/8"		"	0.026
3/8" x 3/4"		"	0.028
1/2" x 1/2"		"	0.031
1/2" x 5/8"		"	0.035
1/2" x 3/4"		"	0.040
1/2" x 7/8"		"	0.041
1/2" x 1"		"	0.044
3/4" x 3/4"		"	0.043
3/4" x 1"		"	0.044
Backer rod, polyethylene			
1/4"		L.F.	0.020
1/2"		"	0.021
3/4"		"	0.022
1"		"	0.024

Metal — 08110.10 Metal Doors

Metal	UNIT	MAN/HOURS
Flush hollow metal, std. duty, 20 ga., 1-3/8" thick		
2-6 x 6-8	EA.	0.889
2-8 x 6-8	"	0.889
3-0 x 6-8	"	0.889
1-3/4" thick		
2-6 x 6-8	EA.	0.889
2-8 x 6-8	"	0.889
3-0 x 6-8	"	0.889
2-6 x 7-0	"	0.889
2-8 x 7-0	"	0.889
3-0 x 7-0	"	0.889
Heavy duty, 20 ga., unrated, 1-3/4"		
2-8 x 6-8	EA.	0.889
3-0 x 6-8	"	0.889
2-8 x 7-0	"	0.889
3-0 x 7-0	"	0.889
3-4 x 7-0	"	0.889
18 ga., 1-3/4", unrated door		
2-0 x 7-0	EA.	0.889
2-4 x 7-0	"	0.889
2-6 x 7-0	"	0.889
2-8 x 7-0	"	0.889
3-0 x 7-0	"	0.889
3-4 x 7-0	"	0.889
2", unrated door		
2-0 x 7-0	EA.	1.000
2-4 x 7-0	"	1.000
2-6 x 7-0	"	1.000
2-8 x 7-0	"	1.000
3-0 x 7-0	"	1.000
3-4 x 7-0	"	1.000
Galvanized metal door		
3-0 x 7-0	EA.	1.000

08110.40 Metal Door Frames

Metal	UNIT	MAN/HOURS
Hollow metal, stock, 18 ga., 4-3/4" x 1-3/4"		
2-0 x 7-0	EA.	1.000
2-4 x 7-0	"	1.000
2-6 x 7-0	"	1.000
2-8 x 7-0	"	1.000
3-0 x 7-0	"	1.000
4-0 x 7-0	"	1.333
5-0 x 7-0	"	1.333
6-0 x 7-0	"	1.333
16 ga., 6-3/4" x 1-3/4"		
2-0 x 7-0	EA.	1.000
2-4 x 7-0	"	1.000
2-6 x 7-0	"	1.000
2-8 x 7-0	"	1.000
3-0 x 7-0	"	1.000
4-0 x 7-0	"	1.333
6-0 x 7-0	"	1.333
Transom frame		

08110.40 Metal Door Frames (Cont.)

Metal	UNIT	MAN/HOURS
3-4 x 1-6	EA.	1.000
3-8 x 1-6	"	1.000
6-4 x 1-6	"	1.000
Transom sash		
3-0 x 1-4	EA.	1.000
3-4 x 1-4	"	1.000
6-0 x 1-4	"	1.000
Sidelights, complete		
1-0 x 7-2	EA.	1.000
1-4 x 7-2	"	1.000
1-0 x 8-8	"	1.000
1-6 x 8-8	"	1.000
16 ga., 4-3/4" x 1-3/4"		
2-0 x 7-0	EA.	1.000
2-4 x 7-0	"	1.000
2-6 x 7-0	"	1.000
2-8 x 7-0	"	1.000
3-0 x 7-0	"	1.000
4-0 x 7-0	"	1.333
6-0 x 7-0	"	1.333
3-4 x 1-6	"	1.000
3-8 x 1-6	"	1.000
6-4 x 1-6	"	1.000
Transom sash		
3-0 x 1-4	EA.	1.000
3-4 x 1-4	"	1.000
6-0 x 1-4	"	1.000
Sidelights, complete		
1-0 x 7-2	EA.	1.000
1-4 x 7-2	"	1.000
1-0 x 8-8	"	1.000
1-4 x 8-8	"	1.000
16 ga., 5-3/4" x 1-3/4"		
2-0 x 7-0	EA.	1.000
2-4 x 7-0	"	1.000
2-6 x 7-0	"	1.000
2-8 x 7-0	"	1.000
3-0 x 7-0	"	1.000
4-0 x 7-0	"	1.333
5-0 x 7-0	"	1.333
6-0 x 7-0	"	1.333
Mullions, vertical		
5-1/4" x 1-3/4"	L.F.	0.100
5-1/4" x 2"	"	0.100
Horizontal		
5-1/4" x 1-3/4"	L.F.	0.100
5-1/4" x 2"	"	0.100

Metal	UNIT	MAN/HOURS
08120.10 **Aluminum Doors**		
Aluminum doors, commercial		
Narrow stile		
2-6 x 7-0	EA.	4.000
3-0 x 7-0	"	4.000
3-6 x 7-0	"	4.000
Pair		
5-0 x 7-0	EA.	8.000
6-0 x 7-0	"	8.000
7-0 x 7-0	"	8.000
Wide stile		
2-6 x 7-0	EA.	4.000
3-0 x 7-0	"	4.000
3-6 x 7-0	"	4.000
Pair		
5-0 x 7-0	EA.	8.000
6-0 x 7-0	"	8.000
7-0 x 7-0	"	8.000

Wood And Plastic	UNIT	MAN/HOURS
08210.10 **Wood Doors**		
Solid core, 1-3/8" thick		
Birch faced		
2-4 x 7-0	EA.	1.000
2-8 x 7-0	"	1.000
3-0 x 7-0	"	1.000
3-4 x 7-0	"	1.000
2-4 x 6-8	"	1.000
2-6 x 6-8	"	1.000
2-8 x 6-8	"	1.000
3-0 x 6-8	"	1.000
Lauan faced		
2-4 x 6-8	EA.	1.000
2-8 x 6-8	"	1.000
3-0 x 6-8	"	1.000
3-4 x 6-8	"	1.000
Tempered hardboard faced		
2-4 x 7-0	EA.	1.000
2-8 x 7-0	"	1.000
3-0 x 7-0	"	1.000
3-4 x 7-0	"	1.000
Hollow core, 1-3/8" thick		
Birch faced		
2-4 x 7-0	EA.	1.000
2-8 x 7-0	"	1.000
3-0 x 7-0	"	1.000

Wood And Plastic	UNIT	MAN/HOURS
08210.10 **Wood Doors** *(Cont.)*		
3-4 x 7-0	EA.	1.000
Lauan faced		
2-4 x 6-8	EA.	1.000
2-6 x 6-8	"	1.000
2-8 x 6-8	"	1.000
3-0 x 6-8	"	1.000
3-4 x 6-8	"	1.000
Tempered hardboard faced		
2-4 x 7-0	EA.	1.000
2-6 x 7-0	"	1.000
2-8 x 7-0	"	1.000
3-0 x 7-0	"	1.000
3-4 x 7-0	"	1.000
Solid core, 1-3/4" thick		
Birch faced		
2-4 x 7-0	EA.	1.000
2-6 x 7-0	"	1.000
2-8 x 7-0	"	1.000
3-0 x 7-0	"	1.000
3-4 x 7-0	"	1.000
Lauan faced		
2-4 x 7-0	EA.	1.000
2-6 x 7-0	"	1.000
2-8 x 7-0	"	1.000
3-4 x 7-0	"	1.000
3-0 x 7-0	"	1.000
Tempered hardboard faced		
2-4 x 7-0	EA.	1.000
2-6 x 7-0	"	1.000
2-8 x 7-0	"	1.000
3-0 x 7-0	"	1.000
3-4 x 7-0	"	1.000
Hollow core, 1-3/4" thick		
Birch faced		
2-4 x 7-0	EA.	1.000
2-6 x 7-0	"	1.000
2-8 x 7-0	"	1.000
3-0 x 7-0	"	1.000
3-4 x 7-0	"	1.000
Lauan faced		
2-4 x 6-8	EA.	1.000
2-6 x 6-8	"	1.000
2-8 x 6-8	"	1.000
3-0 x 6-8	"	1.000
3-4 x 6-8	"	1.000
Tempered hardboard		
2-4 x 7-0	EA.	1.000
2-6 x 7-0	"	1.000
2-8 x 7-0	"	1.000
3-0 x 7-0	"	1.000
3-4 x 7-0	"	1.000
Add-on, louver	"	0.800
Glass	"	0.800

Wood And Plastic

08210.10 — Wood Doors (Cont.)

	UNIT	MAN/HOURS
Exterior doors, 3-0 x 7-0 x 2-1/2", solid core		
Carved		
One face	EA.	2.000
Two faces	"	2.000
Closet doors, 1-3/4" thick		
Bi-fold or bi-passing, includes frame and trim		
Paneled		
4-0 x 6-8	EA.	1.333
6-0 x 6-8	"	1.333
Louvered		
4-0 x 6-8	EA.	1.333
6-0 x 6-8	"	1.333
Flush		
4-0 x 6-8	EA.	1.333
6-0 x 6-8	"	1.333
Primed		
4-0 x 6-8	EA.	1.333
6-0 x 6-8	"	1.333

08210.90 — Wood Frames

	UNIT	MAN/HOURS
Frame, interior, pine		
2-6 x 6-8	EA.	1.143
2-8 x 6-8	"	1.143
3-0 x 6-8	"	1.143
5-0 x 6-8	"	1.143
6-0 x 6-8	"	1.143
2-6 x 7-0	"	1.143
2-8 x 7-0	"	1.143
3-0 x 7-0	"	1.143
5-0 x 7-0	"	1.600
6-0 x 7-0	"	1.600
Exterior, custom, with threshold, including trim		
Walnut		
3-0 x 7-0	EA.	2.000
6-0 x 7-0	"	2.000
Oak		
3-0 x 7-0	EA.	2.000
6-0 x 7-0	"	2.000
Pine		
2-4 x 7-0	EA.	1.600
2-6 x 7-0	"	1.600
2-8 x 7-0	"	1.600
3-0 x 7-0	"	1.600
3-4 x 7-0	"	1.600
6-0 x 7-0	"	2.667

08300.10 — Special Doors

	UNIT	MAN/HOURS
Vault door and frame, class 5, steel	EA.	8.000
Overhead door, coiling insulated		
Chain gear, no frame, 12' x 12'	EA.	10.000
Aluminum, bronze glass panels, 12-9 x 13-0	"	8.000
Garage door, flush insulated metal, primed, 9-0 x 7-0	"	2.667
Sliding metal fire doors, motorized, fusible link, 3 hr.		

Wood And Plastic

08300.10 — Special Doors (Cont.)

	UNIT	MAN/HOURS
3-0 x 6-8	EA.	16.000
3-8 x 6-8	"	16.000
4-0 x 8-0	"	16.000
5-0 x 8-0	"	16.000
Metal clad doors, including electric motor		
Light duty		
Minimum	S.F.	0.133
Maximum	"	0.320
Heavy duty		
Minimum	S.F.	0.400
Maximum	"	0.500
Hangar doors, based on 150' openings		
To 20' high	S.F.	0.096
20' to 40' high	"	0.060
40' to 60' high	"	0.040
60' to 80' high	"	0.024
Over 80' high	"	0.019
Counter doors, (roll-up shutters), standard, manual		
Opening, 4' high		
4' wide	EA.	6.667
6' wide	"	6.667
8' wide	"	7.273
10' wide	"	10.000
14' wide	"	10.000
6' high		
4' wide	EA.	6.667
6' wide	"	7.273
8' wide	"	8.000
10' wide	"	10.000
14' wide	"	11.429
Service doors, (roll up shutters), standard, manual		
Opening		
8' high x 8' wide	EA.	4.444
10' high x 10' wide	"	6.667
12' high x 12' wide	"	10.000
14' high x 14' wide	"	13.333
16' high x 14' wide	"	13.333
20' high x 14' wide	"	20.000
24' high x 16' wide	"	17.778
Roll-up doors		
13-0 high x 14-0 wide	EA.	11.429
12-0 high x 14-0 wide	"	11.429
Top coiling grilles, manually operated, steel or aluminum		
Opening, 4' high x		
4' wide	EA.	3.200
6' wide	"	3.200
8' wide	"	4.444
12' wide	"	4.444
16' wide	"	6.667
6' high x		
4' wide	EA.	6.667
6' wide	"	7.273
8' wide	"	8.000

Wood And Plastic	UNIT	MAN/ HOURS
08300.10 **Special Doors** *(Cont.)*		
12' wide	EA.	8.889
16' wide	"	11.429
Side coiling grilles, manually operated, aluminum		
Opening, 8' high x		
18' wide	EA.	60.000
24' wide	"	68.571
12' high x		
12' wide	EA.	60.000
18' wide	"	68.571
24' wide	"	80.000
Accordion folding doors, tracks and fittings included		
Vinyl covered, 2 layers	S.F.	0.320
Woven mahogany and vinyl	"	0.320
Economy vinyl	"	0.320
Rigid polyvinyl chloride	"	0.320
Sectional wood overhead doors, frames not included		
Commercial grade, heavy duty, 1-3/4" thick, manual		
8' x 8'	EA.	6.667
10' x 10'	"	7.273
12' x 12'	"	8.000
Chain hoist		
12' x 16' high	EA.	13.333
14' x 14' high	"	10.000
20' x 8' high	"	16.000
16' high	"	20.000
Sectional metal overhead doors, complete		
Residential grade, manual		
9' x 7'	EA.	3.200
16' x 7'	"	4.000
Commercial grade		
8' x 8'	EA.	6.667
10' x 10'	"	7.273
12' x 12'	"	8.000
20' x 14', with chain hoist	"	16.000
Sliding glass doors		
Tempered plate glass, 1/4" thick		
6' wide		
Economy grade	EA.	2.667
Premium grade	"	2.667
12' wide		
Economy grade	EA.	4.000
Premium grade	"	4.000
Insulating glass, 5/8" thick		
6' wide		
Economy grade	EA.	2.667
Premium grade	"	2.667
12' wide		
Economy grade	EA.	4.000
Premium grade	"	4.000
1" thick		
6' wide		
Economy grade	EA.	2.667
Premium grade	"	2.667

Wood And Plastic	UNIT	MAN/ HOURS
08300.10 **Special Doors** *(Cont.)*		
12' wide		
Economy grade	EA.	4.000
Premium grade	"	4.000
Vertical lift doors, channel frame construction		
20' high x		
10' wide	EA.	9.600
15' wide	"	9.600
20' wide	"	17.143
25' wide	"	17.143
25' high x		
20' wide	EA.	17.143
25' wide	"	20.000
30' high x		
25' wide	EA.	20.000
30' wide	"	20.000
35' wide	"	20.000
Residential storm door		
Minimum	EA.	1.333
Average	"	1.333
Maximum	"	2.000

Storefronts	UNIT	MAN/ HOURS
08410.10 **Storefronts**		
Storefront, aluminum and glass		
Minimum	S.F.	0.100
Average	"	0.114
Maximum	"	0.133
Entrance drs, prem., incl. glass, closers, panic dev.,etc.		
1/2" thick glass		
3' x 7'	EA.	6.667
6' x 7'	"	10.000
3/4" thick glass		
3' x 7'	EA.	6.667
6' x 7'	"	10.000
1" thick glass		
3' x 7'	EA.	6.667
6' x 7'	"	10.000
Revolving doors		
7'diameter, 7' high		
Minimum	EA.	60.000
Average	"	96.000
Maximum	"	120.000

Metal Windows

08510.10 — Steel Windows

	UNIT	MAN/HOURS
Steel windows, primed		
Casements		
Operable		
Minimum	S.F.	0.047
Maximum	"	0.053
Fixed sash	"	0.040
Double hung	"	0.044
Industrial windows		
Horizontally pivoted sash	S.F.	0.053
Fixed sash	"	0.044
Security sash		
Operable	S.F.	0.053
Fixed	"	0.044
Picture window	"	0.044
Projecting sash		
Minimum	S.F.	0.050
Maximum	"	0.050
Mullions	L.F.	0.040

08520.10 — Aluminum Windows

	UNIT	MAN/HOURS
Jalousie		
3-0 x 4-0	EA.	1.000
3-0 x 5-0	"	1.000
Fixed window		
6 sf to 8 sf	S.F.	0.114
12 sf to 16 sf	"	0.089
Projecting window		
6 sf to 8 sf	S.F.	0.200
12 sf to 16 sf	"	0.133
Horizontal sliding		
6 sf to 8 sf	S.F.	0.100
12 sf to 16 sf	"	0.080
Double hung		
6 sf to 8 sf	S.F.	0.160
10 sf to 12 sf	"	0.133
Storm window, 0.5 cfm, up to		
60 u.i. (united inches)	EA.	0.400
70 u.i.	"	0.400
80 u.i.	"	0.400
90 u.i.	"	0.444
100 u.i.	"	0.444
2.0 cfm, up to		
60 u.i.	EA.	0.400
70 u.i.	"	0.400
80 u.i.	"	0.400
90 u.i.	"	0.444
100 u.i.	"	0.444

Wood And Plastic

08600.10 — Wood Windows

	UNIT	MAN/HOURS
Double hung		
24" x 36"		
Minimum	EA.	0.800
Average	"	1.000
Maximum	"	1.333
24" x 48"		
Minimum	EA.	0.800
Average	"	1.000
Maximum	"	1.333
30" x 48"		
Minimum	EA.	0.889
Average	"	1.143
Maximum	"	1.600
30" x 60"		
Minimum	EA.	0.889
Average	"	1.143
Maximum	"	1.600
Casement		
1 leaf, 22" x 38" high		
Minimum	EA.	0.800
Average	"	1.000
Maximum	"	1.333
2 leaf, 50" x 50" high		
Minimum	EA.	1.000
Average	"	1.333
Maximum	"	2.000
3 leaf, 71" x 62" high		
Minimum	EA.	1.000
Average	"	1.333
Maximum	"	2.000
4 leaf, 95" x 75" high		
Minimum	EA.	1.143
Average	"	1.600
Maximum	"	2.667
5 leaf, 119" x 75" high		
Minimum	EA.	1.143
Average	"	1.600
Maximum	"	2.667
Picture window, fixed glass, 54" x 54" high		
Minimum	EA.	1.000
Average	"	1.143
Maximum	"	1.333
68" x 55" high		
Minimum	EA.	1.000
Average	"	1.143
Maximum	"	1.333
Sliding, 40" x 31" high		
Minimum	EA.	0.800
Average	"	1.000
Maximum	"	1.333
52" x 39" high		
Minimum	EA.	1.000
Average	"	1.143

Wood And Plastic

08600.10 — Wood Windows (Cont.)

	UNIT	MAN/HOURS
Maximum	EA.	1.333
64" x 72" high		
Minimum	EA.	1.000
Average	"	1.333
Maximum	"	1.600
Awning windows		
34" x 21" high		
Minimum	EA.	0.800
Average	"	1.000
Maximum	"	1.333
40" x 21" high		
Minimum	EA.	0.889
Average	"	1.143
Maximum	"	1.600
48" x 27" high		
Minimum	EA.	0.889
Average	"	1.143
Maximum	"	1.600
60" x 36" high		
Minimum	EA.	1.000
Average	"	1.333
Maximum	"	1.600
Window frame, milled		
Minimum	L.F.	0.160
Average	"	0.200
Maximum	"	0.267

Hardware

08710.20 — Locksets

	UNIT	MAN/HOURS
Latchset, heavy duty		
Cylindrical	EA.	0.500
Mortise	"	0.800
Lockset, heavy duty		
Cylindrical	EA.	0.500
Mortise	"	0.800
Mortise locks and latchsets, chrome		
Latchset passage or closet latch	EA.	0.667
Privacy (bath or bedroom)	"	0.667
Entry lockset	"	0.667
Classroom lockset (outside key operated)	"	0.667
Storeroom lock	"	0.667
Front door lock	"	0.667
Dormitory or exit lock	"	0.667
Preassembled locks and latches, brass		
Latchset, passage or closet latch	EA.	0.667

Hardware

08710.20 — Locksets (Cont.)

	UNIT	MAN/HOURS
Lockset		
Privacy (bath or bathroom)	EA.	0.667
Entry lock	"	0.667
Classroom lock (outside key, operated)	"	0.667
Storeroom lock	"	0.667
Bored locks and latches, satin chrome plated		
Latchset passage or closet latch	EA.	0.667
Lockset		
Privacy (bath or bedroom)	EA.	0.667
Entry lock	"	0.667
Classroom lock	"	0.667
Corridor lock	"	0.667
Miscellaneous locks		
Exit lock with alarm, single door	EA.	3.200
Electric strike		
Rim mounted wrought steel	EA.	2.000
Mortised, wrought steel with bronze plating	"	3.200
Dead bolt		
Bored, wrought brass, keyed both sides	EA.	1.333
Mortised, cast brass	"	1.333
Lockset, cipher, mechanical	"	0.800

08710.30 — Closers

	UNIT	MAN/HOURS
Door closers		
Surface mounted, traditional type, parallel arm		
Standard	EA.	1.000
Heavy duty	"	1.000
Modern type, parallel arm, standard duty	"	1.000
Overhead, concealed, pivot hung, single acting		
Interior	EA.	1.000
Exterior	"	1.000
Floor concealed, single acting, offset, pivoted		
Interior	EA.	2.667
Exterior	"	2.667

08710.40 — Door Trim

	UNIT	MAN/HOURS
Door bumper, bronze, wall type	EA.	0.160
Wall type, 4" dia. with convex rubber pad, aluminum	"	0.160
Floor type		
Aluminum	EA.	0.160
Brass	"	0.160
Door holders		
Wall type, bronze	EA.	0.160
Overhead	"	0.400
Floor type	"	0.400
Plunger type	"	0.400
Wall type, aluminum	"	0.400
Surface bolt	"	0.160
Panic device		
Rim type with thumb piece	EA.	2.000
Mortise	"	2.000
Vertical rod	"	2.000
Labeled, rim type	"	2.000

Hardware

08710.40 — Door Trim (Cont.)

	UNIT	MAN/HOURS
Mortise	EA.	2.000
Vertical rod	"	2.000
Silencers, rubber type	"	0.016
Dust proof strike with plate, brass	"	0.267
Flush bolt, lever extension, brass, rated	"	0.160
Surface bolt with strike, brass, 6" long	"	0.160
Door coordinator, labeled, brass, satin chrome	"	0.571
Door plates		
Kick plate, aluminum, 3 beveled edges		
10" x 28"	EA.	0.400
10" x 30"	"	0.400
10" x 34"	"	0.400
10" x 38"	"	0.400
Push plate, 4" x 16"		
Aluminum	EA.	0.160
Bronze	"	0.160
Stainless steel	"	0.160
Armor plate, 40" x 34"	"	0.320
Pull handle, 4" x 16"		
Aluminum	EA.	0.160
Bronze	"	0.160
Stainless steel	"	0.160
Hasp assembly		
3"	EA.	0.133
4-1/2"	"	0.178
6"	"	0.229
Electro-magnetic door holder		
Wall mounted	EA.	2.667
Floor mounted	"	2.667
Smoke detector door holder		
Photo electric type	EA.	2.667
Ionization type	"	2.667
Pneumatic operators, activated by rubber mats		
Swing		
Single	EA.	6.667
Double	"	10.000
Sliding		
Single	EA.	6.667
Double	"	10.000

08710.60 — Weatherstripping

	UNIT	MAN/HOURS
Weatherstrip, head and jamb, metal strip, neoprene bulb		
Standard duty	L.F.	0.044
Heavy duty	"	0.050
Spring type		
Metal doors	EA.	2.000
Wood doors	"	2.667
Sponge type with adhesive backing	"	0.800
Astragal		
1-3/4" x 13 ga., aluminum	L.F.	0.067
1-3/8" x 5/8", oak	"	0.053
Thresholds		
Bronze	L.F.	0.200

Hardware

08710.60 — Weatherstripping (Cont.)

	UNIT	MAN/HOURS
Aluminum		
Plain	L.F.	0.200
Vinyl insert	"	0.200
Aluminum with grit	"	0.200
Steel		
Plain	L.F.	0.200
Interlocking	"	0.667

Glazing

08810.10 — Glazing

	UNIT	MAN/HOURS
Sheet glass, 1/8" thick	S.F.	0.044
Plate glass, bronze or grey, 1/4" thick	"	0.073
Clear	"	0.073
Polished	"	0.073
Plexiglass		
1/8" thick	S.F.	0.073
1/4" thick	"	0.044
Float glass, clear		
3/16" thick	S.F.	0.067
1/4" thick	"	0.073
5/16" thick	"	0.080
3/8" thick	"	0.100
1/2" thick	"	0.133
5/8" thick	"	0.160
3/4" thick	"	0.200
1" thick	"	0.267
Tinted glass, polished plate, twin ground		
3/16" thick	S.F.	0.067
1/4" thick	"	0.073
3/8" thick	"	0.100
1/2" thick	"	0.133
Total, full vision, all glass window system		
To 10' high		
Minimum	S.F.	0.200
Average	"	0.200
Maximum	"	0.200
10' to 20' high		
Minimum	S.F.	0.200
Average	"	0.200
Maximum	"	0.200
Insulated glass, bronze or gray		
1/2" thick	S.F.	0.133
1" thick	"	0.200
Spandrel glass, polished bronze/grey, 1 side, 1/4" thick	"	0.073
Tempered glass (safety)		

Glazing	UNIT	MAN/HOURS
08810.10 Glazing *(Cont.)*		
Clear sheet glass		
1/8" thick	S.F.	0.044
3/16" thick	"	0.062
Clear float glass		
1/4" thick	S.F.	0.067
5/16" thick	"	0.080
3/8" thick	"	0.100
1/2" thick	"	0.133
5/8" thick	"	0.160
3/4" thick	"	0.267
Tinted float glass		
3/16" thick	S.F.	0.062
1/4" thick	"	0.067
3/8" thick	"	0.100
1/2" thick	"	0.133
Laminated glass		
Float safety glass with polyvinyl plastic interlayer		
1/4", sheet or float		
Two lites, 1/8" thick, clear glass	S.F.	0.067
1/2" thick, float glass		
Two lites, 1/4" thick, clear glass	S.F.	0.133
Tinted glass	"	0.133
Insulating glass, two lites, clear float glass		
1/2" thick	S.F.	0.133
5/8" thick	"	0.160
3/4" thick	"	0.200
7/8" thick	"	0.229
1" thick	"	0.267
Glass seal edge		
3/8" thick	S.F.	0.133
Tinted glass		
1/2" thick	S.F.	0.133
1" thick	"	0.267
Tempered, clear		
1" thick	S.F.	0.267
Wire reinforced	"	0.267
Plate mirror glass		
1/4" thick		
15 sf	S.F.	0.080
Over 15 sf	"	0.073
Door type, 1/4" thick	"	0.080
Transparent, one way vision, 1/4" thick	"	0.080
Sheet mirror glass		
3/16" thick	S.F.	0.080
1/4" thick	"	0.067
Wall tiles, 12" x 12"		
Clear glass	S.F.	0.044
Veined glass	"	0.044
Wire glass, 1/4" thick		
Clear	S.F.	0.267
Hammered	"	0.267
Obscure	"	0.267
Bullet resistant glass, plate glass with inter-leaved vinyl		

Glazing	UNIT	MAN/HOURS
08810.10 Glazing *(Cont.)*		
1-3/16" thick		
To 15 sf	S.F.	0.400
Over 15 sf	"	0.400
2" thick		
To 15 sf	S.F.	0.667
Over 15 sf	"	0.667
Glazing accessories		
Neoprene glazing gaskets		
1/4" glass	L.F.	0.032
3/8" glass	"	0.033
1/2" glass	"	0.035
3/4" glass	"	0.036
1" glass	"	0.040
Mullion section		
1/4" glass	L.F.	0.016
3/8" glass	"	0.020
1/2" glass	"	0.023
3/4" glass	"	0.027
1" glass	"	0.032
Molded corners	EA.	0.533

Glazed Curtain Walls	UNIT	MAN/HOURS
08910.10 Glazed Curtain Walls		
Curtain wall, aluminum system, framing sections		
2" x 3"		
Jamb	L.F.	0.067
Horizontal	"	0.067
Mullion	"	0.067
2" x 4"		
Jamb	L.F.	0.100
Horizontal	"	0.100
Mullion	"	0.100
3" x 5-1/2"		
Jamb	L.F.	0.100
Horizontal	"	0.100
Mullion	"	0.100
4" corner mullion	"	0.133
Coping sections		
1/8" x 8"	L.F.	0.133
1/8" x 9"	"	0.133
1/8" x 12-1/2"	"	0.160
Sill section		
1/8" x 6"	L.F.	0.080
1/8" x 7"	"	0.080
1/8" x 8-1/2"	"	0.080

Glazed Curtain Walls	UNIT	MAN/HOURS
08910.10 **Glazed Curtain Walls** *(Cont.)*		
Column covers, aluminum		
1/8" x 26"	L.F.	0.200
1/8" x 34"	"	0.211
1/8" x 38"	"	0.211
Doors		
Aluminum framed, standard hardware		
Narrow stile		
2-6 x 7-0	EA.	4.000
3-0 x 7-0	"	4.000
3-6 x 7-0	"	4.000
Wide stile		
2-6 x 7-0	EA.	4.000
3-0 x 7-0	"	4.000
3-6 x 7-0	"	4.000
Flush panel doors, to match adjacent wall panels		
2-6 x 7-0	EA.	5.000
3-0 x 7-0	"	5.000
3-6 x 7-0	"	5.000
Wall panel, insulated		
"U"=.08	S.F.	0.067
"U"=.10	"	0.067
"U"=.15	"	0.067
Window wall system, complete		
Minimum	S.F.	0.080
Average	"	0.089
Maximum	"	0.114

Support Systems

09110.10 — Metal Studs

Support Systems	UNIT	MAN/HOURS
Studs, non load bearing, galvanized		
2-1/2", 20 ga.		
12" o.c.	S.F.	0.017
16" o.c.	"	0.013
25 ga.		
12" o.c.	S.F.	0.017
16" o.c.	"	0.013
24" o.c.	"	0.011
3-5/8", 20 ga.		
12" o.c.	S.F.	0.020
16" o.c.	"	0.016
24" o.c.	"	0.013
25 ga.		
12" o.c.	S.F.	0.020
16" o.c.	"	0.016
24" o.c.	"	0.013
4", 20 ga.		
12" o.c.	S.F.	0.020
16" o.c.	"	0.016
24" o.c.	"	0.013
25 ga.		
12" o.c.	S.F.	0.020
16" o.c.	"	0.016
24" o.c.	"	0.013
6", 20 ga.		
12" o.c.	S.F.	0.025
16" o.c.	"	0.020
24" o.c.	"	0.017
25 ga.		
12" o.c.	S.F.	0.025
16" o.c.	"	0.020
24" o.c.	"	0.017
Load bearing studs, galvanized		
3-5/8", 16 ga.		
12" o.c.	S.F.	0.020
16" o.c.	"	0.016
18 ga.		
12" o.c.	S.F.	0.013
16" o.c.	"	0.016
4", 16 ga.		
12" o.c.	S.F.	0.020
16" o.c.	"	0.016
6", 16 ga.		
12" o.c.	S.F.	0.025
16" o.c.	"	0.020
Furring		
On beams and columns		
7/8" channel	L.F.	0.053
1-1/2" channel	"	0.062
On ceilings		
3/4" furring channels		
12" o.c.	S.F.	0.033
16" o.c.	"	0.032

09110.10 — Metal Studs (Cont.)

Support Systems	UNIT	MAN/HOURS
24" o.c.	S.F.	0.029
1-1/2" furring channels		
12" o.c.	S.F.	0.036
16" o.c.	"	0.033
24" o.c.	"	0.031
On walls		
3/4" furring channels		
12" o.c.	S.F.	0.027
16" o.c.	"	0.025
24" o.c.	"	0.024
1-1/2" furring channels		
12" o.c.	S.F.	0.029
16" o.c.	"	0.027
24" o.c.	"	0.025

Lath And Plaster

09205.10 — Gypsum Lath

Lath And Plaster	UNIT	MAN/HOURS
Gypsum lath, 1/2" thick		
Clipped	S.Y.	0.044
Nailed	"	0.050

09205.20 — Metal Lath

Lath And Plaster	UNIT	MAN/HOURS
Diamond expanded, galvanized		
2.5 lb., on walls		
Nailed	S.Y.	0.100
Wired	"	0.114
On ceilings		
Nailed	S.Y.	0.114
Wired	"	0.133
3.4 lb., on walls		
Nailed	S.Y.	0.100
Wired	"	0.114
On ceilings		
Nailed	S.Y.	0.114
Wired	"	0.133
Flat rib		
2.75 lb., on walls		
Nailed	S.Y.	0.100
Wired	"	0.114
On ceilings		
Nailed	S.Y.	0.114
Wired	"	0.133
3.4 lb., on walls		
Nailed	S.Y.	0.100
Wired	"	0.114

09205.20 — Metal Lath (Cont.)

Lath And Plaster	UNIT	MAN/HOURS
On ceilings		
Nailed	S.Y.	0.114
Wired	"	0.133
Stucco lath		
1.8 lb.	S.Y.	0.100
3.6 lb.	"	0.100
Paper backed		
Minimum	S.Y.	0.080
Maximum	"	0.114

09205.60 — Plaster Accessories

Lath And Plaster	UNIT	MAN/HOURS
Expansion joint, 3/4", 26 ga., galvanized, one piece	L.F.	0.020
Plaster corner beads, 3/4", galvanized	"	0.023
Casing bead, expanded flange, galvanized	"	0.020
Expanded wing, 1-1/4" wide, galvanized	"	0.020
Joint clips for lath	EA.	0.004
Metal base, galvanized, 2-1/2" high	L.F.	0.027
Stud clips for gypsum lath	EA.	0.004
Sound deadening board, 1/4", nailed or clipped	S.F.	0.013

09210.10 — Plaster

Lath And Plaster	UNIT	MAN/HOURS
Gypsum plaster, trowel finish, 2 coats		
Ceilings	S.Y.	0.250
Walls	"	0.235
3 coats		
Ceilings	S.Y.	0.348
Walls	"	0.308
Vermiculite plaster		
2 coats		
Ceilings	S.Y.	0.381
Walls	"	0.348
3 coats		
Ceilings	S.Y.	0.471
Walls	"	0.421
Keenes cement plaster		
2 coats		
Ceilings	S.Y.	0.308
Walls	"	0.267
3 coats		
Ceilings	S.Y.	0.348
Walls	"	0.308
On columns, add to installation, 50%	"	
Chases, fascia, and soffits, add to installation, 50%	"	
Beams, add to installation, 50%	"	
Patch holes, average size holes		
1 sf to 5 sf		
Minimum	S.F.	0.133
Average	"	0.160
Maximum	"	0.200
Over 5 sf		
Minimum	S.F.	0.080

09210.10 — Plaster (Cont.)

Lath And Plaster	UNIT	MAN/HOURS
Average	S.F.	0.114
Maximum	"	0.133
Patch cracks		
Minimum	S.F.	0.027
average	"	0.040
Maximum	"	0.080

09220.10 — Portland Cement Plaster

Lath And Plaster	UNIT	MAN/HOURS
Stucco, portland, gray, 3 coat, 1" thick		
Sand finish	S.Y.	0.348
Trowel finish	"	0.364
White cement		
Sand finish	S.Y.	0.364
Trowel finish	"	0.400
Scratch coat		
For ceramic tile	S.Y.	0.080
For quarry tile	"	0.080
Portland cement plaster		
2 coats, 1/2"	S.Y.	0.160
3 coats, 7/8"	"	0.200

09250.10 — Gypsum Board

Lath And Plaster	UNIT	MAN/HOURS
Drywall, plasterboard, 3/8" clipped to		
Metal furred ceiling	S.F.	0.009
Columns and beams	"	0.020
Walls	"	0.008
Nailed or screwed to		
Wood framed ceiling	S.F.	0.008
Columns and beams	"	0.018
Walls	"	0.007
1/2", clipped to		
Metal furred ceiling	S.F.	0.009
Columns and beams	"	0.020
Walls	"	0.008
Nailed or screwed to		
Wood framed ceiling	S.F.	0.008
Columns and beams	"	0.018
Walls	"	0.007
5/8", clipped to		
Metal furred ceiling	S.F.	0.010
Columns and beams	"	0.022
Walls	"	0.009
Nailed or screwed to		
Wood framed ceiling	S.F.	0.010
Columns and beams	"	0.022
Walls	"	0.009
Vinyl faced, clipped to metal studs		
1/2"	S.F.	0.010
5/8"	"	0.010
Taping and finishing joints		
Minimum	S.F.	0.005
Average	"	0.007
Maximum	"	0.008

Lath And Plaster

	UNIT	MAN/HOURS
09250.10 **Gypsum Board** *(Cont.)*		
Casing bead		
Minimum	L.F.	0.023
Average	"	0.027
Maximum	"	0.040
Corner bead		
Minimum	L.F.	0.023
Average	"	0.027
Maximum	"	0.040

Tile

	UNIT	MAN/HOURS
09310.10 **Ceramic Tile**		
Glazed wall tile, 4-1/4" x 4-1/4"		
Minimum	S.F.	0.057
Average	"	0.067
Maximum	"	0.080
Base, 4-1/4" high		
Minimum	L.F.	0.100
Average	"	0.100
Maximum	"	0.100
Unglazed floor tile		
Portland cement bed, cushion edge, face mounted		
1" x 1"	S.F.	0.073
1" x 2"	"	0.070
2" x 2"	"	0.067
Adhesive bed, with white grout		
1" x 1"	S.F.	0.073
1" x 2"	"	0.070
2" x 2"	"	0.067
Organic adhesive bed, thin set, back mounted		
1" x 1"	S.F.	0.073
1" x 2"	"	0.070
2" x 2"	"	0.067
Unglazed wall tile		
Organic adhesive, face mounted cushion edge		
1" x 1"		
Minimum	S.F.	0.067
Average	"	0.073
Maximum	"	0.080
1" x 2"		
Minimum	S.F.	0.064
Average	"	0.070
Maximum	"	0.076
2" x 2"		
Minimum	S.F.	0.062
Average	"	0.067

Tile

	UNIT	MAN/HOURS
09310.10 **Ceramic Tile** *(Cont.)*		
Maximum	S.F.	0.073
Back mounted		
1" x 1"		
Minimum	S.F.	0.067
Average	"	0.073
Maximum	"	0.080
1" x 2"		
Minimum	S.F.	0.064
Average	"	0.070
Maximum	"	0.076
2" x 2"		
Minimum	S.F.	0.062
Average	"	0.067
Maximum	"	0.073
Conductive floor tile, unglazed square edged		
Portland cement bed		
1 x 1	S.F.	0.100
1-9/16 x 1-9/16	"	0.100
Dry set		
1 x 1	S.F.	0.100
1-9/16 x 1-9/16	"	0.100
Epoxy bed with epoxy joints		
1 x 1	S.F.	0.100
1-9/16 x 1-9/16	"	0.100
Ceramic accessories		
Towel bar, 24" long		
Minimum	EA.	0.320
Average	"	0.400
Maximum	"	0.533
Soap dish		
Minimum	EA.	0.533
Average	"	0.667
Maximum	"	0.800
09330.10 **Quarry Tile**		
Floor		
4 x 4 x 1/2"	S.F.	0.107
6 x 6 x 1/2"	"	0.100
6 x 6 x 3/4"	"	0.100
Wall, applied to 3/4" portland cement bed		
4 x 4 x 1/2"	S.F.	0.160
6 x 6 x 3/4"	"	0.133
Cove base		
5 x 6 x 1/2" straight top	L.F.	0.133
6 x 6 x 3/4" round top	"	0.133
Stair treads 6 x 6 x 3/4"	"	0.200
Window sill 6 x 8 x 3/4"	"	0.160
For abrasive surface, add to material, 25%		

Tile	UNIT	MAN/ HOURS
09410.10 **Terrazzo**		
Floors on concrete, 1-3/4" thick, 5/8" topping		
Gray cement	S.F.	0.114
White cement	"	0.114
Sand cushion, 3" thick, 5/8" top, 1/4"		
Gray cement	S.F.	0.133
White cement	"	0.133
Monolithic terrazzo, 3-1/2" base slab, 5/8" topping	"	0.100
Terrazzo wainscot, cast-in-place, 1/2" thick	"	0.200
Base, cast in place, terrazzo cove type, 6" high	L.F.	0.114
Curb, cast in place, 6" wide x 6" high, polished top	"	0.400
Stairs, cast-in-place, topping on concrete or metal		
1-1/2" thick treads, 12" wide	L.F.	0.400
Combined tread and riser	"	1.000
Precast terrazzo, thin set		
Terrazzo tiles, non-slip surface		
9" x 9" x 1" thick	S.F.	0.114
12" x 12"		
1" thick	S.F.	0.107
1-1/2" thick	"	0.114
18" x 18" x 1-1/2" thick	"	0.114
24" x 24" x 1-1/2" thick	"	0.094
Terrazzo wainscot		
12" x 12" x 1" thick	S.F.	0.200
18" x 18" x 1-1/2" thick	"	0.229
Base		
6" high		
Straight	L.F.	0.062
Coved	"	0.062
8" high		
Straight	L.F.	0.067
Coved	"	0.067
Terrazzo curbs		
8" wide x 8" high	L.F.	0.320
6" wide x 6" high	"	0.267
Precast terrazzo stair treads, 12" wide		
1-1/2" thick		
Diamond pattern	L.F.	0.145
Non-slip surface	"	0.145
2" thick		
Diamond pattern	L.F.	0.145
Non-slip surface	"	0.160
Stair risers, 1" thick to 6" high		
Straight sections	L.F.	0.080
Cove sections	"	0.080
Combined tread and riser		
Straight sections		
1-1/2" tread, 3/4" riser	L.F.	0.229
3" tread, 1" riser	"	0.229
Curved sections		
2" tread, 1" riser	L.F.	0.267
3" tread, 1" riser	"	0.267
Stair stringers, notched for treads and risers		
1" thick	L.F.	0.200

Tile	UNIT	MAN/ HOURS
09410.10 **Terrazzo** *(Cont.)*		
2" thick	L.F.	0.267
Landings, structural, nonslip		
1-1/2" thick	S.F.	0.133
3" thick	"	0.160
Conductive terrazzo, spark proof industrial floor		
Epoxy terrazzo		
Floor	S.F.	0.050
Base	"	0.067
Polyacrylate		
Floor	S.F.	0.050
Base	"	0.067
Polyester		
Floor	S.F.	0.032
Base	"	0.040
Synthetic latex mastic		
Floor	S.F.	0.050
Base	"	0.067

Acoustical Treatment	UNIT	MAN/ HOURS
09510.10 **Ceilings And Walls**		
Acoustical panels, suspension system not included		
Fiberglass panels		
5/8" thick		
2' x 2'	S.F.	0.011
2' x 4'	"	0.009
3/4" thick		
2' x 2'	S.F.	0.011
2' x 4'	"	0.009
Glass cloth faced fiberglass panels		
3/4" thick	S.F.	0.013
1" thick	"	0.013
Mineral fiber panels		
5/8" thick		
2' x 2'	S.F.	0.011
2' x 4'	"	0.009
3/4" thick		
2' x 2'	S.F.	0.011
2' x 4'	"	0.009
Wood fiber panels		
2" thick		
2' x 2'	S.F.	0.013
2' x 4'	"	0.010
Air distributing panels		
3/4" thick	S.F.	0.020
5/8" thick	"	0.016

Acoustical Treatment	UNIT	MAN/HOURS

09510.10   Ceilings And Walls *(Cont.)*

	UNIT	MAN/HOURS
Acoustical tiles, suspension system not included		
Fiberglass tile, 12" x 12"		
5/8" thick	S.F.	0.015
3/4" thick	"	0.018
Glass cloth faced fiberglass tile		
3/4" thick	S.F.	0.018
3" thick	"	0.020
Mineral fiber tile, 12" x 12"		
5/8" thick		
Standard	S.F.	0.016
Vinyl faced	"	0.016
3/4" thick		
Standard	S.F.	0.016
Vinyl faced	"	0.016
Fire rated	"	0.016
Aluminum or mylar faced	"	0.016
Wood fiber tile, 12" x 12"		
1/2" thick	S.F.	0.016
3/4" thick	"	0.016
Metal pan units, 24 ga. steel		
12" x 12"	S.F.	0.032
12" x 24"	"	0.027
Aluminum, .025" thick		
12" x 12"	S.F.	0.032
12" x 24"	"	0.027
Anodized aluminum, 0.25" thick		
12" x 12"	S.F.	0.032
12" x 24"	"	0.027
Stainless steel, 24 ga.		
12" x 12"	S.F.	0.032
12" x 24"	"	0.027
Metal ceiling systems		
.020" thick panels		
10', 12', and 16' lengths	S.F.	0.023
Custom lengths, 3' to 20'	"	0.023
.025" thick panels		
32 sf, 38 sf, and 52 sf pieces	S.F.	0.027
Custom lengths, 10 sf to 65 sf	"	0.027
Sound absorption walls, with fabric cover		
2-6" x 9' x 3/4"	S.F.	0.027
2' x 9' x 1"	"	0.027
Starter spline	L.F.	0.020
Internal spline	"	0.020
Acoustical treatment		
Barriers for plenums		
Leaded vinyl		
0.48 lb per sf	S.F.	0.038
0.87 lb per sf	"	0.040
Aluminum foil, fiberglass reinforcement		
Minimum	S.F.	0.027
Maximum	"	0.040
Aluminum mesh, paper backed	"	0.027
Fibered cement sheet, 3/16" thick	"	0.029

Acoustical Treatment	UNIT	MAN/HOURS

09510.10   Ceilings And Walls *(Cont.)*

	UNIT	MAN/HOURS
Sheet lead, 1/64" thick	S.F.	0.020
Sound attenuation blanket		
1" thick	S.F.	0.080
1-1/2" thick	"	0.080
2" thick	"	0.080
3" thick	"	0.089
Ceiling suspension systems		
T bar system		
2' x 4'	S.F.	0.008
2' x 2'	"	0.009
Concealed Z bar suspension system, 12" module	"	0.013

Flooring	UNIT	MAN/HOURS

09550.10   Wood Flooring

	UNIT	MAN/HOURS
Wood strip flooring, unfinished		
Fir floor		
C and better		
Vertical grain	S.F.	0.027
Flat grain	"	0.027
Oak floor		
Minimum	S.F.	0.038
Average	"	0.038
Maximum	"	0.038
Maple floor		
25/32" x 2-1/4"		
Minimum	S.F.	0.038
Maximum	"	0.038
33/32" x 3-1/4"		
Minimum	S.F.	0.038
Maximum	"	0.038
Wood block industrial flooring		
Creosoted		
2" thick	S.F.	0.021
2-1/2" thick	"	0.025
3" thick	"	0.027
Parquet, 5/16", white oak		
Finished	S.F.	0.040
Unfinished	"	0.040
Gym floor, 2 ply felt, 25/32" maple, finished, in mastic	"	0.044
Over wood sleepers	"	0.050
Finishing, sand, fill, finish, and wax	"	0.020
Refinish sand, seal, and 2 coats of polyurethane	"	0.027
Clean and wax floors	"	0.004

Flooring

09630.10 — Unit Masonry Flooring

	UNIT	MAN/HOURS
Clay brick		
9 x 4-1/2 x 3" thick		
Glazed	S.F.	0.067
Unglazed	"	0.067
8 x 4 x 3/4" thick		
Glazed	S.F.	0.070
Unglazed	"	0.070

09660.10 — Resilient Tile Flooring

	UNIT	MAN/HOURS
Solid vinyl tile, 1/8" thick, 12" x 12"		
Marble patterns	S.F.	0.020
Solid colors	"	0.020
Travertine patterns	"	0.020
Conductive resilient flooring, vinyl tile		
1/8" thick, 12" x 12"	S.F.	0.023

09665.10 — Resilient Sheet Flooring

	UNIT	MAN/HOURS
Vinyl sheet flooring		
Minimum	S.F.	0.008
Average	"	0.010
Maximum	"	0.013
Cove, to 6"	L.F.	0.016
Fluid applied resilient flooring		
Polyurethane, poured in place, 3/8" thick	S.F.	0.067
Vinyl sheet goods, backed		
0.070" thick	S.F.	0.010
0.093" thick	"	0.010
0.125" thick	"	0.010
0.250" thick	"	0.010

09678.10 — Resilient Base And Accessories

	UNIT	MAN/HOURS
Wall base, vinyl		
Group 1		
4" high	L.F.	0.027
6" high	"	0.027
Group 2		
4" high	L.F.	0.027
6" high	"	0.027
Group 3		
4" high	L.F.	0.027
6" high	"	0.027
Stair accessories		
Treads, 1/4" x 12", rubber diamond surface		
Marbled	L.F.	0.067
Plain	"	0.067
Grit strip safety tread, 12" wide, colors		
3/16" thick	L.F.	0.067
5/16" thick	"	0.067
Risers, 7" high, 1/8" thick, colors		
Flat	L.F.	0.040
Coved	"	0.040
Nosing, rubber		

Flooring

09678.10 — Resilient Base And Accessories

	UNIT	MAN/HOURS
3/16" thick, 3" wide		
Black	L.F.	0.040
Colors	"	0.040
6" wide		
Black	L.F.	0.067
Colors	"	0.067

Carpet

09680.10 — Floor Leveling

	UNIT	MAN/HOURS
Repair and level floors to receive new flooring		
Minimum	S.Y.	0.027
Average	"	0.067
Maximum	"	0.080

09682.10 — Carpet Padding

	UNIT	MAN/HOURS
Carpet padding		
Foam rubber, waffle type, 0.3" thick	S.Y.	0.040
Jute padding		
Minimum	S.Y.	0.036
Average	"	0.040
Maximum	"	0.044
Sponge rubber cushion		
Minimum	S.Y.	0.036
Average	"	0.040
Maximum	"	0.044
Urethane cushion, 3/8" thick		
Minimum	S.Y.	0.036
Average	"	0.040
Maximum	"	0.044

09685.10 — Carpet

	UNIT	MAN/HOURS
Carpet, acrylic		
24 oz., light traffic	S.Y.	0.089
28 oz., medium traffic	"	0.089
Residential		
Nylon		
15 oz., light traffic	S.Y.	0.089
28 oz., medium traffic	"	0.089
Commercial		
Nylon		
28 oz., medium traffic	S.Y.	0.089
35 oz., heavy traffic	"	0.089
Wool		
30 oz., medium traffic	S.Y.	0.089
36 oz., medium traffic	"	0.089

Carpet

Carpet *(Cont.)*	UNIT	MAN/HOURS
09685.10		
42 oz., heavy traffic	S.Y.	0.089
Carpet tile		
Foam backed		
Minimum	S.F.	0.016
Average	"	0.018
Maximum	"	0.020
Tufted loop or shag		
Minimum	S.F.	0.016
Average	"	0.018
Maximum	"	0.020
Clean and vacuum carpet		
Minimum	S.Y.	0.004
Average	"	0.005
Maximum	"	0.008

Special Flooring

09700.10	UNIT	MAN/HOURS
Epoxy flooring, marble chips		
Epoxy with colored quartz chips in 1/4" base	S.F.	0.044
Heavy duty epoxy topping, 3/16" thick	"	0.044
Epoxy terrazzo		
1/4" thick chemical resistant	S.F.	0.050

Painting

Painting Preparation	UNIT	MAN/HOURS
09905.10		
Dropcloths		
Minimum	S.F.	0.001
Average	"	0.001
Maximum	"	0.001
Masking		
Paper and tape		
Minimum	L.F.	0.008
Average	"	0.010
Maximum	"	0.013
Doors		
Minimum	EA.	0.100
Average	"	0.133
Maximum	"	0.178
Windows		
Minimum	EA.	0.100
Average	"	0.133
Maximum	"	0.178
Sanding		
Walls and flat surfaces		
Minimum	S.F.	0.005
Average	"	0.007

Painting

Painting Preparation *(Cont.)*	UNIT	MAN/HOURS
09905.10		
Maximum	S.F.	0.008
Doors and windows		
Minimum	EA.	0.133
Average	"	0.200
Maximum	"	0.267
Trim		
Minimum	L.F.	0.010
Average	"	0.013
Maximum	"	0.018
Puttying		
Minimum	S.F.	0.012
Average	"	0.016
Maximum	"	0.020
Chemical Preparation		
Concrete floors		
Acid Etch		
Minimum	S.F.	0.002
Average	"	0.003
Maximum	"	0.004
Chemical Stripping		
Minimum	S.F.	0.013
Average	"	0.020
Maximum	"	0.032
Stone		
Chemical cleaning		
Minimum	S.F.	0.040
Average	"	0.053
Maximum	"	0.067
Wood		
Bleaching		
Minimum	S.F.	0.009
Average	"	0.011
Maximum	"	0.015
Chemical stripping		
Minimum	S.F.	0.032
Average	"	0.080
Maximum	"	0.133
Water cleaning/preparation		
Washing (General)		
Minimum	S.F.	0.001
Average	"	0.001
Maximum	"	0.001
Mildew eradication		
Minimum	S.F.	0.001
Average	"	0.002
Maximum	"	0.003
Remove loose paint		
Minimum	S.F.	0.002

Painting	UNIT	MAN/ HOURS
09905.10 Painting Preparation *(Cont.)*		
Average	S.F.	0.003
Maximum	"	0.004
Steam clean		
Minimum	S.F.	0.002
Average	"	0.003
Maximum	"	0.004
09910.05 Ext. Painting, Sitework		
Benches		
Brush		
First Coat		
Minimum	S.F.	0.008
Average	"	0.010
Maximum	"	0.013
Second Coat		
Minimum	S.F.	0.005
Average	"	0.006
Maximum	"	0.007
Roller		
First Coat		
Minimum	S.F.	0.004
Average	"	0.004
Maximum	"	0.005
Second Coat		
Minimum	S.F.	0.003
Average	"	0.003
Maximum	"	0.004
Brickwork		
Brush		
First Coat		
Minimum	S.F.	0.005
Average	"	0.007
Maximum	"	0.010
Second Coat		
Minimum	S.F.	0.004
Average	"	0.005
Maximum	"	0.007
Roller		
First Coat		
Minimum	S.F.	0.004
Average	"	0.005
Maximum	"	0.007
Second Coat		
Minimum	S.F.	0.003
Average	"	0.004
Maximum	"	0.005
Spray		
First Coat		
Minimum	S.F.	0.002
Average	"	0.003
Maximum	"	0.004
Second Coat		
Minimum	S.F.	0.002

Painting	UNIT	MAN/ HOURS
09910.05 Ext. Painting, Sitework *(Cont.)*		
Average	S.F.	0.003
Maximum	"	0.003
Concrete Block		
Roller		
First Coat		
Minimum	S.F.	0.004
Average	"	0.005
Maximum	"	0.008
Second Coat		
Minimum	S.F.	0.003
Average	"	0.004
Maximum	"	0.007
Spray		
First Coat		
Minimum	S.F.	0.002
Average	"	0.003
Maximum	"	0.003
Second Coat		
Minimum	S.F.	0.001
Average	"	0.002
Maximum	"	0.003
Fences, Chain Link		
Brush		
First Coat		
Minimum	S.F.	0.008
Average	"	0.009
Maximum	"	0.010
Second Coat		
Minimum	S.F.	0.005
Average	"	0.006
Maximum	"	0.007
Roller		
First Coat		
Minimum	S.F.	0.006
Average	"	0.007
Maximum	"	0.008
Second Coat		
Minimum	S.F.	0.003
Average	"	0.004
Maximum	"	0.005
Spray		
First Coat		
Minimum	S.F.	0.003
Average	"	0.003
Maximum	"	0.003
Second Coat		
Minimum	S.F.	0.002
Average	"	0.002
Maximum	"	0.003
Fences, Wood or Masonry		
Brush		
First Coat		
Minimum	S.F.	0.008

Painting	UNIT	MAN/HOURS
09910.05 **Ext. Painting, Sitework** (Cont.)		
Average	S.F.	0.010
Maximum	"	0.013
Second Coat		
Minimum	S.F.	0.005
Average	"	0.006
Maximum	"	0.008
Roller		
First Coat		
Minimum	S.F.	0.004
Average	"	0.005
Maximum	"	0.006
Second Coat		
Minimum	S.F.	0.003
Average	"	0.004
Maximum	"	0.005
Spray		
First Coat		
Minimum	S.F.	0.003
Average	"	0.004
Maximum	"	0.005
Second Coat		
Minimum	S.F.	0.002
Average	"	0.003
Maximum	"	0.003
Storage Tanks		
Roller		
First Coat		
Minimum	S.F.	0.003
Average	"	0.004
Maximum	"	0.005
Second Coat		
Minimum	S.F.	0.003
Average	"	0.003
Maximum	"	0.004
Spray		
First Coat		
Minimum	S.F.	0.002
Average	"	0.002
Maximum	"	0.003
Second Coat		
Minimum	S.F.	0.002
Average	"	0.002
Maximum	"	0.002
09910.15 **Ext. Painting, Buildings**		
Decks, Metal		
Spray		
First Coat		
Minimum	S.F.	0.004
Average	"	0.004
Maximum	"	0.004
Second Coat		
Minimum	S.F.	0.003

Painting	UNIT	MAN/HOURS
09910.15 **Ext. Painting, Buildings** (Cont.)		
Average	S.F.	0.003
Maximum	"	0.003
Decks, Wood, Stained		
Brush		
First Coat		
Minimum	S.F.	0.004
Average	"	0.004
Maximum	"	0.005
Second Coat		
Minimum	S.F.	0.003
Average	"	0.003
Maximum	"	0.003
Roller		
First Coat		
Minimum	S.F.	0.003
Average	"	0.003
Maximum	"	0.003
Second Coat		
Minimum	S.F.	0.003
Average	"	0.003
Maximum	"	0.003
Spray		
First Coat		
Minimum	S.F.	0.003
Average	"	0.003
Maximum	"	0.003
Second Coat		
Minimum	S.F.	0.002
Average	"	0.002
Maximum	"	0.003
Doors, Metal		
Roller		
First Coat		
Minimum	S.F.	0.006
Average	"	0.007
Maximum	"	0.008
Second Coat		
Minimum	S.F.	0.004
Average	"	0.004
Maximum	"	0.005
Spray		
First Coat		
Minimum	S.F.	0.005
Average	"	0.006
Maximum	"	0.007
Second Coat		
Minimum	S.F.	0.004
Average	"	0.004
Maximum	"	0.004
Door Frames, Metal		
Brush		
First Coat		
Minimum	L.F.	0.010

Painting

09910.15 — Ext. Painting, Buildings (Cont.)	UNIT	MAN/HOURS
Average	L.F.	0.013
Maximum	"	0.015
Second Coat		
Minimum	L.F.	0.006
Average	"	0.007
Maximum	"	0.008
Spray		
First Coat		
Minimum	L.F.	0.004
Average	"	0.006
Maximum	"	0.008
Second Coat		
Minimum	L.F.	0.004
Average	"	0.004
Maximum	"	0.004
Doors, Wood		
Brush		
First Coat		
Minimum	S.F.	0.012
Average	"	0.016
Maximum	"	0.020
Second Coat		
Minimum	S.F.	0.010
Average	"	0.011
Maximum	"	0.013
Roller		
First Coat		
Minimum	S.F.	0.005
Average	"	0.007
Maximum	"	0.010
Second Coat		
Minimum	S.F.	0.004
Average	"	0.004
Maximum	"	0.007
Spray		
First Coat		
Minimum	S.F.	0.003
Average	"	0.003
Maximum	"	0.004
Second Coat		
Minimum	S.F.	0.002
Average	"	0.002
Maximum	"	0.003
Gutters and Downspouts		
Brush		
First Coat		
Minimum	L.F.	0.010
Average	"	0.011
Maximum	"	0.013
Second Coat		
Minimum	L.F.	0.007
Average	"	0.008
Maximum	"	0.010

Painting

09910.15 — Ext. Painting, Buildings (Cont.)	UNIT	MAN/HOURS
Siding, Metal		
Roller		
First Coat		
Minimum	S.F.	0.003
Average	"	0.004
Maximum	"	0.004
Second Coat		
Minimum	S.F.	0.003
Average	"	0.003
Maximum	"	0.004
Spray		
First Coat		
Minimum	S.F.	0.003
Average	"	0.003
Maximum	"	0.003
Second Coat		
Minimum	S.F.	0.002
Average	"	0.002
Maximum	"	0.003
Siding, Wood		
Roller		
First Coat		
Minimum	S.F.	0.003
Average	"	0.003
Maximum	"	0.004
Second Coat		
Minimum	S.F.	0.003
Average	"	0.004
Maximum	"	0.004
Spray		
First Coat		
Minimum	S.F.	0.003
Average	"	0.003
Maximum	"	0.003
Second Coat		
Minimum	S.F.	0.002
Average	"	0.003
Maximum	"	0.004
Stucco		
Roller		
First Coat		
Minimum	S.F.	0.004
Average	"	0.004
Maximum	"	0.005
Second Coat		
Minimum	S.F.	0.003
Average	"	0.003
Maximum	"	0.004
Spray		
First Coat		
Minimum	S.F.	0.003
Average	"	0.003
Maximum	"	0.003

Painting

09910.15   Ext. Painting, Buildings *(Cont.)*

	UNIT	MAN/HOURS
Second Coat		
Minimum	S.F.	0.002
Average	"	0.002
Maximum	"	0.003
Trim		
Brush		
First Coat		
Minimum	L.F.	0.003
Average	"	0.004
Maximum	"	0.005
Second Coat		
Minimum	L.F.	0.003
Average	"	0.003
Maximum	"	0.005
Walls		
Roller		
First Coat		
Minimum	S.F.	0.003
Average	"	0.003
Maximum	"	0.003
Second Coat		
Minimum	S.F.	0.003
Average	"	0.003
Maximum	"	0.003
Spray		
First Coat		
Minimum	S.F.	0.001
Average	"	0.002
Maximum	"	0.002
Second Coat		
Minimum	S.F.	0.001
Average	"	0.001
Maximum	"	0.002
Windows		
Brush		
First Coat		
Minimum	S.F.	0.013
Average	"	0.016
Maximum	"	0.020
Second Coat		
Minimum	S.F.	0.011
Average	"	0.013
Maximum	"	0.016

09910.25   Ext. Painting, Misc.

	UNIT	MAN/HOURS
Gratings, Metal		
Roller		
First Coat		
Minimum	S.F.	0.023
Average	"	0.027
Maximum	"	0.032
Second Coat		
Minimum	S.F.	0.016

09910.25   Ext. Painting, Misc. *(Cont.)*

	UNIT	MAN/HOURS
Average	S.F.	0.020
Maximum	"	0.027
Spray		
First Coat		
Minimum	S.F.	0.011
Average	"	0.013
Maximum	"	0.016
Second Coat		
Minimum	S.F.	0.009
Average	"	0.010
Maximum	"	0.011
Ladders		
Brush		
First Coat		
Minimum	L.F.	0.020
Average	"	0.023
Maximum	"	0.027
Second Coat		
Minimum	L.F.	0.016
Average	"	0.018
Maximum	"	0.020
Spray		
First Coat		
Minimum	L.F.	0.013
Average	"	0.015
Maximum	"	0.016
Second Coat		
Minimum	L.F.	0.011
Average	"	0.012
Maximum	"	0.013
Shakes		
Spray		
First Coat		
Minimum	S.F.	0.003
Average	"	0.004
Maximum	"	0.004
Second Coat		
Minimum	S.F.	0.003
Average	"	0.003
Maximum	"	0.004
Shingles, Wood		
Roller		
First Coat		
Minimum	S.F.	0.004
Average	"	0.005
Maximum	"	0.006
Second Coat		
Minimum	S.F.	0.003
Average	"	0.003
Maximum	"	0.004
Spray		
First Coat		
Minimum	L.F.	0.003

09910.25 — Ext. Painting, Misc. *(Cont.)*

Painting	UNIT	MAN/HOURS
Average	L.F.	0.003
Maximum	"	0.004
Second Coat		
Minimum	L.F.	0.002
Average	"	0.003
Maximum	"	0.003
Shutters and Louvres		
Brush		
First Coat		
Minimum	EA.	0.160
Average	"	0.200
Maximum	"	0.267
Second Coat		
Minimum	EA.	0.100
Average	"	0.123
Maximum	"	0.160
Spray		
First Coat		
Minimum	EA.	0.053
Average	"	0.064
Maximum	"	0.080
Second Coat		
Minimum	EA.	0.040
Average	"	0.053
Maximum	"	0.064
Stairs, metal		
Brush		
First Coat		
Minimum	S.F.	0.009
Average	"	0.010
Maximum	"	0.011
Second Coat		
Minimum	S.F.	0.005
Average	"	0.006
Maximum	"	0.007
Spray		
First Coat		
Minimum	S.F.	0.004
Average	"	0.006
Maximum	"	0.006
Second Coat		
Minimum	S.F.	0.003
Average	"	0.004
Maximum	"	0.005
Steel, Structural, Light		
Brush		
First Coat		
Minimum	S.F.	0.016
Average	"	0.020
Maximum	"	0.027
Second Coat		
Minimum	S.F.	0.011
Average	"	0.013

09910.25 — Ext. Painting, Misc. *(Cont.)*

Painting	UNIT	MAN/HOURS
Maximum	S.F.	0.016
Roller		
First Coat		
Minimum	S.F.	0.011
Average	"	0.013
Maximum	"	0.016
Second Coat		
Minimum	S.F.	0.007
Average	"	0.008
Maximum	"	0.010
Spray		
First Coat		
Minimum	S.F.	0.007
Average	"	0.008
Maximum	"	0.010
Second Coat		
Minimum	S.F.	0.006
Average	"	0.007
Maximum	"	0.008
Steel, Medium to Heavy		
Brush		
First Coat		
Minimum	S.F.	0.008
Average	"	0.009
Maximum	"	0.010
Second Coat		
Minimum	S.F.	0.007
Average	"	0.008
Maximum	"	0.009
Roller		
First Coat		
Minimum	S.F.	0.007
Average	"	0.008
Maximum	"	0.009
Second Coat		
Minimum	S.F.	0.005
Average	"	0.006
Maximum	"	0.007
Spray		
First Coat		
Minimum	S.F.	0.004
Average	"	0.005
Maximum	"	0.006
Second Coat		
Minimum	S.F.	0.004
Average	"	0.004
Maximum	"	0.004

Painting	UNIT	MAN/HOURS
09910.35 **Int. Painting, Buildings**		
Acoustical Ceiling		
Roller		
First Coat		
Minimum	S.F.	0.005
Average	"	0.007
Maximum	"	0.010
Second Coat		
Minimum	S.F.	0.004
Average	"	0.005
Maximum	"	0.007
Spray		
First Coat		
Minimum	S.F.	0.002
Average	"	0.003
Maximum	"	0.003
Second Coat		
Minimum	S.F.	0.002
Average	"	0.002
Maximum	"	0.002
Cabinets and Casework		
Brush		
First Coat		
Minimum	S.F.	0.008
Average	"	0.009
Maximum	"	0.010
Second Coat		
Minimum	S.F.	0.007
Average	"	0.007
Maximum	"	0.008
Spray		
First Coat		
Minimum	S.F.	0.004
Average	"	0.005
Maximum	"	0.006
Second Coat		
Minimum	S.F.	0.003
Average	"	0.003
Maximum	"	0.004
Ceilings		
Roller		
First Coat		
Minimum	S.F.	0.003
Average	"	0.004
Maximum	"	0.004
Second Coat		
Minimum	S.F.	0.003
Average	"	0.003
Maximum	"	0.003
Spray		
First Coat		
Minimum	S.F.	0.002
Average	"	0.002
Maximum	"	0.003

Painting	UNIT	MAN/HOURS
09910.35 **Int. Painting, Buildings** *(Cont.)*		
Second Coat		
Minimum	S.F.	0.002
Average	"	0.002
Maximum	"	0.002
Doors, Metal		
Roller		
First Coat		
Minimum	L.F.	0.005
Average	"	0.006
Maximum	"	0.007
Second Coat		
Minimum	L.F.	0.004
Average	"	0.004
Maximum	"	0.005
Spray		
First Coat		
Minimum	L.F.	0.004
Average	"	0.005
Maximum	"	0.006
Second Coat		
Minimum	L.F.	0.003
Average	"	0.004
Maximum	"	0.004
Doors, Wood		
Brush		
First Coat		
Minimum	S.F.	0.011
Average	"	0.015
Maximum	"	0.018
Second Coat		
Minimum	S.F.	0.009
Average	"	0.010
Maximum	"	0.011
Spray		
First Coat		
Minimum	S.F.	0.002
Average	"	0.003
Maximum	"	0.004
Second Coat		
Minimum	S.F.	0.002
Average	"	0.002
Maximum	"	0.003
Ductwork		
Brush		
Minimum	L.F.	0.010
Average	"	0.011
Maximum	"	0.013
Roller		
Minimum	S.F.	0.007
Average	"	0.007
Maximum	"	0.008
Spray		
Minimum	S.F.	0.003

Painting	UNIT	MAN/HOURS
09910.35 **Int. Painting, Buildings** (Cont.)		
Average	S.F.	0.003
Maximum	"	0.003
Floors		
Roller		
First Coat		
Minimum	S.F.	0.003
Average	"	0.003
Maximum	"	0.003
Second Coat		
Minimum	S.F.	0.002
Average	"	0.002
Maximum	"	0.002
Spray		
First Coat		
Minimum	S.F.	0.002
Average	"	0.002
Maximum	"	0.002
Second Coat		
Minimum	S.F.	0.002
Average	"	0.002
Maximum	"	0.002
Pipes to 6" diameter		
Brush		
Minimum	L.F.	0.010
Average	"	0.011
Maximum	"	0.013
Spray		
Minimum	L.F.	0.003
Average	"	0.004
Maximum	"	0.005
Pipes to 12" diameter		
Brush		
Minimum	L.F.	0.020
Average	"	0.023
Maximum	"	0.027
Spray		
Minimum	L.F.	0.007
Average	"	0.008
Maximum	"	0.010
Trim		
Brush		
First Coat		
Minimum	L.F.	0.003
Average	"	0.004
Maximum	"	0.004
Second Coat		
Minimum	L.F.	0.002
Average	"	0.003
Maximum	"	0.004
Walls		
Roller		
First Coat		
Minimum	S.F.	0.003

Painting	UNIT	MAN/HOURS
09910.35 **Int. Painting, Buildings** (Cont.)		
Average	S.F.	0.003
Maximum	"	0.003
Second Coat		
Minimum	S.F.	0.003
Average	"	0.003
Maximum	"	0.003
Spray		
First Coat		
Minimum	S.F.	0.001
Average	"	0.002
Maximum	"	0.002
Second Coat		
Minimum	S.F.	0.001
Average	"	0.001
Maximum	"	0.002
09955.10 **Wall Covering**		
Vinyl wall covering		
Medium duty	S.F.	0.011
Heavy duty	"	0.013
Over pipes and irregular shapes		
Lightweight, 13 oz.	S.F.	0.016
Medium weight, 25 oz.	"	0.018
Heavy weight, 34 oz.	"	0.020
Cork wall covering		
1' x 1' squares		
1/4" thick	S.F.	0.020
1/2" thick	"	0.020
3/4" thick	"	0.020
Wall fabrics		
Natural fabrics, grass cloths		
Minimum	S.F.	0.012
Average	"	0.013
Maximum	"	0.016
Flexible gypsum coated wall fabric, fire resistant	"	0.008
Vinyl corner guards		
3/4" x 3/4" x 8'	EA.	0.100
2-3/4" x 2-3/4" x 4'	"	0.100

Specialties	UNIT	MAN/HOURS
10110.10 — Chalkboards		
Chalkboard, metal frame, 1/4" thick		
48"x60"	EA.	0.800
48"x96"	"	0.889
48"x144"	"	1.000
48"x192"	"	1.143
Liquid chalkboard		
48"x60"	EA.	0.800
48"x96"	"	0.889
48"x144"	"	1.000
48"x192"	"	1.143
Map rail, deluxe	L.F.	0.040
10165.10 — Toilet Partitions		
Toilet partition, plastic laminate		
Ceiling mounted	EA.	2.667
Floor mounted	"	2.000
Metal		
Ceiling mounted	EA.	2.667
Floor mounted	"	2.000
Wheel chair partition, plastic laminate		
Ceiling mounted	EA.	2.667
Floor mounted	"	2.000
Painted metal		
Ceiling mounted	EA.	2.667
Floor mounted	"	2.000
Urinal screen, plastic laminate		
Wall hung	EA.	1.000
Floor mounted	"	1.000
Porcelain enameled steel, floor mounted	"	1.000
Painted metal, floor mounted	"	1.000
Stainless steel, floor mounted	"	1.000
Metal toilet partitions		
Front door and side divider, floor mounted		
Porcelain enameled steel	EA.	2.000
Painted steel	"	2.000
Stainless steel	"	2.000
10185.10 — Shower Stalls		
Shower receptors		
Precast, terrazzo		
32" x 32"	EA.	0.667
32" x 48"	"	0.800
Concrete		
32" x 32"	EA.	0.667
48" x 48"	"	0.889
Shower door, trim and hardware		
Economy, 24" wide, chrome frame, tempered glass	EA.	0.800
Porcelain enameled steel, flush	"	0.800
Baked enameled steel, flush	"	0.800
Aluminum frame, tempered glass, 48" wide, sliding	"	1.000
Folding	"	1.000
Aluminum frame and tempered glass, molded plastic		
Complete with receptor and door		

Specialties	UNIT	MAN/HOURS
10185.10 — Shower Stalls (Cont.)		
32" x 32"	EA.	2.000
36" x 36"	"	2.000
40" x 40"	"	2.286
Shower compartment, precast concrete receptor		
Single entry type		
Porcelain enameled steel	EA.	8.000
Baked enameled steel	"	8.000
Stainless steel	"	8.000
Double entry type		
Porcelain enameled steel	EA.	10.000
Baked enameled steel	"	10.000
Stainless steel	"	10.000
10190.10 — Cubicles		
Hospital track		
Ceiling hung	L.F.	0.089
Suspended	"	0.114
Hospital metal dividers, galvanized steel		
Baked enamel finish		
54" high		
10" glass light	L.F.	0.400
14" glass light	"	0.400
24" glass light	"	0.400
60" high		
10" glass light	L.F.	0.444
14" glass light	"	0.444
24" glass light	"	0.444
Stainless steel		
54" high		
10" glass light	L.F.	0.444
14" glass light	"	0.444
24" glass light	"	0.444
60" high		
14" glass light	L.F.	0.500
14" glass light	"	0.500
24" glass light	"	0.500
10210.10 — Vents And Wall Louvres		
Block vent, 8"x16"x4" aluminum, w/screen, mill finish	EA.	0.267
Standard	"	0.250
Vents w/screen, 4" deep, 8" wide, 5" high		
Modular	EA.	0.250
Aluminum gable louvers	S.F.	0.133
Vent screen aluminum, 4" wide, continuous	L.F.	0.027
Louvers, aluminum, anodized, fixed blade		
Horizontal line	S.F.	0.200
Vertical line	"	0.200
Wall louvre, aluminum mill finish		
Under, 2 sf	S.F.	0.100

Specialties		UNIT	MAN/HOURS
10210.10	**Vents And Wall Louvres** *(Cont.)*		
2 to 4 sf		S.F.	0.089
5 to 10 sf		"	0.089
Galvanized steel			
Under 2 sf		S.F.	0.100
2 to 4 sf		"	0.089
5 to 10 sf		"	0.089
10225.10	**Door Louvres**		
Fixed, 1" thick, enameled steel			
8"x8"		EA.	0.100
12"x8"		"	0.100
12"x12"		"	0.114
16"x12"		"	0.123
18"x12"		"	0.200
20"x8"		"	0.114
20"x12"		"	0.229
20"x16"		"	0.267
20"x20"		"	0.320
24"x12"		"	0.267
24"x16"		"	0.286
24"x18"		"	0.308
24"x20"		"	0.333
24"x24"		"	0.364
26"x26"		"	0.500
10270.40	**Access & Pedestal Floor**		
Panels, no covering, 2'x2'			
Plain		S.F.	0.010
Perforated		"	0.400
Pedestals			
For 6" to 12" clearance		EA.	0.080
Stringers			
2'		L.F.	0.038
6'		"	0.027
Accessories			
Ramp assembly		S.F.	0.032
Elevated floor assembly		"	0.030
Handrail		L.F.	0.400
Fascia plate		"	0.200
RF shielding components, floor liner			
Hot rolled steel sheet			
14 ga.		S.F.	0.020
11 ga.		"	0.062
10290.10	**Pest Control**		
Termite control			
Under slab spraying			
Minimum		S.F.	0.002
Average		"	0.004
Maximum		"	0.008

Specialties		UNIT	MAN/HOURS
10350.10	**Flagpoles**		
Installed in concrete base			
Fiberglass			
25' high		EA.	5.333
50' high		"	13.333
Aluminum			
25' high		EA.	5.333
50' high		"	13.333
Bonderized steel			
25' high		EA.	6.154
50' high		"	16.000
Freestanding tapered, fiberglass			
30' high		EA.	5.714
40' high		"	7.273
50' high		"	8.000
60' high		"	9.412
Wall mounted, with collar, brushed aluminum finish			
15' long		EA.	4.000
18' long		"	4.000
20' long		"	4.211
24' long		"	4.706
Outrigger, wall, including base			
10' long		EA.	5.333
20' long		"	6.667
10400.10	**Identifying Devices**		
Directory and bulletin boards			
Open face boards			
Chrome plated steel frame		S.F.	0.400
Aluminum framed		"	0.400
Bronze framed		"	0.400
Stainless steel framed		"	0.400
Tack board, aluminum framed		"	0.400
Visual aid board, aluminum framed		"	0.400
Glass encased boards, hinged and keyed			
Aluminum framed		S.F.	1.000
Bronze framed		"	1.000
Stainless steel framed		"	1.000
Chrome plated steel framed		"	1.000
Metal plaque			
Cast bronze		S.F.	0.667
Aluminum		"	0.667
Metal engraved plaque			
Porcelain steel		S.F.	0.667
Stainless steel		"	0.667
Brass		"	0.667
Aluminum		"	0.667
Metal built-up plaque			
Bronze		S.F.	0.800
Copper and bronze		"	0.800
Copper and aluminum		"	0.800
Metal nameplate plaques			
Cast bronze		S.F.	0.500
Cast aluminum		"	0.500

Specialties

10400.10 — Identifying Devices *(Cont.)*

Specialties	UNIT	MAN/HOURS
Engraved, 1-1/2" x 6"		
Bronze	EA.	0.500
Aluminum	"	0.500
Letters, on masonry or concrete, aluminum, satin finish		
1/2" thick		
2" high	EA.	0.320
4" high	"	0.400
6" high	"	0.444
3/4" thick		
8" high	EA.	0.500
10" high	"	0.571
1" thick		
12" high	EA.	0.667
14" high	"	0.800
16" high	"	1.000
3/8" thick		
2" high	EA.	0.320
4" high	"	0.400
1/2" thick, 6" high	"	0.444
5/8" thick, 8" high	"	0.500
1" thick		
10" high	EA.	0.571
12" high	"	0.667
14" high	"	0.800
16" high	"	1.000
Interior door signs, adhesive, flexible		
2" x 8"	EA.	0.200
4" x 4"	"	0.200
6" x 7"	"	0.200
6" x 9"	"	0.200
10" x 9"	"	0.200
10" x 12"	"	0.200
Hard plastic type, no frame		
3" x 8"	EA.	0.200
4" x 4"	"	0.200
4" x 12"	"	0.200
Hard plastic type, with frame		
3" x 8"	EA.	0.200
4" x 4"	"	0.200
4" x 12"	"	0.200

10450.10 — Control

Specialties	UNIT	MAN/HOURS
Access control, 7' high, indoor or outdoor impenetrability		
Remote or card control, type B	EA.	10.667
Free passage, type B	"	10.667
Remote or card control, type AA	"	10.667
Free passage, type AA	"	10.667

10500.10 — Lockers

Specialties	UNIT	MAN/HOURS
Locker bench, floor mounted, laminated maple		
4'	EA.	0.667
6'	"	0.667
Wardrobe locker, single tier type		
12" x 15" x 72"	EA.	0.800
18" x 15" x 72"	"	0.842
12" x 18" x 72"	"	0.889
18" x 18" x 72"	"	0.941
Double tier type		
12" x 15" x 36"	EA.	0.400
18" x 15" x 36"	"	0.400
12" x 18" x 36"	"	0.400
18" x 18" x 36"	"	0.400
Two person unit		
18" x 15" x 72"	EA.	1.333
18" x 18" x 72"	"	1.600
Duplex unit		
15" x 15" x 72"	EA.	0.800
15" x 21" x 72"	"	0.800
Basket lockers, basket sets with baskets		
24 basket set	SET	4.000
30 basket set	"	5.000
36 basket set	"	6.667
42 basket set	"	8.000

10520.10 — Fire Protection

Specialties	UNIT	MAN/HOURS
Portable fire extinguishers		
Water pump tank type		
2.5 gal.		
Red enameled galvanized	EA.	0.533
Red enameled copper	"	0.533
Polished copper	"	0.533
Carbon dioxide type, red enamel steel		
Squeeze grip with hose and horn		
2.5 lb	EA.	0.533
5 lb	"	0.615
10 lb	"	0.800
15 lb	"	1.000
20 lb	"	1.000
Wheeled type		
125 lb	EA.	1.600
250 lb	"	1.600
500 lb	"	1.600
Dry chemical, pressurized type		
Red enameled steel		
2.5 lb	EA.	0.533
5 lb	"	0.615
10 lb	"	0.800
20 lb	"	1.000
30 lb	"	1.000
Chrome plated steel, 2.5 lb	"	0.533
Other type extinguishers		
2.5 gal, stainless steel, pressurized water tanks	EA.	0.533

Left Column

Specialties	UNIT	MAN/HOURS
10520.10 **Fire Protection** *(Cont.)*		
Soda and acid type	EA.	0.533
Cartridge operated, water type	"	0.533
Loaded stream, water type	"	0.533
Foam type	"	0.533
40 gal, wheeled foam type	"	1.600
Fire extinguisher cabinets		
Enameled steel		
8" x 12" x 27"	EA.	1.600
8" x 16" x 38"	"	1.600
Aluminum		
8" x 12" x 27"	EA.	1.600
8" x 16" x 38"	"	1.600
8" x 12" x 27"	"	1.600
Stainless steel		
8" x 16" x 38"	EA.	1.600
10550.10 **Postal Specialties**		
Mail chutes		
Single mail chute		
Finished aluminum	L.F.	2.000
Bronze	"	2.000
Single mail chute receiving box		
Finished aluminum	EA.	4.000
Bronze	"	4.000
Twin mail chute, double parallel		
Finished aluminum	FLR	4.000
Bronze	"	4.000
Receiving box, 36" x 20" x 12"		
Finished aluminum	EA.	6.667
Bronze	"	6.667
Locked receiving mail box		
Finished aluminum	EA.	4.000
Bronze	"	4.000
Commercial postal accessories for mail chutes		
Letter slot, brass	EA.	1.333
Bulk mail slot, brass	"	1.333
Mail boxes		
Residential postal accessories		
Letter slot	EA.	0.400
Rural letter box	"	1.000
Apartment house, keyed, 3.5" x 4.5" x 16"	"	0.267
Ranch style	"	0.400
Commercial postal accessories		
Letter box, with combination lock	EA.	0.286
Key lock	"	0.286
Mail box, aluminum w/glass front, 4x5		
Horizontal rear load	EA.	0.229
Vertical front load	"	0.229

Right Column

Specialties	UNIT	MAN/HOURS
10600.10 **Movable Partitions**		
Partition, movable office type, 2-1/2" thick, vinyl-gypsum	S.F.	0.040
Enameled steel frame, with 1/4" thick clear glass	"	0.040
Door frame and hardware for movable partitions	EA.	2.667
Add for acoustic movable partition	S.F.	
Accordion partition, 12' high		
Vinyl	S.F.	0.133
Acoustical	"	0.133
Standard office cubicles, 8' high, steel framed		
Baked enamel finish		
100% flush	L.F.	0.200
75% flush and 25% glass	"	0.222
50% flush and 50% glass	"	0.222
100% glass	"	0.267
Natural hardwood panels		
100% flush	L.F.	0.267
50% flush and 50% glass	"	0.286
Plastic laminated panels		
100% flush	L.F.	0.267
75% flush and 25% glass	"	0.286
50% flush and 50% glass	"	0.286
Vinyl covered panels		
100% flush	L.F.	0.276
75% flush and 25% glass	"	0.296
50% and 50% glass	"	0.296
Aluminum framed		
Enameled or anodized aluminum panels		
100% flush	L.F.	0.200
75% flush and 25% glass	"	0.222
50% flush and 50% glass	"	0.222
Vinyl covered panels		
100% flush	L.F.	0.276
75% flush and 25% glass	"	0.296
50% flush and 50% glass	"	0.296
60" high partitions, steel framed		
Enameled panels	L.F.	0.178
Natural hardwood panels, two sides	"	0.186
Plastic laminated panels	"	0.186
Vinyl covered panels	"	0.178
Aluminum framed		
Anodized or baked enamel panels	L.F.	0.178
Natural hardwood panels	"	0.186
Plastic laminated panels	"	0.186
Vinyl covered panels	"	0.178
Wire mesh partitions		
Wall panels		
4' x 7'	EA.	0.500
4' x 8'	"	0.533
4' x 10'	"	0.615
Wall filler panels		
1' x 7'	EA.	0.500
1' x 8'	"	0.533
1' x 10'	"	0.571
2' x 7'	"	0.500

Left column

Specialties	UNIT	MAN/HOURS
10600.10 — Movable Partitions *(Cont.)*		
2' x 8'	EA.	0.533
2' x 10'	"	0.571
3' x 7'	"	0.500
3' x 8'	"	0.533
3' x 10'	"	0.571
Ceiling panels		
10' x 2'	EA.	1.143
10' x 4'	"	1.600
Wall panel with service window		
5' wide		
7' high	EA.	0.500
8' high	"	0.533
10' high	"	0.571
Doors		
Sliding		
3' x 7'	EA.	2.000
3' x 8'	"	2.286
3' x 10'	"	3.200
4' x 7'	"	2.286
4' x 8'	"	3.200
4' x 10'	"	4.000
5' x 7'	"	3.200
5' x 8'	"	4.000
5' x 10'	"	4.000
Swing door		
3' x 7'	EA.	2.000
4' x 7'	"	2.286
Swing door, with 1' transom		
3' x 7'	EA.	2.286
4' x 7'	"	3.200
Swing door, with 3' transom		
3' x 7'	EA.	3.200
4' x 7'	"	4.000
10670.10 — Shelving		
Shelving, enamel, closed side and back, 12" x 36"		
5 shelves	EA.	1.333
8 shelves	"	1.778
Open		
5 shelves	EA.	1.333
8 shelves	"	1.778
Metal storage shelving, baked enamel		
7 shelf unit, 72" or 84" high		
10" shelf	L.F.	0.800
12" shelf	"	0.842
15" shelf	"	0.889
18" shelf	"	0.941
24" shelf	"	1.000
30" shelf	"	1.067
36" shelf	"	1.143
4 shelf unit, 40" high		
10" shelf	L.F.	0.667
12" shelf	"	0.727

Right column

Specialties	UNIT	MAN/HOURS
10670.10 — Shelving *(Cont.)*		
15" shelf	L.F.	0.800
18" shelf	"	0.842
24" shelf	"	0.889
3 shelf unit, 32" high		
10" shelf	L.F.	0.400
12" shelf	"	0.421
15" shelf	"	0.444
18" shelf	"	0.471
24" shelf	"	0.500
Single shelf unit, attached to masonry		
10" shelf	L.F.	0.133
12" shelf	"	0.145
15" shelf	"	0.154
18" shelf	"	0.163
24" shelf	"	0.174
Built-in wood shelves		
Posts and trimmed plywood	L.F.	0.114
Solid clear pine	"	0.123
Closet shelf, pine with rod	"	0.123
10750.10 — Telephone Enclosures		
Telephone enclosure, wall mounted, shelf, 28" x 30" x 15"	EA.	2.000
Directory shelf, stainless steel, 3 binders	"	1.333
10800.10 — Bath Accessories		
Ash receiver, wall mounted, aluminum	EA.	0.400
Grab bar, 1-1/2" dia., stainless steel, wall mounted		
24" long	EA.	0.400
36" long	"	0.421
42" long	"	0.444
48" long	"	0.471
52" long	"	0.500
1" dia., stainless steel		
12" long	EA.	0.348
18" long	"	0.364
24" long	"	0.400
30" long	"	0.421
36" long	"	0.444
48" long	"	0.471
Hand dryer, surface mounted, 110 volt	"	1.000
Medicine cabinet, 16 x 22, baked enamel, steel, lighted	"	0.320
With mirror, lighted	"	0.533
Mirror, 1/4" plate glass, up to 10 sf	S.F.	0.080
Mirror, stainless steel frame		
18"x24"	EA.	0.267
18"x32"	"	0.320
18"x36"	"	0.400
24"x30"	"	0.400
24"x36"	"	0.444
24"x48"	"	0.667
24"x60"	"	0.800
30"x30"	"	0.800
30"x72"	"	1.000

Specialties	UNIT	MAN/HOURS
10800.10 **Bath Accessories** *(Cont.)*		
48"x72"	EA.	1.333
With shelf, 18"x24"	"	0.320
Sanitary napkin dispenser, stainless steel, wall mounted	"	0.533
Shower rod, 1" diameter		
Chrome finish over brass	EA.	0.400
Stainless steel	"	0.400
Soap dish, stainless steel, wall mounted	"	0.533
Toilet tissue dispenser, stainless, wall mounted		
Single roll	EA.	0.200
Double roll	"	0.229
Towel dispenser, stainless steel		
Flush mounted	EA.	0.444
Surface mounted	"	0.400
Combination towel dispenser and waste receptacle	"	0.533
Towel bar, stainless steel		
18" long	EA.	0.320
24" long	"	0.364
30" long	"	0.400
36" long	"	0.444
Toothbrush and tumbler holder	"	0.267
Waste receptacle, stainless steel, wall mounted	"	0.667
10900.10 **Wardrobe Specialties**		
Hospital wardrobe units, 24" x 24" x 76", with door		
Baked enameled steel	EA.	4.444
Hardwood	"	4.444
Stainless steel	"	4.444
Plastic laminated	"	4.444
Dormitory wardrobe units, 24" x 76", with door		
Hardwood	EA.	4.444
Plastic laminated	"	4.444
Hat and coat rack		
Single tier		
Baked enameled steel	L.F.	0.200
Stainless steel	"	0.200
Aluminum	"	0.200
Double tier		
Baked enameled steel	L.F.	0.229
Stainless steel	"	0.229
Aluminum	"	0.229

Architectural Equipment

11010.10 — Maintenance Equipment

	UNIT	MAN/HOURS
Vacuum cleaning system		
3 valves		
1.5 hp	EA.	8.889
2.5 hp	"	11.429
5 valves	"	16.000
7 valves	"	20.000

11020.10 — Security Equipment

	UNIT	MAN/HOURS
Bulletproof teller window		
4' x 4'	EA.	13.333
5' x 4'	"	16.000
Bulletproof partitions		
Up to 12' high, 2.5" thick	S.F.	0.053
Counter for banks		
Minimum	L.F.	1.600
Maximum	"	2.667
Drive-up window		
Minimum	EA.	11.429
Maximum	"	26.667
Night depository		
Minimum	EA.	11.429
Maximum	"	26.667
Office safes, 30" x 20" x 20", 1 hr rating	"	2.000
30" x 16" x 15", 2 hr rating	"	1.600
30" x 28" x 20", H&G rating	"	1.000
Service windows, pass through painted steel		
24" x 36"	EA.	8.000
48" x 40"	"	10.000
72" x 40"	"	16.000
Special doors and windows		
3' x 7' bulletproof door with frame	EA.	11.429
12" x 12" vision panel	"	5.714
Surveillance system		
Minimum	EA.	16.000
Maximum	"	80.000
Vault door, 3' wide, 6'6" high		
3-1/2" thick	EA.	100.000
7" thick	"	133.333
10" thick	"	160.000
Insulated vault door		
2 hr rating		
32" wide	EA.	8.000
40" wide	"	8.421
4 hr rating		
32" wide	EA.	8.889
40" wide	"	10.000
6 hr rating		
32" wide	EA.	8.889
40" wide	"	10.000
Insulated file room door		
1 hr rating		
32" wide	EA.	8.000
40" wide	"	8.889

Architectural Equipment

11060.10 — Theater Equipment

	UNIT	MAN/HOURS
Roll out stage, steel frame, wood floor		
Manual	S.F.	0.050
Electric	"	0.080
Portable stages		
8" high	S.F.	0.040
18" high	"	0.044
36" high	"	0.047
48" high	"	0.050
Band risers		
Minimum	S.F.	0.040
Maximum	"	0.040
Chairs for risers		
Minimum	EA.	0.036
Maximum	"	0.036

11080.10 — Police Equipment

	UNIT	MAN/HOURS
Firing range equipment, rifle		
3 position	EA.	26.667
4 position	"	40.000
5 position	"	44.444
6 position	"	47.059

11090.10 — Checkroom Equipment

	UNIT	MAN/HOURS
Motorized checkroom equipment		
No shelf system, 6'4" height		
7'6" length	EA.	8.000
14'6" length	"	8.000
28' length	"	8.000
One shelf, 6'8" height		
7'6" length	EA.	8.000
14'6" length	"	8.000
28' length	"	8.000
Two shelves, 7'5" height		
7'6" length	EA.	8.000
14'6" length	"	8.000
28' length	"	8.000
Three shelves, 8' height		
7'6" length	EA.	16.000
14'6" length	"	16.000
28' length	"	16.000
Four shelves, 8'7" height		
7'6" length	EA.	16.000
14'6" length	"	16.000
28' length	"	16.000

11110.10 — Laundry Equipment

	UNIT	MAN/HOURS
High capacity, heavy duty		
Washer extractors		
135 lb		
Standard	EA.	6.667
Pass through	"	6.667
200 lb		
Standard	EA.	6.667

Architectural Equipment	UNIT	MAN/HOURS
11110.10 — Laundry Equipment *(Cont.)*		
Pass through	EA.	6.667
110 lb dryer	"	6.667
Hand operated presser	"	8.889
Mushroom press	"	8.889
Spreader feeders		
2 station	EA.	8.889
4 station	"	16.000
Delivery carts		
12 bushel	EA.	0.100
16 bushel	"	0.107
18 bushel	"	0.114
30 bushel	"	0.133
40 bushel	"	0.160
Low capacity		
Pressers		
Air operated	EA.	3.200
Hand operated	"	3.200
Extractor, low capacity	"	3.200
Ironer, 48"	"	1.600
Coin washers		
10 lb capacity	EA.	1.600
20 lb capacity	"	1.600
Coin dryer	"	1.000
Coin dry cleaner, 20 lb	"	3.200
11161.10 — Loading Dock Equipment		
Dock leveler, 10 ton capacity		
6' x 8'	EA.	8.000
7' x 8'	"	8.000
Bumpers, laminated rubber		
4-1/2" thick		
6" x 14"	EA.	0.160
6" x 36"	"	0.178
10" x 14"	"	0.200
10" x 24"	"	0.229
10" x 36"	"	0.267
12" x 14"	"	0.211
12" x 24"	"	0.250
12" x 36"	"	0.296
6" thick		
10" x 14"	EA.	0.229
10" x 24"	"	0.276
10" x 36"	"	0.400
Extruded rubber bumpers		
T-section, 22" x 22" x 3"	EA.	0.160
Molded rubber bumpers		
24" x 12" x 3" thick	EA.	0.400
Door seal, 12" x 12", vinyl covered	L.F.	0.200
Dock boards, heavy duty, 5' x 5'		
5000 lb		
Minimum	EA.	6.667
Maximum	"	6.667
9000 lb		

Architectural Equipment	UNIT	MAN/HOURS
11161.10 — Loading Dock Equipment *(Cont.)*		
Minimum	EA.	6.667
Maximum	"	7.273
15,000 lb	"	7.273
Truck shelters		
Minimum	EA.	6.154
Maximum	"	11.429
11170.10 — Waste Handling		
Incinerator, electric		
100 lb/hr		
Minimum	EA.	8.000
Maximum	"	8.000
400 lb/hr		
Minimum	EA.	16.000
Maximum	"	16.000
1000 lb/hr		
Minimum	EA.	24.242
Maximum	"	24.242
Incinerator, medical-waste		
25 lb/hr, 2-7 x 4-0	EA.	16.000
50 lb/hr, 2-11 x 4-11	"	16.000
75 lb/hr, 3-8 x 5-0	"	32.000
100 lb/hr, 3-8 x 6-0	"	32.000
Industrial compactor		
1 cy	EA.	8.889
3 cy	"	11.429
5 cy	"	16.000
Trash chutes steel, including sprinklers		
18" dia.	L.F.	4.000
24" dia.	"	4.211
30" dia.	"	4.444
36" dia.	"	4.706
Refuse bottom hopper	EA.	4.444
11400.10 — Food Service Equipment		
Unit kitchens		
30" compact kitchen		
Refrigerator, with range, sink	EA.	4.000
Sink only	"	2.667
Range only	"	2.000
Cabinet for upper wall section	"	1.143
Stainless shield, for rear wall	"	0.320
Side wall	"	0.320
42" compact kitchen		
Refrigerator with range, sink	EA.	4.444
Sink only	"	4.000
Cabinet for upper wall section	"	1.333
Stainless shield, for rear wall	"	0.333
Side wall	"	0.333
54" compact kitchen		
Refrigerator, oven, range, sink	EA.	5.714
Cabinet for upper wall section	"	1.600
Stainless shield, for		

Architectural Equipment	UNIT	MAN/ HOURS
11400.10 **Food Service Equipment** *(Cont.)*		
Rear wall	EA.	0.364
Side wall	"	0.364
60" compact kitchen		
Refrigerator, oven, range, sink	EA.	5.714
Cabinet for upper wall section	"	1.600
Stainless shield, for		
Rear wall	EA.	0.364
Side wall	"	0.364
72" compact kitchen		
Refrigerator, oven, range, sink	EA.	6.667
Cabinet for upper wall section	"	1.600
Stainless shield for		
Rear wall	EA.	0.400
Side wall	"	0.400
Bake oven		
Single deck		
Minimum	EA.	1.000
Maximum	"	2.000
Double deck		
Minimum	EA.	1.333
Maximum	"	2.000
Triple deck		
Minimum	EA.	1.333
Maximum	"	2.667
Convection type oven, electric, 40" x 45" x 57"		
Minimum	EA.	1.000
Maximum	"	2.000
Broiler, without oven, 69" x 26" x 39"		
Minimum	EA.	1.000
Maximum	"	1.333
Coffee urns, 10 gallons		
Minimum	EA.	2.667
Maximum	"	4.000
Fryer, with submerger		
Single		
Minimum	EA.	1.600
Maximum	"	2.667
Double		
Minimum	EA.	2.000
Maximum	"	2.667
Griddle, counter		
3' long		
Minimum	EA.	1.333
Maximum	"	1.600
5' long		
Minimum	EA.	2.000
Maximum	"	2.667
Kettles, steam, jacketed		
20 gallons		
Minimum	EA.	2.000
Maximum	"	4.000
40 gallons		
Minimum	EA.	2.000

Architectural Equipment	UNIT	MAN/ HOURS
11400.10 **Food Service Equipment** *(Cont.)*		
Maximum	EA.	4.000
60 gallons		
Minimum	EA.	2.000
Maximum	"	4.000
Range		
Heavy duty, single oven, open top		
Minimum	EA.	1.000
Maximum	"	2.667
Fry top		
Minimum	EA.	1.000
Maximum	"	2.667
Steamers, electric		
27 kw		
Minimum	EA.	2.000
Maximum	"	2.667
18 kw		
Minimum	EA.	2.000
Maximum	"	2.667
Dishwasher, rack type		
Single tank, 190 racks/hr	EA.	4.000
Double tank		
234 racks/hr	EA.	4.444
265 racks/hr	"	5.333
Dishwasher, automatic 100 meals/hr	"	2.667
Disposals		
100 gal/hr	EA.	2.667
120 gal/hr	"	2.759
250 gal/hr	"	2.857
Exhaust hood for dishwasher, gutter 4 sides, s-steel		
4'x4'x2'	EA.	2.963
4'x7'x2'	"	3.200
Food preparation machines		
Vertical cutter mixers		
25 quart	EA.	2.667
40 quart	"	2.667
80 quart	"	4.000
130 quart	"	6.667
Choppers		
5 lb	EA.	2.000
16 lb	"	2.667
40 lb	"	4.000
Mixers, floor models		
20 quart	EA.	1.000
60 quart	"	1.000
80 quart	"	1.143
140 quart	"	1.600
Ice cube maker		
50 lb per day		
Minimum	EA.	8.000
Maximum	"	8.000
500 lb per day		
Minimum	EA.	13.333
Maximum	"	13.333

Architectural Equipment

11400.10 Food Service Equipment (Cont.)

	UNIT	MAN/HOURS
Ice flakers		
300 lb per day	EA.	8.000
600 lb per day	"	13.333
1000 lb per day	"	17.778
2000 lb per day	"	20.000
Refrigerated cases		
Dairy products		
Multi deck type	L.F.	0.533
Delicatessen case, service deli		
Single deck	L.F.	4.000
Multi deck	"	5.000
Meat case		
Single deck	L.F.	4.706
Multi deck	"	5.000
Produce case		
Single deck	L.F.	4.706
Multi deck	"	5.000
Bottle coolers		
6' long		
Minimum	EA.	16.000
Maximum	"	16.000
10' long		
Minimum	EA.	26.667
Maximum	"	26.667
Frozen food cases		
Chest type	L.F.	4.706
Reach-in, glass door	"	5.000
Island case, single	"	4.706
Multi deck	"	5.000
Ice storage bins		
500 lb capacity	EA.	11.429
1000 lb capacity	"	22.857

11450.10 Residential Equipment

	UNIT	MAN/HOURS
Compactor, 4 to 1 compaction	EA.	2.000
Dishwasher, built-in		
2 cycles	EA.	4.000
4 or more cycles	"	4.000
Disposal		
Garbage disposer	EA.	2.667
Heaters, electric, built-in		
Ceiling type	EA.	2.667
Wall type		
Minimum	EA.	2.000
Maximum	"	2.667
Hood for range, 2-speed, vented		
30" wide	EA.	2.667
42" wide	"	2.667
Ice maker, automatic		
30 lb per day	EA.	1.143
50 lb per day	"	4.000
Folding access stairs, disappearing metal stair		
8' long	EA.	1.143

Architectural Equipment

11450.10 Residential Equipment (Cont.)

	UNIT	MAN/HOURS
11' long	EA.	1.143
12' long	"	1.143
Wood frame, wood stair		
22" x 54" x 8'9" long	EA.	0.800
25" x 54" x 10' long	"	0.800
Ranges electric		
Built-in, 30", 1 oven	EA.	2.667
2 oven	"	2.667
Counter top, 4 burner, standard	"	2.000
With grill	"	2.000
Free standing, 21", 1 oven	"	2.667
30", 1 oven	"	1.600
2 oven	"	1.600
Water softener		
30 grains per gallon	EA.	2.667
70 grains per gallon	"	4.000

11470.10 Darkroom Equipment

	UNIT	MAN/HOURS
Dryers		
36" x 25" x 68"	EA.	4.000
48" x 25" x 68"	"	4.000
Processors, film		
Black and white	EA.	4.000
Color negatives	"	4.000
Prints	"	4.000
Transparencies	"	4.000
Sinks with cabinet and/or stand		
5" sink with stand		
24" x 48"	EA.	2.000
32" x 64"	"	2.667
38" x 52"	"	2.667
42" x 132"	"	4.000
48" x 52"	"	4.000
5" sink with cabinet		
24" x 48"	EA.	2.000
32" x 64"	"	2.667
38" x 52"	"	2.667
42" x 132"	"	4.000
48" x 52"	"	4.000
10" sink with stand		
24" x 48"	EA.	2.000
32" x 64"	"	2.667
38" x 52"	"	2.667
10" sink with cabinet		
24" x 48"	EA.	2.000
38" x 52"	"	2.667

11480.10 Athletic Equipment

	UNIT	MAN/HOURS
Basketball backboard		
Fixed	EA.	10.000
Swing-up	"	16.000
Portable, hydraulic	"	4.000
Suspended type, standard	"	16.000

Architectural Equipment	UNIT	MAN/HOURS
11480.10 Athletic Equipment *(Cont.)*		
Bleacher, telescoping, manual		
15 tier, minimum	SEAT	0.160
Maximum	"	0.160
20 tier, minimum	"	0.178
Maximum	"	0.178
30 tier, minimum	"	0.267
Maximum	"	0.267
Boxing ring elevated, complete, 22' x 22'	EA.	114.286
Gym divider curtain		
Minimum	S.F.	0.011
Maximum	"	0.011
Scoreboards, single face		
Minimum	EA.	8.000
Maximum	"	40.000
Parallel bars		
Minimum	EA.	8.000
Maximum	"	13.333
11500.10 Industrial Equipment		
Vehicular paint spray booth, solid back, 14'4" x 9'6"		
24' deep	EA.	8.000
26'6" deep	"	8.000
28'6" deep	"	8.000
Drive through, 14'9" x 9'6"		
24' deep	EA.	8.000
26'6" deep	"	8.000
28'6" deep	"	8.000
Water wash, paint spray booth		
5' x 11'2" x 10'8"	EA.	8.000
6' x 11'2" x 10'8"	"	8.000
8' x 11'2" x 10'8"	"	8.000
10' x 11'2" x 11'2"	"	8.000
12' x 12'2" x 11'2"	"	8.000
14' x 12'2" x 11'2"	"	8.000
16' x 12'2" x 11'2"	"	8.000
20' x 12'2" x 11'2"	"	8.000
Dry type spray booth, with paint arrestors		
5'4" x 7'2" x 6'8"	EA.	8.000
6'4" x 7'2" x 6'8"	"	8.000
8'4" x 7'2" x 9'2"	"	8.000
10'4" x 7'2" x 9'2"	"	8.000
12'4" x 7'6" x 9'2"	"	8.000
14'4" x 7'6" x 9'8"	"	8.000
16'4" x 7'7" x 9'8"	"	8.000
20'4" x 7'7" x 10'8"	"	8.000
Air compressor, electric		
1 hp		
115 volt	EA.	5.333
7.5 hp		
115 volt	EA.	8.000
230 volt	"	8.000
Hydraulic lifts		
8,000 lb capacity	EA.	20.000

Architectural Equipment	UNIT	MAN/HOURS
11500.10 Industrial Equipment *(Cont.)*		
11,000 lb capacity	EA.	32.000
24,000 lb capacity	"	53.333
Power tools		
Band saws		
10"	EA.	0.667
14"	"	0.800
Motorized shaper	"	0.615
Motorized lathe	"	0.667
Bench saws		
9" saw	EA.	0.533
10" saw	"	0.571
12" saw	"	0.667
Electric grinders		
1/3 hp	EA.	0.320
1/2 hp	"	0.348
3/4 hp	"	0.348
11600.10 Laboratory Equipment		
Cabinets, base		
Minimum	L.F.	0.667
Maximum	"	0.667
Full storage, 7' high		
Minimum	L.F.	0.667
Maximum	"	0.667
Wall		
Minimum	L.F.	0.800
Maximum	"	0.800
Counter tops		
Minimum	S.F.	0.100
Average	"	0.114
Maximum	"	0.133
Tables		
Open underneath	S.F.	0.400
Doors underneath	"	0.500
Medical laboratory equipment		
Analyzer		
Chloride	EA.	0.400
Blood	"	0.667
Bath, water, utility, countertop unit	"	0.800
Hot plate, lab, countertop	"	0.727
Stirrer	"	0.727
Incubator, anaerobic, 23x23x36"	"	4.000
Dry heat bath	"	1.333
Incinerator, for sterilizing	"	0.080
Meter, serum protein	"	0.100
Ph analog, general purpose	"	0.114
Refrigerator, blood bank, undercounter type 153 litres	"	1.333
5.4 cf, undercounter type	"	1.333
Refrigerator/freezer, 4.4 cf, undercounter type	"	1.333
Sealer, impulse, free standing, 20x12x4"	"	0.267

Architectural Equipment	UNIT	MAN/HOURS
11600.10 Laboratory Equipment *(Cont.)*		
Timer, electric, 1-60 minutes, bench or wall mounted	EA.	0.444
Glassware washer - dryer, undercounter	"	10.000
Balance, torsion suspension, tabletop, 4.5 lb capacity	"	0.444
Binocular microscope, with in base illuminator	"	0.308
Centrifuge, table model, 19x16x13"	"	0.320
Clinical model, with four place head	"	0.178
11700.10 Medical Equipment		
Hospital equipment, lights		
Examination, portable	EA.	0.667
Meters		
Air flow meter	EA.	0.444
Oxygen flow meters	"	0.333
Racks		
40 chart, revolving open frame; mobile caddy	EA.	0.667
Scales.		
Clinical, metric with measure rod, 350 lb	EA.	0.727
Physical therapy		
Chair, hydrotherapy	EA.	0.133
Diathermy, shortwave, portable, on casters	"	0.320
Exercise bicycle, floor standing, 35" x 15"	"	0.267
Hydrocollator, 4 pack, portable, 129 x 90 x 160"	"	0.114
Lamp, infrared, mobile with variable heat control	"	0.615
Ultra violet, base mounted	"	0.615
Mirror, posture training, 27" wide and 72" high	"	0.200
Parallel bars, adjustable	"	1.000
Platform mat 10'x6', 1" thick	"	0.200
Pulley, duplex, wall mounted	"	2.667
Rack, crutch, wall mounted, 66 x 16 x 13"	"	0.800
Stimulator, galvanic-faradic, hand held	"	0.053
Ultrasound, muscle stimulator, portable, 13x13x8"	"	0.067
Sandbag set, velcro straps, saddle bag type	"	0.114
Whirlpool, 85 gallon	"	4.000
65 gallon capacity	"	4.000
Radiology		
Radiographic table, motor driven tilting table	EA.	80.000
Fluoroscope image/tv system	"	160.000
Processor for washing and drying radiographs		
Water filter unit, 30" x 48-1/2" x 37-1/2"	EA.	13.333
Cassette transfer cabinet	"	0.667
Base storage cabinets, sectional design		
With back splash, 24" deep and 35" high	L.F.	0.667
Wall storage cabinets	"	1.000
Steam sterilizers		
For heat and moisture stable materials	EA.	0.800
For fast drying after sterilization	"	1.000
Compact unit	"	1.000
Semi-automatic	"	4.000
Floor loading		
Single door	EA.	6.667
Double door	"	8.000
Utensil washer, sanitizer	"	6.154
Automatic washer/sterilizer	"	16.000

Architectural Equipment	UNIT	MAN/HOURS
11700.10 Medical Equipment *(Cont.)*		
16 x 16 x 26", including generator & accessories	EA.	26.667
Steam generator, elec., 10 kw to 180 kw	"	16.000
Surgical scrub		
Minimum	EA.	2.667
Maximum	"	2.667
Gas sterilizers		
Automatic, free standing, 21x19x29"	EA.	8.000
Surgical tables		
Minimum	EA.	11.429
Maximum	"	16.000
Surgical lights, ceiling mounted		
Minimum	EA.	13.333
Maximum	"	16.000
Water stills		
4 liters/hr	EA.	2.667
8 liters/hr	"	2.667
19 liters/hr	"	6.667
X-ray equipment		
Mobile unit		
Minimum	EA.	4.000
Maximum	"	8.000
Film viewers		
Minimum	EA.	1.333
Maximum	"	2.667
Autopsy table		
Minimum	EA.	8.000
Maximum	"	8.000
Incubators		
15 cf	EA.	4.000
29 cf	"	6.667
Infant transport, portable	"	4.211
Beds		
Stretcher, with pad, 30" x 78"	EA.	2.000
Transfer, for patient transport	"	2.000
Headwall		
Aluminum, with back frame and console	EA.	4.000
Hospital ground detection system		
Power ground module	EA.	2.286
Ground slave module	"	1.739
Master ground module	"	1.509
Remote indicator	"	1.600
X-ray indicator	"	1.739
Micro ammeter	"	2.000
Supervisory module	"	1.739
Ground cords	"	0.296
Hospital isolation monitors, 5 ma		
120v	EA.	3.478
208v	"	3.478
240v	"	3.478

Architectural Equipment	UNIT	MAN/HOURS
11700.10 **Medical Equipment** *(Cont.)*		
Digital clock-timers separate display	EA.	1.600
One display	"	1.600
Remote control	"	1.250
Battery pack	"	1.250
Surgical chronometer clock and 3 timers	"	2.500
Auxilary control	"	1.159
11700.20 **Dental Equipment**		
Dental care equipment		
Drill console with accessories	EA.	13.333
Amalgamator	"	0.400
Lathe	"	0.267
Finish polisher	"	0.533
Model trimmer	"	0.364
Motor, wall mounted	"	0.364
Cleaner, ultrasonic	"	0.800
Curing unit, bench mounted	"	1.333
Oral evacuation system, dual pump	"	1.000
Sterilizer, table top, self contained	"	0.444
Dental lights		
Light, floor or ceiling mounted	EA.	4.000
X-ray unit		
Portable	EA.	2.000
Wall mounted with remote control	"	6.667
Illuminator, single panel	"	11.429
X-ray film processor	"	6.667
Shield, portable x-ray, lead lined	"	0.533

Interior	UNIT	MAN/HOURS
12302.10 **Casework**		
Kitchen base cabinet, prefinished, 24" deep, 35" high		
12"wide	EA.	0.800
18" wide	"	0.800
24" wide	"	0.889
27" wide	"	0.889
36" wide	"	1.000
48" wide	"	1.000
Corner cabinet, 36" wide	"	1.000
Wall cabinet, 12" deep, 12" high		
30" wide	EA.	0.800
36" wide	"	0.800
15" high		
30" wide	EA.	0.889
36" wide	"	0.889
24" high		
30" wide	EA.	0.889
36" wide	"	0.889
30" high		
12" wide	EA.	1.000
18" wide	"	1.000
24" wide	"	1.000
27" wide	"	1.000
30" wide	"	1.143
36" wide	"	1.143
Corner cabinet, 30" high		
24" wide	EA.	1.333
30" wide	"	1.333
36" wide	"	1.333
Wardrobe	"	2.000
Vanity with top, laminated plastic		
24" wide	EA.	2.000
30" wide	"	2.000
36" wide	"	2.667
48" wide	"	3.200
12390.10 **Counter Tops**		
Stainless steel, counter top, with backsplash	S.F.	0.200
Acid-proof, kemrock surface	"	0.133
12500.10 **Window Treatment**		
Drapery tracks, wall or ceiling mounted		
Basic traverse rod		
50 to 90"	EA.	0.400
84 to 156"	"	0.444
136 to 250"	"	0.444
165 to 312"	"	0.500
Traverse rod with stationary curtain rod		
30 to 50"	EA.	0.400
50 to 90"	"	0.400
84 to 156"	"	0.444

Interior	UNIT	MAN/HOURS
12500.10 **Window Treatment** *(Cont.)*		
136 to 250"	EA.	0.500
Double traverse rod		
30 to 50"	EA.	0.400
50 to 84"	"	0.400
84 to 156"	"	0.444
136 to 250"	"	0.500
12510.10 **Blinds**		
Venetian blinds		
2" slats	S.F.	0.020
1" slats	"	0.020
12690.40 **Floor Mats**		
Recessed entrance mat, 3/8" thick, aluminum link	S.F.	0.400
Steel, flexible	"	0.400

13056.10 — Vaults

Construction	UNIT	MAN/HOURS
Floor safes		
Class C		
1.0 cf	EA.	0.667
1.3 cf	"	1.000
1.9 cf	"	1.333
5.2 cf	"	1.333

13121.10 — Pre-engineered Buildings

Construction	UNIT	MAN/HOURS
Pre-engineered metal building, 40'x100'		
14' eave height	S.F.	0.032
16' eave height	"	0.037
20' eave height	"	0.048
60'x100'		
14' eave height	S.F.	0.032
16' eave height	"	0.037
20' eave height	"	0.048
80'x100'		
14' eave height	S.F.	0.032
16' eave height	"	0.037
20' eave height	"	0.048
100'x100'		
14' eave height	S.F.	0.032
16' eave height	"	0.037
20' eave height	"	0.048
100'x150'		
14' eave height	S.F.	0.032
16' eave height	"	0.037
20' eave height	"	0.048
120'x150'		
14' eave height	S.F.	0.032
16' eave height	"	0.037
20' eave height	"	0.048
140'x150'		
14' eave height	S.F.	0.032
16' eave height	"	0.037
20' eave height	"	0.048
160'x200'		
14' eave height	S.F.	0.032
16' eave height	"	0.037
20' eave height	"	0.048
200'x200'		
14' eave height	S.F.	0.032
16' eave height	"	0.037
20' eave height	"	0.048
Liner panel, 26 ga, painted steel	"	0.020
Wall panel insulated, 26 ga. steel, foam core	"	0.020
Roof panel, 26 ga. painted steel	"	0.011
Plastic (sky light)	"	0.011
Insulation, 3-1/2" thick blanket, R11	"	0.005

13152.10 — Swimming Pool Equipment

Construction	UNIT	MAN/HOURS
Diving boards		
14' long		
Aluminum	EA.	4.444
Fiberglass	"	4.444
Ladders, heavy duty		
2 steps		
Minimum	EA.	1.600
Maximum	"	1.600
4 steps		
Minimum	EA.	2.000
Maximum	"	2.000
Lifeguard chair		
Minimum	EA.	8.000
Maximum	"	8.000
Lights, underwater		
12 volt, with transformer	EA.	2.000
110 volt		
Minimum	EA.	2.000
Maximum	"	2.000
Ground fault interrupter for 110 volt, each light	"	0.667
Pool covers		
Reinforced polyethylene	S.F.	0.062
Vinyl water tube		
Minimum	S.F.	0.062
Maximum	"	0.062
Slides with water tube		
Minimum	EA.	6.667
Maximum	"	6.667

Elevators

14210.10 — Elevators

Elevators	UNIT	MAN/HOURS
Passenger elevators, electric, geared		
Based on a shaft of 6 stops and 6 openings		
50 fpm, 2000 lb	EA.	24.000
100 fpm, 2000 lb	"	26.667
150 fpm		
2000 lb	EA.	30.000
3000 lb	"	34.286
4000 lb	"	40.000
200 fpm		
2500 lb	EA.	34.286
3000 lb	"	36.923
4000 lb	"	40.000
250 fpm		
2500 lb	EA.	34.286
3000 lb	"	36.923
4000 lb	"	40.000
300 fpm		
2500 lb	EA.	34.286
3000 lb	"	36.923
4000 lb	"	24.000
Based on a shaft of 8 stops and 8 openings		
300 fpm		
3000 lb	EA.	48.000
3500 lb	"	48.000
4000 lb	"	53.333
5000 lb	"	57.143
400 fpm		
3000 lb	EA.	48.000
3500 lb	"	48.000
4000 lb	"	53.333
5000 lb	"	57.143
600 fpm		
3000 lb	EA.	53.333
3500 lb	"	57.143
4000 lb	"	58.537
5000 lb	"	60.000
800 fpm		
3000 lb	EA.	53.333
3500 lb	"	57.143
4000 lb	"	58.537
5000 lb	"	60.000
Hydraulic, based on a shaft of 3 stops, 3 openings		
50 fpm		
2000 lb	EA.	20.000
2500 lb	"	20.000
3000 lb	"	20.870
100 fpm		
2000 lb	EA.	20.000
2500 lb	"	20.870
3000 lb	"	21.818
150 fpm		
2000 lb	EA.	20.000
2500 lb	"	20.870

14210.10 — Elevators (Cont.)

Elevators	UNIT	MAN/HOURS
3000 lb	EA.	22.857
Small elevators, 4 to 6 passenger capacity		
Electric, push		
2 stops	EA.	20.000
3 stops	"	21.818
4 stops	"	24.000
Freight elevators, electric		
Based on a shaft of 6 stops and 6 openings		
50 fpm		
3500 lb	EA.	26.667
4000 lb	"	26.667
5000 lb	"	30.000
100 fpm		
3500 lb	EA.	30.000
4000 lb	"	30.000
5000 lb	"	34.286
200 fpm		
3500 lb	EA.	34.286
4000 lb	"	34.286
5000 lb	"	40.000
Based on shaft of 8 stops and 8 openings		
100 fpm		
4000 lb	EA.	30.000
6000 lb	"	30.769
8000 lb	"	32.432
150 fpm		
4000 lb	EA.	34.286
6000 lb	"	35.294
8000 lb	"	37.500
200 fpm		
4000 lb	EA.	40.000
6000 lb	"	41.379
8000 lb	"	43.636
Hydraulic, based on 3 stops and 3 openings		
50 fpm		
3000 lb	EA.	17.143
4000 lb	"	17.778
6000 lb	"	18.462
100 fpm		
3000 lb	EA.	17.143
4000 lb	"	17.778
6000 lb	"	18.462
150 fpm		
3000 lb	EA.	17.143
4000 lb	"	17.778
6000 lb	"	18.462

14300.10 — Escalators

Escalators	UNIT	MAN/HOURS
Escalators		
32" wide, floor to floor		
12' high	EA.	40.000
15' high	"	48.000
18' high	"	60.000

Elevators

14300.10 Escalators *(Cont.)*	UNIT	MAN/HOURS
22' high	EA.	80.000
25' high	"	96.000
48" wide		
12' high	EA.	41.379
15' high	"	50.000
18' high	"	63.158
22' high	"	85.714
25' high	"	96.000

Lifts

14410.10 Personnel Lifts	UNIT	MAN/HOURS
Residential stair climber, per story	EA.	6.667

14450.10 Vehicle Lifts	UNIT	MAN/HOURS
Automotive hoist, one post, semi-hydraulic, 8,000 lb	EA.	24.000
Full hydraulic, 8,000 lb	"	24.000
2 post, semi-hydraulic, 10,000 lb	"	34.286
Full hydraulic		
10,000 lb	EA.	34.286
13,000 lb	"	60.000
18,500 lb	"	60.000
24,000 lb	"	60.000
26,000 lb	"	60.000
Pneumatic hoist, fully hydraulic		
11,000 lb	EA.	80.000
24,000 lb	"	80.000

Material Handling

14560.10 Chutes	UNIT	MAN/HOURS
Linen chutes, stainless steel, with supports		
18" dia.	L.F.	0.057
24" dia.	"	0.062

Material Handling

14560.10 Chutes *(Cont.)*	UNIT	MAN/HOURS
30" dia.	L.F.	0.067
Hopper	EA.	0.533
Skylight	"	0.800
Sprinkler unit at top	"	0.889

14580.10 Pneumatic Systems		
Pneumatic message tube system		
Average, 20 station job		
3" round system	E.A.	72.727
4" round system	"	80.000
6" round system	"	88.889
4" x 7" oval system	"	160.000
Trash and linen tube system		
10 stations	EA.	120.000
15 stations	"	160.000
20 stations	"	184.615
30 stations	"	218.182

Hoists And Cranes

14600.10 Industrial Hoists	UNIT	MAN/HOURS
Industrial hoists, electric, light to medium duty		
500 lb	EA.	4.000
1000 lb	"	4.211
2000 lb	"	4.444
3000 lb	"	4.706
4000 lb	"	5.000
5000 lb	"	5.333
6000 lb	"	5.517
7500 lb	"	5.714
10,000 lb	"	5.926
15,000 lb	"	6.154
20,000 lb	"	6.667
25,000 lb	"	7.273
30,000 lb	"	8.000
Heavy duty		
500 lb	EA.	4.000
1000 lb	"	4.211
2000 lb	"	4.444
3000 lb	"	4.706
4000 lb	"	5.000
5000 lb	"	5.333
6000 lb	"	5.517
7500 lb	"	5.714
10,000 lb	"	5.926
15,000 lb	"	6.154

Hoists And Cranes	UNIT	MAN/ HOURS
14600.10 **Industrial Hoists** *(Cont.)*		
20,000 lb	EA.	6.667
25,000 lb	"	7.273
30,000 lb	"	8.000
Air powered hoists		
500 lb	EA.	4.000
1000 lb	"	4.000
2000 lb	"	4.211
4000 lb	"	4.706
6000 lb	"	6.154
Overhead traveling bridge crane		
Single girder, 20' span		
3 ton	EA.	12.000
5 ton	"	12.000
7.5 ton	"	12.000
10 ton	"	15.000
15 ton	"	15.000
30' span		
3 ton	EA.	12.000
5 ton	"	12.000
10 ton	"	15.000
15 ton	"	15.000
Double girder, 40' span		
3 ton	EA.	26.667
5 ton	"	26.667
7.5 ton	"	26.667
10 ton	"	34.286
15 ton	"	34.286
25 ton	"	34.286
50' span		
3 ton	EA.	26.667
5 ton	"	26.667
7.5 ton	"	26.667
10 ton	"	34.286
15 ton	"	34.286
25 ton	"	34.286
14650.10 **Jib Cranes**		
Self supporting, swinging 8' boom, 200 deg rotation		
1000 lb	EA.	6.667
2000 lb	"	6.667
3000 lb	"	13.333
4000 lb	"	13.333
6000 lb	"	13.333
10,000 lb	"	13.333
Wall mounted, 180 deg rotation		
2000 lb	EA.	6.667
3000 lb	"	6.667
4000 lb	"	13.333
6000 lb	"	13.333
10,000 lb	"	13.333

Basic Materials	UNIT	MAN/ HOURS
15100.10 Specialties		
Wall penetration		
Concrete wall, 6" thick		
2" dia.	EA.	0.267
4" dia.	"	0.400
8" dia.	"	0.571
12" thick		
2" dia.	EA.	0.364
4" dia.	"	0.571
8" dia.	"	0.889
Non-destructive testing, piping systems		
X-ray of welds		
3" dia. pipe	EA.	0.800
4" dia. pipe	"	0.800
6" dia. pipe	"	0.800
8" dia. pipe	"	1.000
10" dia. pipe	"	1.000
Liquid penetration of welds		
2" dia. pipe	EA.	0.500
3" dia. pipe	"	0.500
4" dia. pipe	"	0.500
6" dia. pipe	"	0.500
8" dia. pipe	"	0.500
10" dia. pipe	"	0.500
15120.10 Backflow Preventers		
Backflow preventer, flanged, cast iron, with valves		
3" pipe	EA.	4.000
4" pipe	"	4.444
6" pipe	"	6.667
8" pipe	"	8.000
Threaded		
3/4" pipe	EA.	0.500
2" pipe	"	0.800
Reduced pressure assembly, bronze, threaded		
3/4"	EA.	0.500
1"	"	0.571
1-1/4"	"	0.667
1-1/2"	"	0.800
15140.10 Pipe Hangers, Heavy		
Hangers		
1/2" pipe, clevis pipe hanger		
Black steel	EA.	0.267
Galvanized	"	0.267
U bolt	"	0.080
3/4" pipe, clevis pipe hanger		
Black steel	EA.	0.267
Galvanized	"	0.267
U bolt	"	0.080
1" pipe, clevis pipe hanger		
Black steel	EA.	0.267
Galvanized	"	0.267
U bolt	"	0.080

Basic Materials	UNIT	MAN/ HOURS
15140.10 Pipe Hangers, Heavy *(Cont.)*		
1-1/4" pipe, clevis pipe hanger		
Black steel	EA.	0.267
Galvanized	"	0.267
U bolt	"	0.080
1-1/2" pipe, clevis pipe hanger		
Black steel	EA.	0.267
Galvanized	"	0.267
U bolt	"	0.080
2" pipe, clevis pipe hanger		
Black steel	EA.	0.267
Galvanized	"	0.267
U bolt	"	0.080
2-1/2" pipe, clevis pipe hanger		
Black steel	EA.	0.267
Galvanized	"	0.267
3" pipe, clevis pipe hanger		
Black steel	EA.	0.267
Galvanized	"	0.267
U bolt	"	0.080
3-1/2" pipe, clevis pipe hanger		
Black steel	EA.	0.267
Galvanized	"	0.267
U bolt	"	0.080
4" pipe, clevis pipe hanger		
Black steel	EA.	0.267
Galvanized	"	0.267
U bolt	"	0.080
5" pipe, clevis pipe hanger		
Black steel	EA.	0.320
Galvanized	"	0.320
U bolt	"	0.080
6" pipe, clevis pipe hanger		
Black steel	EA.	0.320
Galvanized	"	0.320
U bolt	"	0.100
8" pipe, clevis pipe hanger		
Black steel	EA.	0.320
Galvanized	"	0.320
U bolt	"	0.100
10" clevis pipe hanger		
Black steel	EA.	0.320
Galvanized	"	0.320
12" pipe, clevis pipe hanger		
Black steel	EA.	0.320
Galvanized	"	0.320
Threaded rod, galvanized		
C-clamp, steel, with lock nut		
3/8"	EA.	0.100
1/2"	"	0.100
5/8"	"	0.100
3/4"	"	0.100
7/8"	"	0.100
Angle support, medium, welded steel		

Basic Materials	UNIT	MAN/HOURS
15140.10 Pipe Hangers, Heavy (Cont.)		
12"x18"	EA.	0.800
18"x24"	"	0.800
24"x30"	"	0.800
Heavy, welded steel		
12"x18"	EA.	0.800
18"x24"	"	0.800
24"x30"	"	0.800
15140.11 Pipe Hangers, Light		
A band, black iron		
1/2"	EA.	0.057
1"	"	0.059
1-1/4"	"	0.062
1-1/2"	"	0.067
2"	"	0.073
2-1/2"	"	0.080
3"	"	0.089
4"	"	0.100
5"	"	0.107
6"	"	0.114
8"	"	0.133
Copper		
1/2"	EA.	0.057
3/4"	"	0.059
1"	"	0.059
1-1/4"	"	0.062
1-1/2"	"	0.067
2"	"	0.073
2-1/2"	"	0.080
3"	"	0.089
4"	"	0.100
Black riser friction hangers		
3/4"	EA.	0.067
1"	"	0.070
1-1/4"	"	0.073
1-1/2"	"	0.076
2"	"	0.080
2-1/2"	"	0.089
3"	"	0.100
4"	"	0.114
5"	"	0.123
6"	"	0.133
8"	"	0.145
10"	"	0.160
2 hole clips, galvanized		
3/4"	EA.	0.053
1"	"	0.055
1-1/4"	"	0.057
1-1/2"	"	0.059
2"	"	0.062
2-1/2"	"	0.064
3"	"	0.067
4"	"	0.073

Basic Materials	UNIT	MAN/HOURS
15140.11 Pipe Hangers, Light (Cont.)		
Perforated strap		
3/4"		
Galvanized, 20 ga.	L.F.	0.040
Copper, 22 ga.	"	0.040
J-Hooks		
1/2"	EA.	0.036
3/4"	"	0.036
1"	"	0.038
1-1/4"	"	0.039
1-1/2"	"	0.040
2"	"	0.040
3"	"	0.042
4"	"	0.042
PVC coated hangers, galvanized, 28 ga.		
1-1/2" x 12"	EA.	0.053
2" x 12"	"	0.057
3" x 12"	"	0.062
4" x 12"	"	0.067
Copper, 30 ga.		
1-1/2" x 12"	EA.	0.053
2" x 12"	"	0.057
3" x 12"	"	0.062
4" x 12"	"	0.067
2" x 24"	"	0.062
3" x 24"	"	0.067
4" x 24"	"	0.073
Wire hook hangers		
Black wire, 1/2" x		
4"	EA.	0.040
6"	"	0.042
8"	"	0.044
10"	"	0.044
12"	"	0.047
3/4" x		
4"	EA.	0.042
6"	"	0.044
8"	"	0.047
10"	"	0.050
12"	"	0.053
1" x		
4"	EA.	0.044
6"	"	0.047
8"	"	0.050
10"	"	0.053
12"	"	0.057
1-1/4" x		
4"	EA.	0.047
6"	"	0.050
8"	"	0.053
10"	"	0.057
12"	"	0.062
1-1/2" x		
6"	EA.	0.053

Basic Materials	UNIT	MAN/HOURS
15140.11 Pipe Hangers, Light *(Cont.)*		
8"	EA.	0.057
10"	"	0.062
12"	"	0.067
2" x		
6"	EA.	0.057
8"	"	0.062
10"	"	0.067
12"	"	0.073
Copper wire hooks		
1/2" x		
4"	EA.	0.040
6"	"	0.042
8"	"	0.044
10"	"	0.047
12"	"	0.050
3/4" x		
4"	EA.	0.042
6"	"	0.044
8"	"	0.047
10"	"	0.050
12"	"	0.053
1" x		
4"	EA.	0.044
6"	"	0.047
8"	"	0.050
10"	"	0.053
12"	"	0.057
1-1/4" x		
6"	EA.	0.047
8"	"	0.050
10"	"	0.053
12"	"	0.057
1-1/2" x		
6"	EA.	0.053
8"	"	0.057
10"	"	0.062
12"	"	0.067
2" x		
6"	EA.	0.057
8"	"	0.062
10"	"	0.067
12"	"	0.073
15175.60 Expansion Tanks		
Expansion tank, 125 psi, steel		
20 gallon	EA.	1.000
30 gallon	"	1.333
65 gallon	"	2.000
80 gallon	"	2.286

Basic Materials	UNIT	MAN/HOURS
15240.10 Vibration Control		
Vibration isolator, in-line, stainless connector, screwed		
1/2"	EA.	0.444
3/4"	"	0.471
1"	"	0.500
1-1/4"	"	0.533
1-1/2"	"	0.571
2"	"	0.615
2-1/2"	"	0.667
3"	"	0.727
4"	"	0.800
6"	"	0.889
Flanged		
8"	EA.	1.000
10"	"	1.143
12"	"	1.333

Insulation	UNIT	MAN/HOURS
15260.10 Fiberglass Pipe Insulation		
Fiberglass insulation on 1/2" pipe		
1" thick	L.F.	0.027
1-1/2" thick	"	0.033
3/4" pipe		
1" thick	L.F.	0.027
1-1/2" thick	"	0.033
1" pipe		
1" thick	L.F.	0.027
1-1/2" thick	"	0.033
2" pipe		
1" thick	L.F.	0.033
1-1/2" thick	"	0.036
3" pipe		
1" thick	L.F.	0.038
1-1/2" thick	"	0.040
6" pipe		
1" thick	L.F.	0.042
2" thick	"	0.044
10" pipe		
2" thick	L.F.	0.042
3" thick	"	0.044

Insulation

15260.20 — Calcium Silicate

	UNIT	MAN/HOURS
Calcium silicate insulation, 6" pipe		
2" thick	L.F.	0.057
2-1/2" thick	"	0.062
3" thick	"	0.067
4" thick	"	0.073
6" thick	"	0.080
12" pipe		
2" thick	L.F.	0.062
2-1/2" thick	"	0.067
3" thick	"	0.073
4" thick	"	0.080
6" thick	"	0.089

15260.60 — Exterior Pipe Insulation

	UNIT	MAN/HOURS
Fiberglass insulation, aluminum jacket		
1/2" pipe		
1" thick	L.F.	0.062
1-1/2" thick	"	0.067
3/4" pipe		
1" thick	L.F.	0.062
1-1/2" thick	"	0.067
1" pipe		
1" thick	L.F.	0.062
1-1/2" thick	"	0.067
2" pipe		
1" thick	L.F.	0.073
1-1/2" thick	"	0.076
3" pipe		
1" thick	L.F.	0.080
1-1/2" thick	"	0.084
6" pipe		
1" thick	L.F.	0.089
2" thick	"	0.094
10" pipe		
2" thick	L.F.	0.089
3" thick	"	0.094

15260.90 — Pipe Insulation Fittings

	UNIT	MAN/HOURS
Insulation protection saddle		
1" thick covering		
1/2" pipe	EA.	0.320
3/4" pipe	"	0.320
1" pipe	"	0.320
2" pipe	"	0.320
3" pipe	"	0.364
6" pipe	"	0.500
1-1/2" thick covering		
3/4" pipe	EA.	0.320
1" pipe	"	0.320
2" pipe	"	0.320
3" pipe	"	0.320
6" pipe	"	0.500
10" pipe	"	0.667

Insulation

15280.10 — Equipment Insulation

	UNIT	MAN/HOURS
Equipment insulation, 2" thick, cellular glass	S.F.	0.050
Urethane, rigid, field applied jacket, plastered finish	"	0.100
Fiberglass, rigid, with vapor barrier	"	0.044

15290.10 — Ductwork Insulation

	UNIT	MAN/HOURS
Fiberglass duct insulation, plain blanket		
1-1/2" thick	S.F.	0.010
2" thick	"	0.013
With vapor barrier		
1-1/2" thick	S.F.	0.010
2" thick	"	0.013
Rigid with vapor barrier		
2" thick	S.F.	0.027
3" thick	"	0.032
4" thick	"	0.040
6" thick	"	0.053
Weatherproof, polystyrene, 3" thick, w/vapor barrier	"	0.080
Urethane board with vapor barrier	"	0.100

Fire Protection

15330.10 — Wet Sprinkler System

	UNIT	MAN/HOURS
Sprinkler head, 212 deg, brass, exposed piping	EA.	0.320
Chrome, concealed piping	"	0.444
Water motor alarm	"	1.333
Fire department inlet connection	"	1.600
Wall plate for fire dept connection	"	0.667
Swing check valve flanged iron body, 4"	"	2.667
Check valve, 6"	"	4.000
Wet pipe valve, flange to groove, 4"	"	0.889
Flange to flange		
6"	EA.	1.333
8"	"	2.667
Alarm valve, flange to flange, (wet valve)		
4"	EA.	0.889
8"	"	6.667
Inspector's test connection	"	0.667
Wall hydrant, polished brass, 2-1/2" x 2-1/2", single	"	0.571
2-way	"	0.571
3-way	"	0.571
Wet valve trim, includes retard chamber & gauges, 4"-6"	"	0.667
Retard pressure switch for wet systems	"	1.600
Air maintenance device	"	0.667
Wall hydrant non-freeze, 8" thick wall, vacuum breaker	"	0.400
12" thick wall	"	0.400

Fire Protection

15330.50 — Dry Sprinkler System

	UNIT	MAN/HOURS
Dry pipe valve, flange to flange		
4"	EA.	1.600
6"	"	2.000
Trim, 4" and 6", includes gauges	"	0.667
Field testing and flushing	"	6.667
Disinfection	"	6.667
Pressure switch double circuit, open/close contacts	"	2.000
Low air		
Supervisory unit	EA.	1.333
Pressure switch	"	0.667

15330.70 — Co2 System

	UNIT	MAN/HOURS
CO2 system, high pressure, 75# cylinder with		
Valve assemblies	EA.	1.600
Storage rack	"	1.143
Manifold	"	5.714
Flexible loops	"	0.100
Beam scale for cylinders	"	1.333
Mechanically control head	"	0.533
Electrically control head	"	0.533
Stop valves	"	0.800
Check valves	"	1.000
Activation station	"	0.800
Nozzles	"	0.667
Hose reel with 75' of 3/4" hose	"	4.000
Main/reserve transfer switch	"	1.333
Pressure switch	"	0.800
Heat responsive device	"	1.333
Battery and charger	"	4.000
Low pressure		
Battery and charger	EA.	4.000
Pressure switch	"	0.889
Nozzles	"	0.727
Master selector valve	"	1.333
Selector valve	"	1.333
Low pressure hose reel with 75' of 3/4" hose	"	4.000
Tank fill lines	"	1.000
Activation stations	"	0.667
Electro manual pilot panels	"	1.000

Plumbing

15410.05 — C.i. Pipe, Above Ground

	UNIT	MAN/HOURS
No hub pipe		
1-1/2" pipe	L.F.	0.057
2" pipe	"	0.067
3" pipe	"	0.080
4" pipe	"	0.133
6" pipe	"	0.160
8" pipe	"	0.267
10" pipe	"	0.320
No hub fittings, 1-1/2" pipe		
1/4 bend	EA.	0.267
1/8 bend	"	0.267
Sanitary tee	"	0.400
Sanitary cross	"	0.400
Wye	"	0.400
2" pipe		
1/4 bend	EA.	0.320
1/8 bend	"	0.320
Sanitary tee	"	0.533
Sanitary cross	"	0.533
Wye	"	0.667
3" pipe		
1/4 bend	EA.	0.400
1/8 bend	"	0.400
Sanitary tee	"	0.500
3"x2" sanitary tee	"	0.500
3"x1-1/2" sanitary tee	"	0.500
Sanitary cross	"	0.667
3x2" sanitary cross	"	0.667
Wye	"	0.667
4" pipe		
1/4 bend	EA.	0.400
1/8 bend	"	0.400
Sanitary tee	"	0.667
4x3" sanitary tee	"	0.667
4x2" sanitary tee	"	0.667
Sanitary cross	"	0.800
4x3" sanitary cross	"	0.800
4x2" sanitary cross	"	0.800
Wye	"	0.667
8" deep	"	0.400
6" pipe		
1/4 bend	EA.	0.667
1/8 bend	"	0.667
Sanitary tee	"	0.800
6x4" sanitary tee	"	0.800
Wye	"	0.800
8" pipe		
1/4 bend	EA.	0.667
1/8 bend	"	0.667
Sanitary tee	"	1.000
8x6" sanitary tee	"	1.000
Wye	"	0.800

Plumbing	UNIT	MAN/HOURS
15410.05 **C.i. Pipe, Above Ground** *(Cont.)*		
10" pipe		
1/4 bend	EA.	0.667
1/8 bend	"	0.667
Wye	"	1.333
15410.06 **C.i. Pipe, Below Ground**		
No hub pipe		
1-1/2" pipe	L.F.	0.040
2" pipe	"	0.044
3" pipe	"	0.050
4" pipe	"	0.067
6" pipe	"	0.073
8" pipe	"	0.089
10" pipe	"	0.100
Fittings, 1-1/2"		
1/4 bend	EA.	0.229
1/8 bend	"	0.229
Wye	"	0.320
Wye & 1/8 bend	"	0.229
P-trap	"	0.229
2"		
1/4 bend	EA.	0.267
1/8 bend	"	0.267
Double wye	"	0.500
Wye & 1/8 bend	"	0.400
Double wye & 1/8 bend	"	0.500
P-trap	"	0.267
3"		
1/4 bend	EA.	0.320
1/8 bend	"	0.320
Wye	"	0.500
3x2" wye	"	0.500
Wye & 1/8 bend	"	0.500
Double wye & 1/8 bend	"	0.500
3x2" double wye & 1/8 bend	"	0.500
3x2" reducer	"	0.320
P-trap	"	0.320
4"		
1/4 bend	EA.	0.320
1/8 bend	"	0.320
Wye	"	0.500
4x3" wye	"	0.500
4x2" wye	"	0.500
Double wye	"	0.667
4x3" double wye	"	0.667
4x2" double wye	"	0.667
Wye & 1/8 bend	"	0.500
Double wye & 1/8 bend	"	0.667
6"		
1/4 bend	EA.	0.500
1/8 bend	"	0.500
Wye & 1/8 bend	"	0.667
P-trap	"	0.400

Plumbing	UNIT	MAN/HOURS
15410.06 **C.i. Pipe, Below Ground** *(Cont.)*		
8"		
1/4 bend	EA.	0.500
1/8 bend	"	0.500
Wye	"	0.667
10"		
1/4 bend	EA.	0.500
1/8 bend	"	0.500
Wye	"	1.000
15410.09 **Service Weight Pipe**		
Service weight pipe, single hub		
3" x 5'	EA.	0.170
4" x 5'	"	0.178
6" x 5'	"	0.200
1/8 bend		
3"	EA.	0.320
4"	"	0.364
6"	"	0.400
1/4 bend		
3"	EA.	0.320
4"	"	0.364
6"	"	0.400
Sweep		
3"	EA.	0.320
4"	"	0.364
6"	"	0.400
Sanitary T		
3"	EA.	0.571
4"	"	0.667
6"	"	0.727
Wye		
3"	EA.	0.444
4"	"	0.471
6"	"	0.571
15410.10 **Copper Pipe**		
Type "K" copper		
1/2"	L.F.	0.025
3/4"	"	0.027
1"	"	0.029
1-1/4"	"	0.031
1-1/2"	"	0.033
2"	"	0.036
2-1/2"	"	0.040
3"	"	0.042
4"	"	0.044
DWV, copper		
1-1/4"	L.F.	0.033
1-1/2"	"	0.036
2"	"	0.040
3"	"	0.044
4"	"	0.050
6"	"	0.057

Plumbing	UNIT	MAN/HOURS
15410.10 **Copper Pipe** *(Cont.)*		
Refrigeration tubing, copper, sealed		
1/8"	L.F.	0.032
3/16"	"	0.033
1/4"	"	0.035
5/16"	"	0.036
3/8"	"	0.038
1/2"	"	0.040
7/8"	"	0.046
1-1/8"	"	0.053
1-3/8"	"	0.062
Type "L" copper		
1/4"	L.F.	0.024
3/8"	"	0.024
1/2"	"	0.025
3/4"	"	0.027
1"	"	0.029
1-1/4"	"	0.031
1-1/2"	"	0.033
2"	"	0.036
Type "M" copper		
1/2"	L.F.	0.025
3/4"	"	0.027
1"	"	0.029
1-1/4"	"	0.031
2"	"	0.036
15410.11 **Copper Fittings**		
Coupling, with stop		
1/4"	EA.	0.267
3/8"	"	0.320
1/2"	"	0.348
5/8"	"	0.400
3/4"	"	0.444
1"	"	0.471
3"	"	0.800
4"	"	1.000
Reducing coupling		
1/4" x 1/8"	EA.	0.320
3/8" x 1/4"	"	0.348
1/2" x		
3/8"	EA.	0.400
1/4"	"	0.400
1/8"	"	0.400
3/4" x		
3/8"	EA.	0.444
1/2"	"	0.444
1" x		
3/8"	EA.	0.500
1" x 1/2"	"	0.500
1" x 3/4"	"	0.500
1-1/4" x		
1/2"	EA.	0.533
3/4"	"	0.533

Plumbing	UNIT	MAN/HOURS
15410.11 **Copper Fittings** *(Cont.)*		
1"	EA.	0.533
1-1/2" x		
1/2"	EA.	0.571
3/4"	"	0.571
1"	"	0.571
1-1/4"	"	0.571
2" x		
1/2"	EA.	0.667
3/4"	"	0.667
1"	"	0.667
1-1/4"	"	0.667
1-1/2"	"	0.667
2-1/2" x		
1"	EA.	0.800
1-1/4"	"	0.800
1-1/2"	"	0.800
2"	"	0.800
3" x		
1-1/2"	EA.	1.000
2"	"	1.000
2-1/2"	"	1.000
4" x		
2"	EA.	1.143
2-1/2"	"	1.143
3"	"	1.143
Slip coupling		
1/4"	EA.	0.267
1/2"	"	0.320
3/4"	"	0.400
1"	"	0.444
1-1/4"	"	0.500
1-1/2"	"	0.533
2"	"	0.667
2-1/2"	"	0.667
3"	"	0.800
4"	"	1.000
Coupling with drain		
1/2"	EA.	0.400
3/4"	"	0.444
1"	"	0.500
Reducer		
3/8" x 1/4"	EA.	0.320
1/2" x 3/8"	"	0.320
3/4" x		
1/4"	EA.	0.364
3/8"	"	0.364
1/2"	"	0.364
1" x		
1/2"	EA.	0.400
3/4"	"	0.400
1-1/4" x		
1/2"	EA.	0.444
3/4"	"	0.444

15410.11 — Copper Fittings (Cont.)

Plumbing	UNIT	MAN/HOURS
1"	EA.	0.444
1-1/2" x		
1/2"	EA.	0.500
3/4"	"	0.500
1"	"	0.500
1-1/4"	"	0.500
2" x		
1/2"	EA.	0.571
3/4"	"	0.571
1"	"	0.571
1-1/4"	"	0.571
1-1/2"	"	0.571
2-1/2" x		
1"	EA.	0.667
1-1/4"	"	0.667
1-1/2"	"	0.667
2"	"	0.667
3" x		
1-1/4"	EA.	0.800
1-1/2"	"	0.800
2"	"	0.800
2-1/2"	"	0.800
4" x		
2"	EA.	1.000
3"	"	1.000
Female adapters		
1/4"	EA.	0.320
3/8"	"	0.364
1/2"	"	0.400
3/4"	"	0.444
1"	"	0.444
1-1/4"	"	0.500
1-1/2"	"	0.500
2"	"	0.533
2-1/2"	"	0.571
3"	"	0.667
4"	"	0.800
Increasing female adapters		
1/8" x		
3/8"	EA.	0.320
1/2"	"	0.320
1/4" x 1/2"	"	0.348
3/8" x 1/2"	"	0.364
1/2" X		
3/4"	EA.	0.400
1"	"	0.400
3/4" X		
1"	EA.	0.444
1-1/4"	"	0.444
1" x		
1-1/4"	EA.	0.444
1-1/2"	"	0.444
1-1/4" x		

15410.11 — Copper Fittings (Cont.)

Plumbing	UNIT	MAN/HOURS
1-1/2"	EA.	0.500
2"	"	0.500
1-1/2" x 2"	"	0.533
Reducing female adapters		
3/8" x 1/4"	EA.	0.364
1/2" x		
1/4"	EA.	0.400
3/8"	"	0.400
3/4" x 1/2"	"	0.444
1" x		
1/2"	EA.	0.444
3/4"	"	0.444
1-1/4" x		
1/2"	EA.	0.500
3/4"	"	0.500
1"	"	0.500
1-1/2" x		
1"	EA.	0.533
1-1/4"	"	0.533
2" x		
1"	EA.	0.571
1-1/4"	"	0.571
1-1/2"	"	0.571
Female fitting adapters		
1/2"	EA.	0.400
3/4"	"	0.400
3/4" x 1/2"	"	0.421
1"	"	0.444
1-1/4"	"	0.471
1-1/2"	"	0.500
2"	"	0.533
Male adapters		
1/4"	EA.	0.364
3/8"	"	0.364
3"	"	0.667
4"	"	0.800
Increasing male adapters		
3/8" x 1/2"	EA.	0.364
1/2" x		
3/4"	EA.	0.400
1"	"	0.400
3/4" x		
1"	EA.	0.421
1-1/4"	"	0.421
1" x 1-1/4"	"	0.444
1-1/2" x		
3/4"	EA.	0.471
1"	"	0.471
1-1/4"	"	0.471
2" x		
1"	EA.	0.500
1-1/4"	"	0.500
1-1/2"	"	0.500

Plumbing	UNIT	MAN/HOURS
15410.11 Copper Fittings *(Cont.)*		
2" x 2-1/2"	EA.	0.533
Copper pipe fittings		
1/2"		
90 deg ell	EA.	0.178
45 deg ell	"	0.178
Tee	"	0.229
Cap	"	0.089
Coupling	"	0.178
Union	"	0.200
3/4"		
90 deg ell	EA.	0.200
45 deg ell	"	0.200
Tee	"	0.267
Cap	"	0.094
Coupling	"	0.200
Union	"	0.229
1"		
90 deg ell	EA.	0.267
45 deg ell	"	0.267
Tee	"	0.320
Cap	"	0.133
Coupling	"	0.267
Union	"	0.267
1-1/4"		
90 deg ell	EA.	0.229
45 deg ell	"	0.229
Tee	"	0.400
Cap	"	0.133
Union	"	0.286
1-1/2"		
90 deg ell	EA.	0.286
45 deg ell	"	0.286
Tee	"	0.444
Cap	"	0.133
Coupling	"	0.267
Union	"	0.364
2"		
90 deg ell	EA.	0.320
45 deg ell	"	0.500
Tee	"	0.500
Cap	"	0.160
Coupling	"	0.320
Union	"	0.400
2-1/2"		
90 deg ell	EA.	0.400
45 deg ell	"	0.400
Tee	"	0.571
Cap	"	0.200
Coupling	"	0.400
Union	"	0.444

Plumbing	UNIT	MAN/HOURS
15410.15 Brass Fittings		
Compression fittings, union		
3/8"	EA.	0.133
1/2"	"	0.133
5/8"	"	0.133
Union elbow		
3/8"	EA.	0.133
1/2"	"	0.133
5/8"	"	0.133
Union tee		
3/8"	EA.	0.133
1/2"	"	0.133
5/8"	"	0.133
Male connector		
3/8"	EA.	0.133
1/2"	"	0.133
5/8"	"	0.133
Female connector		
3/8"	EA.	0.133
1/2"	"	0.133
5/8"	"	0.133
Brass flare fittings, union		
3/8"	EA.	0.129
1/2"	"	0.129
5/8"	"	0.129
90 deg elbow union		
3/8"	EA.	0.129
1/2"	"	0.129
5/8"	"	0.129
Three way tee		
3/8"	EA.	0.216
1/2"	"	0.216
5/8"	"	0.216
Cross		
3/8"	EA.	0.286
1/2"	"	0.286
5/8"	"	0.286
Male connector, half union		
3/8"	EA.	0.129
1/2"	"	0.129
5/8"	"	0.129
Female connector, half union		
3/8"	EA.	0.129
1/2"	"	0.129
5/8"	"	0.129
Long forged nut		
3/8"	EA.	0.129
1/2"	"	0.129
5/8"	"	0.129
Short forged nut		
3/8"	EA.	0.129
1/2"	"	0.129
5/8"	"	0.129

Plumbing	UNIT	MAN/HOURS
15410.18 — Glass Pipe		
Glass pipe		
1-1/2" dia.	L.F.	0.160
2" dia.	"	0.178
3" dia.	"	0.200
4" dia.	"	0.229
6" dia.	"	0.267
15410.30 — Pvc/cpvc Pipe		
PVC schedule 40		
1/2" pipe	L.F.	0.033
3/4" pipe	"	0.036
1" pipe	"	0.040
1-1/4" pipe	"	0.044
1-1/2" pipe	"	0.050
2" pipe	"	0.057
2-1/2" pipe	"	0.067
3" pipe	"	0.080
4" pipe	"	0.100
6" pipe	"	0.200
8" pipe	"	0.267
Fittings, 1/2"		
90 deg ell	EA.	0.100
45 deg ell	"	0.100
Tee	"	0.114
3/4"		
90 deg elbow	EA.	0.133
45 deg elbow	"	0.133
Tee	"	0.160
1"		
90 deg elbow	EA.	0.160
45 deg elbow	"	0.160
Tee	"	0.178
1-1/4"		
90 deg elbow	EA.	0.229
45 deg elbow	"	0.229
Tee	"	0.267
1-1/2"		
90 deg elbow	EA.	0.229
45 deg elbow	"	0.229
Tee	"	0.267
2"		
90 deg elbow	EA.	0.267
45 deg elbow	"	0.267
Tee	"	0.320
2-1/2"		
90 deg elbow	EA.	0.500
45 deg elbow	"	0.500
Tee	"	0.533
3"		
90 deg elbow	EA.	0.667
45 deg elbow	"	0.667
Tee	"	0.727
4"		

Plumbing	UNIT	MAN/HOURS
15410.30 — Pvc/cpvc Pipe (Cont.)		
90 deg elbow	EA.	0.800
45 deg elbow	"	0.800
Tee	"	0.889
PVC schedule 80 pipe		
1-1/2" pipe	L.F.	0.050
2" pipe	"	0.057
3" pipe	"	0.080
4" pipe	"	0.100
Fittings, 1-1/2"		
90 deg elbow	EA.	0.267
45 deg elbow	"	0.267
Tee	"	0.400
2"		
90 deg elbow	EA.	0.320
45 deg elbow	"	0.320
Tee	"	0.500
2-1/2"		
90 deg elbow	EA.	0.500
45 deg elbow	"	0.500
Tee	"	0.667
3"		
90 deg elbow	EA.	0.667
45 deg elbow	"	0.667
Tee	"	0.800
4"		
90 deg elbow	EA.	0.800
45 deg elbow	"	0.800
Tee	"	1.000
CPVC schedule 40		
1/2" pipe	L.F.	0.033
3/4" pipe	"	0.036
1" pipe	"	0.040
1-1/4" pipe	"	0.044
1-1/2" pipe	"	0.050
2" pipe	"	0.057
Fittings, CPVC, schedule 80		
1/2", 90 deg ell	EA.	0.080
Tee	"	0.133
3/4", 90 deg ell	"	0.080
Tee	"	0.133
1", 90 deg ell	"	0.089
Tee	"	0.145
1-1/4", 90 deg ell	"	0.089
Tee	"	0.145
1-1/2", 90 deg ell	"	0.160
Tee	"	0.200
2", 90 deg ell	"	0.160
Tee	"	0.200
Polypropylene, acid resistant, DWV pipe		
Schedule 40		
1-1/2" pipe	L.F.	0.057
2" pipe	"	0.067
3" pipe	"	0.080

15410.30 — Pvc/cpvc Pipe (Cont.)

Plumbing	UNIT	MAN/HOURS
4" pipe	L.F.	0.100
6" pipe	"	0.200
Fittings		
1-1/2"		
1/4 bend	EA.	0.200
1/8 bend	"	0.200
Sanitary tee	"	0.400
Cleanout with plug	"	0.400
Wye	"	0.400
2"		
1/4 bend	EA.	0.229
1/8 bend	"	0.229
Sanitary tee	"	0.471
Cleanout with plug	"	0.471
Wye	"	0.471
3"		
1/4 bend	EA.	0.267
1/8 bend	"	0.267
Sanitary tee	"	0.533
Cleanout with plug	"	0.533
Wye	"	0.533
4"		
1/4 bend	EA.	0.400
1/8 bend	"	0.400
Sanitary tee	"	0.800
Cleanout with plug	"	0.800
Wye	"	0.800
6"		
1/4 bend	EA.	0.667
1/8 bend	"	0.667
Sanitary tee	"	1.333
Cleanout with plug	"	1.333
Wye	"	1.333
Polyethylene pipe and fittings		
SDR-21		
3" pipe	L.F.	0.100
4" pipe	"	0.133
6" pipe	"	0.200
8" pipe	"	0.229
10" pipe	"	0.267
12" pipe	"	0.320
14" pipe	"	0.400
16" pipe	"	0.500
18" pipe	"	0.615
20" pipe	"	0.800
22" pipe	"	0.889
24" pipe	"	1.000
Fittings, 3"		
90 deg elbow	EA.	0.400
45 deg elbow	"	0.400
Tee	"	0.667
4"		
90 deg elbow	EA.	0.500

15410.30 — Pvc/cpvc Pipe (Cont.)

Plumbing	UNIT	MAN/HOURS
45 deg elbow	EA.	0.500
Tee	"	0.800
8"		
90 deg elbow	EA.	1.000
45 deg elbow	"	1.000
Tee	"	1.600
10"		
90 deg elbow	EA.	1.333
45 deg elbow	"	1.333
Tee	"	2.000
12"		
90 deg elbow	EA.	1.600
45 deg elbow	"	1.600
Tee	"	2.667
14"		
90 deg elbow	EA.	2.000
45 deg elbow	"	2.000
Tee	"	3.200
16"		
90 deg elbow	EA.	2.000
45 deg elbow	"	2.000
Tee	"	3.200
18"		
90 deg elbow	EA.	2.667
45 deg elbow	"	2.667
Tee	"	4.000
20"		
90 deg elbow	EA.	2.667
45 deg elbow	"	2.667

15410.33 — Abs Dwv Pipe

Plumbing	UNIT	MAN/HOURS
Schedule 40 ABS		
1-1/2" pipe	L.F.	0.040
2" pipe	"	0.044
3" pipe	"	0.057
4" pipe	"	0.080
6" pipe	"	0.100
Fittings		
1/8 bend		
1-1/2"	EA.	0.160
2"	"	0.200
3"	"	0.267
4"	"	0.320
6"	"	0.400
Tee, sanitary		
1-1/2"	EA.	0.267
2"	"	0.320
3"	"	0.400
4"	"	0.500
6"	"	0.667
Tee, sanitary reducing		
2 x 1-1/2 x 1-1/2	EA.	0.320
2 x 1-1/2 x 2	"	0.333

Left Column

Plumbing	UNIT	MAN/HOURS
15410.33 — Abs Dwv Pipe *(Cont.)*		
2 x 2 x 1-1/2	EA.	0.364
3 x 3 x 1-1/2	"	0.400
3 x 3 x 2	"	0.444
4 x 4 x 1-1/2	"	0.500
4 x 4 x 2	"	0.571
4 x 4 x 3	"	0.615
6 x 6 x 4	"	0.667
Wye		
1-1/2"	EA.	0.229
2"	"	0.320
3"	"	0.400
4"	"	0.500
6"	"	0.667
Reducer		
2 x 1-1/2	EA.	0.200
3 x 1-1/2	"	0.267
3 x 2	"	0.267
4 x 2	"	0.320
4 x 3	"	0.320
6 x 4	"	0.400
P-trap		
1-1/2"	EA.	0.267
2"	"	0.296
3"	"	0.348
4"	"	0.400
6"	"	0.500
Double sanitary, tee		
1-1/2"	EA.	0.320
2"	"	0.400
3"	"	0.500
4"	"	0.667
Long sweep, 1/4 bend		
1-1/2"	EA.	0.160
2"	"	0.200
3"	"	0.267
4"	"	0.400
Wye, standard		
1-1/2"	EA.	0.267
2"	"	0.320
3"	"	0.400
4"	"	0.500
15410.35 — Plastic Pipe		
Fiberglass reinforced pipe		
2" pipe	L.F.	0.062
3" pipe	"	0.067
4" pipe	"	0.073
6" pipe	"	0.080
8" pipe	"	0.133
10" pipe	"	0.160
12" pipe	"	0.200
Fittings		
90 deg elbow, flanged		

Right Column

Plumbing	UNIT	MAN/HOURS
15410.35 — Plastic Pipe *(Cont.)*		
2"	EA.	0.800
3"	"	0.889
4"	"	1.000
6"	"	1.333
8"	"	1.600
10"	"	2.000
12"	"	2.667
45 deg elbow, flanged		
2"	EA.	0.667
3"	"	0.800
4"	"	1.000
6"	"	1.333
8"	"	1.600
10"	"	2.000
12"	"	2.667
Tee, flanged		
2"	EA.	1.000
3"	"	1.143
4"	"	1.333
6"	"	1.600
8"	"	2.000
10"	"	2.667
12"	"	4.000
Wye, flanged		
2"	EA.	1.000
3"	"	1.143
4"	"	1.333
6"	"	1.600
8"	"	2.000
10"	"	2.667
12"	"	4.000
15410.70 — Stainless Steel Pipe		
Stainless steel, schedule 40, threaded		
1/2" pipe	L.F.	0.114
1" pipe	"	0.123
1-1/2" pipe	"	0.133
2" pipe	"	0.145
2-1/2" pipe	"	0.160
3" pipe	"	0.178
4" pipe	"	0.200
Fittings, 1/2"		
90 deg ell	EA.	1.000
45 deg ell	"	1.000
Tee	"	1.333
3/4"		
90 deg ell	EA.	1.000
45 deg ell	"	1.000
Tee	"	1.333
1"		
90 deg ell	EA.	1.000
45 deg ell	"	1.000
Tee	"	1.333

Plumbing	UNIT	MAN/HOURS	Plumbing	UNIT	MAN/HOURS
15410.70 Stainless Steel Pipe *(Cont.)*			**15410.70** Stainless Steel Pipe *(Cont.)*		
1-1/4"			1/4"	L.F.	0.044
90 deg ell	EA.	1.000	3/8"	"	0.050
45 deg ell	"	1.000	1/2"	"	0.057
Tee	"	1.333	5/8"	"	0.067
1-1/2"			3/4"	"	0.080
90 deg ell	EA.	1.333	7/8"	"	0.089
45 deg ell	"	1.333	1"	"	0.100
Tee	"	1.600	.049 wall		
2"			1/4"	L.F.	0.047
90 deg ell	EA.	1.600	3/8"	"	0.053
45 deg ell	"	1.600	1/2"	"	0.062
Tee	"	2.667	5/8"	"	0.073
Type 304, sch 10 pipe			3/4"	"	0.089
1" pipe	L.F.	0.100	7/8"	"	0.100
1-1/4" pipe	"	0.133	1"	"	0.114
1-1/2" pipe	"	0.145	.065 wall		
2" pipe	"	0.178	1/4"	L.F.	0.053
2-1/2" pipe	"	0.200	3/8"	"	0.067
3" pipe	"	0.229	1/2"	"	0.073
4" pipe	"	0.267	5/8"	"	0.089
6" pipe	"	0.320	3/4"	"	0.114
Fittings, 1"			7/8"	"	0.133
90 deg elbow	EA.	1.600	1"	"	0.160
45 deg elbow	"	1.600	Type 316 tubing		
Tee	"	2.667	.035 wall		
1-1/4"			1/4"	L.F.	0.044
90 deg elbow	EA.	1.600	3/8"	"	0.050
45 deg elbow	"	1.600	1/2"	"	0.057
Tee	"	2.667	5/8"	"	0.067
1-1/2"			3/4"	"	0.080
90 deg elbow	EA.	1.333	7/8"	"	0.089
45 deg elbow	"	1.333	1"	"	0.100
Tee	"	2.667	.049 wall		
2"			1/4"	L.F.	0.053
90 deg elbow	EA.	1.600	3/8"	"	0.067
45 deg elbow	"	1.600	1/2"	"	0.073
Tee	"	4.000	5/8"	"	0.089
2-1/2"			3/4"	"	0.114
90 deg elbow	EA.	1.600	7/8"	"	0.133
45 deg elbow	"	1.600	1"	"	0.160
Tee	"	4.000	.065 wall		
3"			1/4"	L.F.	0.053
90 deg elbow	EA.	2.000	3/8"	"	0.067
45 deg elbow	"	2.000	1/2"	"	0.073
Tee	"	5.333	5/8"	"	0.089
4"			3/4"	"	0.114
90 deg elbow	EA.	2.000	7/8"	"	0.133
45 deg elbow	"	2.000	1"	"	0.160
6"			Fittings, 1/4"		
90 deg elbow	EA.	2.667	90 deg elbow	EA.	0.160
45 deg elbow	"	2.667	Union tee	"	0.267
Type 304 tubing			Union	"	0.267
.035 wall			Male connector	"	0.200

Plumbing	UNIT	MAN/HOURS
15410.70 Stainless Steel Pipe *(Cont.)*		
3/8"		
90 deg elbow	EA.	0.200
Union tee	"	0.308
Union	"	0.308
Male connector	"	0.200
1/2"		
90 deg elbow	EA.	0.211
Union tee	"	0.333
Union	"	0.333
Male connector	"	0.200
5/8"		
90 deg elbow	EA.	0.267
Union tee	"	0.400
Union	"	0.400
Male connector	"	0.267
3/4"		
90 deg elbow	EA.	0.267
Union tee	"	0.400
Union	"	0.400
Male connector	"	0.267
7/8"		
90 deg elbow	EA.	0.286
Union tee	"	0.444
Union	"	0.444
Male connector	"	0.286
1"		
90 deg elbow	EA.	0.364
Union tee	"	0.500
Union	"	0.500
Male connector	"	0.400
Type 316 valves		
Gate valves		
1/4"	EA.	0.267
3/8"	"	0.320
1/2"	"	0.348
3/4"	"	0.400
1"	"	0.533
Globe valves		
1/4"	EA.	0.267
3/8"	"	0.320
1/2"	"	0.348
3/4"	"	0.400
1"	"	0.533
Check valves		
1/4"	EA.	0.267
3/8"	"	0.320
1/2"	"	0.348
3/4"	"	0.400
1"	"	0.533

Plumbing	UNIT	MAN/HOURS
15410.80 Steel Pipe		
Black steel, extra heavy pipe, threaded		
1/2" pipe	L.F.	0.032
3/4" pipe	"	0.032
1" pipe	"	0.040
1-1/2" pipe	"	0.044
2-1/2" pipe	"	0.100
3" pipe	"	0.133
4" pipe	"	0.160
5" pipe	"	0.200
6" pipe	"	0.200
8" pipe	"	0.267
10" pipe	"	0.320
12" pipe	"	0.400
Fittings, malleable iron, threaded, 1/2" pipe		
90 deg ell	EA.	0.267
45 deg ell	"	0.267
Tee	"	0.400
Reducing tee	"	0.400
Cap	"	0.160
Coupling	"	0.320
Union	"	0.267
Nipple, 4" long	"	0.267
3/4" pipe		
90 deg ell	EA.	0.267
45 deg ell	"	0.400
Tee	"	0.400
Reducing tee	"	0.267
Cap	"	0.160
Coupling	"	0.267
Union	"	0.267
Nipple, 4" long	"	0.267
1" pipe		
90 deg ell	EA.	0.320
45 deg ell	"	0.320
Tee	"	0.444
Reducing tee	"	0.444
Cap	"	0.160
Coupling	"	0.320
Union	"	0.320
Nipple, 4" long	"	0.320
1-1/2" pipe		
90 deg ell	EA.	0.400
45 deg ell	"	0.400
Tee	"	0.571
Reducing tee	"	0.571
Cap	"	0.200
Coupling	"	0.400
Union	"	0.400
Nipple, 4" long	"	0.400
2-1/2" pipe		
90 deg ell	EA.	1.000
45 deg ell	"	1.000
Tee	"	1.333

15410.80 — Steel Pipe (Cont.)

Plumbing	UNIT	MAN/HOURS
Reducing tee	EA.	1.333
Cap	"	0.500
Coupling	"	1.333
Union	"	1.333
Nipple, 4" long	"	1.333
3" pipe		
90 deg ell	EA.	1.333
45 deg ell	"	1.333
Tee	"	2.000
Reducing tee	"	2.000
Cap	"	0.667
Coupling	"	1.333
Union	"	1.333
Nipple, 4" long	"	1.333
4" pipe		
90 deg ell	EA.	1.600
45 deg ell	"	1.600
Tee	"	2.667
Reducing tee	"	2.667
Cap	"	2.667
Coupling	"	0.800
Union	"	2.667
Nipple, 4" long	"	2.667
6" pipe		
90 deg ell	EA.	1.600
45 deg ell	"	1.600
Tee	"	2.667
Reducing tee	"	2.667
Cap	"	0.800
8" pipe		
90 deg ell	EA.	3.200
45 deg ell	"	3.200
Tee	"	5.000
Reducing tee	"	4.211
Cap	"	1.600
10" pipe		
90 deg ell	EA.	4.000
45 deg ell	"	4.000
Tee	"	5.000
Reducing tee	"	2.000
Cap	"	2.000
12" pipe		
90 deg ell	EA.	5.000
45 deg ell	"	5.000
Tee	"	6.667
Reducing tee	"	6.667
Cap	"	2.667

15410.82 — Galvanized Steel Pipe

Plumbing	UNIT	MAN/HOURS
Galvanized pipe		
1/2" pipe	L.F.	0.080
3/4" pipe	"	0.100
1" pipe	"	0.114
1-1/4" pipe	"	0.133
1-1/2" pipe	"	0.160
2" pipe	"	0.200
2-1/2" pipe	"	0.267
3" pipe	"	0.286
4" pipe	"	0.333
6" pipe	"	0.667
90 degree ell, 150 lb malleable iron, galvanized		
1/2"	EA.	0.160
3/4"	"	0.200
1"	"	0.211
1-1/4"	"	0.235
1-1/2"	"	0.267
2"	"	0.320
2-1/2"	"	0.500
3"	"	0.615
4"	"	0.667
5"	"	0.800
6"	"	0.800
45 degree ell, 150 lb m.i., galv.		
1/2"	EA.	0.160
3/4"	"	0.200
1"	"	0.211
1-1/4"	"	0.235
1-1/2"	"	0.267
2"	"	0.320
2-1/2"	"	0.500
3"	"	0.615
4"	"	0.800
5"	"	0.800
6"	"	1.000
Tees, straight, 150 lb m.i., galv.		
1/2"	EA.	0.200
3/4"	"	0.229
1"	"	0.267
1-1/4"	"	0.320
1-1/2"	"	0.400
2"	"	0.500
2-1/2"	"	0.667
3"	"	0.800
4"	"	1.000
5"	"	1.143
6"	"	1.333
Couplings, straight, 150 lb m.i., galv.		
1/2"	EA.	0.160
3/4"	"	0.178
1"	"	0.200
1-1/4"	"	0.229
1-1/2"	"	0.267

Plumbing	UNIT	MAN/HOURS
15410.82 Galvanized Steel Pipe *(Cont.)*		
2"	EA.	0.320
2-1/2"	"	0.500
3"	"	0.667
4"	"	0.727
5"	"	0.800
6"	"	0.800
Caps, 150 lb m.i., galv.		
1/2"	EA.	0.080
3/4"	"	0.084
1"	"	0.089
1-1/4"	"	0.094
1-1/2"	"	0.100
2"	"	0.114
2-1/2"	"	0.145
3"	"	0.200
4"	"	0.250
5"	"	0.308
6"	"	0.400
Unions, 150 lb m.i., galv.		
1/2"	EA.	0.200
3/4"	"	0.229
1"	"	0.267
1-1/4"	"	0.320
1-1/2"	"	0.400
2"	"	0.444
2-1/2"	"	0.533
3"	"	0.667
Nipples, galvanized steel, 4" long		
1/2"	EA.	0.100
3/4"	"	0.107
1"	"	0.114
1-1/4"	"	0.123
1-1/2"	"	0.133
2"	"	0.145
2-1/2"	"	0.160
3"	"	0.200
4"	"	0.267
Square head plug (C.I.)		
1/2"	EA.	0.089
3/4"	"	0.100
1"	"	0.107
1-1/4"	"	0.114
1-1/2"	"	0.123
2"	"	0.133
2-1/2"	"	0.178
3"	"	0.200
4"	"	0.267
5"	"	0.320
6"	"	0.400
Screwed flanges, galv.		
1"	EA.	0.400
1-1/4"	"	0.444
1-1/2"	"	0.500

Plumbing	UNIT	MAN/HOURS
15410.82 Galvanized Steel Pipe *(Cont.)*		
2"	EA.	0.500
2-1/2"	"	0.533
3"	"	0.727
4"	"	1.000
5"	"	1.333
6"	"	1.333
15430.23 Cleanouts		
Cleanout, wall		
2"	EA.	0.533
3"	"	0.533
4"	"	0.667
6"	"	0.800
8"	"	1.000
Floor		
2"	EA.	0.667
3"	"	0.667
4"	"	0.800
6"	"	1.000
8"	"	1.143
15430.24 Grease Traps		
Grease traps, cast iron, 3" pipe		
35 gpm, 70 lb capacity	EA.	8.000
50 gpm, 100 lb capacity	"	10.000
15430.25 Hose Bibbs		
Hose bibb		
1/2"	EA.	0.267
3/4"	"	0.267
15430.60 Valves		
Gate valve, 125 lb, bronze, soldered		
1/2"	EA.	0.200
3/4"	"	0.200
1"	"	0.267
1-1/2"	"	0.320
2"	"	0.400
2-1/2"	"	0.500
Threaded		
1/4", 125 lb	EA.	0.320
1/2"		
125 lb	EA.	0.320
150 lb	"	0.320
300 lb	"	0.320
3/4"		
125 lb	EA.	0.320
150 lb	"	0.320
300 lb	"	0.320
1"		
125 lb	EA.	0.320
150 lb	"	0.320
300 lb	"	0.400

Plumbing	UNIT	MAN/HOURS
15430.60 **Valves** *(Cont.)*		
1-1/2"		
125 lb	EA.	0.400
150 lb	"	0.400
300 lb	"	0.444
2"		
125 lb	EA.	0.571
150 lb	"	0.571
300 lb	"	0.667
Cast iron, flanged		
2", 150 lb	EA.	0.667
2-1/2"		
125 lb	EA.	0.667
150 lb	"	0.667
250 lb	"	0.667
3"		
125 lb	EA.	0.800
150 lb	"	0.800
250 lb	"	0.800
4"		
125 lb	EA.	1.143
150 lb	"	1.143
250 lb	"	1.143
6"		
125 lb	EA.	1.600
250 lb	"	1.600
8"		
125 lb	EA.	2.000
250 lb	"	2.000
OS&Y, flanged		
2"		
125 lb	EA.	0.667
250 lb	"	0.667
2-1/2"		
125 lb	EA.	0.667
250 lb	"	0.800
3"		
125 lb	EA.	0.800
250 lb	"	0.800
4"		
125 lb	EA.	1.333
250 lb	"	1.333
6"		
125 lb	EA.	1.600
250 lb	"	1.600
Check valve, bronze, soldered, 125 lb		
1/2"	EA.	0.200
3/4"	"	0.200
1"	"	0.267
1-1/4"	"	0.320
1-1/2"	"	0.320
2"	"	0.400
Threaded		
1/2"		

Plumbing	UNIT	MAN/HOURS
15430.60 **Valves** *(Cont.)*		
125 lb	EA.	0.267
150 lb	"	0.267
200 lb	"	0.267
3/4"		
125 lb	EA.	0.320
150 lb	"	0.320
200 lb	"	0.320
1"		
125 lb	EA.	0.400
150 lb	"	0.400
200 lb	"	0.400
Flow check valve, cast iron, threaded		
1"	EA.	0.320
1-1/4"	"	0.400
1-1/2"		
125 lb	EA.	0.400
150 lb	"	0.400
200 lb	"	0.444
2"		
125 lb	EA.	0.444
150 lb	"	0.444
200 lb	"	0.500
2-1/2"		
125 lb	EA.	0.667
250 lb	"	0.800
3"		
125 lb	EA.	0.800
250 lb	"	1.000
4"		
125 lb	EA.	1.143
250 lb	"	1.333
6"		
125 lb	EA.	1.600
250 lb	"	1.600
Vertical check valve, bronze, 125 lb, threaded		
1/2"	EA.	0.320
3/4"	"	0.364
1"	"	0.400
1-1/4"	"	0.444
1-1/2"	"	0.500
2"	"	0.571
Cast iron, flanged		
2-1/2"	EA.	0.800
3"	"	1.000
4"	"	1.333
6	"	1.600
8"	"	2.000
10"	"	2.667
12"	"	3.200
Globe valve, bronze, soldered, 125 lb		
1/2"	EA.	0.229
3/4"	"	0.250
1"	"	0.267

Plumbing	UNIT	MAN/HOURS
15430.60 Valves *(Cont.)*		
1-1/4"	EA.	0.286
1-1/2"	"	0.333
2"	"	0.400
Threaded		
1/2"		
125 lb	EA.	0.267
150 lb	"	0.267
300 lb	"	0.267
3/4"		
125 lb	EA.	0.320
150 lb	"	0.320
300 lb	"	0.320
1"		
125 lb	EA.	0.400
150 lb	"	0.400
300 lb	"	0.400
1-1/4"		
125 lb	EA.	0.400
150 lb	"	0.400
300 lb	"	0.400
1-1/2"		
125 lb	EA.	0.444
150 lb	"	0.444
300 lb	"	0.444
2"		
125 lb	EA.	0.533
150 lb	"	0.533
300 lb	"	0.533
Cast iron flanged		
2-1/2"		
125 lb	EA.	0.800
250 lb	"	0.800
3"		
125 lb	EA.	1.000
250 lb	"	1.000
4"		
125 lb	EA.	1.333
250 lb	"	1.333
6"		
125 lb	EA.	1.600
250 lb	"	1.600
8"		
125 lb	EA.	2.000
250 lb	"	2.000
Butterfly valve, cast iron, wafer type		
2"		
150 lb	EA.	0.571
200 lb	"	0.667
2-1/2"		
150 lb	EA.	0.667
200 lb	"	0.727
3"		
150 lb	EA.	0.800

Plumbing	UNIT	MAN/HOURS
15430.60 Valves *(Cont.)*		
200 lb	EA.	0.889
4"		
150 lb	EA.	1.143
200 lb	"	1.333
6"		
150 lb	EA.	1.600
200 lb	"	1.600
8"		
150 lb	EA.	1.778
200 lb	"	2.000
10"		
150 lb	EA.	2.000
200 lb	"	2.667
Ball valve, bronze, 250 lb, threaded		
1/2"	EA.	0.320
3/4"	"	0.320
1"	"	0.400
1-1/4"	"	0.444
1-1/2"	"	0.500
2"	"	0.571
Angle valve, bronze, 150 lb, threaded		
1/2"	EA.	0.286
3/4"	"	0.320
1"	"	0.320
1-1/4"	"	0.400
1-1/2"	"	0.444
Balancing valve, with meter connections, circuit setter		
1/2"	EA.	0.320
3/4"	"	0.364
1"	"	0.400
1-1/4"	"	0.444
1-1/2"	"	0.533
2"	"	0.667
2-1/2"	"	0.800
3"	"	1.000
4"	"	1.333
Balancing valve, straight type		
1/2"	EA.	0.320
3/4"	"	0.320
Angle type		
1/2"	EA.	0.320
3/4"	"	0.320
Square head cock, 125 lb, bronze body		
1/2"	EA.	0.267
3/4"	"	0.320
1"	"	0.364
1-1/4"	"	0.400
Radiator temp control valve, with control and sensor		
1/2" valve	EA.	0.500
1" valve	"	0.500
Pressure relief valve, 1/2", bronze		
Low pressure	EA.	0.320
High pressure	"	0.320

Plumbing

15430.60	Valves *(Cont.)*	UNIT	MAN/HOURS
Pressure and temperature relief valve			
Bronze, 3/4"		EA.	0.320
Cast iron, 3/4"			
High pressure		EA.	0.320
Temperature relief		"	0.320
Pressure & temp relief valve		"	0.320
Pressure reducing valve, bronze, threaded, 250 lb			
1/2"		EA.	0.500
3/4"		"	0.500
1"		"	0.500
1-1/4"		"	0.571
1-1/2"		"	0.667
Pressure regulating valve, bronze, class 300			
1"		EA.	0.500
1-1/2"		"	0.615
2"		"	0.800
3"		"	1.143
4"		"	1.600
5"		"	2.000
6"		"	2.667
Solar water temperature regulating valve			
3/4"		EA.	0.667
1"		"	0.800
1-1/4"		"	0.889
1-1/2"		"	1.000
2"		"	1.143
2-1/2"		"	2.000
Tempering valve, threaded			
3/4"		EA.	0.267
1"		"	0.320
1-1/4"		"	0.400
1-1/2"		"	0.400
2"		"	0.500
2-1/2"		"	0.667
3"		"	0.800
4"		"	1.143
Thermostatic mixing valve, threaded			
1/2"		EA.	0.286
3/4"		"	0.320
1"		"	0.348
1-1/2"		"	0.400
2"		"	0.500
Sweat connection			
1/2"		EA.	0.286
3/4"		"	0.320
Mixing valve, sweat connection			
1/2"		EA.	0.286
3/4"		"	0.320
Liquid level gauge, aluminum body			
3/4"		EA.	0.320

Plumbing

15430.60	Valves *(Cont.)*	UNIT	MAN/HOURS
4125 psi, pvc body			
3/4"		EA.	0.320
150 psi, crs body			
3/4"		EA.	0.320
1"		"	0.320
175 psi, bronze body, 1/2"		"	0.286

15430.65	Vacuum Breakers	UNIT	MAN/HOURS
Vacuum breaker, atmospheric, threaded connection			
3/4"		EA.	0.320
1"		"	0.320
Anti-siphon, brass			
3/4"		EA.	0.320
1"		"	0.320
1-1/4"		"	0.400
1-1/2"		"	0.444
2"		"	0.500

15430.68	Strainers	UNIT	MAN/HOURS
Strainer, Y pattern, 125 psi, cast iron body, threaded			
3/4"		EA.	0.286
1"		"	0.320
1-1/4"		"	0.400
1-1/2"		"	0.400
2"		"	0.500
250 psi, brass body, threaded			
3/4"		EA.	0.320
1"		"	0.320
1-1/4"		"	0.400
1-1/2"		"	0.400
2"		"	0.500
Cast iron body, threaded			
3/4"		EA.	0.320
1"		"	0.320
1-1/4"		"	0.400
1-1/2"		"	0.400
2"		"	0.500

15430.70	Drains, Roof & Floor	UNIT	MAN/HOURS
Floor drain, cast iron, with cast iron top			
2"		EA.	0.667
3"		"	0.667
4"		"	0.667
6"		"	0.800
Roof drain, cast iron			
2"		EA.	0.667
3"		"	0.667
4"		"	0.667
5"		"	0.800
6"		"	0.800

Plumbing	UNIT	MAN/HOURS
15430.80 — Traps		
Bucket trap, threaded		
3/4"	EA.	0.500
1"	"	0.533
1-1/4"	"	0.615
1-1/2"	"	0.727
Inverted bucket steam trap, threaded		
3/4"	EA.	0.500
1"	"	0.500
1-1/4"	"	0.444
1-1/2"	"	0.667
Float trap, 15 psi		
3/4"	EA.	0.500
1"	"	0.533
1-1/4"	"	0.571
1-1/2"	"	0.667
2"	"	0.800
Float and thermostatic trap, 15 psi		
3/4"	EA.	0.500
1"	"	0.533
1-1/4"	"	0.571
1-1/2"	"	0.667
2"	"	0.800
Steam trap, cast iron body, threaded, 125 psi		
3/4"	EA.	0.500
1"	"	0.533
1-1/4"	"	0.571
1-1/2"	"	0.667
Thermostatic trap, low pressure, angle type, 25 psi		
1/2"	EA.	0.500
3/4"	"	0.500
1"	"	0.533
50 psi		
1/2"	EA.	0.500
3/4"	"	0.500
1"	"	0.533
Cast iron body, threaded, 125 psi		
3/4"	EA.	0.500
1"	"	0.571
1-1/4"	"	0.615
1-1/2"	"	0.667

Plumbing Fixtures	UNIT	MAN/HOURS
15440.10 — Baths		
Bath tub, 5' long		
Minimum	EA.	2.667
Average	"	4.000
Maximum	"	8.000
6' long		
Minimum	EA.	2.667
Average	"	4.000
Maximum	"	8.000
Square tub, whirlpool, 4'x4'		
Minimum	EA.	4.000
Average	"	8.000
Maximum	"	10.000
5'x5'		
Minimum	EA.	4.000
Average	"	8.000
Maximum	"	10.000
6'x6'		
Minimum	EA.	4.000
Average	"	8.000
Maximum	"	10.000
For trim and rough-in		
Minimum	EA.	2.667
Average	"	4.000
Maximum	"	8.000
15440.12 — Disposals & Accessories		
Continuous feed		
Minimum	EA.	1.600
Average	"	2.000
Maximum	"	2.667
Batch feed, 1/2 hp		
Minimum	EA.	1.600
Average	"	2.000
Maximum	"	2.667
Hot water dispenser		
Minimum	EA.	1.600
Average	"	2.000
Maximum	"	2.667
Epoxy finish faucet	"	1.600
Lock stop assembly	"	1.000
Mounting gasket	"	0.667
Tailpipe gasket	"	0.667
Stopper assembly	"	0.800
Switch assembly, on/off	"	1.333
Tailpipe gasket washer	"	0.400
Stop gasket	"	0.444
Tailpipe flange	"	0.400
Tailpipe	"	0.500

Plumbing Fixtures

15440.15 — Faucets

Plumbing Fixtures	UNIT	MAN/HOURS
Kitchen		
Minimum	EA.	1.333
Average	"	1.600
Maximum	"	2.000
Bath		
Minimum	EA.	1.333
Average	"	1.600
Maximum	"	2.000
Lavatory, domestic		
Minimum	EA.	1.333
Average	"	1.600
Maximum	"	2.000
Hospital, patient rooms		
Minimum	EA.	2.000
Average	"	2.667
Maximum	"	4.000
Operating room		
Minimum	EA.	2.000
Average	"	2.667
Maximum	"	4.000
Washroom		
Minimum	EA.	1.333
Average	"	1.600
Maximum	"	2.000
Handicapped		
Minimum	EA.	1.600
Average	"	2.000
Maximum	"	2.667
Shower		
Minimum	EA.	1.333
Average	"	1.600
Maximum	"	2.000
For trim and rough-in		
Minimum	EA.	1.600
Average	"	2.000
Maximum	"	4.000

15440.18 — Hydrants

Plumbing Fixtures	UNIT	MAN/HOURS
Wall hydrant		
8" thick	EA.	1.333
12" thick	"	1.600
18" thick	"	1.778
24" thick	"	2.000
Ground hydrant		
2' deep	EA.	1.000
4' deep	"	1.143
6' deep	"	1.333
8' deep	"	2.000

15440.20 — Lavatories

Plumbing Fixtures	UNIT	MAN/HOURS
Lavatory, counter top, porcelain enamel on cast iron		
Minimum	EA.	1.600
Average	"	2.000
Maximum	"	2.667
Wall hung, china		
Minimum	EA.	1.600
Average	"	2.000
Maximum	"	2.667
Handicapped		
Minimum	EA.	2.000
Average	"	2.667
Maximum	"	4.000
For trim and rough-in		
Minimum	EA.	2.000
Average	"	2.667
Maximum	"	4.000

15440.30 — Showers

Plumbing Fixtures	UNIT	MAN/HOURS
Shower, fiberglass, 36"x34"x84"		
Minimum	EA.	5.714
Average	"	8.000
Maximum	"	8.000
Steel, 1 piece, 36"x36"		
Minimum	EA.	5.714
Average	"	8.000
Maximum	"	8.000
Receptor, molded stone, 36"x36"		
Minimum	EA.	2.667
Average	"	4.000
Maximum	"	6.667
For trim and rough-in		
Minimum	EA.	3.636
Average	"	4.444
Maximum	"	8.000

15440.40 — Sinks

Plumbing Fixtures	UNIT	MAN/HOURS
Service sink, 24"x29"		
Minimum	EA.	2.000
Average	"	2.667
Maximum	"	4.000
Kitchen sink, single, stainless steel, single bowl		
Minimum	EA.	1.600
Average	"	2.000
Maximum	"	2.667
Double bowl		
Minimum	EA.	2.000
Average	"	2.667
Maximum	"	4.000
Porcelain enamel, cast iron, single bowl		
Minimum	EA.	1.600
Average	"	2.000
Maximum	"	2.667
Double bowl		

15440.40 — Sinks (Cont.)

Plumbing Fixtures	UNIT	MAN/HOURS
Minimum	EA.	2.000
Average	"	2.667
Maximum	"	4.000
Mop sink, 24"x36"x10"		
Minimum	EA.	1.600
Average	"	2.000
Maximum	"	2.667
Washing machine box		
Minimum	EA.	2.000
Average	"	2.667
Maximum	"	4.000
For trim and rough-in		
Minimum	EA.	2.667
Average	"	4.000
Maximum	"	5.333

15440.50 — Urinals

Plumbing Fixtures	UNIT	MAN/HOURS
Urinal, flush valve, floor mounted		
Minimum	EA.	2.000
Average	"	2.667
Maximum	"	4.000
Wall mounted		
Minimum	EA.	2.000
Average	"	2.667
Maximum	"	4.000
For trim and rough-in		
Minimum	EA.	2.000
Average	"	4.000
Maximum	"	5.333

15440.60 — Water Closets

Plumbing Fixtures	UNIT	MAN/HOURS
Water closet flush tank, floor mounted		
Minimum	EA.	2.000
Average	"	2.667
Maximum	"	4.000
Handicapped		
Minimum	EA.	2.667
Average	"	4.000
Maximum	"	8.000
Bowl, with flush valve, floor mounted		
Minimum	EA.	2.000
Average	"	2.667
Maximum	"	4.000
Wall mounted		
Minimum	EA.	2.000
Average	"	2.667
Maximum	"	4.000
For trim and rough-in		
Minimum	EA.	2.000
Average	"	2.667
Maximum	"	4.000

15440.70 — Water Heaters

Plumbing Fixtures	UNIT	MAN/HOURS
Water heater, electric		
6 gal	EA.	1.333
10 gal	"	1.333
15 gal	"	1.333
20 gal	"	1.600
30 gal	"	1.600
40 gal	"	1.600
52 gal	"	2.000
66 gal	"	2.000
80 gal	"	2.000
100 gal	"	2.667
120 gal	"	2.667
Oil fired		
20 gal	EA.	4.000
50 gal	"	5.714

15440.90 — Miscellaneous Fixtures

Plumbing Fixtures	UNIT	MAN/HOURS
Electric water cooler		
Floor mounted	EA.	2.667
Wall mounted	"	2.667
Wash fountain		
Wall mounted	EA.	4.000
Circular, floor supported	"	8.000
Deluge shower and eye wash	"	4.000

15440.95 — Fixture Carriers

Plumbing Fixtures	UNIT	MAN/HOURS
Water fountain, wall carrier		
Minimum	EA.	0.800
Average	"	1.000
Maximum	"	1.333
Lavatory, wall carrier		
Minimum	EA.	0.800
Average	"	1.000
Maximum	"	1.333
Sink, industrial, wall carrier		
Minimum	EA.	0.800
Average	"	1.000
Maximum	"	1.333
Toilets, water closets, wall carrier		
Minimum	EA.	0.800
Average	"	1.000
Maximum	"	1.333
Floor support		
Minimum	EA.	0.667
Average	"	0.800
Maximum	"	1.000
Urinals, wall carrier		
Minimum	EA.	0.800

Plumbing Fixtures	UNIT	MAN/ HOURS
15440.95 Fixture Carriers *(Cont.)*		
Average	EA.	1.000
Maximum	"	1.333
Floor support		
Minimum	EA.	0.667
Average	"	0.800
Maximum	"	1.000
15450.30 Pumps		
In-line pump, bronze, centrifugal		
5 gpm, 20' head	EA.	0.500
20 gpm, 40' head	"	0.500
50 gpm		
50' head	EA.	1.000
100' head	"	1.000
70 gpm, 100' head	"	1.333
100 gpm, 80' head	"	1.333
250 gpm, 150' head	"	2.000
Cast iron, centrifugal		
50 gpm, 200' head	EA.	1.000
100 gpm		
100' head	EA.	1.333
200' head	"	1.333
200 gpm		
100' head	EA.	2.000
200' head	"	2.000
Centrifugal, close coupled, c.i., single stage		
50 gpm, 100' head	EA.	1.000
100 gpm, 100' head	"	1.333
Base mounted		
50 gpm, 100' head	EA.	1.000
100 gpm, 50' head	"	1.333
200 gpm, 100' head	"	2.000
300 gpm, 175' head	"	2.000
Sump pump, bronze, 1750 rpm, 25 gpm		
20' head	EA.	10.000
150' head	"	13.333
50 gpm		
100' head	EA.	10.000
100 gpm		
50' head	EA.	10.000
Condensate pump, simplex		
1000 sf EDR, 2 gpm	EA.	6.667
2000 sf EDR, 3 gpm	"	6.667
4000 sf EDR, 6 gpm	"	7.273
6000 sf EDR, 9 gpm	"	7.273
Duplex, bronze		
8000 sf EDR, 12 gpm	EA.	7.273
10,000 sf EDR, 15 gpm	"	10.000
15,000 sf EDR, 23 gpm	"	11.429
20,000 sf EDR, 30 gpm	"	16.000
25,000 sf EDR, 38 gpm	"	16.000

Plumbing Fixtures	UNIT	MAN/ HOURS
15450.40 Storage Tanks		
Hot water storage tank, cement lined		
10 gallon	EA.	2.667
70 gallon	"	4.000
200 gallon	"	5.714
900 gallon	"	10.000
1100 gallon	"	10.000
2000 gallon	"	10.000
15480.10 Special Systems		
Air compressor, air cooled, two stage		
5.0 cfm, 175 psi	EA.	16.000
10 cfm, 175 psi	"	17.778
20 cfm, 175 psi	"	19.048
50 cfm, 125 psi	"	21.053
80 cfm, 125 psi	"	22.857
Single stage, 125 psi		
1.0 cfm	EA.	11.429
1.5 cfm	"	11.429
2.0 cfm	"	11.429
Automotive, hose reel, air and water, 50' hose	"	6.667
Lube equipment, 3 reel, with pumps	"	32.000
Tire changer		
Truck	EA.	11.429
Passenger car	"	6.154
Air hose reel, includes, 50' hose	"	6.154
Hose reel, 5 reel, motor oil, gear oil, lube, air & water	"	32.000
Water hose reel, 50' hose	"	6.154
Pump, air operated, for motor or gear oil, fits 55 gal drum	"	0.800
For chassis lube	"	0.800
Fuel dispensing pump, lighted dial, one product		
One hose	EA.	6.667
Two hose	"	6.667
Two products, two hose	"	6.667

Heating & Ventilating	UNIT	MAN/ HOURS
15555.10 Boilers		
Cast iron, gas fired, hot water		
115 mbh	EA.	20.000
175 mbh	"	21.818
235 mbh	"	24.000
940 mbh	"	48.000
1600 mbh	"	60.000
3000 mbh	"	80.000
6000 mbh	"	120.000
Steam		

Heating & Ventilating

15555.10 — Boilers (Cont.)

	UNIT	MAN/HOURS
115 mbh	EA.	20.000
175 mbh	"	21.818
235 mbh	"	24.000
940 mbh	"	48.000
1600 mbh	"	60.000
3000 mbh	"	80.000
6000 mbh	"	120.000
Electric, hot water		
115 mbh	EA.	12.000
175 mbh	"	12.000
235 mbh	"	12.000
940 mbh	"	24.000
1600 mbh	"	48.000
3000 mbh	"	60.000
6000 mbh	"	80.000
Steam		
115 mbh	EA.	12.000
175 mbh	"	12.000
235 mbh	"	12.000
940 mbh	"	24.000
1600 mbh	"	48.000
3000 mbh	"	60.000
6000 mbh	"	80.000
Oil fired, hot water		
115 mbh	EA.	16.000
175 mbh	"	18.462
235 mbh	"	21.818
940 mbh	"	40.000
1600 mbh	"	48.000
3000 mbh	"	60.000
6000 mbh	"	120.000
Steam		
115 mbh	EA.	16.000
175 mbh	"	18.462
235 mbh	"	21.818
940 mbh	"	40.000
1600 mbh	"	48.000
3000 mbh	"	60.000
6000 mbh	"	120.000

15610.10 — Furnaces

	UNIT	MAN/HOURS
Electric, hot air		
40 mbh	EA.	4.000
60 mbh	"	4.211
80 mbh	"	4.444
100 mbh	"	4.706
125 mbh	"	4.848
160 mbh	"	5.000
200 mbh	"	5.161
400 mbh	"	5.333
Gas fired hot air		
40 mbh	EA.	4.000
60 mbh	"	4.211

Heating & Ventilating

15610.10 — Furnaces (Cont.)

	UNIT	MAN/HOURS
80 mbh	EA.	4.444
100 mbh	"	4.706
125 mbh	"	4.848
160 mbh	"	5.000
200 mbh	"	5.161
400 mbh	"	5.333
Oil fired hot air		
40 mbh	EA.	4.000
60 mbh	"	4.211
80 mbh	"	4.444
100 mbh	"	4.706
125 mbh	"	4.848
160 mbh	"	5.000
200 mbh	"	5.161
400 mbh	"	5.333

Refrigeration

15670.10 — Condensing Units

	UNIT	MAN/HOURS
Air cooled condenser, single circuit		
3 ton	EA.	1.333
5 ton	"	1.333
7.5 ton	"	3.810
20 ton	"	4.000
25 ton	"	4.000
30 ton	"	4.000
With low ambient dampers		
3 ton	EA.	2.000
5 ton	"	2.000
7.5 ton	"	4.000
20 ton	"	5.333
25 ton	"	5.333
30 ton	"	5.333
Dual circuit		
10 ton	EA.	4.000
15 ton	"	5.714
20 ton	"	5.714
25 ton	"	5.714
30 ton	"	5.714
With low ambient dampers		
15 ton	EA.	5.714
20 ton	"	5.714
25 ton	"	5.714
30 ton	"	5.714

Refrigeration

Refrigeration	UNIT	MAN/HOURS
15680.10 — Chillers		
Chiller, reciprocal		
Air cooled, remote condenser, starter		
20 ton	EA.	8.000
25 ton	"	8.000
30 ton	"	8.000
40 ton	"	12.000
Water cooled, with starter		
20 ton	EA.	8.000
25 ton	"	8.000
30 ton	"	12.000
40 ton	"	12.000
Packaged, air cooled, with starter		
20 ton	EA.	6.000
25 ton	"	6.000
30 ton	"	6.000
40 ton	"	6.000
15710.10 — Cooling Towers		
Cooling tower, propeller type		
100 ton	EA.	8.000
200 ton	"	12.000
300 ton	"	20.000
400 ton	"	24.000
600 ton	"	34.286
800 ton	"	48.000
1000 ton	"	60.000
Centrifugal		
100 ton	EA.	8.000
200 ton	"	12.000
300 ton	"	20.000
400 ton	"	24.000
600 ton	"	34.286
800 ton	"	48.000
1000 ton	"	60.000

Heat Transfer	UNIT	MAN/HOURS
15780.10 — Computer Room A/c		
Air cooled, alarm, high efficiency filter, elec. heat		
3 ton	EA.	6.154
5 ton	"	6.667
7.5 ton	"	8.000
10 ton	"	10.000
15 ton	"	11.429
Steam heat		
3 ton	EA.	6.154

Heat Transfer	UNIT	MAN/HOURS
15780.10 — Computer Room A/c (Cont.)		
5 ton	EA.	6.667
7.5 ton	"	8.000
10 ton	"	10.000
15 ton	"	11.429
Hot water heat		
3 ton	EA.	6.154
5 ton	"	6.667
7.5 ton	"	8.000
10 ton	"	10.000
15 ton	"	11.429
Air cooled condenser, low ambient damper		
3 ton	EA.	1.600
5 ton	"	2.000
7.5 ton	"	4.000
10 ton	"	5.714
15 ton	"	4.706
Water cooled, high efficiency filter, alarm, elec. heat		
3 ton	EA.	5.714
5 ton	"	6.667
7.5 ton	"	10.000
10 ton	"	11.429
15 ton	"	13.333
Steam heat		
3 ton	EA.	5.714
5 ton	"	6.667
7.5 ton	"	10.000
10 ton	"	11.429
15 ton	"	13.333
Hot water heat		
3 ton	EA.	5.714
5 ton	"	6.667
7.5 ton	"	10.000
10 ton	"	11.429
15 ton	"	13.333
Chilled water, alarm, high eff. filter, elec. heat		
7.5 ton	EA.	7.273
10 ton	"	8.889
15 ton	"	10.000
Steam heat		
7.5 ton	EA.	7.273
10 ton	"	8.889
15 ton	"	10.000
Hot water heat		
7.5 ton	EA.	7.273
10 ton	"	8.889
15 ton	"	10.000
15780.20 — Rooftop Units		

Heat Transfer

15780.20 — Rooftop Units (Cont.)

	UNIT	MAN/HOURS
Packaged, single zone rooftop unit, with roof curb		
2 ton	EA.	8.000
3 ton	"	8.000
4 ton	"	10.000
5 ton	"	13.333
7.5 ton	"	16.000

15830.10 — Radiation Units

	UNIT	MAN/HOURS
Baseboard radiation unit		
1.7 mbh/lf	L.F.	0.320
2.1 mbh/lf	"	0.400
Enclosure only		
Two tier	L.F.	0.133
Three tier	"	0.133
Copper element only, 3/4" dia.		
Two tier	L.F.	0.200
Three tier	"	0.267
Fin-tube, 16 ga, sloping cover, 1-1/4" steel		
One tier	L.F.	0.267
Two tier	"	0.320
2" steel		
Two tier	L.F.	0.320
Three tier	"	0.400
1-1/4" copper		
Two tier	L.F.	0.267
18 ga flat cover, 1-1/4" steel		
One tier	L.F.	0.267
Two tier	"	0.320
Three tier	"	0.400
2" steel		
One tier	L.F.	0.267
Two tier	"	0.320
Three tier	"	0.400
1-1/4" copper		
One tier	L.F.	0.267
Two tier	"	0.320
Three tier	"	0.400

15830.20 — Fan Coil Units

	UNIT	MAN/HOURS
Fan coil unit, 2 pipe, complete		
200 cfm ceiling hung	EA.	2.667
Floor mounted	"	2.000
300 cfm, ceiling hung	"	3.200
Floor mounted	"	2.667
400 cfm, ceiling hung	"	3.810
Floor mounted	"	2.667
500 cfm, ceiling hung	"	4.000
Floor mounted	"	3.077
600 cfm, ceiling hung	"	4.420
Floor mounted	"	3.636

Heat Transfer

15830.20 — Fan Coil Units (Cont.)

	UNIT	MAN/HOURS
800 cfm, ceiling hung	EA.	5.000
Floor mounted	"	3.810
1000 cfm, ceiling hung	"	5.714
Floor mounted	"	4.211
1200 cfm ceiling hung	"	6.667
Floor mounted	"	5.000

15830.70 — Unit Heaters

	UNIT	MAN/HOURS
Steam unit heater, horizontal		
12,500 btuh, 200 cfm	EA.	1.333
17,000 btuh, 300 cfm	"	1.333
40,000 btuh, 500 cfm	"	1.333
60,000 btuh, 700 cfm	"	1.333
70,000 btuh, 1000 cfm	"	2.000
Vertical		
12,500 btuh, 200 cfm	EA.	1.333
17,000 btuh, 300 cfm	"	1.333
40,000 btuh, 500 cfm	"	1.333
60,000 btuh, 700 cfm	"	1.333
70,000 btuh, 1000 cfm	"	1.333
Gas unit heater, horizontal		
27,400 btuh	EA.	3.200
38,000 btuh	"	3.200
56,000 btuh	"	3.200
82,200 btuh	"	3.200
103,900 btuh	"	5.000
125,700 btuh	"	5.000
133,200 btuh	"	5.000
149,000 btuh	"	5.000
172,000 btuh	"	5.000
190,000 btuh	"	5.000
225,000 btuh	"	5.000
Hot water unit heater, horizontal		
12,500 btuh, 200 cfm	EA.	1.333
17,000 btuh, 300 cfm	"	1.333
25,000 btuh, 500 cfm	"	1.333
30,000 btuh, 700 cfm	"	1.333
50,000 btuh, 1000 cfm	"	2.000
60,000 btuh, 1300 cfm	"	2.000
Vertical		
12,500 btuh, 200 cfm	EA.	1.333
17,000 btuh, 300 cfm	"	1.333
25,000 btuh, 500 cfm	"	1.333
30,000 btuh, 700 cfm	"	1.333
50,000 btuh, 1000 cfm	"	1.333
60,000 btuh, 1300 cfm	"	1.333
Cabinet unit heaters, ceiling, exposed, hot water		
200 cfm	EA.	2.667
300 cfm	"	3.200

Heat Transfer

15830.70 — Unit Heaters *(Cont.)*

	UNIT	MAN/HOURS
400 cfm	EA.	3.810
600 cfm	"	4.211
800 cfm	"	5.000
1000 cfm	"	5.714
1200 cfm	"	6.667
2000 cfm	"	8.889

Air Handling

15855.10 — Air Handling Units

	UNIT	MAN/HOURS
Air handling unit, medium pressure, single zone		
1500 cfm	EA.	5.000
3000 cfm	"	8.889
4000 cfm	"	10.000
5000 cfm	"	10.667
6000 cfm	"	11.429
7000 cfm	"	12.308
8500 cfm	"	13.333
10,500 cfm	"	16.000
12,500 cfm	"	17.778
15,500 cfm	"	22.857
17,500 cfm	"	26.667
20,500 cfm	"	32.000
25,000 cfm	"	40.000
31,500 cfm	"	53.333
Rooftop air handling units		
4950 cfm	EA.	8.889
7370 cfm	"	11.429
9790 cfm	"	13.333
14,300 cfm	"	11.429
21,725 cfm	"	11.429
33,000 cfm	"	13.333

15870.20 — Exhaust Fans

	UNIT	MAN/HOURS
Belt drive roof exhaust fans		
640 cfm, 2618 fpm	EA.	1.000
940 cfm, 2604 fpm	"	1.000
1050 cfm, 3325 fpm	"	1.000
1170 cfm, 2373 fpm	"	1.000
2440 cfm, 4501 fpm	"	1.000
2760 cfm, 4950 fpm	"	1.000
3890 cfm, 6769 fpm	"	1.000
2380 cfm, 3382 fpm	"	1.000
2880 cfm, 3859 fpm	"	1.000
3200 cfm, 4173 fpm	"	1.333
3660 cfm, 3437 fpm	"	1.333

Air Handling

15870.20 — Exhaust Fans *(Cont.)*

	UNIT	MAN/HOURS
Direct drive fans		
60 to 390 cfm	EA.	1.000
145 to 590 cfm	"	1.000
295 to 860 cfm	"	1.000
235 to 1300 cfm	"	1.000
415 to 1630 cfm	"	1.000
590 to 2045 cfm	"	1.000

Air Distribution

15890.10 — Metal Ductwork

	UNIT	MAN/HOURS
Rectangular duct		
Galvanized steel		
Minimum	Lb.	0.073
Average	"	0.089
Maximum	"	0.133
Aluminum		
Minimum	Lb.	0.160
Average	"	0.200
Maximum	"	0.267
Fittings		
Minimum	EA.	0.267
Average	"	0.400
Maximum	"	0.800

15890.30 — Flexible Ductwork

	UNIT	MAN/HOURS
Flexible duct, 1.25" fiberglass		
5" dia.	L.F.	0.040
6" dia.	"	0.044
7" dia.	"	0.047
8" dia.	"	0.050
10" dia.	"	0.057
12" dia.	"	0.062
14" dia.	"	0.067
16" dia.	"	0.073
Flexible duct connector, 3" wide fabric	"	0.133

15895.10 — Roof Curbs

	UNIT	MAN/HOURS
8" high, insulated, with liner and raised can		
15" x 15"	EA.	0.400
17" x 17"	"	0.400
19" x 19"	"	0.400
21" x 21"	"	0.400
25" x 25"	"	0.500
28" x 28"	"	0.533
32" x 32"	"	0.571

Air Distribution

15895.10 — Roof Curbs (Cont.)

Size	UNIT	MAN/HOURS
36" x 36"	EA.	0.571
40" x 40"	"	0.571
44" x 44"	"	0.615
48" x 48"	"	0.615
52" x 52"	"	0.667
56" x 56"	"	0.667
60" x 60"	"	0.800
64" x 64"	"	0.800
68" x 68"	"	0.889
72" x 72"	"	1.000

15910.10 — Dampers

Size	UNIT	MAN/HOURS
Horizontal parallel aluminum backdraft damper		
12" x 12"	EA.	0.200
16" x 16"	"	0.229
20" x 20"	"	0.286
24" x 24"	"	0.400
28" x 28"	"	0.444
32" x 32"	"	0.500
36" x 36"	"	0.571
40" x 40"	"	0.667
44" x 44"	"	0.727
48" x 48"	"	0.800
"Up", parallel dampers		
12" x 12"	EA.	0.200
16" x 16"	"	0.229
20" x 20"	"	0.286
24" x 24"	"	0.400
28" x 28"	"	0.444
32" x 32"	"	0.500
36" x 36"	"	0.571
40" x 40"	"	0.667
44" x 44"	"	0.727
48" x 48"	"	0.800
"Down", parallel dampers		
12" x 12"	EA.	0.200
16" x 16"	"	0.229
20" x 20"	"	0.286
24" x 24"	"	0.400
28" x 28"	"	0.444
32" x 32"	"	0.500
36" x 36"	"	0.571
40" x 40"	"	0.667
44" x 44"	"	0.727
48" x 48"	"	0.800
Fire damper, 1.5 hr rating		
12" x 12"	EA.	0.400
16" x 16"	"	0.400
20" x 20"	"	0.400
24" x 24"	"	0.400

Air Distribution

15910.10 — Dampers (Cont.)

Size	UNIT	MAN/HOURS
28" x 28"	EA.	0.571
32" x 32"	"	0.667
36" x 36"	"	0.800
40" x 40"	"	0.889
44" x 44"	"	1.000
48" x 48"	"	1.143

15940.10 — Diffusers

Size	UNIT	MAN/HOURS
Ceiling diffusers, round, baked enamel finish		
6" dia.	EA.	0.267
8" dia.	"	0.333
10" dia.	"	0.333
12" dia.	"	0.333
14" dia.	"	0.364
16" dia.	"	0.364
18" dia.	"	0.400
20" dia.	"	0.400
Rectangular		
6x6"	EA.	0.267
9x9"	"	0.400
12x12"	"	0.400
15x15"	"	0.400
18x18"	"	0.400
21x21"	"	0.500
24x24"	"	0.500
Lay in, flush mounted, perforated face, with grid		
6x6/24x24	EA.	0.320
8x8/24x24	"	0.320
9x9/24x24	"	0.320
10x10/24x24	"	0.320
12x12/24x24	"	0.320
15x15/24x24	"	0.320
18x6/24x24	"	0.320
18x18/24x24	"	0.320
Two-way slot diffuser with balancing damper, 4'	"	0.800

15940.20 — Relief Ventilators

Size	UNIT	MAN/HOURS
Intake ventilator, aluminum, with screen, no curbs		
12" x 12"	EA.	0.667
16" x 16"	"	0.800
20" x 20"	"	0.800
30" x 30"	"	1.143
36" x 36"	"	1.333
42" x 42"	"	1.333
48" x 48"	"	1.600

15940.40 — Registers And Grilles

Size	UNIT	MAN/HOURS
Lay in flush mounted, perforated face, return		
6x6/24x24	EA.	0.320
8x8/24x24	"	0.320
9x9/24x24	"	0.320
10x10/24x24	"	0.320
12x12/24x24	"	0.320

Air Distribution		UNIT	MAN/ HOURS
15940.40	**Registers And Grilles** (Cont.)		
Rectangular, ceiling return, single deflection			
10x10		EA.	0.400
12x12		"	0.400
14x14		"	0.400
16x8		"	0.400
16x16		"	0.400
18x8		"	0.400
20x20		"	0.400
24x12		"	0.400
24x18		"	0.400
36x24		"	0.444
36x30		"	0.444
Wall, return air register			
12x12		EA.	0.200
16x16		"	0.200
18x18		"	0.200
20x20		"	0.200
24x24		"	0.200
Ceiling, return air grille			
6x6		EA.	0.267
8x8		"	0.320
10x10		"	0.320
Ceiling, exhaust grille, aluminum egg crate			
6x6		EA.	0.267
8x8		"	0.320
10x10		"	0.320
12x12		"	0.400
14x14		"	0.400
16x16		"	0.400
18x18		"	0.400
15940.80	**Penthouse Louvers**		
Penthouse louvers			
12" high, extruded aluminum, 4" louver			
6' perimeter		EA.	2.000
8' perimeter		"	2.000
10' perimeter		"	2.000
12' perimeter		"	2.000
14' perimeter		"	2.667
16' perimeter		"	3.200
18' perimeter		"	4.444
20' perimeter		"	5.333
16" high x 4' perimeter		"	2.000
6' perimeter		"	2.000
8' perimeter		"	2.000
10' perimeter		"	2.000
12' perimeter		"	2.000
14' perimeter		"	2.667
16' perimeter		"	3.200
18' perimeter		"	4.444
20' perimeter		"	5.333
22' perimeter		"	6.667
24' perimeter		"	8.889

Air Distribution		UNIT	MAN/ HOURS
15940.80	**Penthouse Louvers** (Cont.)		
20" high x 4' perimeter		EA.	2.000
6' perimeter		"	2.000
8' perimeter		"	2.000
10' perimeter		"	2.000
12' perimeter		"	2.000
14' perimeter		"	2.667
16' perimeter		"	3.200
18' perimeter		"	4.444
20' perimeter		"	5.333
22' perimeter		"	6.667
24' perimeter		"	8.889
24" high x 4' perimeter		"	2.000
6' perimeter		"	2.000
8' perimeter		"	2.000
10' perimeter		"	2.000
12' perimeter		"	2.000
16' perimeter		"	3.200
18' perimeter		"	4.444
20' perimeter		"	5.333
22' perimeter		"	6.667
24' perimeter		"	8.889

Controls		UNIT	MAN/ HOURS
15950.10	**Hvac Controls**		
Pressure gauge, direct reading gage cock and siphon		EA.	0.500
Control valve, 1", modulating			
2-way		EA.	0.667
3-way		"	1.000
Self contained control valve w/ sensing elmnt, 3/4"		"	0.500
Control dampers, round			
6" dia.		EA.	0.320
8" dia		"	0.320
10" dia		"	0.320
12" dia		"	0.320
12" dia		"	0.400
18" dia		"	0.400
20" dia		"	0.400
Rectangular, parallel blade standard leakage			
12" x 12"		EA.	0.400
16" x 16"		"	0.400
20" x 20"		"	0.400
28" x 28"		"	0.500
32" x 32"		"	0.500
36" x 36"		"	0.667
40" x 40"		"	0.800

Controls	UNIT	MAN/HOURS
15950.10 **Hvac Controls** *(Cont.)*		
44" x 44"	EA.	1.000
48" x 48"	"	1.143
48" x 52"	"	1.333
48" x 56"	"	1.333
48" x 60"	"	1.333
48" x 64"	"	1.333
48" x 68"	"	1.333
48" x 72"	"	1.333
Low leakage		
12" x 12"	EA.	0.400
16" x 16"	"	0.400
20" x 20"	"	0.400
24" x 24"	"	0.400
28" x 28"	"	0.500
32" x 32"	"	0.571
36" x 36"	"	0.667
40" x 40"	"	0.800
44" x 44"	"	1.000
48" x 48"	"	1.143
48" x 56"	"	1.333
48" x 60"	"	1.333
48" x 64"	"	1.333
48" x 68"	"	1.333
48" x 72"	"	1.333
Rectangular, opposed horizontal blade		
12" x 12"	EA.	0.400
16" x 16"	"	0.400
20" x 20"	"	0.400
24" x 24"	"	0.400
28" x 28"	"	0.500
32" x 32"	"	0.533
36" x 36"	"	0.667
40" x 40"	"	0.800
44" x 44"	"	1.000
48" x 48"	"	1.143
48" x 52"	"	1.143
48" x 56"	"	1.333
48" x 60"	"	1.333
48" x 64"	"	1.333
48" x 68"	"	1.333
48" x 72"	"	1.333

Basic Materials

16110.20 — Conduit Specialties

Basic Materials	UNIT	MAN/HOURS
Rod beam clamp, 1/2"	EA.	0.050
Hanger rod		
3/8"	L.F.	0.040
1/2"	"	0.050
All thread rod		
1/4"	L.F.	0.030
3/8"	"	0.040
1/2"	"	0.050
5/8"	"	0.080
Hanger channel, 1-1/2"		
No holes	EA.	0.030
Holes	"	0.030
Channel strap		
1/2"	EA.	0.050
3/4"	"	0.050
1"	"	0.050
1-1/4"	"	0.080
1-1/2"	"	0.080
2"	"	0.080
2-1/2"	"	0.123
3"	"	0.123
3-1/2"	"	0.123
4"	"	0.145
5"	"	0.145
6"	"	0.145
Conduit penetrations, roof and wall, 8" thick		
1/2"	EA.	0.615
3/4"	"	0.615
1"	"	0.800
1-1/4"	"	0.800
1-1/2"	"	0.800
2"	"	1.600
2-1/2"	"	1.600
3"	"	1.600
3-1/2"	"	2.000
4"	"	2.000
Fireproofing, for conduit penetrations		
1/2"	EA.	0.500
3/4"	"	0.500
1"	"	0.500
1-1/4"	"	0.727
1-1/2"	"	0.727
2"	"	0.727
2-1/2"	"	0.899
3"	"	0.899
3-1/2"	"	1.250
4"	"	1.509

16110.21 — Aluminum Conduit

Basic Materials	UNIT	MAN/HOURS
Aluminum conduit		
1/2"	L.F.	0.030
3/4"	"	0.040
1"	"	0.050
1-1/4"	"	0.059
1-1/2"	"	0.080
2"	"	0.089
2-1/2"	"	0.100
3"	"	0.107
3-1/2"	"	0.123
4"	"	0.145
5"	"	0.182
6"	"	0.200
90 deg. elbow		
1/2"	EA.	0.190
3/4"	"	0.250
1"	"	0.308
1-1/4"	"	0.381
1-1/2"	"	0.400
2"	"	0.444
2-1/2"	"	0.571
3"	"	0.667
3-1/2"	"	0.800
4"	"	0.889
5"	"	1.143
6"	"	2.222
Coupling		
1/2"	EA.	0.050
3/4"	"	0.059
1"	"	0.080
1-1/4"	"	0.089
1-1/2"	"	0.100
2"	"	0.107
2-1/2"	"	0.123
3"	"	0.123
3-1/2"	"	0.145
4"	"	0.160
5"	"	0.160
6"	"	0.190

16110.22 — Emt Conduit

Basic Materials	UNIT	MAN/HOURS
EMT conduit		
1/2"	L.F.	0.030
3/4"	"	0.040
1"	"	0.050
1-1/4"	"	0.059
1-1/2"	"	0.080
2"	"	0.089
2-1/2"	"	0.100
3"	"	0.123
3-1/2"	"	0.145
4"	"	0.182
90 deg. elbow		

Basic Materials	UNIT	MAN/HOURS
16110.22 — Emt Conduit *(Cont.)*		
1/2"	EA.	0.089
3/4"	"	0.100
1"	"	0.107
1-1/4"	"	0.123
1-1/2"	"	0.145
2"	"	0.190
2-1/2"	"	0.211
3"	"	0.242
3-1/2"	"	0.258
4"	"	0.286
Connector, die cast set screw		
1/2"	EA.	0.059
3/4"	"	0.059
1"	"	0.059
1-1/4"	"	0.089
1-1/2"	"	0.107
2"	"	0.145
2-1/2"	"	0.200
3"	"	0.222
3-1/2"	"	0.250
4"	"	0.286
Coupling, steel set screw		
1/2"	EA.	0.040
3/4"	"	0.040
1"	"	0.040
1-1/4"	"	0.050
1-1/2"	"	0.080
2"	"	0.107
2-1/2"	"	0.160
3"	"	0.190
3-1/2"	"	0.222
4"	"	0.250
16110.23 — Flexible Conduit		
Flexible conduit, steel		
3/8"	L.F.	0.030
1/2	"	0.030
3/4"	"	0.040
1"	"	0.040
1-1/4"	"	0.050
1-1/2"	"	0.059
2"	"	0.080
2-1/2"	"	0.089
3"	"	0.107
Flexible conduit, liquid tight		
3/8"	L.F.	0.030
1/2"	"	0.030
3/4"	"	0.040
1"	"	0.040
1-1/4"	"	0.050
1-1/2"	"	0.059
2"	EA.	0.080
2-1/2"	"	0.089

Basic Materials	UNIT	MAN/HOURS
16110.23 — Flexible Conduit *(Cont.)*		
3"	EA.	0.107
4"	"	0.145
Connector, straight		
3/8"	EA.	0.080
1/2"	"	0.080
3/4"	"	0.089
1"	"	0.100
1-1/4"	"	0.107
1-1/2"	"	0.123
2"	"	0.145
2-1/2"	"	0.182
3"	"	0.190
Flexible aluminum conduit		
3/8"	L.F.	0.030
1/2"	"	0.030
3/4"	"	0.040
1"	"	0.040
1-1/4"	"	0.050
1-1/2"	"	0.059
2"	"	0.080
2-1/2"	"	0.089
3"	"	0.107
3-1/2"	"	0.123
4"	"	0.145
16110.24 — Galvanized Conduit		
Galvanized rigid steel conduit		
1/2"	L.F.	0.040
3/4"	"	0.050
1"	"	0.059
1-1/4"	"	0.080
1-1/2"	"	0.089
2"	"	0.100
2-1/2"	"	0.145
3"	"	0.182
3-1/2"	"	0.190
4"	"	0.211
5"	"	0.286
6"	"	0.381
90 degree ell		
1/2"	EA.	0.250
3/4"	"	0.308
1"	"	0.381
1-1/4"	"	0.444
1-1/2"	"	0.500
2"	"	0.533
2-1/2"	"	0.667
3"	"	0.889
3-1/2"	"	1.000
4"	"	1.333
5"	"	2.222
6"	"	3.333
Couplings, with set screws		

Basic Materials	UNIT	MAN/HOURS
16110.24 — **Galvanized Conduit** *(Cont.)*		
1/2"	EA.	0.050
3/4"	"	0.059
1"	"	0.080
1-1/4"	"	0.100
1-1/2"	"	0.123
2"	"	0.145
2-1/2"	"	0.190
3"	"	0.250
3-1/2"	"	0.286
4"	"	0.308
5"	"	0.444
6"	"	0.500
Split couplings		
1/2"	EA.	0.190
3/4"	"	0.250
1"	"	0.276
1-1/4"	"	0.308
1-1/2"	"	0.381
2"	"	0.571
2-1/2"	"	0.571
3"	"	0.727
3-1/2"	"	1.000
4"	"	1.333
5"	"	1.633
6"	"	2.051
Erickson couplings		
1/2"	EA.	0.444
3/4"	"	0.500
1"	"	0.615
1-1/4"	"	0.889
1-1/2"	"	1.000
2"	"	1.333
2-1/2"	"	1.860
3"	"	2.105
3-1/2"	"	2.500
4"	"	2.667
5"	"	2.963
6"	"	3.200
Seal fittings		
1/2"	EA.	0.667
3/4"	"	0.800
1"	"	1.000
1-1/4"	"	1.143
1-1/2"	"	1.333
2"	"	1.600
2-1/2"	"	1.905
3"	"	2.105
3-1/2"	"	2.500
4"	"	2.963
5"	"	4.444
6"	"	5.000
Entrance fitting, (weather head), threaded		
1/2"	EA.	0.444

Basic Materials	UNIT	MAN/HOURS
16110.24 — **Galvanized Conduit** *(Cont.)*		
3/4"	EA.	0.500
1"	"	0.571
1-1/4"	"	0.727
1-1/2"	"	0.800
2"	"	0.889
2-1/2"	"	1.000
3"	"	1.333
3-1/2"	"	1.739
4"	"	2.500
5"	"	3.478
6"	"	4.444
Locknuts		
1/2"	EA.	0.050
3/4"	"	0.050
1"	"	0.050
1-1/4"	"	0.050
1-1/2"	"	0.059
2"	"	0.059
2-1/2"	"	0.080
3"	"	0.080
3-1/2"	"	0.080
4"	"	0.089
5"	"	0.089
6"	"	0.089
Plastic conduit bushings		
1/2"	EA.	0.123
3/4"	"	0.145
1"	"	0.190
1-1/4"	"	0.222
1-1/2"	"	0.250
2"	"	0.308
2-1/2"	"	0.500
3"	"	0.667
3-1/2"	"	0.800
4"	"	0.889
5"	"	1.143
6"	"	1.600
Conduit bushings, steel		
1/2"	EA.	0.123
3/4"	"	0.145
1"	"	0.190
1-1/4"	"	0.222
1-1/2"	"	0.250
2"	"	0.308
2-1/2"	"	0.500
3"	"	0.667
3-1/2"	"	0.800
4"	"	0.889
5"	"	1.143
6"	"	1.600
Pipe cap		
1/2"	EA.	0.050
3/4"	"	0.050

Basic Materials

16110.24 — Galvanized Conduit (Cont.)

Basic Materials	UNIT	MAN/HOURS
1"	EA.	0.050
1-1/4"	"	0.080
1-1/2"	"	0.080
2"	"	0.080
2-1/2"	"	0.089
3"	"	0.089
3-1/2"	"	0.089
4"	"	0.107
5"	"	0.145
6"	"	0.200

16110.25 — Plastic Conduit

Basic Materials	UNIT	MAN/HOURS
PVC conduit, schedule 40		
1/2"	L.F.	0.030
3/4"	"	0.030
1"	"	0.040
1-1/4"	"	0.040
1-1/2"	"	0.050
2"	"	0.050
2-1/2"	"	0.059
3"	"	0.059
3-1/2"	"	0.080
4"	"	0.080
5"	"	0.089
6"	"	0.100
Couplings		
1/2"	EA.	0.050
3/4"	"	0.050
1"	"	0.050
1-1/4"	"	0.059
1-1/2"	"	0.059
2"	"	0.059
2-1/2"	"	0.059
3"	"	0.080
3-1/2"	"	0.080
4"	"	0.100
5"	"	0.100
6"	"	0.100
90 degree elbows		
1/2"	EA.	0.100
3/4"	"	0.123
1"	"	0.123
1-1/4"	"	0.145
1-1/2"	"	0.190
2"	"	0.222
2-1/2"	"	0.250
3"	"	0.308
3-1/2"	"	0.381
4"	"	0.500
5"	"	0.615
6"	"	0.727
Terminal adapters		
1/2"	EA.	0.100

16110.25 — Plastic Conduit (Cont.)

Basic Materials	UNIT	MAN/HOURS
3/4"	EA.	0.100
1"	"	0.100
1-1/4"	"	0.160
1-1/2"	"	0.160
2"	"	0.160
2-1/2"	"	0.222
3"	"	0.222
3-1/2"	"	0.222
4"	"	0.381
5"	"	0.381
6"	"	0.381
End bells		
1"	EA.	0.100
1-1/4"	"	0.160
1-1/2"	"	0.160
2"	"	0.160
2-1/2"	"	0.222
3"	"	0.222
3-1/2"	"	0.222
4"	"	0.381
5"	"	0.381
6"	"	0.381
LB conduit body		
1/2"	EA.	0.190
3/4"	"	0.190
1	"	0.190
1-1/4"	"	0.308
1-1/2"	"	0.308
2"	"	0.308
Direct burial, conduit		
2"	L.F.	0.050
3"	"	0.059
4"	"	0.080
5"	"	0.089
6"	"	0.100
Encased burial conduit		
2"	L.F.	0.050
3"	"	0.059
4"	"	0.080
5"	"	0.089
6"	"	0.100

16110.27 — Plastic Coated Conduit

Basic Materials	UNIT	MAN/HOURS
Rigid steel conduit, plastic coated		
1/2"	L.F.	0.050
3/4"	"	0.059
1"	"	0.080
1-1/4"	"	0.100
1-1/2"	"	0.123
2"	"	0.145
2-1/2"	"	0.190
3"	"	0.222
3-1/2"	"	0.250

Basic Materials		UNIT	MAN/HOURS
16110.27	**Plastic Coated Conduit** *(Cont.)*		
4"		L.F.	0.308
5"		"	0.381
90 degree elbows			
1/2"		EA.	0.308
3/4"		"	0.381
1"		"	0.444
1-1/4"		"	0.500
1-1/2"		"	0.615
2"		"	0.800
2-1/2"		"	1.143
3"		"	1.333
3-1/2"		"	1.633
4"		"	2.000
5"		"	2.500
Couplings			
1/2"		EA.	0.059
3/4"		"	0.080
1"		"	0.089
1-1/4"		"	0.107
1-1/2"		"	0.123
2"		"	0.145
2-1/2"		"	0.182
3"		"	0.190
3-1/2"		"	0.200
4"		"	0.222
5"		"	0.250
16110.28	**Steel Conduit**		
Intermediate metal conduit (IMC)			
1/2"		L.F.	0.030
3/4"		"	0.040
1"		"	0.050
1-1/4"		"	0.059
1-1/2"		"	0.080
2"		"	0.089
2-1/2"		"	0.119
3"		"	0.145
3-1/2"		"	0.182
4"		"	0.190
90 degree ell			
1/2"		EA.	0.250
3/4"		"	0.308
1"		"	0.381
1-1/4"		"	0.444
1-1/2"		"	0.500
2"		"	0.571
2-1/2"		"	0.667
3"		"	0.889
3-1/2"		"	1.143
4"		"	1.333

Basic Materials		UNIT	MAN/HOURS
16110.35	**Surface Mounted Raceway**		
Single Raceway			
3/4" x 17/32" Conduit		L.F.	0.040
Mounting Strap		EA.	0.053
Connector		"	0.053
Elbow			
45 degree		EA.	0.050
90 degree		"	0.050
internal		"	0.050
external		"	0.050
Switch		"	0.400
Utility Box		"	0.400
Receptacle		"	0.400
3/4" x 21/32" Conduit		L.F.	0.040
Mounting Strap		EA.	0.053
Connector		"	0.053
Elbow			
45 degree		EA.	0.050
90 degree		"	0.050
internal		"	0.050
external		"	0.050
Switch		"	0.400
Utility Box		"	0.400
Receptacle		"	0.400
16110.80	**Wireways**		
Wireway, hinge cover type			
2-1/2" x 2-1/2"			
1' section		EA.	0.154
2'		"	0.190
3'		"	0.250
5'		"	0.381
10'		"	0.667
4" x 4"			
1'		EA.	0.250
2'		"	0.250
3'		"	0.308
4'		"	0.308
10'		"	0.800
6" x 6"			
1'		EA.	0.381
2'		"	0.381
3'		"	0.444
4'		"	0.444
5'		"	0.571
10'		"	0.889
8" x 8"			
1'		EA.	0.444
2'		"	0.444
3'		"	0.500
4'		"	0.500
5'		"	0.615

Basic Materials	UNIT	MAN/HOURS
16110.80 **Wireways** *(Cont.)*		
12" x 12"		
1'	EA.	0.615
2'	"	0.615
3'	"	0.727
4'	"	0.727
5'	"	0.889
16120.43 **Copper Conductors**		
Copper conductors, type THW, solid		
#14	L.F.	0.004
#12	"	0.005
#10	"	0.006
Stranded		
#14	L.F.	0.004
#12	"	0.005
#10	"	0.006
#8	"	0.008
#6	"	0.009
#4	"	0.010
#3	"	0.010
#2	"	0.012
#1	"	0.014
1/0	"	0.016
2/0	"	0.020
3/0	"	0.025
4/0	"	0.028
250 MCM	"	0.030
300 MCM	"	0.033
350 MCM	"	0.040
400 MCM	"	0.044
500 MCM	"	0.052
600 MCM	"	0.059
750 MCM	"	0.067
1000 MCM	"	0.076
THHN-THWN, solid		
#14	L.F.	0.004
#12	"	0.005
#10	"	0.006
Stranded		
#14	L.F.	0.004
#12	"	0.005
#10	"	0.006
#8	"	0.008
#6	"	0.009
#4	"	0.010
#2	"	0.012
#1	"	0.014
1/0	"	0.016
2/0	"	0.020
3/0	"	0.025
4/0	"	0.028
250 MCM	"	0.030
350 MCM	"	0.040

Basic Materials	UNIT	MAN/HOURS
16120.43 **Copper Conductors** *(Cont.)*		
XHHW		
#14	L.F.	0.004
#10	"	0.006
#8	"	0.008
#6	"	0.009
#4	"	0.009
#2	"	0.011
#1	"	0.014
1/0	"	0.016
2/0	"	0.019
3/0	"	0.025
XLP, 600v		
#12	L.F.	0.005
#10	"	0.006
#8	"	0.008
#6	"	0.009
#4	"	0.010
#3	"	0.011
#2	"	0.012
#1	"	0.014
1/0	"	0.016
2/0	"	0.020
3/0	"	0.026
4/0	"	0.028
250 MCM	"	0.030
300 MCM	"	0.033
350 MCM	"	0.039
400 MCM	"	0.044
500 MCM	"	0.052
600 MCM	"	0.059
750 MCM	"	0.067
1000 MCM	"	0.076
16120.47 **Sheathed Cable**		
Non-metallic sheathed cable		
Type NM cable with ground		
#14/2	L.F.	0.015
#12/2	"	0.016
#10/2	"	0.018
#8/2	"	0.020
#6/2	"	0.025
#14/3	"	0.026
#12/3	"	0.027
#10/3	"	0.027
#8/3	"	0.028
#6/3	"	0.028
#4/3	"	0.032
#2/3	"	0.035
Type U.F. cable with ground		
#14/2	L.F.	0.016
#12/2	"	0.019
#10/2	"	0.020
#8/2	"	0.023

Basic Materials	UNIT	MAN/HOURS
16120.47 **Sheathed Cable** *(Cont.)*		
#6/2	L.F.	0.027
#14/3	"	0.020
#12/3	"	0.022
#10/3	"	0.025
#8/3	"	0.028
#6/3	"	0.032
16130.60 **Pull And Junction Boxes**		
4"		
Octagon box	EA.	0.114
Box extension	"	0.059
Plaster ring	"	0.059
Cover blank	"	0.059
Square box	"	0.114
Box extension	"	0.059
Plaster ring	"	0.059
Cover blank	"	0.059
4-11/16"		
Square box	EA.	0.114
Box extension	"	0.059
Plaster ring	"	0.059
Cover blank	"	0.059
Switch and device boxes		
2 gang	EA.	0.114
3 gang	"	0.114
4 gang	"	0.160
Device covers		
2 gang	EA.	0.059
3 gang	"	0.059
4 gang	"	0.059
Handy box	"	0.114
Extension	"	0.059
Switch cover	"	0.059
Switch box with knockout	"	0.145
Weatherproof cover, spring type	"	0.080
Cover plate, dryer receptacle 1 gang plastic	"	0.100
For 4" receptacle, 2 gang	"	0.100
Duplex receptacle cover plate, plastic	"	0.059
4", vertical bracket box, 1-1/2" with		
RMX clamps	EA.	0.145
BX clamps	"	0.145
4", octagon device cover		
1 switch	EA.	0.059
1 duplex recept	"	0.059
4", square face bracket boxes, 1-1/2"		
RMX	EA.	0.145
BX	"	0.145
4" square to round plaster rings	"	0.059
2 gang device plaster rings	"	0.059
Surface covers		
1 gang switch	EA.	0.059
2 gang switch	"	0.059
1 single recept	"	0.059

Basic Materials	UNIT	MAN/HOURS
16130.60 **Pull And Junction Boxes** *(Cont.)*		
1 20a twist lock recept	EA.	0.059
1 30a twist lock recept	"	0.059
1 duplex recept	"	0.059
2 duplex recept	"	0.059
Switch and duplex recept	"	0.059
4-11/16" square to round plaster rings	"	0.059
2 gang device plaster rings	"	0.059
Surface covers		
1 gang switch	EA.	0.059
2 gang switch	"	0.059
1 single recept	"	0.059
1 20a twist lock recept	"	0.059
1 30a twist lock recept	"	0.059
1 duplex recept	"	0.059
2 duplex recept	"	0.059
Switch and duplex recept	"	0.059
4" plastic round boxes, ground straps		
Box only	EA.	0.145
Box w/clamps	"	0.200
Box w/16" bar	"	0.229
Box w/24" bar	"	0.250
4" plastic round box covers		
Blank cover	EA.	0.059
Plaster ring	"	0.059
4" plastic square boxes		
Box only	EA.	0.145
Box w/clamps	"	0.200
Box w/hanger	"	0.250
Box w/nails and clamp	"	0.250
4" plastic square box covers		
Blank cover	EA.	0.059
1 gang ring	"	0.059
2 gang ring	"	0.059
Round ring	"	0.059
16130.65 **Pull Boxes And Cabinets**		
Galvanized pull boxes, screw cover		
4x4x4	EA.	0.190
4x6x4	"	0.190
6x6x4	"	0.190
6x8x4	"	0.190
8x8x4	"	0.250
8x10x4	"	0.242
8x12x4	"	0.250
Screw cover		
10x10x4	EA.	0.308
12x12x6	"	0.444
12x15x6	"	0.444
12x18x6	"	0.500
15x18x6	"	0.571
18x24x6	"	0.615
18x30x6	"	0.727
24x36x6	"	0.727

16130.65   Pull Boxes And Cabinets *(Cont.)*

Basic Materials	UNIT	MAN/HOURS
Cast iron junction box, unflanged		
6x6x4		
3/4" tap	EA.	0.500
1" tap	"	0.500
Two 1/2" taps	"	0.500
3/4" taps	"	0.500
6" adapter plate	"	0.348
6" exterior collar	"	0.348
Screw cover cabinet		
12x12x4	EA.	0.615
12x16x4	"	0.615
12x16x6	"	0.615
12x18x4	"	0.667
12x18x6	"	0.667
18x18x4	"	1.000
18x18x6	"	1.000
18x24x6	"	1.143
24x24x6	"	1.333
24x36x6	"	1.667
36x48x6	"	2.500

16130.80   Receptacles

Basic Materials	UNIT	MAN/HOURS
Contractor grade duplex receptacles, 15a 120v		
Duplex	EA.	0.200
125 volt, 20a, duplex, grounding type, standard grade	"	0.200
Ground fault interrupter type	"	0.296
250 volt, 20a, 2 pole, single receptacle, ground type	"	0.200
120/208v, 4 pole, single receptacle, twist lock		
20a	EA.	0.348
50a	"	0.348
125/250v, 3 pole, flush receptacle		
30a	EA.	0.296
50a	"	0.296
60a	"	0.348
277v, 20a, 2 pole, grounding type, twist lock	"	0.200
Dryer receptacle, 250v, 30a/50a, 3 wire	"	0.296
Clock receptacle, 2 pole, grounding type	"	0.200
125v, 20a single recept. grounding type		
Standard grade	EA.	0.200
Specification	"	0.200
Hospital	"	0.200
Isolated ground orange	"	0.250
Duplex		
Specification grade	EA.	0.200
Hospital	"	0.200
Isolated ground orange	"	0.250
250v, 20a, duplex, 2 pole, grounding, spec. grade	"	0.200
Combination recepts, 20a, 125v and 250v, duplex	"	0.200
GFI hospital grade recepts, 20a, 125v, duplex	"	0.296
125/250v, 3 pole, 3 wire surface recepts		

16130.80   Receptacles *(Cont.)*

Basic Materials	UNIT	MAN/HOURS
30a	EA.	0.296
50a	"	0.296
60a	"	0.348
Cord set, 3 wire, 6' cord		
30a	EA.	0.296
50a	"	0.296

16198.10   Electric Manholes

Basic Materials	UNIT	MAN/HOURS
Precast, handhole, 4' deep		
2'x2'	EA.	3.478
3'x3'	"	5.556
4'x4'	"	10.256
Power manhole, complete, precast, 8' deep		
4'x4'	EA.	14.035
6'x6'	"	20.000
8'x8'	"	21.053
6' deep, 9' x 12'	"	25.000
Cast in place, power manhole, 8' deep		
4'x4'	EA.	14.035
6'x6'	"	20.000
8'x8'	"	21.053

16199.10   Utility Poles & Fittings

Basic Materials	UNIT	MAN/HOURS
Wood pole, creosoted		
25'	EA.	2.353
30'	"	2.963
35'	"	3.478
40'	"	3.791
45'	"	6.957
50'	"	7.207
55'	"	7.547
Treated, wood preservative, 6"x6"		
8'	EA.	0.500
10'	"	0.800
12'	"	0.889
14'	"	1.333
16'	"	1.600
18'	"	2.000
20'	"	2.000
Aluminum, brushed, no base		
8'	EA.	2.000
10'	"	2.667
15'	"	2.759
20'	"	3.200
25'	"	3.810
30'	"	4.396
35'	"	5.000
40'	"	6.250
Steel, no base		
10'	EA.	2.500
15'	"	2.963
20'	"	3.810
25'	"	4.520

Basic Materials

	UNIT	MAN/ HOURS

16199.10 — Utility Poles & Fittings (Cont.)

	UNIT	MAN/ HOURS
30'	EA.	5.096
35'	"	6.250
Concrete, no base		
13'	EA.	5.517
16'	"	7.273
18'	"	8.791
25'	"	10.000
30'	"	12.121
35'	"	14.035
40'	"	16.000
45'	"	17.021
50'	"	18.182
55'	"	19.048
60'	"	20.000
Pole line hardware		
Wood crossarm		
4'	EA.	1.333
8'	"	1.667
10'	"	2.051
Angle steel brace		
1 piece	EA.	0.250
2 piece	"	0.348
Eye nut, 5/8"	"	0.050
Bolt (14-16"), 5/8"	"	0.200
Transformer, ground connection	"	0.250
Stirrup	"	0.308
Secondary lead support	"	0.400
Spool insulator	"	0.200
Guy grip, preformed		
7/16"	EA.	0.145
1/2"	"	0.145
Hook	"	0.250
Strain insulator	"	0.364
Wire		
5/16"	L.F.	0.005
7/16"	"	0.006
1/2"	"	0.008
Soft drawn ground, copper, #8	"	0.008
Ground clamp	EA.	0.308
Perforated strapping for conduit, 1-1/2"	L.F.	0.145
Hot line clamp	EA.	0.800
Lightning arrester		
3kv	EA.	1.000
10kv	"	1.600
30kv	"	2.000
36kv	"	2.500

Power Generation

	UNIT	MAN/ HOURS

16210.10 — Generators

	UNIT	MAN/ HOURS
Diesel generator, with auto transfer switch		
30kw	EA.	30.769
50kw	"	30.769
75kw	"	42.105
100kw	"	47.059
125kw	"	50.000
150kw	"	57.143
175kw	"	66.667
200kw	"	80.000
250kw	"	88.889
300kw	"	100.000
350kw	"	114.286
400kw	"	133.333
450kw	"	145.455
500kw	"	160.000
600kw	"	200.000
750kw	"	200.000

16320.10 — Transformers

	UNIT	MAN/ HOURS
Floor mounted, single phase, int. dry, 480v-120/240v		
3 kva	EA.	1.818
5 kva	"	3.077
7.5 kva	"	3.478
10 kva	"	3.810
15 kva	"	4.301
25 kva	"	7.547
37.5 kva	"	9.412
50 kva	"	10.256
75 kva	"	10.667
100 kva	"	11.594
Three phase, 480v-120/208v		
15 kva	EA.	6.015
30 kva	"	9.412
45 kva	"	10.811
75 kva	"	10.959
112.5 kva	"	12.698
150 kva	"	13.559
225 kva	"	15.385

16350.10 — Circuit Breakers

	UNIT	MAN/ HOURS
Molded case, 240v, 15-60a, bolt-on		
1 pole	EA.	0.250
2 pole	"	0.348
70-100a, 2 pole	"	0.533
15-60a, 3 pole	"	0.400
70-100a, 3 pole	"	0.615
480v, 2 pole		
15-60a	EA.	0.296
70-100a	"	0.400
3 pole		
15-60a	EA.	0.400
70-100a	"	0.444
70-225a	"	0.615

Power Generation

16350.10 — Circuit Breakers *(Cont.)*

	UNIT	MAN/HOURS
Draw out air circuit breakers		
600a	EA.	16.000
800a	"	18.182
1600a	"	24.242
2000a	"	27.586
3000a	"	32.000
4000a	"	38.095
Load center circuit breakers, 240v		
1 pole, 10-60a	EA.	0.250
2 pole		
10-60a	EA.	0.400
70-100a	"	0.667
110-150a	"	0.727
3 pole		
10-60a	EA.	0.500
70-100a	"	0.727
Load center, G.F.I. breakers, 240v		
1 pole, 15-30a	EA.	0.296
2 pole, 15-30a	"	0.400
Key operated breakers, 240v, 1 pole, 10-30a	"	0.296
Tandem breakers, 240v		
1 pole, 15-30a	EA.	0.400
2 pole, 15-30a	"	0.533
Bolt-on, G.F.I. breakers, 240v, 1 pole, 15-30a	"	0.348

16360.10 — Safety Switches

	UNIT	MAN/HOURS
Fused, 3 phase, 30 amp, 600v, heavy duty		
NEMA 1	EA.	1.143
NEMA 3r	"	1.143
NEMA 4	"	1.600
NEMA 12	"	1.739
60a		
NEMA 1	EA.	1.143
NEMA 3r	"	1.143
NEMA 4	"	1.600
NEMA 12	"	1.739
100a		
NEMA 1	EA.	1.739
NEMA 3r	"	1.739
NEMA 4	"	2.000
NEMA 12	"	2.500
200a		
NEMA 1	EA.	2.500
NEMA 3r	"	2.500
NEMA 4	"	2.759
NEMA 12	"	3.478
400a		
NEMA 1	EA.	5.517
NEMA 3r	"	5.517
NEMA 4	"	5.755
NEMA 12	"	7.018
NEMA 4	"	8.989
NEMA 12	"	12.308

Power Generation

16360.10 — Safety Switches *(Cont.)*

	UNIT	MAN/HOURS
Non-fused, 240-600v, heavy duty, 3 phase, 30 amp		
NEMA 1	EA.	1.143
NEMA 3r	"	1.143
NEMA 4	"	1.739
NEMA 12	"	1.739
60a		
NEMA1	EA.	1.143
NEMA 3r	"	1.143
NEMA 4	"	1.739
NEMA 12	"	1.739
100a		
NEMA 1	EA.	1.739
NEMA 3r	"	1.739
NEMA 4	"	2.500
NEMA 12	"	2.500
200a, NEMA 1	"	2.500
600a, NEMA 12	"	12.308

16365.10 — Fuses

	UNIT	MAN/HOURS
Fuse, one-time, 250v		
30a	EA.	0.050
60a	"	0.050
100a	"	0.050
200a	"	0.050
400a	"	0.050
600a	"	0.050
600v		
30a	EA.	0.050
60a	"	0.050
100a	"	0.050
200a	"	0.050
400a	"	0.050
Fusetron, 600v		
200a	EA.	0.050
400a	"	0.050
Fuse, amp-trap, K1, 250v		
30a	EA.	0.050
60a	"	0.050
100a	"	0.050
200a	"	0.050
400a	"	0.050
600a	"	0.050
600v		
30a	EA.	0.050
60a	"	0.050
100a	"	0.050
200a	"	0.050
400a	"	0.050
K5, 250v		
30a	EA.	0.050
60a	"	0.050
100a	"	0.050
200a	"	0.050

Power Generation

	UNIT	MAN/HOURS

16365.10 — Fuses (Cont.)

	UNIT	MAN/HOURS
400a	EA.	0.050
600a	"	0.050
600v		
30a	EA.	0.050
60a	"	0.050
100a	"	0.050
200a	"	0.050
400a	"	0.050
600a	"	0.050
J, 600v		
30a	EA.	0.050
60a	"	0.050
100a	"	0.050
200a	"	0.050
400a	"	0.050
L, 600v		
1200a	EA.	0.400
1600a	"	0.400
2000a	"	0.400
2500a	"	0.400
3000a	"	0.400
4000a	"	0.400
5000a	"	0.400
Fuse cl-ay 250v		
600a	EA.	0.296
1200a	"	0.296
1600a	"	0.296
2000a	"	0.296
600v		
1200a	EA.	0.296
1600a	"	0.296
2000a	"	0.296

16395.10 — Grounding

	UNIT	MAN/HOURS
Ground rods, copper clad, 1/2" x		
6'	EA.	0.667
8'	"	0.727
10'	"	1.000
5/8" x		
5'	EA.	0.615
6'	"	0.727
8'	"	1.000
10'	"	1.250
3/4" x		
8'	EA.	0.727
10'	"	0.800
Ground rod clamp		
5/8"	EA.	0.123
3/4"	"	0.123

Service And Distribution

	UNIT	MAN/HOURS

16425.10 — Switchboards

	UNIT	MAN/HOURS
Switchboard, 90" high, no main disconnect, 208/120v		
400a	EA.	7.921
600a	"	8.000
1000a	"	8.000
1200a	"	10.000
1600a	"	11.940
2000a	"	14.035
2500a	"	16.000
277/480v		
600a	EA.	8.163
800a	"	8.163
1600a	"	11.940
2000a	"	14.035
2500a	"	16.000
3000a	"	27.586
4000a	"	29.630
1500 kva	"	57.143

16470.10 — Panelboards

	UNIT	MAN/HOURS
3 phase, 480/277v, main lugs only, 120a, 30 circuits	EA.	3.478
277/480v, 4 wire, flush surface		
225a, 30 circuits	EA.	4.000
400a, 30 circuits	"	5.000
600a, 42 circuits	"	6.015
208/120v, main circuit breaker, 3 phase, 4 wire		
100a		
12 circuits	EA.	5.096
20 circuits	"	6.299
30 circuits	"	7.018
225a		
30 circuits	EA.	7.767
42 circuits	"	9.524
400a		
30 circuits	EA.	14.815
42 circuits	"	16.000
600a, 42 circuits	"	18.182
120/208v, flush, 3 ph., 4 wire, main only		
100a		
12 circuits	EA.	5.096
20 circuits	"	6.299
30 circuits	"	7.018
225a		
30 circuits	EA.	7.767
42 circuits	"	9.524
400a		
30 circuits	EA.	14.815
42 circuits	"	16.000
600a, 42 circuits	"	18.182
Panelboard accessories		
Grounding bus	EA.	0.348
Handle lock device	"	0.145
Factory assembled panel		
1 pole space	EA.	0.348

Service And Distribution

16470.10 Panelboards (Cont.)

	UNIT	MAN/HOURS
2 pole space	EA.	0.145
3 pole space	"	0.133
Panelboards 1 phase, 240/120v main circuit breaker		
Single phase, 3 wire, 120/240v flush		
100a, 20 circuits	EA.	3.478
225a, 30 circuits	"	4.000
240/120v, main lugs only		
100a		
8 circuits	EA.	2.963
12 circuits	"	2.963
20 circuits	"	2.963
225a		
24 circuits	EA.	3.478
30 circuits	"	3.810
42 circuits	"	3.810

16480.10 Motor Controls

	UNIT	MAN/HOURS
Motor generator set, 3 phase, 480/277v, w/controls		
10kw	EA.	27.586
15kw	"	30.769
20kw	"	32.000
25kw	"	34.783
30kw	"	36.364
40kw	"	38.095
50kw	"	40.000
60kw	"	44.444
75kw	"	50.000
100kw	"	61.538
125kw	"	66.667
150kw	"	66.667
200kw	"	72.727
250kw	"	72.727
300kw	"	80.000
2 pole, 230 volt starter, w/NEMA-1		
1 hp, 9a, size 00	EA.	1.000
2 hp, 18a, size 0	"	1.000
3 hp, 27a, size 1	"	1.000
5 hp, 45a, size 1p	"	1.000
7-1/2 hp, 45a, size 2	"	1.000
15 hp, 90a, size 3	"	1.000
2 pole, w/NEMA-4 enclosure		
2 hp, 18a, size 1	EA.	1.600
5 hp, 45a, size 1p	"	1.600
7-1/2 hp, 45a, size 2	"	1.600
3 pole, 2 hp, 9a, 200-575v starter		
W/NEMA-1, size 00	EA.	1.333
W/NEMA-4 enclosure, size 00	"	1.739
5hp, 18a		
W/NEMA-1 enclosure, size 0	EA.	1.333
W/NEMA-4 enclosure, size 0	"	1.739
7.5-10hp, 27a		
7.5-10hp 27a, w/NEMA-1 enclosure, size 1	EA.	1.333
W/NEMA-4 enclosure size 1	"	1.739

16480.10 Motor Controls (Cont.)

	UNIT	MAN/HOURS
10-25hp, 45a		
W/NEMA-1 enclosure, size 2	EA.	1.333
W/NEMA-4 enclosure, size 2	"	1.739
25-50hp, 90a		
W/NEMA-1 enclosure, size 3	EA.	1.739
W/NEMA-4 enclosure, size 3	"	2.500
40-100hp, 135a		
W/NEMA-1 enclosure, size 4	EA.	2.500
W/NEMA-4 enclosure, size 4	"	3.478
75-200hp, 270a		
W/NEMA-1 enclosure, size 5	EA.	5.517
W/NEMA-4 enclosure, size 5	"	7.018
Magnetic starter accessories		
On-off-auto selector switch kit	EA.	0.320
With pilot light	"	0.348

16490.10 Switches

	UNIT	MAN/HOURS
Oil switches, medium voltage, bus components		
Switches, 277/120v, toggle device only	EA.	1.600
With oil 35kv, g&w gram 44, 4 way switch	"	8.000
Weatherproof enclosure		
3 way switch	EA.	10.000
4 way switch	"	10.959
Fused interrupter load, 35kv		
20A		
1 pole	EA.	16.000
2 pole	"	17.021
3 way	"	17.021
4 way	"	18.182
30a, 1 pole	"	16.000
3 way	"	17.021
4 way	"	18.182
Weatherproof switch, including box & cover, 20a		
1 pole	EA.	16.000
2 pole	"	17.021
3 way	"	18.182
4 way	"	18.182
3 way, oil switch, 15kv enclosure	"	11.940
Pedestal for 35kv double breaker switch	"	5.000
Bus terminal connector, 2	"	2.500
2 to 3	"	2.500
Support connector, 3	"	1.600
Tee connector, 2 to 3	"	2.000
Flexible bus stud connector	"	1.739
End cap 3	"	1.333
Weldment connection, 3	"	1.000
Plate switch, 1 gang	"	0.050
Start stop stations, manual motor starters	"	0.727
Lockout switch	"	0.250
Forward-reverse switch	"	0.727
On-off switch	"	0.727
Open-close switch	"	0.727
Forward-reverse-stop switch	"	1.000

Service And Distribution	UNIT	MAN/HOURS
16490.10 — Switches *(Cont.)*		
Standard 3 button switch any standard legend	EA.	1.000
Standard 3 button with lockout	"	1.000
Manual motor starters, tog, 115/230v		
Size 1 gp	EA.	1.000
Size 2	"	1.000
Button		
Size 0	EA.	1.000
Size 1	"	1.000
Size 2	"	1.000
3-phase		
Size 0	EA.	1.333
Size 1	"	1.333
Time & float switches	"	1.600
Astronomical time switch, 40a, 240v	"	1.000
Timer switch 0-5 minute, with box	"	0.500
Single pole/single throw time, 277v, NEMA-1	"	0.727
Single toggle switch, 20a, 120v, with pilot	"	0.250
3-way toggle	"	0.296
Photo electric switches		
1000 watt		
105-135v	EA.	0.727
208-277v	"	0.727
3000 watt, 105-130v		
Double throw	EA.	1.000
Single throw	"	1.000
Double pole/single throw, 210-250v	"	1.333
Dimmer switch and switch plate		
600w	EA.	0.308
1000w	"	0.348
Dimmer switch incandescent		
1500w	EA.	0.702
2000w	"	0.748
Fluorescent		
12 lamps	EA.	0.500
20 lamps	"	0.552
30 lamps	"	0.602
40 lamps	"	0.702
Specification grade toggle switches, 20a, 120-277v		
Single pole	EA.	0.200
Double pole	"	0.296
3 way	"	0.250
4 way	"	0.296
Switch plates, plastic ivory		
1 gang	EA.	0.080
2 gang	"	0.100
3 gang	"	0.119
4 gang	"	0.145
5 gang	"	0.160
6 gang	"	0.182
Stainless steel		
1 gang	EA.	0.080
2 gang	"	0.100
3 gang	"	0.123

Service And Distribution	UNIT	MAN/HOURS
16490.10 — Switches *(Cont.)*		
4 gang	EA.	0.145
5 gang	"	0.160
6 gang	"	0.182
Brass		
1 gang	EA.	0.080
2 gang	"	0.100
3 gang	"	0.123
4 gang	"	0.145
5 gang	"	0.160
6 gang	"	0.182
16490.20 — Transfer Switches		
Automatic transfer switch 600v, 3 pole		
30a	EA.	3.478
60a	"	3.478
100a	"	4.762
150a	"	6.015
225a	"	8.000
260a	"	8.000
400a	"	10.000
600a	"	15.094
800a	"	18.182
1000a	"	21.053
1200a	"	22.857
1600a	"	25.000
2000a	"	29.630
2600a	"	42.105
3000a	"	50.000
16490.80 — Safety Switches		
Safety switch, 600v, 3 pole, heavy duty, NEMA-1		
30a	EA.	1.000
60a	"	1.143
100a	"	1.600
200a	"	2.500
400a	"	5.517
600a	"	8.000
800a	"	10.526
1200a	"	14.286

Lighting	UNIT	MAN/HOURS	Lighting	UNIT	MAN/HOURS
16510.05 **Interior Lighting**			**16510.05** **Interior Lighting** *(Cont.)*		
Recessed fluorescent fixtures, 2'x2'			Starter kit	EA.	0.250
2 lamp	EA.	0.727	Conduit feed	"	0.145
4 lamp	"	0.727	Straight connector	"	0.145
2 lamp w/flange	"	1.000	Center feed	"	0.145
4 lamp w/flange	"	1.000	L-connector	"	0.145
1'x4'			T-connector	"	0.145
2 lamp	EA.	0.667	X-connector	"	0.200
3 lamp	"	0.667	Cord and plug	"	0.100
2 lamp w/flange	"	0.727	Rigid corner	"	0.145
3 lamp w/flange	"	0.727	Flex connector	"	0.145
2'x4'			2 way connector	"	0.200
2 lamp	EA.	0.727	Spacer clip	"	0.050
3 lamp	"	0.727	Grid box	"	0.145
4 lamp	"	0.727	T-bar clip	"	0.050
2 lamp w/flange	"	1.000	Utility hook	"	0.145
3 lamp w/flange	"	1.000	Fixtures, square		
4 lamp w/flange	"	1.000	R-20	EA.	0.145
4'x4'			R-30	"	0.145
4 lamp	EA.	1.000	40w flood	"	0.145
6 lamp	"	1.000	40w spot	"	0.145
8 lamp	"	1.000	100w flood	"	0.145
4 lamp w/flange	"	1.509	100w spot	"	0.145
6 lamp w/flange	"	1.509	Mini spot	"	0.145
8 lamp, w/flange	"	1.509	Mini flood	"	0.145
Surface mounted incandescent fixtures			Quartz, 500w	"	0.145
40w	EA.	0.667	R-20 sphere	"	0.145
75w	"	0.667	R-30 sphere	"	0.145
100w	"	0.667	R-20 cylinder	"	0.145
150w	"	0.667	R-30 cylinder	"	0.145
Pendant			R-40 cylinder	"	0.145
40w	EA.	0.800	R-30 wall wash	"	0.145
75w	"	0.800	R-40 wall wash	"	0.145
100w	"	0.800	Energy saving rapid start fluor. lamps		
150w	"	0.800	F30 cw	EA.	0.100
Recessed incandescent fixtures			F40 cw	"	0.100
40w	EA.	1.509	F40 cwx	"	0.100
75w	"	1.509	F30 ww	"	0.100
100w	"	1.509	F40 ww	"	0.100
150w	"	1.509	F40 wwx	"	0.100
Exit lights, 120v			Incandescent lamps		
Recessed	EA.	1.250	200w	EA.	0.100
Back mount	"	0.727	300w	"	0.100
Universal mount	"	0.727	500w	"	0.100
Emergency battery units, 6v-120v, 50 unit	"	1.509	750w	"	0.145
With 1 head	"	1.509	1000w	"	0.200
With 2 heads	"	1.509	1500w	"	0.200
Light track single circuit			Energy saving reflector floodlight lamps		
2'	EA.	0.500	25w	EA.	0.100
4'	"	0.500	30w	"	0.100
8'	"	1.000	50w	"	0.100
12'	"	1.509	75w	"	0.100
Fittings and accessories			120w	"	0.100
Dead end	EA.	0.145	150w	"	0.100

16510.05 — Interior Lighting (Cont.)

Lighting	UNIT	MAN/HOURS
200w	EA.	0.100
300w	"	0.100
500w	"	0.145
750w	"	0.145
Reflector spotlight		
75w	EA.	0.100
100w	"	0.100
125w	"	0.100
150w	"	0.100
250w	"	0.100
300w	"	0.100
400w	"	0.145
500w	"	0.145
1000w	"	0.200
Medium par flood lamps		
75w	EA.	0.100
100w	"	0.100
150w	"	0.100
200w	"	0.100
300w	"	0.100
500w	"	0.145
Medium par spot lamps		
75w	EA.	0.100
120w	"	0.100
150w	"	0.100
Ballast replacements rapid start fluor		
1f-40-120v	EA.	0.727
1f-40-277v	"	0.727
1f-96-120v	"	0.727
1f-96-277v	"	0.727
2f-40-120v	"	0.727
2f-40-277v	"	0.727
2f-96-120v	"	0.727
2f-96-277v	"	0.727

16510.10 — Lighting Industrial

Lighting	UNIT	MAN/HOURS
Surface mounted fluorescent, wrap around lens		
1 lamp	EA.	0.800
2 lamps	"	0.889
4 lamps	"	1.000
Wall mounted fluorescent		
2-20w lamps	EA.	0.500
2-30w lamps	"	0.500
2-40w lamps	"	0.667
Indirect, with wood shielding, 2049w lamps		
4'	EA.	1.000
8'	"	1.600
Industrial fluorescent, 2 lamp		
4'	EA.	0.727
8'	"	1.333
Strip fluorescent		
4'		
1 lamp	EA.	0.667

16510.10 — Lighting Industrial (Cont.)

Lighting	UNIT	MAN/HOURS
2 lamps	EA.	0.667
8'		
1 lamp	EA.	0.727
2 lamps	"	0.889
Wire guard for strip fixture, 4' long	"	0.348
Strip fluorescent, 8' long, two 4' lamps	"	1.333
With four 4' lamps	"	1.600
Wet location fluorescent, plastic housing		
4' long		
1 lamp	EA.	1.000
2 lamps	"	1.333
8' long		
2 lamps	EA.	1.600
4 lamps	"	1.739
Parabolic troffer, 2'x2'		
With 2 "U" lamps	EA.	1.000
With 3 "U" lamps	"	1.143
2'x4'		
With 2 40w lamps	EA.	1.143
With 3 40w lamps	"	1.333
With 4 40w lamps	"	1.333
1'x4'		
With 1 T-12 lamp, 9 cell	EA.	0.727
With 2 T-12 lamps	"	0.889
With 1 T-12 lamp, 20 cell	"	0.727
With 2 T-12 lamps	"	0.889
Steel sided surface fluorescent, 2'x4'		
3 lamps	EA.	1.333
4 lamps	"	1.333
Outdoor sign fluor., 1 lamp, remote ballast		
4' long	EA.	6.015
6' long	"	8.000
Recess mounted, commercial, 2'x2', 13" high		
100w	EA.	4.000
250w	"	4.494
High pressure sodium, hi-bay open		
400w	EA.	1.739
1000w	"	2.424
Enclosed		
400w	EA.	2.424
1000w	"	2.963
Metal halide hi-bay, open		
400w	EA.	1.739
1000w	"	2.424
Enclosed		
400w	EA.	2.424
1000w	"	2.963
High pressure sodium, low bay, surface mounted		
100w	EA.	1.000
150w	"	1.143
250w	"	1.333
400w	"	1.600
Metal halide, low bay, pendant mounted		

Lighting	UNIT	MAN/ HOURS
16510.10 — **Lighting Industrial** *(Cont.)*		
175w	EA.	1.333
250w	"	1.600
400w	"	2.222
Indirect luminare, square, metal halide, freestanding		
175w	EA.	1.000
250w	"	1.000
400w	"	1.000
High pressure sodium		
150w	EA.	1.000
250w	"	1.000
400w	"	1.000
Round, metal halide		
175w	EA.	1.000
250w	"	1.000
400w	"	1.000
High pressure sodium		
150w	EA.	1.000
250w	"	1.000
400w	"	1.000
Wall mounted, metal halide		
175w	EA.	2.500
250w	"	2.500
400w	"	3.200
High pressure sodium		
150w	EA.	2.500
250w	"	2.500
400w	"	3.200
Wall pack lithonia, high pressure sodium		
35w	EA.	0.889
55w	"	1.000
150w	"	1.600
250w	"	1.739
Low pressure sodium		
35w	EA.	1.739
55w	"	2.000
Wall pack hubbell, high pressure sodium		
35w	EA.	0.889
150w	"	1.600
250w	"	1.739
Compact fluorescent		
2-7w	EA.	1.000
2-13w	"	1.333
1-18w	"	1.333
Handball & racquet ball court, 2'x2', metal halide		
250w	EA.	2.500
400w	"	2.759
High pressure sodium		
250w	EA.	2.500
400w	"	2.759
Bollard light, 42" w/found., high pressure sodium		
70w	EA.	2.581
100w	"	2.581
150w	"	2.581

Lighting	UNIT	MAN/ HOURS
16510.10 — **Lighting Industrial** *(Cont.)*		
Light fixture lamps		
Lamp		
20w med. bipin base, cool white, 24"	EA.	0.145
30w cool white, rapid start, 36"	"	0.145
40w cool white "U", 3"	"	0.145
40w cool white, rapid start, 48"	"	0.145
70w high pressure sodium, mogul base	"	0.200
75w slimline, 96"	"	0.200
100w		
Incandescent, 100a, inside frost	EA.	0.100
Mercury vapor, clear, mogul base	"	0.200
High pressure sodium, mogul base	"	0.200
150w		
Par 38 flood or spot, incandescent	EA.	0.100
High pressure sodium, 1/2 mogul base	"	0.200
175w		
Mercury vapor, clear, mogul base	EA.	0.200
Metal halide, clear, mogul base	"	0.200
High pressure sodium, mogul base	"	0.200
250w		
Mercury vapor, clear, mogul base	EA.	0.200
Metal halide, clear, mogul base	"	0.200
High pressure sodium, mogul base	"	0.200
400w		
Mercury vapor, clear, mogul base	EA.	0.200
Metal halide, clear, mogul base	"	0.200
High pressure sodium, mogul base	"	0.200
1000w		
Mercury vapor, clear, mogul base	EA.	0.250
High pressure sodium, mogul base	"	0.250
16510.30 — **Exterior Lighting**		
Exterior light fixtures		
Rectangle, high pressure sodium		
70w	EA.	2.500
100w	"	2.581
150w	"	2.581
250w	"	2.759
400w	"	3.478
Flood, rectangular, high pressure sodium		
70w	EA.	2.500
100w	"	2.581
150w	"	2.581
400w	"	3.478
1000w	"	4.494
Round		
400w	EA.	3.478
1000w	"	4.494
Round, metal halide		
400w	EA.	3.478
1000w	"	4.494
Light fixture arms, cobra head, 6', high press. sodium		
100w	EA.	2.000

Lighting	UNIT	MAN/ HOURS
16510.30 Exterior Lighting (Cont.)		
150w	EA.	2.500
250w	"	2.500
400w	"	2.963
Flood, metal halide		
400w	EA.	3.478
1000w	"	4.494
1500w	"	6.015
Mercury vapor		
250w	EA.	2.759
400w	"	3.478
Incandescent		
300w	EA.	1.739
500w	"	2.000
1000w	"	3.200
16510.90 Power Line Filters		
Heavy duty power line filter, 240v		
100a	EA.	10.000
300a	"	16.000
600a	"	24.242
16610.30 Uninterruptible Power		
Uninterruptible power systems, (U.P.S.), 3kva	EA.	8.000
5 kva	"	11.004
7.5 kva	"	16.000
10 kva	"	21.978
15 kva	"	22.857
20 kva	"	24.024
25 kva	"	25.000
30 kva	"	25.974
35 kva	"	27.027
40 kva	"	27.972
45 kva	"	28.986
50 kva	"	29.963
62.5 kva	"	32.000
75 kva	"	34.934
100 kva	"	36.036
150 kva	"	50.000
200 kva	"	55.172
300 kva	"	74.766
400 kva	"	89.888
500 kva	"	109.589
16670.10 Lightning Protection		
Lightning protection		
Copper point, nickel plated, 12'		
1/2" dia.	EA.	1.000
5/8" dia.	"	1.000

Communications	UNIT	MAN/ HOURS
16720.10 Fire Alarm Systems		
Master fire alarm box, pedestal mounted	EA.	16.000
Master fire alarm box	"	6.015
Box light	"	0.500
Ground assembly for box	"	0.667
Bracket for pole type box	"	0.727
Pull station		
Waterproof	EA.	0.500
Manual	"	0.400
Horn, waterproof	"	1.000
Interior alarm	"	0.727
Coded transmitter, automatic	"	2.000
Control panel, 8 zone	"	8.000
Battery charger and cabinet	"	2.000
Batteries, nickel cadmium or lead calcium	"	5.000
CO2 pressure switch connection	"	0.727
Annunciator panels		
Fire detection annunciator, remote type, 8 zone	EA.	1.818
12 zone	"	2.000
16 zone	"	2.500
Fire alarm systems		
Bell	EA.	0.615
Weatherproof bell	"	0.667
Horn	"	0.727
Siren	"	2.000
Chime	"	0.615
Audio/visual	"	0.727
Strobe light	"	0.727
Smoke detector	"	0.667
Heat detection	"	0.500
Thermal detector	"	0.500
Ionization detector	"	0.533
Duct detector	"	2.759
Test switch	"	0.500
Remote indicator	"	0.571
Door holder	"	0.727
Telephone jack	"	0.296
Fireman phone	"	1.000
Speaker	"	0.800
Remote fire alarm annunciator panel		
24 zone	EA.	6.667
48 zone	"	13.008
Control panel		
12 zone	EA.	2.963
16 zone	"	4.444
24 zone	"	6.667
48 zone	"	16.000
Power supply	"	1.509
Status command	"	5.000
Printer	"	1.509
Transponder	"	0.899
Transformer	"	0.667
Transceiver	"	0.727
Relays	"	0.500

Communications

16720.10   Fire Alarm Systems *(Cont.)*

	UNIT	MAN/HOURS
Flow switch	EA.	2.000
Tamper switch	"	2.963
End of line resistor	"	0.348
Printed ckt. card	"	0.500
Central processing unit	"	6.154
UPS backup to c.p.u.	"	8.999
Smoke detector, fixed temp. & rate of rise comb.	"	1.600

16720.50   Security Systems

	UNIT	MAN/HOURS
Sensors		
Balanced magnetic door switch, surface mounted	EA.	0.500
With remote test	"	1.000
Flush mounted	"	1.860
Mounted bracket	"	0.348
Mounted bracket spacer	"	0.348
Photoelectric sensor, for fence		
6 beam	EA.	2.759
9 beam	"	4.255
Photoelectric sensor, 12 volt dc		
500' range	EA.	1.600
800' range	"	2.000
Capacitance wire grid kit		
Surface	EA.	1.000
Duct	"	1.600
Tube grid kit	"	0.500
Vibration sensor, 30 max per zone	"	0.500
Audio sensor, 30 max per zone	"	0.500
Inertia sensor		
Outdoor	EA.	0.727
Indoor	"	0.500
Ultrasonic transmitter, 20 max per zone		
Omni-directional	EA.	1.600
Directional	"	1.333
Transceiver		
Omni-directional	EA.	1.000
Directional	"	1.000
Passive infra-red sensor, 20 max per zone	"	1.600
Access/secure control unit, balanced magnetic switch	"	1.600
Photoelectric sensor	"	1.600
Photoelectric fence sensor	"	1.600
Capacitance sensor	"	1.739
Audio and vibration sensor	"	1.600
Inertia sensor	"	1.600
Ultrasonic sensor	"	1.739
Infra-red sensor	"	2.000
Monitor panel, with access/secure tone, standard	"	1.739
High security	"	2.000
Emergency power indicator	"	0.500
Monitor rack with 115v power supply		
1 zone	EA.	1.000
10 zone	"	2.500
Monitor cabinet, wall mounted		
1 zone	EA.	1.000

Communications

16720.50   Security Systems *(Cont.)*

	UNIT	MAN/HOURS
5 zone	EA.	1.600
10 zone	"	1.739
20 zone	"	2.000
Floor mounted, 50 zone	"	4.000
Security system accessories		
Tamper assembly for monitor cabinet	EA.	0.444
Monitor panel blank	"	0.348
Audible alarm	"	0.500
Audible alarm control	"	0.348
Termination screw, terminal cabinet		
25 pair	EA.	1.600
50 pair	"	2.500
150 pair	"	5.000
Universal termination, cabinets & panel		
Remote test	EA.	1.739
No remote test	"	0.727
High security line supervision termination	"	1.000
Door cord for capacitance sensor, 12"	"	0.500
Insulation block kit for capacitance sensor	"	0.348
Termination block for capacitance sensor	"	0.348
Guard alert display	"	0.615
Uninterrupted power supply	"	8.000
Plug-in 40kva transformer		
12 volt	EA.	0.348
18 volt	"	0.348
24 volt	"	0.348
Test relay	"	0.348
Coaxial cable, 50 ohm	L.F.	0.006
Door openers	EA.	0.500
Push buttons		
Standard	EA.	0.348
Weatherproof	"	0.444
Bells	"	0.727
Horns		
Standard	EA.	1.000
Weatherproof	"	1.250
Chimes	"	0.667
Flasher	"	0.615
Motion detectors	"	1.509
Intercom units	"	0.727
Remote annunciator	"	5.000

16740.10   Telephone Systems

	UNIT	MAN/HOURS
Communication cable		
25 pair	L.F.	0.026
100 pair	"	0.029
150 pair	"	0.033
200 pair	"	0.040
300 pair	"	0.042
400 pair	"	0.044
Cable tap in manhole or junction box		
25 pair cable	EA.	3.810
50 pair cable	"	7.547

Communications	UNIT	MAN/HOURS
16740.10 **Telephone Systems** *(Cont.)*		
75 pair cable	EA.	11.268
100 pair cable	"	15.094
150 pair cable	"	22.222
200 pair cable	"	29.630
300 pair cable	"	44.444
400 pair cable	"	61.538
Cable terminations, manhole or junction box		
25 pair cable	EA.	3.756
50 pair cable	"	7.477
100 pair cable	"	15.094
150 pair cable	"	22.222
200 pair cable	"	29.630
300 pair cable	"	44.444
400 pair cable	"	61.538
16780.10 **Antennas And Towers**		
Guy cable, alumaweld		
1x3, 7/32"	L.F.	0.050
1x3, 1/4"	"	0.050
1x3, 25/64"	"	0.059
1x19, 1/2"	"	0.070
1x7, 35/64"	"	0.080
1x19, 13/16"	"	0.100
Preformed alumaweld end grip		
1/4" cable	EA.	0.100
3/8" cable	"	0.100
1/2" cable	"	0.145
9/16" cable	"	0.200
5/8" cable	"	0.250
Fiberglass guy rod, white epoxy coated		
1/4" dia.	L.F.	0.145
3/8" dia	"	0.145
1/2" dia	"	0.200
5/8" dia	"	0.250
Preformed glass grip end grip, guy rod		
1/4" dia.	EA.	0.145
3/8" dia.	"	0.200
1/2" dia.	"	0.250
5/8" dia.	"	0.250
Spelter socket end grip, 1/4" dia. guy rod		
Standard strength	EA.	0.500
High performance	"	0.500
3/8" dia. guy rod		
Standard strength	EA.	0.348
High performance	"	0.500
Timber pole, Douglas Fir		
80-85 ft	EA.	19.512
90-95 ft	"	22.222
Southern yellow pine		
35-45 ft	EA.	10.959
50-55 ft	"	14.035

Communications	UNIT	MAN/HOURS
16780.50 **Television Systems**		
TV outlet, self terminating, w/cover plate	EA.	0.308
Thru splitter	"	1.600
End of line	"	1.333
In line splitter multitap		
4 way	EA.	1.818
2 way	"	1.702
Equipment cabinet	"	1.600
Antenna		
Broad band uhf	EA.	3.478
Lightning arrester	"	0.727
TV cable	L.F.	0.005
Coaxial cable rg	"	0.005
Cable drill, with replacement tip	EA.	0.500
Cable blocks for in-line taps	"	0.727
In-line taps ptu-series 36 tv system	"	1.143
16850.10 **Electric Heating**		
Baseboard heater		
2', 375w	EA.	1.000
3', 500w	"	1.000
4', 750w	"	1.143
5', 935w	"	1.333
6', 1125w	"	1.600
7', 1310w	"	1.818
8', 1500w	"	2.000
9', 1680w	"	2.222
10', 1875w	"	2.286
Unit heater, wall mounted		
1500w	EA.	1.667
2500w	"	1.818
4000w	"	2.286
Thermostat		
Integral	EA.	0.500
Line voltage	"	0.500
Electric heater connection	"	0.250

Supporting Construction Reference Data

This section contains information, text, charts and tables on various aspects of construction. The intent is to provide the user with a better understanding of unfamiliar areas in order to be able to estimate better. This information includes actual takeoff data for some areas and also selected explanations of common construction materials, methods and common practices.

ADDITIONAL COSTS INVOLVED IN REMODELING CONSTRUCTION

SMALL AMOUNTS OF WORK

Material
Since remodeling work often entails a smaller scope of work, many of the materials that are bought at a discount in large quantities for a new construction project have to be purchased at a lesser discount or at the regular retail price. To take this factor into account, it is prudent to add an additional 10 percent to the cost of materials for small remodeling projects.

Labor
In order to obtain maximum labor efficiency and normal production, quantities of work must be sufficient to keep workers busy for long enough periods of time, usually a full day. Even with a full day's worth of a particular task, the start-up and wind-down/cleanup (non-productive time) of a typical day take a significant amount of time. When the task requires less than a full day, the non-productive time increases as a percentage of the work installed. Therefore, the cost of labor per unit increases to a minimum of 2 to 4 hours for the installing function.

PATCHING TO MATCH
Matching and fitting new materials to the pre-existing construction is usually a costly factor in remodeling work, whether it be the exterior siding of a building or interior paint or plaster. Sometimes creative, and often expensive, methods have to be used to match colors, patterns and materials. Adding 10 to 30 percent to the labor costs will insulate against this.

WORKING AROUND ONGOING OPERATIONS
One of the most time-consuming parts of a remodeling job is the distractions caused by working around people. Just normal traffic flow of people through or near a work area affects the way the work gets done. For example, when a contractor has to take precautions during demolition work for the safety of passersby, the job will take longer and cost more. These costs increase even more when work is being done in an office or retail store when the business must continue to function during the remodeling. Adding 5 to 50 percent (depending upon severity) to the labor estimate will provide for this.

DIFFICULT ACCESS
Contractors often find their access to work areas restricted in any number of ways. In remodeling work, the contractor loses the luxury of a logical building process that a new construction project enjoys. Walls have to be cut and patched for the installation of new mechanical or electrical systems; large building materials do not fit through existing doors or windows; even access to the structure's exterior can be limited by added features such as urban locations, landscaping, fences, garages and swimming pools. Delivery and distribution of materials should be approached on a man-hour basis. Add 5 to 20 percent to the labor process for a realistic delivery allowances.

WORKING AROUND SENSITIVE EQUIPMENT
Doing a remodeling job quickly and efficiently usually means being able to make a mess at first and then cleaning up after. Some job locations, however, only allow a minimum amount of dust and debris. When working in close proximity to sensitive machines, computers especially, a job can become up to 100 percent more expensive for labor. Also to be factored in are the hanging of plastic curtains, longer dumpster rental, and laborers cleaning up on a continual basis.

WORKING OFF PEAK HOURS
When a remodeling job involves accommodating a functioning business it is often to the contractor's advantage to work odd hours. Sometimes it is even a requirement. Many retail operations that are open late, or even 24 hours, schedule remodeling work around the second or third shift. While this is not overtime, it is more costly for the contractor to hire subcontractors and tradesmen. However, the distractions and accommodations necessitated by building occupants are much less than if the remodeling work was done during the normal business day. The greater efficiency and speed allowed by working in a quiet environment helps to offset the higher labor costs and reduced productivity. Still, it is justifiable to add 10 to 15 percent to the estimated cost of labor scheduled during off peak hours.

INEFFICIENT WORK METHODS
Working around people and adverse conditions often make the most efficient work methods and the proper tools unusable. The less efficient the method, the more must be spent on additional labor and equipment rental, such as scaffolding and tools a remodeling contractor may not normally have on hand. Depending on the job, another 10 to 40 percent may be added to the labor estimate.

TYPICAL BUILDING COST BROKEN DOWN
BY CSI FORMAT
(Commercial Construction)

Division	New Construction	Remodeling Construction
1. General Requirements	6 to 8%	Up to 30%
2. Sitework	4 to 6%	
3. Concrete	15 to 20%	
4. Masonry	8 to 12%	
5. Metals	5 to 7%	
6. Wood and Plastics	1 to 5 %	
7. Thermal and Moisture Protection	4 to 6%	
8. Doors and Windows	5 to 7%	} Up to 30%
9. Finishes	8 to 12 %	
10. Specialties		
11. Architectural Equipment		
12. Furnishings	} 6 to 10%	
13. Special Construction		
14. Conveying Systems		
15. Mechanical	15 to 25%	} Up to 30%
16. Electrical	8 to 12%	
TOTAL COST	100%	

CONVERSION FACTORS

Change	To	Multiply By
ATMOSPHERES	POUNDS PER SQUARE INCH	14.696
ATMOSPHERES	INCHES OF MERCURY	29.92
ATMOSPHERES	FEET OF WATER	34
BARRELS, OIL	GALLONS, OF OIL	42
BARRELS, CEMENT	POUNDS OF CEMENT	376
BOGS OR SACKS, CEMENT	POUNDS OF CEMENT	94
BTU/MIN	FOOT-POUNDS/SEC	12.96
BTU/MIN	HORSE-POWER	0.02356
BTU/MIN	KILOWATTS	0.01757
BTU/MM	WATTS	17.57
CENTIMETERS	INCHES	0.3937
CENTIMETERS OF MERCURY	ATMOSPHERES	0.01316
CENTIMETERS OF MERCURY	FEET OF WATER	0.4461
CUBIC INCHES	CUBIC FEET	0.00058
CUBIC FEET	CUBIC INCHES	1728
CUBIC FEET	CUBIC YARDS	0.03703
CUBIC YARDS	CUBIC FEET	27
CUBIC INCHES	GALLONS	0.00433
CUBIC FEET	GALLONS	7.48
FEET	INCHES	12
FEET	YARDS	0.3333
YARDS	FEET	3
FEET OF WATER	ATMOSPHERES	0.02950
FEET OF WATER	INCHES OF MERCURY	0.8826
GALLONS	CUBIC INCHES	231
GALLONS	CUBIC FEET	0.1337
GALLONS	POUNDS OF WATER	8.33
GALLONS	QUARTS	4
GALLONS PER MIN	CUBIC FEET SEC	0.002228
GALLONS PER MIN	CUBIC FEET HOUR	8.0208
GALLONS WATER PER MIN	TONS WATER/24 HOURS	6.0086
HORSE-POWER	FOOT-LBS./SEC	550
INCHES	CENTIMETERS	2.540
INCHES	FEET	0.0833
INCHES	MILLIMETERS	25.4
INCHES OF WATER	POUNDS PER SQ. INCH	0.0361
INCHES OF WATER	INCHES OF MERCURY	0.0735
INCHES OF WATER	OUNCES PER SQUARE INCH	0.578
INCHES OF WATER	OUNCES PER SQUARE FOOT	5.2
INCHES OF MERCURY	INCHES OF WATER	13.6
INCHES OF MERCURY	FEET OF WATER	1.1333
INCHES OF MERCURY	POUNDS PER SQUARE INCH	0.4914
KILOMETERS	MILES	0.6214
METERS	INCHES	39.37
MILES	FEET	5280
MILLIMETERS	CENTIMETERS	0.1
MILLIMETERS	INCHES	0.03937
OUNCES (FLUID)	CUBIC INCHES	1.805
OUNCES	POUNDS	0.0625
POUNDS	OUNCES	16
POUNDS PER SQUARE INCH	INCHES OF WATER	27.72
POUNDS PER SQUARE INCH	FEET OF WATER	2.310
POUNDS PER SQUARE INCH	INCHES OF MERCURY	2.04
POUNDS PER SQUARE INCH	ATMOSPHERES	0.0681
QUARTS	CUBIC INCHES	67.20
SQUARE INCHES	SQUARE FEET	0.00694
SQUARE FEET	SQUARE INCHES	144
SQUARE FEET	SQUARE YARDS	0.11111
SQUARE YARDS	SQUARE FEET	9
SQUARE MILES	ACRES	640
SHORT TONS	POUNDS	2000
SHORT TONS	LONG TONS	0.89285
TONS OF WATER/24 HOURS	GALLONS PER MINUTE	0.16643
YARDS	FEET	3
YARDS	CENTIMETERS	91.44
YARDS	INCHES	36

CONVERSION CALCULATIONS

Commercial Measure

16 grams	= 1 ounce
16 ounces	= 1 pound
2,000 pounds	= 1 ton

Long Measure

12 inches	= 1 foot
3 feet	= 1 yard
16 ½ feet	= 1 rod
40 rods	= 1 furlong
8 furlongs (5,280 ft.)	= 1 mile
3 miles	= 1 league

Square Measure

144 square inches	= 1 square foot
9 square feet	= 1 square yard
30 ¼ square yards	= 1 square rod
160 square rods	= 1 acre
4,840 square yards	= 1 acre
640 acres	= 1 square mile
36 square miles	= 1 township

Surveyor's Measure

7.92 inches	= 1 link
25 links	= 1 rod
4 rods (66 ft.)	= 1 chain
10 chains	= 1 furlong
8 furlongs	= 1 mile
1 square mile	= 1 section

Cubic Measure

1728 cubic inches	= 1 cubic foot
27 cubic feet	= 1 cubic yard
128 cubic feet	= 1 cord (wood/stone)
231 cubic inches	= 1 U.S. gallon
7.48 U.S. Gallons	= 1 cubic foot
2150.4 cubic inches	= 1 U.S. bushel

Liquid Measure

4 fluid ounces	= 1 gill
4 gills	= 1 pint
2 pints	= 1 quart
4 quarts	= 1 gallon
9 gallons	= 1 firkin
31 ½ gallons	= 1 barrel
2 barrels	= 1 hogshead

Dry Measure

2 pints	= 1 quart
8 quarts	= 1 peck
4 pecks	= 1 bushel
2150.42 cubic inches	= 1 bushel

SQUARE

$$A = a^2$$

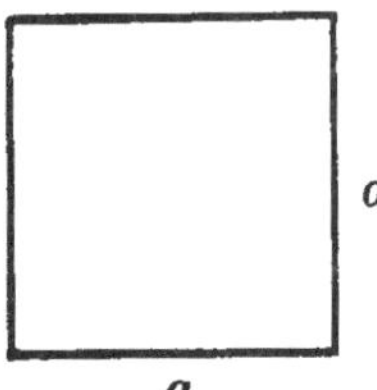

RECTANGLE

$$A = bh$$

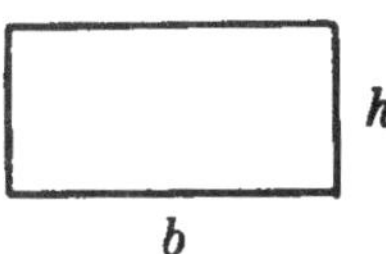

TRIANGLE

$$A = \frac{1}{2} bh$$

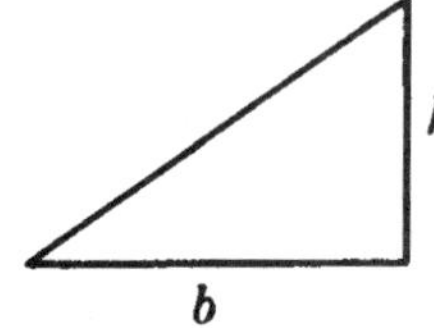

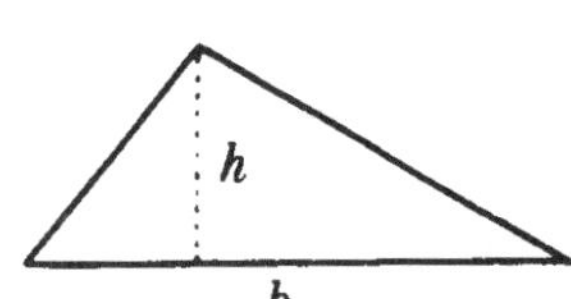

PARALLELOGRAM

$$A = bh = ab\ Sin\emptyset$$

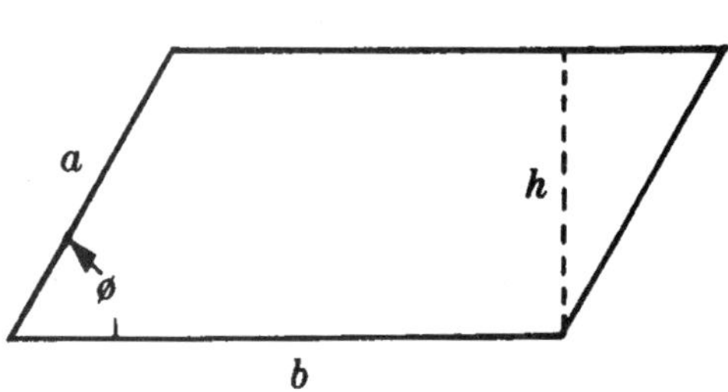

TRAPEZOID

$$A = \left(\frac{a + b}{2}\right) h$$

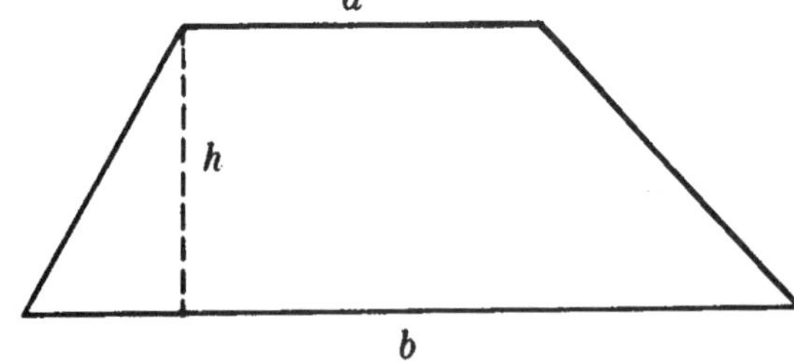

CIRCLE

$$A = \pi r^2 = \frac{\pi d^2}{4}$$

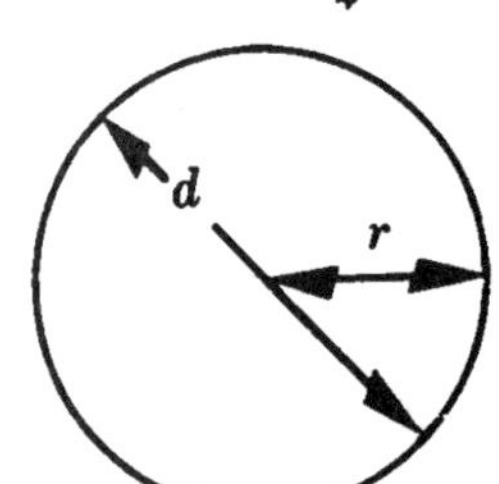

$$Circumference = C = 2\pi r = \pi d$$

ELLIPSE

$A = \pi ab$

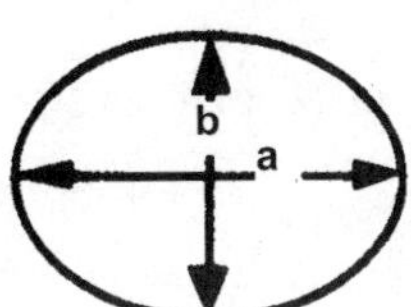

PARABOLA

$A = \dfrac{2}{3} bh$

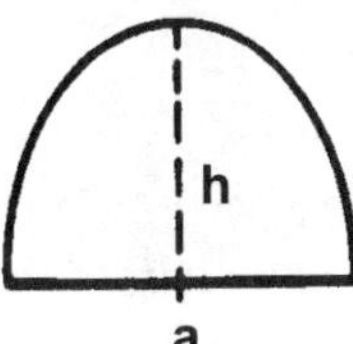

VOLUMES

CUBE

$V = a^3$

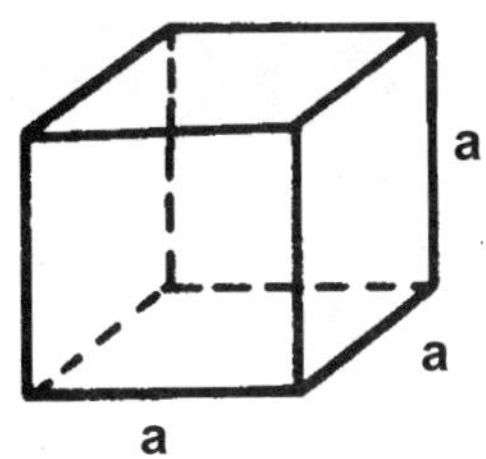

CYLINDER

$V = \pi r^3 h$

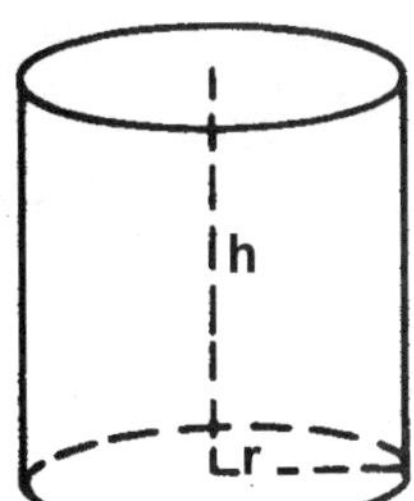

PYRAMID

$V = \dfrac{1}{3}(\text{Base})\, h$

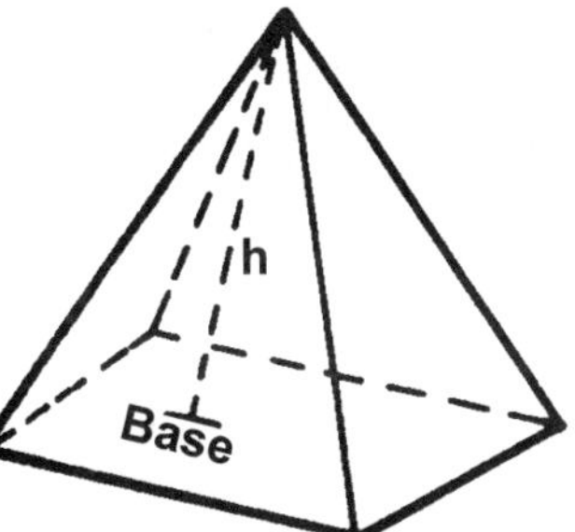

CONE

$V = \dfrac{1}{3}\pi r^2 h$

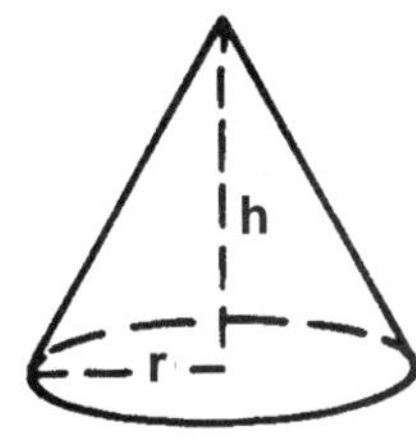

SPHERE

$V = \dfrac{4}{3}\pi r^3 = \dfrac{1}{6}\pi d^3$

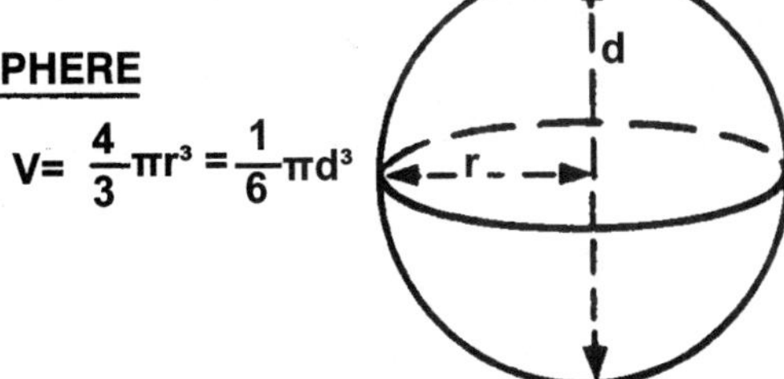

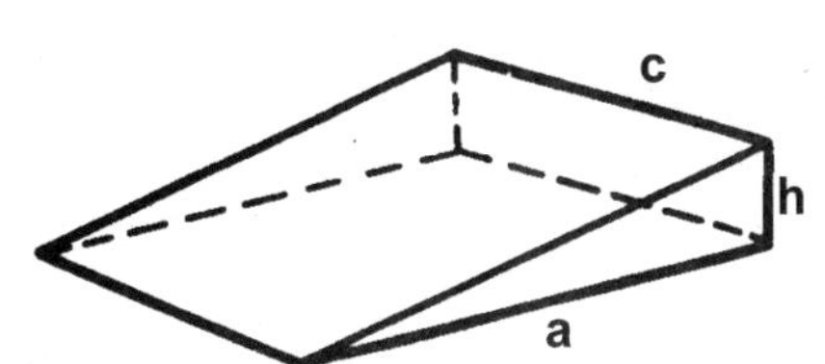

WEDGE

$V = \dfrac{1}{2} abc$

CONVERSION FACTORS
ENGLISH TO SI (SYSTEM INTERNATIONAL)

To Convert from	To	Multiply by
LENGTH		
Inches	Millimetres	25.4[a]
Feet	Metres	0.3048[a]
Yards	Metres	0.9144[a]
Miles (statute)	Kilometres	1.609
AREA		
Square inches	Square millimetres	645.2
Square feet	Square metres	0.0929
Square yards	Square metres	0.8361
VOLUME		
Cubic inches	Cubic millimetres	16.387
Cubic feet	Cubic metres	0.02832
Cubic yards	Cubic metres	0.7646
Gallons (U.S. liquid) [b]	Cubic metres[c]	0.003785
Gallons (Canadian liquid) [b]	Cubic metres[c]	0.004546
Ounces (U.S. liquid) [b]	Millilitres [c,d]	29.57
Quarts (U.S. liquid) [b]	Litres [c,d]	0.9464
Gallons (U.S. liquid) [b]	Litres [c]	3.785
FORCE		
Kilograms force	Newtons	9.807
Pounds force	Newtons	4.448
Pounds force	Kilograms force [d]	0.4536
Kips	Newtons	4448
Kips	Kilograms force [d]	453.6
PRESSURE, STRESS, STRENGTH (FORCE PER UNIT AREA)		
Kilograms force per sq. centimetre	Megapascals	0.09807
Pounds force per square inch (psi)	Megapascals	6895
Kips per square inch	Megapascals	6.895
Pounds force per square inch (psi)	Kilograms force per square centimetre [d]	0.07031
Pounds force per square foot	Pascals	47.88
Pounds force per square foot	Kilograms force per square metre [d]	4.882
SENDING MOMENT OR TORQUE		
Inch-pounds force	Metre-kilog. force [d]	0.01152
Inch-pounds force	Newton-metres	0.1130
Foot-pounds force	Metre-kilog. force [d]	0.1383
Foot-pounds force	Newton-metres	1.356
Metre-kilograms force	Newton-metres	9.807
MASS		
Ounce (avoirdupois)	Grams	28.35
Pounds (avoirdupois)	Kilograms	0.4536
Tons (metric)	Kilograms	1000[a]
Tons, short (2000 pounds)	Kilograms	907.2
Tons, short (2000 pounds)	Megagrams [e]	0.9072
MASS PER UNIT VOLUME		
Pounds mass per cubic foot	Kilog. per cubic metre	16.02
Pounds mass per cubic yard	Kilog. per cubic metre	0.5933
Pds. mass per gallon (U.S. liquid) [b]	Kilog. per cubic metre	119.8
Pds. mass p/gal. (Canadian liquid) [b]	Kilog. per cubic metre	99.78
TEMPERATURE		
Degrees Fahrenheit	Degrees Celsius	$tK = (1F - 32)/1.8$
Degrees Fahrenheit	Degrees Kelvin	$tK = (1F + 459.67)/1.8$
Degree Celsius	Degree Kelvin	$tK = 1C + 273.15$

[a] The factor given is exact.
[b] One U.S. gallon equals 0.8327 Canadian gallon.
[c] 1 litre = 1000 millilitres = 10,000 cubic centimetres = 1 cubic decimetre = 0.001 cubic metre.
[d] Metric but not SI unit.
[e] Called "tonne" in England. Called "metric ton" in other metric systems.

TRENCH BRACING

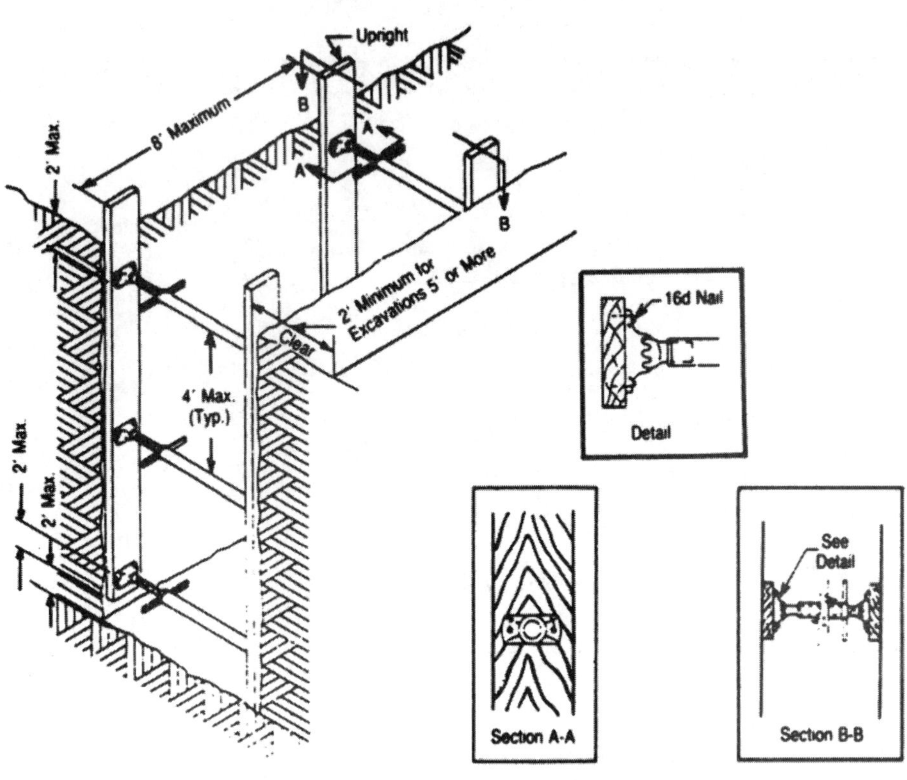

CLOSED VERTICAL SHEETING

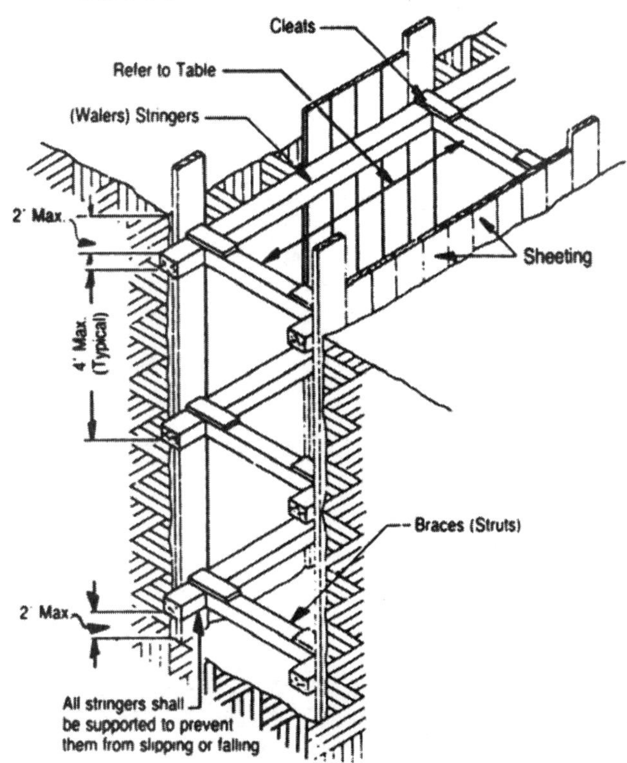

STANDARD NOMENCLATURE FOR STREET CONSTRUCTION

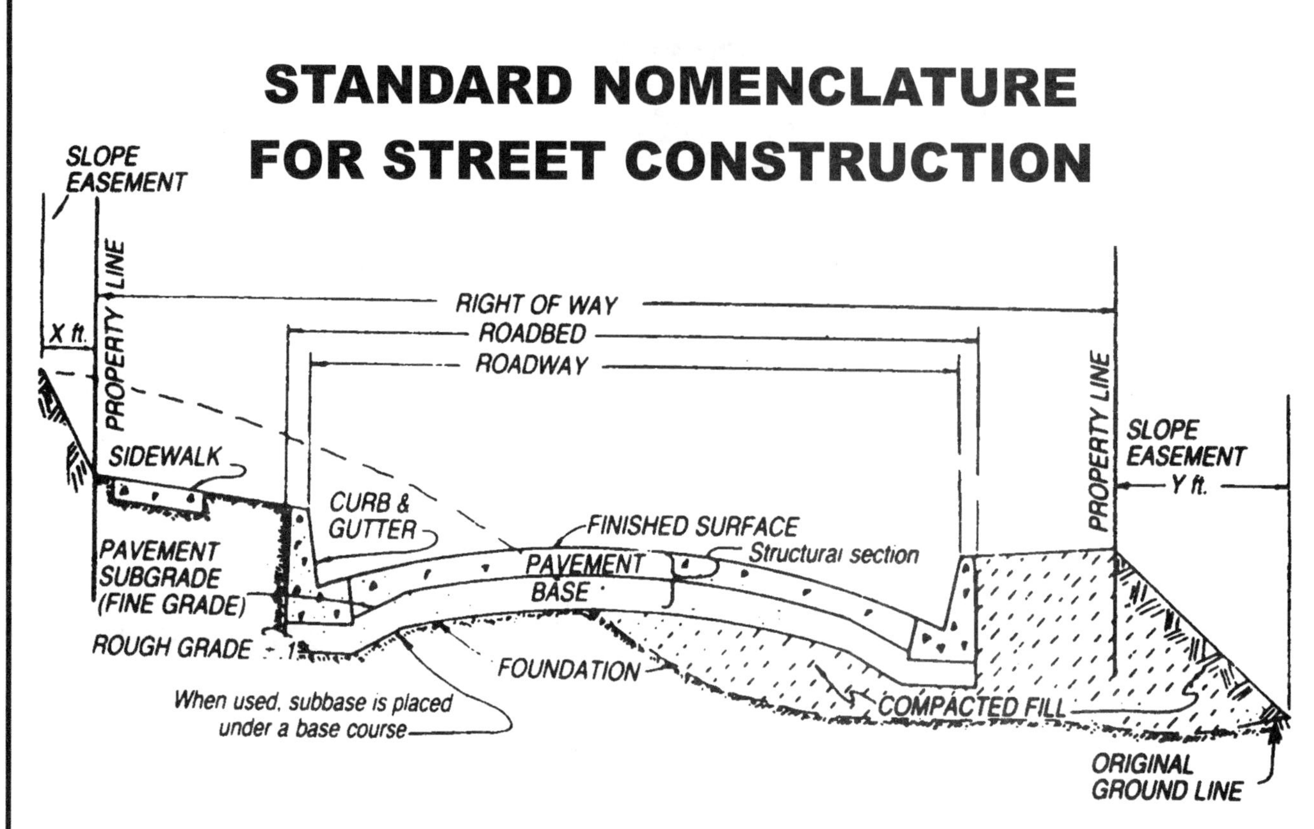

CONCRETE MASONRY PAVING UNITS

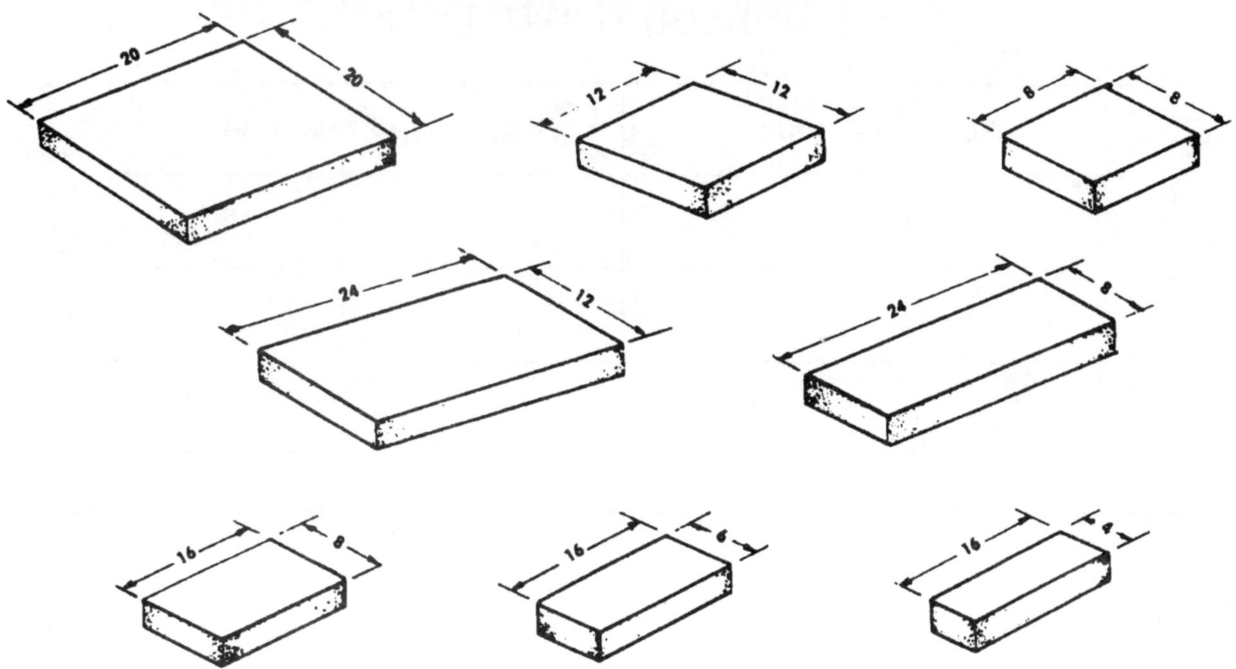

NOTE: Sizes are nominal and will vary by manufacturer.

HEXAGON PAVER UNITS
Various Sizes Available

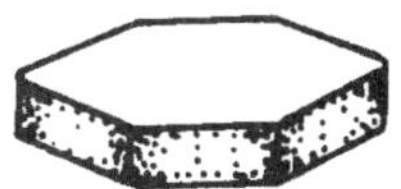

ROUND PAVING UNITS
Various Sizes Available

VEHICULAR PAVING UNITS

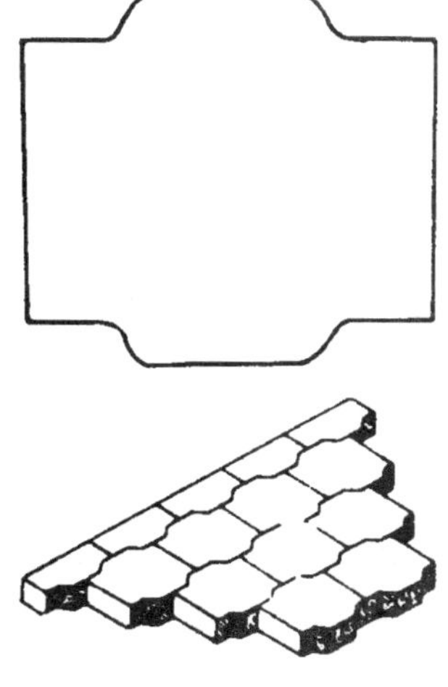

INTERLOCKING PAVER
7¼" x 3" x 8½"

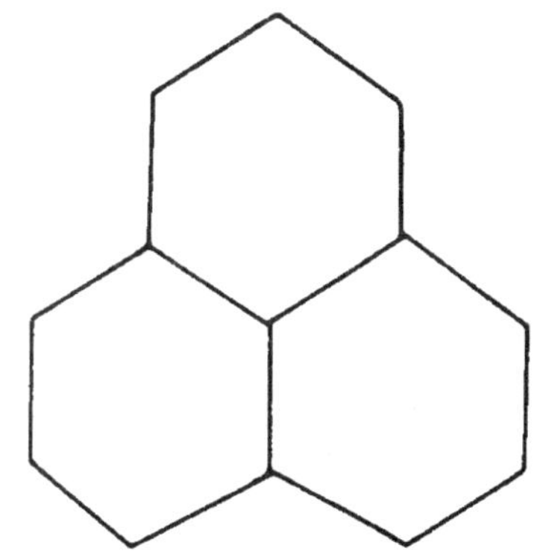

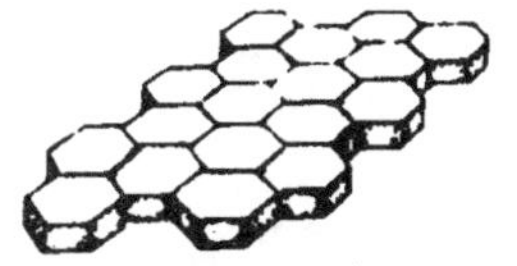

INTERLOCKING PAVER
12" x 3⅜" x 12"

TURF PAVER
24" x 3⅝" x 24"

CAPACITIES FOR SEPTIC TANKS SERVING AN INDIVIDUAL DWELLING

No. of bedrooms	Capacity of tank (gals.)
2 or less	750
3	900
4	1,000

FORM NOMENCLAURE

FALSEWORK NOMENCLATURE

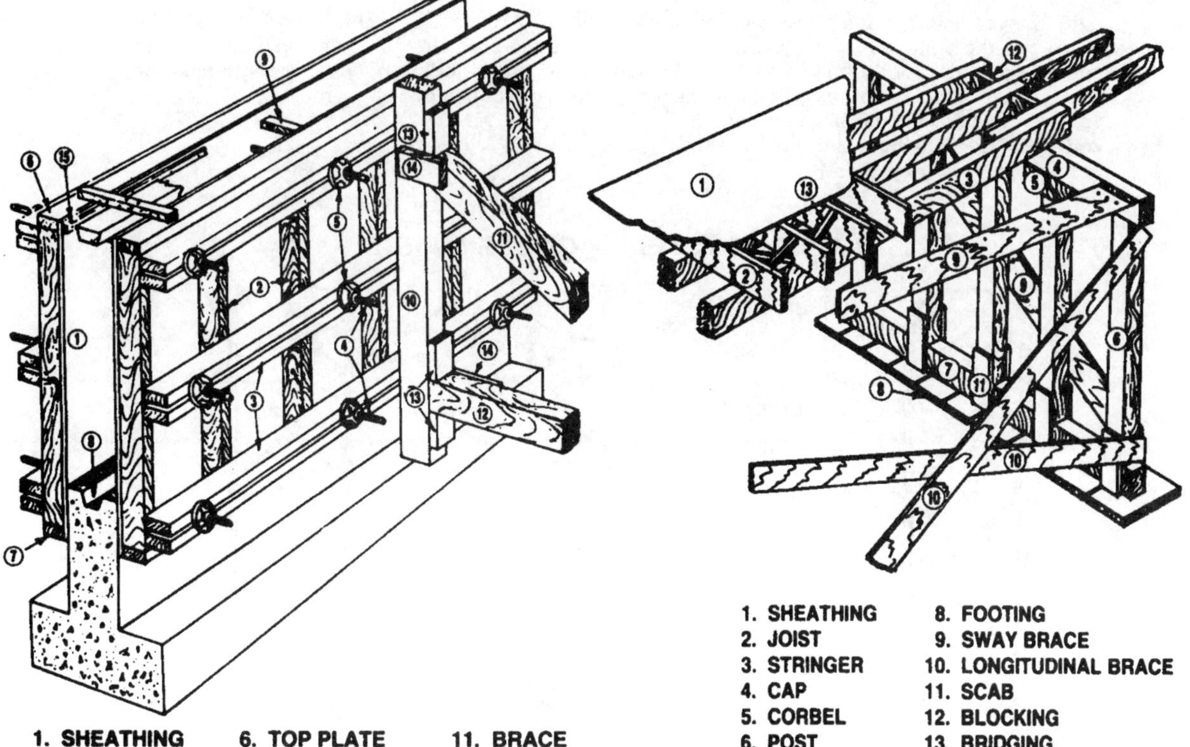

Form Nomenclature legend:

1. SHEATHING
2. STUDS
3. WALES
4. FORM BOLTS
5. NUT WASHER
6. TOP PLATE
7. BOTTOM PLATE
8. KEY-WAY
9. SPREADER
10. STRONGBACK
11. BRACE
12. STRUT
13. CLEATS
14. SCAB
14. POUR STRIP

Falsework Nomenclature legend:

1. SHEATHING
2. JOIST
3. STRINGER
4. CAP
5. CORBEL
6. POST
7. SILL
8. FOOTING
9. SWAY BRACE
10. LONGITUDINAL BRACE
11. SCAB
12. BLOCKING
13. BRIDGING

TYPICAL PAN-JOIST FORM CONSTRUCTION

TYPICAL WAFFLE SLAB FORM CONSTRUCTION

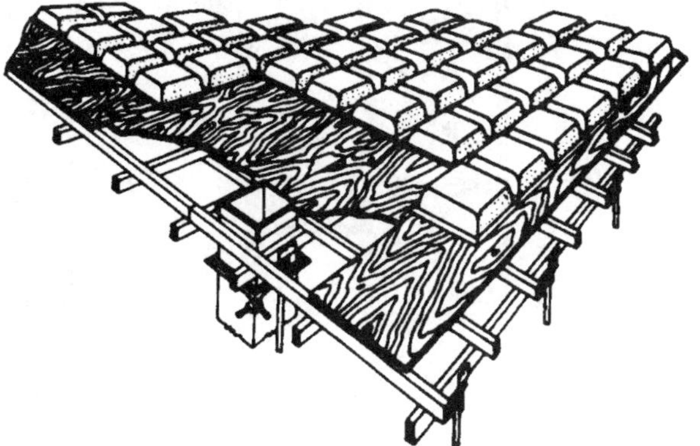

COMMON TYPES OF STEEL
REINFORCEMENT BARS

ASTM specifications for billet steel reinforcing bars (A 615) require identification marks to be rolled into the surface of one side of the bar to denote the producer's mill designation, bar size and type of steel. For Grade 60 and Grade 75 bars, grade marks indicating yield strength must be show. Grade 40 bars show only three marks (no grade mark) in the following order:

1st — Producing Mill (usually an initial)
2nd — Bar Size Number (#3 through # 18)
3rd — Type (N for New Billet)

NUMBER SYSTEM — GRADE MARKS

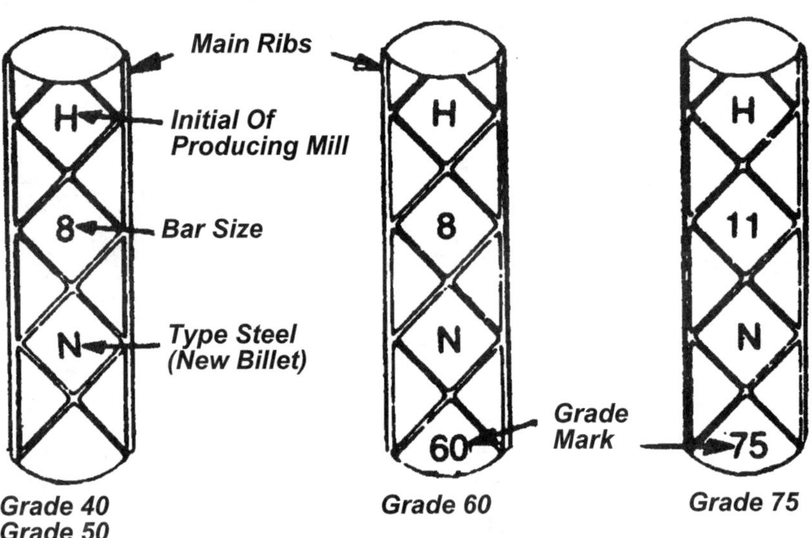

LINE SYSTEM — GRADE MARKS

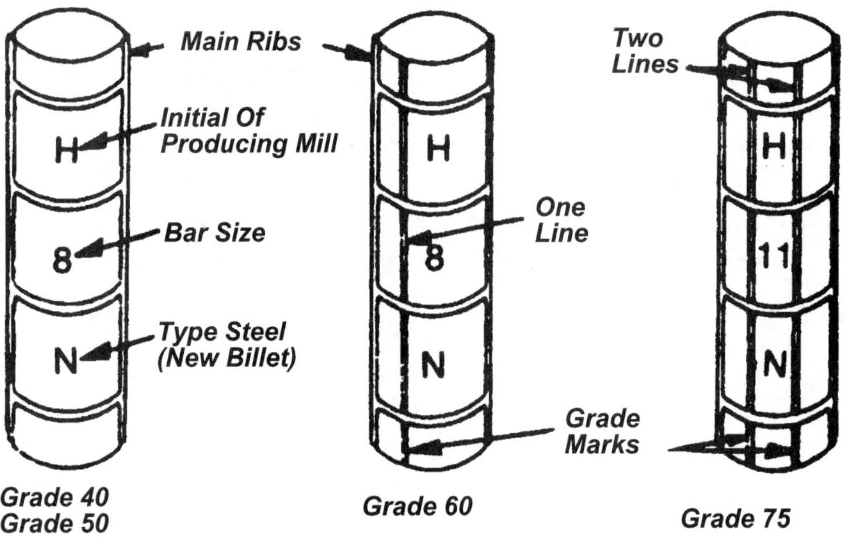

STANDARD SIZES OF STEEL REINFORCEMENT BARS

STANDARD REINFORCEMENT BARS

Bar Designation Number*	Nominal Weight lb. per ft.	Nominal Dimensions		
		Diameter, in.	Cross Sectional Area, sq. in.	Perimeter, in.
3	0.376	0.375	0.11	1.178
4	0.668	0.500	0.20	1.571
5	1.043	0.625	0.31	1.963
6	1.502	0.750	0.44	2.356
7	2.044	0.875	0.60	2.749
8	2.670	1.000	0.79	3.142
9	3.400	1.128	1.00	3.544
10	4.303	1.270	1.27	3.990
11	5.313	1.410	1.56	4.430
14	7.65	1.693	2.25	5.32
18	13.60	2.257	4.00	7.09

*The bar numbers are based on the number of 1/8 inches included in the nominal diameter of the bar.

Type of Steel & ASTM Specification No.	Size Nos. Inclusive	Grade	Tensile Strength Min., psi.	Yield (a) Min., psi
Billet Steel A 615	3-11	40	70,000	40,000
	3-11 14, 18	60	90,000	60,000
	11, 14, 18	75	100,000	75,000

Style Designation	Steel Area sq. in per ft.		Weight Approx. lbs. per 100 sq. ft.
	Longit.	Transv.	
Rolls			
6 x 6 — W1.4 x W1.4	.03	.03	21
6 x 8 — W2 x W2	.04	.04	29
6 x 6 — W2.9 x W2.9	.08	.06	42
6 x 6 — W4 x W4	.08	.08	58
4 x 4 — W1.4 x W1.4	.04	.04	31
4 x 4 — W2 x W2	.06	.06	43
4 x 4 — W2.9 x W2.9	.09	.09	62
4 x 4 — W4 x W4	.12	.12	
Sheets			
6 x 6 — W2.9 x W2.9	.06	.06	42
6 x 6 — W4 x W4	.08	.08	58
6 x 6 — W5.5 x W5.5	.11	.11	80
4 x 4 — W4 x W4	.12	.12	86

Insofar as is possible, the moisture content should be kept uniform to avoid problems in determining the proper amount of water to be added for mixing. Mixing water must be reduced to compensate for moisture in the aggregate in order to control the slump of the concrete and avoid exceeding the specified water-cement ratio.

Handling Concrete by Pumping Methods. Transportation and placement of concrete by pumping is another method gaining increased popularity. Pumps have several advantages, the primary one being that a pump will high-lift concrete without the need for an expensive crane and bucket. Since the concrete is delivered through pipe and hoses, concrete can be conveyed to remote locations in buildings, in tunnels, to locations otherwise inaccessible on steep hillside slopes for anchor walls, pipe bedding or encasement, or for placing concrete for chain link fence post bases. Concrete pumps have been found to be economical and expedient in the placement of concrete, and this has promoted the use and acceptance of this development. The essence of proper concrete pumping is the placement of the concrete in its final location without segregation.

Modern concrete pumps, depending on the mix design and size of line, can pump to a height of 200 feet or a horizontal distance of 1,000 feet. They can handle, economically, structural mixes, standard mixes, low slump mixes, mixes with two-inch maximum size aggregate and light weight concrete. When a special pump mix is required for structural concrete in a major structure, the mix design must be approved by the Engineer and checked and confirmed by the Supervisor of the Materials Control Group. The Inspector should obtain the pump manufacturer's printed information and evaluate its characteristics and ability to handle the concrete mixture specified for the project.

If concrete is being placed for a major reinforced structure, it is important that the placement continue without interruption. The Inspector should be sure that the contractor has ready access to a back-up pump to be used in the event of a breakdown. In order to further insure the success of the concrete placement by the pumping method, the user should be aware of the following points:

(a) A protective grating over the receiving hopper of the pump is necessary to exclude large pieces of aggregate or foreign material.

(b) The pump and lines require lubrication with a grout of cement and water. All of the excess grout is to be wasted prior to pumping the concrete.

(c) All changes in direction must be made by a large radius bend with a maximum bend of 90 degrees. Wye connections induce segregation and shall not be used.

(d) Pump lines should be made of a material capable of resisting abrasion and with a smooth interior surface having a low coefficient of friction. Steel is commonly used

for pump lines, because a chemical reaction occurs between the concrete and the aluminum. Aluminum pipe should not be used for pumping concrete and some of the new plastic or rubber tubing is gaining acceptance. Hydrogen is generated which results in a swelling of the concrete, causing a significant reduction in compressive strength. This reaction is aggravated by any of the following: abrasive coarse aggregate, non-uniformly graded sand, low-slump concrete, low sand-aggregate ratio, high-alkali cement or when no air-entraining agent is used.

(e) During temporary interruptions in pumping, the hopper must remain nearly full, with an occasional turning and pumping to avoid developing a hard slug of concrete in the lines.

(f) Excessive line pressures must be avoided. When this occurs, check these points as the probable cause: segregation caused by too low a slump or too high a slump; large particle contamination caused by large pieces of aggregate or frozen lumps not eliminated by the grating; poor gradation of aggregates or particle shape; rich or lean spots caused by improper mixing.

(g) Corrections must be made to correct excessive slump loss as measured at the transit-mixed concrete truck and as measured at the hose outlet. This may be attributable to porous aggregate, high temperature or rapid setting mixes.

(h) Two transit-mix concrete trucks must be used simultaneously to deliver concrete into the pump hopper. These trucks must be discharged alternately to assure a continuous flow of concrete as trucks are replaced.

(i) Samples of concrete for test specimens prepared to determine the acceptance of the concrete quality are to be taken as required for conventional concrete.

Sampling is done before the concrete is deposited in the pump hopper. However, it is suggested that, where possible, the effect of pumping on the compressive strength be checked by taking companion samples, so identified, from the end of the pump line at the same time. The Record of Test must be properly noted as being a special mix used for pumping purposes. This will enable the Materials Control Group to compile a complete history of mix designs and their respective compressive strengths.

The prudent use of pumped concrete can result in economy and improved quality. However, only the control exercised by the operator will assure continued high standards of quality concrete.

Pump lines must be properly fastened to supports to eliminate excessive vibration. Couplings must be easily and securely fastened in a manner that will prevent mortar leakage. It is preferable to use the flexible hose only at the discharge point. This hose must be moved in such a manner as to avoid kinks or sharp bends. The pump line should be protected from excessive heat during hot weather by water sprinkling or shade.

COMPRESSIVE STRENGTH FOR VARIOUS WATER-CEMENT RATIOS
(The strengths listed are based on the use of normal portland cement)

WATER/CEMENT RATIO		PROBABLE 28-DAY STRENGTH	
WEIGHT	**GALS./100#**	**PSI**	**MEGAPASCALS***
.40	4.8	5000	34
.45	5.4	4500	31
.50	6.0	4000	28
.55	6.6	3500	24
.60	7.2	3000	21
.65	7.8	2500	17
.70	8.4	2000	14

* International system equivalent.

APPROXIMATE CONTENT OF SAND, CEMENT AND WATER PER CUBIC YARD OF CONCRETE

Based on aggregates of average grading and physical characteristics in concrete mixes having a water-cement ratio (W/C) of about .65 by weight (or 7.8 gallons) per sack of cement; 3-in, slump; and a medium natural sand having a fineness modulus of about 2.75.

COURSE AGGREGATE MAX. SIZE	WATER		CEMENT	% SAND
	POUNDS	**GALLONS**		
3/8	385	46	590	57
½	365	44	560	50
¾	340	41	525	43
1	325	39	500	39
1 1/2	300	36	460	37

It can be noted from the above chart that, for a given slump, the amount of mixing water increases as the size of the course aggregate decreases. The size of the course aggregate controls the sand content in the same way; that is, the amount of sand required in the mix increases as the size of the course aggregate decreases.

Other typical examples are contained in the pamphlet published by the Portland Cement Association entitled "Design and Control of Concrete Mixtures."

Effects of Temperature on Concrete. Concrete mixtures gain strength rapidly in the first few days after placement. While the rate of gain in strength diminishes, concrete continues to become stronger with time over a period of many years, so long as drying of the concrete is prevented. Its strength at 28 days is considered to be the compressive strength upon which the Engineer bases his calculations. The temperature of the atmosphere has a significant effect upon the development of strength in concrete. Lower temperatures retard and higher temperatures accelerate the gain in strength.

Most destructive of the natural forces is freezing and thawing action. While the concrete is still wet or moist, expansion of the water as it is converted into ice results in severe damage to the fresh concrete. In situations where freezing may be encountered, high early strength cement may be used. Also, the mixing water or the aggregate (or both) may be preheated before mixing. Covering the concrete, and using steam or salamanders to heat the concrete under the covering, will help prevent freezing. Air-entraining agents help to diminish the effects of freezing of fresh concrete as well as in subsequent freezing and thawing cycles throughout the life of the concrete.

Hot weather will present problems of a different nature in placing concrete. Concrete will set up faster and tend to shrink and crack at the surface. To minimize this problem, the concrete should be placed without delay after mixing. Avoid the use of accelerators (perhaps even use a retarding agent), dampen all subgrade and forms, protect the freshly placed concrete from hot dry winds, and provide for adequate curing. Crushed ice or chilled water can be used as part of the mixing water to reduce the temperature of the mix in extremely hot areas.

ARCHITECTURAL WALL PATTERNS (BONDS)

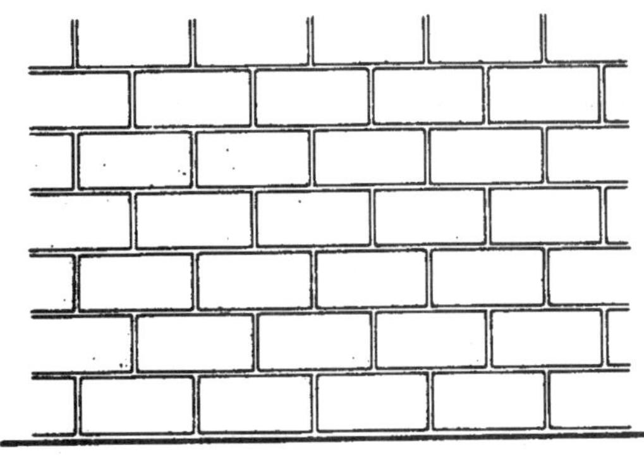

COMMON BOND
8" x 16" UNITS

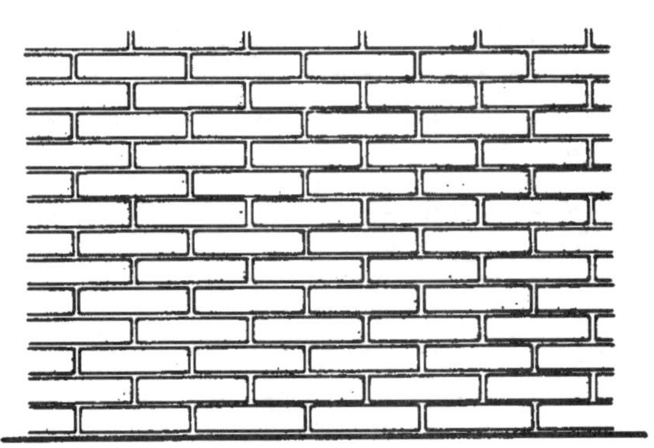

COMMON BOND
4" x 16" UNITS

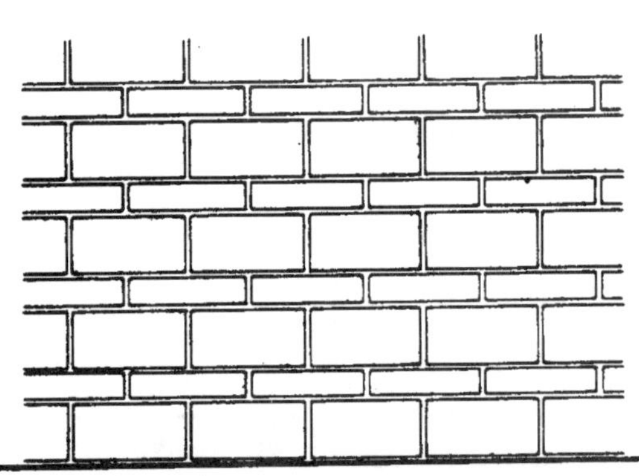

COURSED ASHLAR
8" x 16", 4"x 16" UNITS

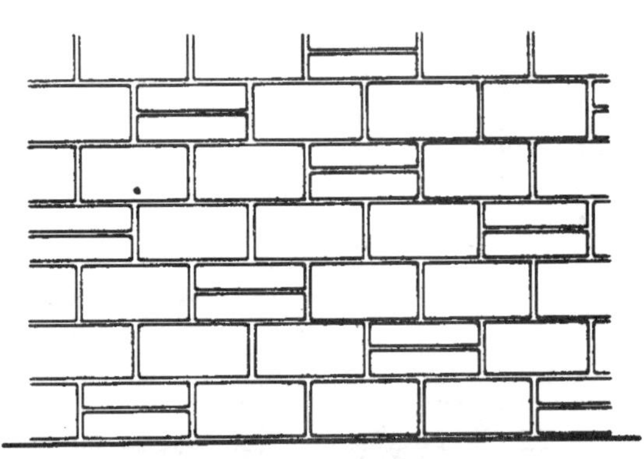

COURSED ASHLAR
8" x 16", 4"x 16" UNITS

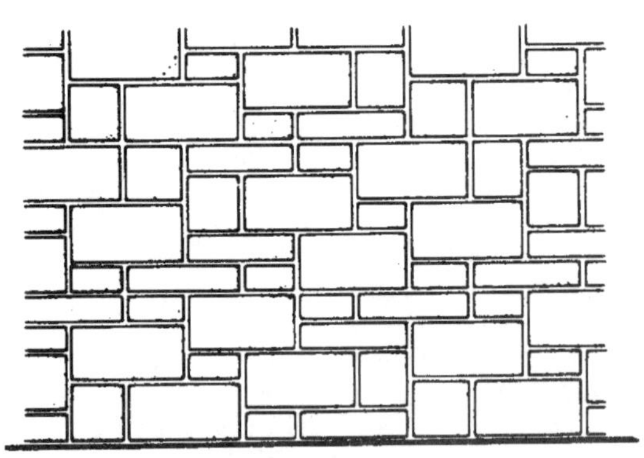

RANDOM ASHLAR
8" x 16", 8"x 8"
AND 4" x 8" UNITS

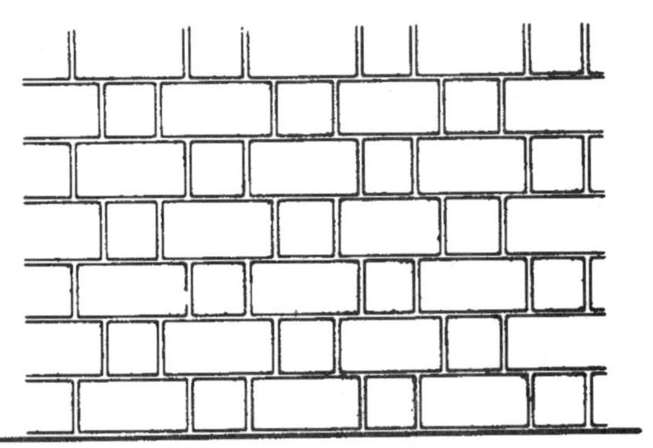

COURSED ASHLAR
8" x 16", 8"x 8" UNITS

ARCHITECTURAL WALL PATTERNS (BONDS) — (Continued)

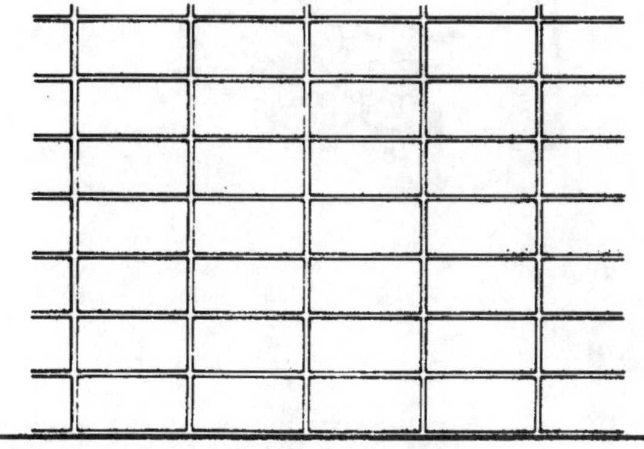

STACKED BOND
8" x 16" UNITS

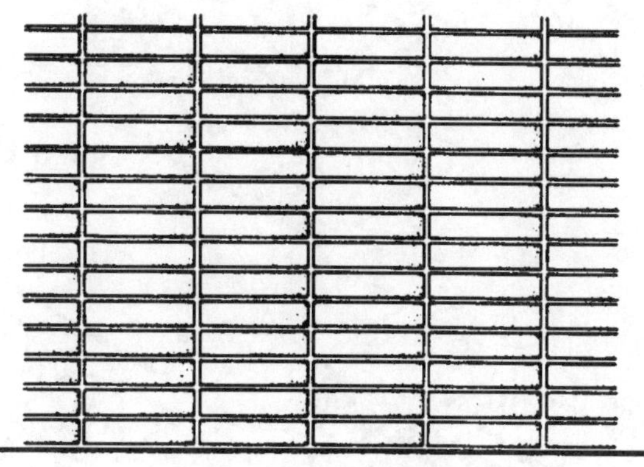

STACKED BOND
8" x 16" UNITS

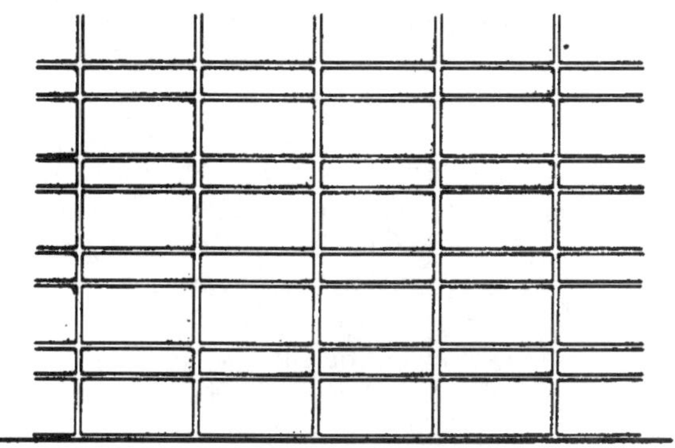

STACKED BOND
8" x 16", 4" x 16" UNITS

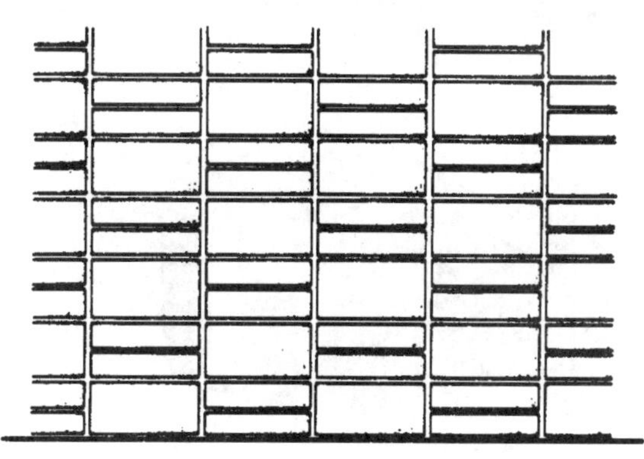

STACKED BOND
8" x 16", 4" x 16" UNITS

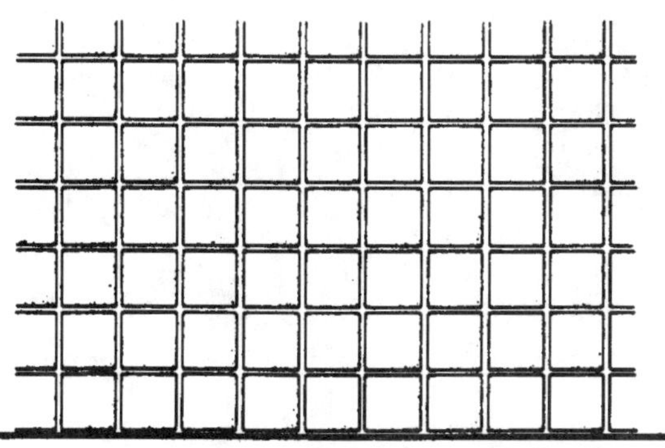

STACKED BOND VERTICAL
SCORED UNITS

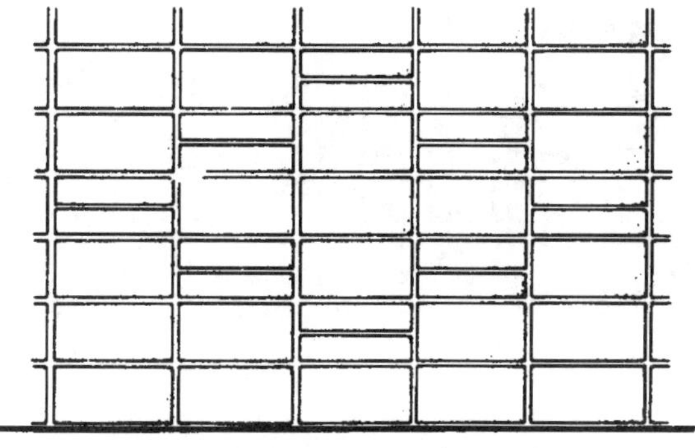

USE OF BLOCK DESIGN
IN STACKED BOND

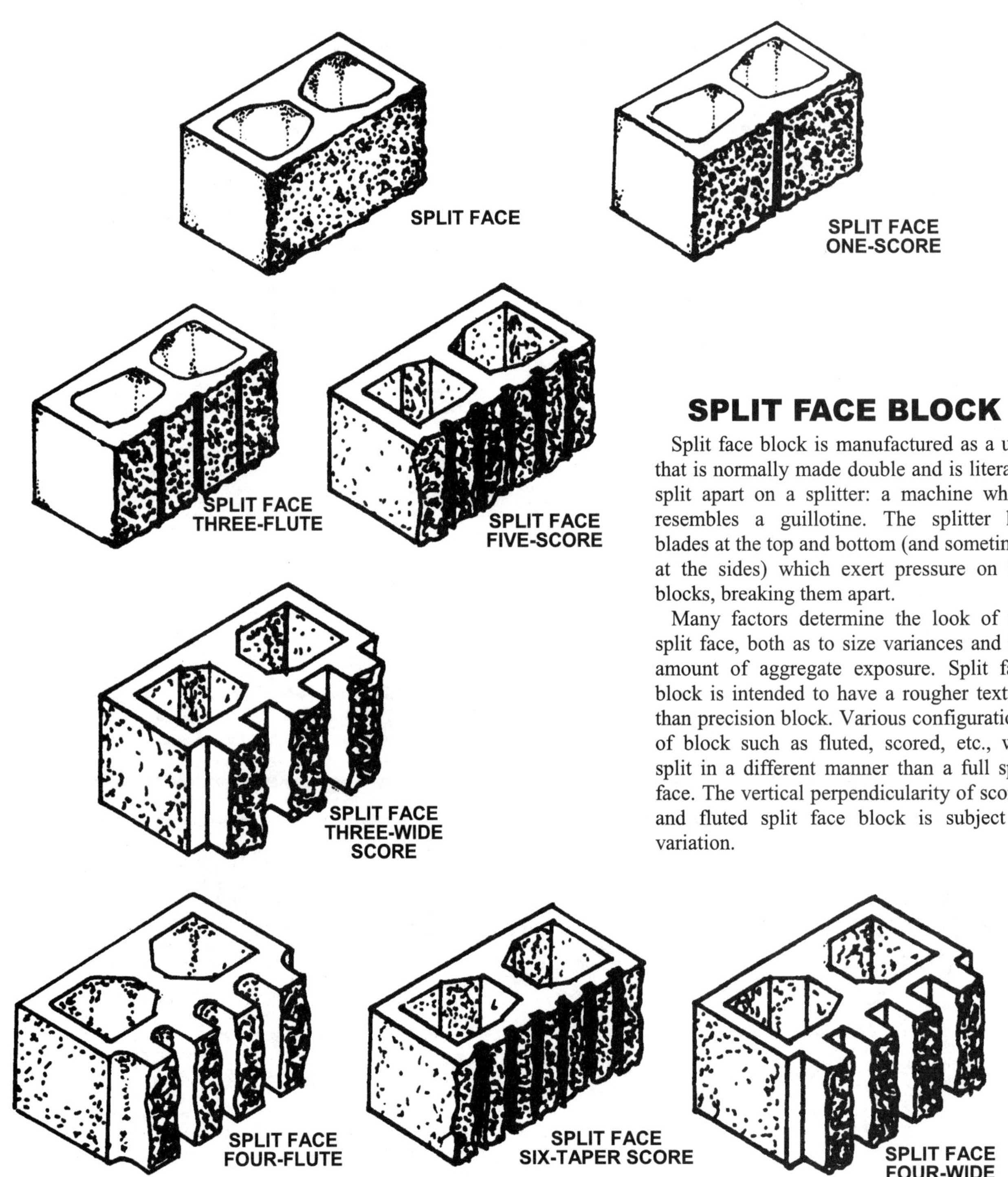

SPLIT FACE BLOCK

Split face block is manufactured as a unit that is normally made double and is literally split apart on a splitter: a machine which resembles a guillotine. The splitter has blades at the top and bottom (and sometimes at the sides) which exert pressure on the blocks, breaking them apart.

Many factors determine the look of the split face, both as to size variances and the amount of aggregate exposure. Split face block is intended to have a rougher texture than precision block. Various configurations of block such as fluted, scored, etc., will split in a different manner than a full split face. The vertical perpendicularity of scored and fluted split face block is subject to variation.

NOTE: Split face units shown in this manual are a small sampling of the broad range of concrete masonry architectural units available from the industry on special order. Depths and widths of scores vary. Consult a local manufacturer for specific information.

TYPICAL DETAILS — LINTELS AND BOND BEAMS

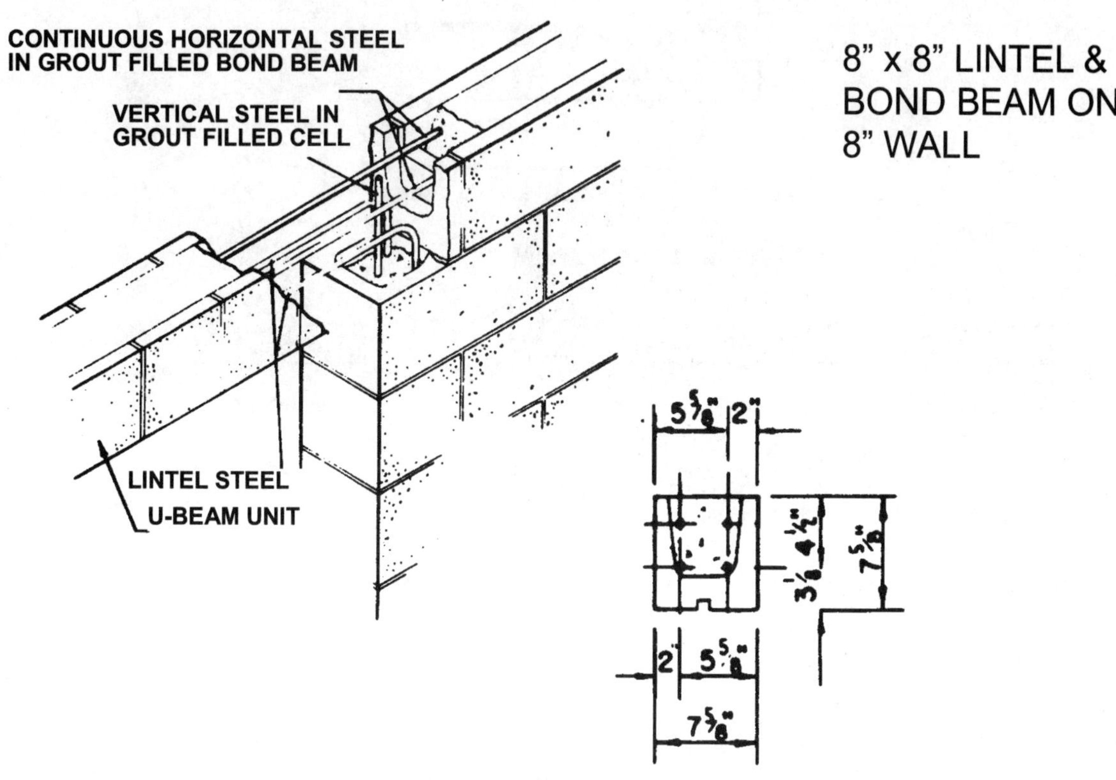

8" x 16"
BOND BEAM ON
8" WALL

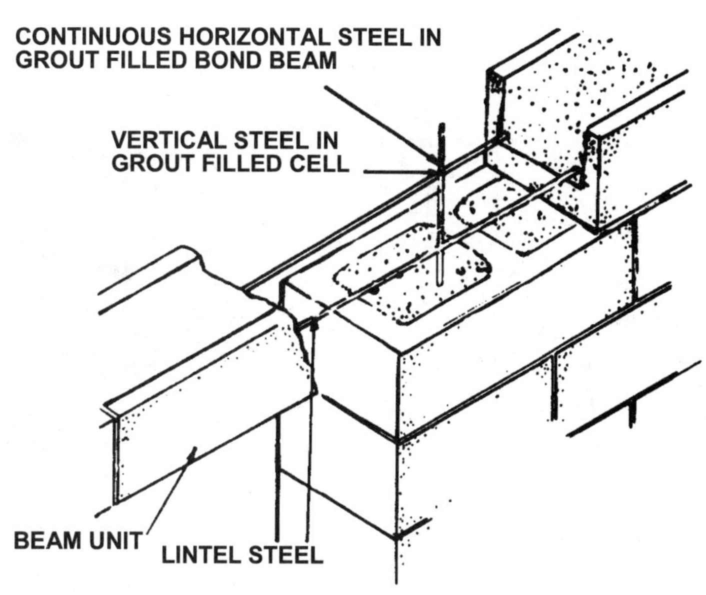

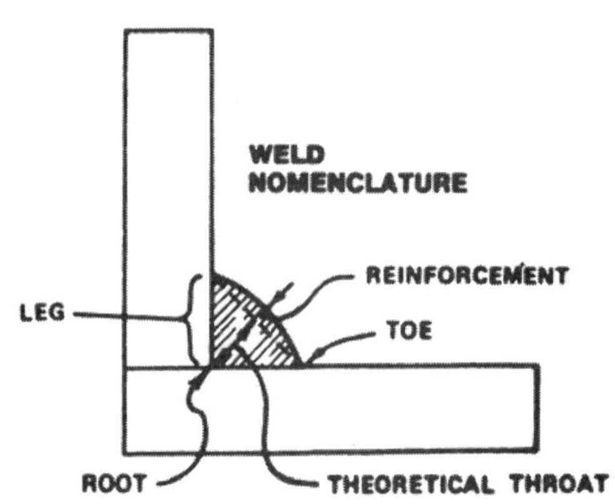

WELDED JOINTS

SQUARE BUTT

SINGLE VEE BUTT

DOUBLE VEE BUTT

SINGLE U BUTT

DOUBLE U BUTT

SINGLE FILLET LAP

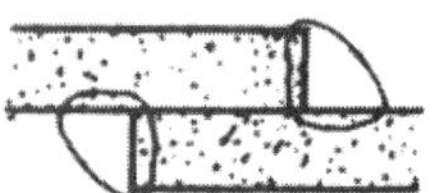

DOUBLE FILLET LAP

STRAP JOINT

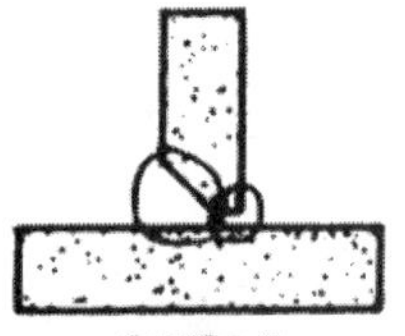

SINGLE
BEVEL TEE

DOUBLE
BEVEL TEE

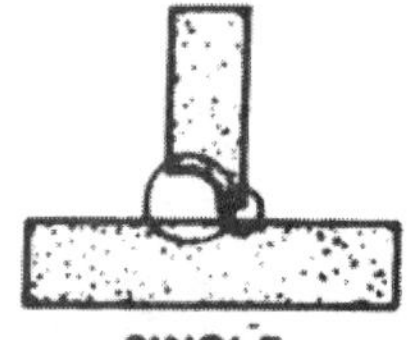

SINGLE
J TEE

SQUARE
TEE

DOUBLE J TEE

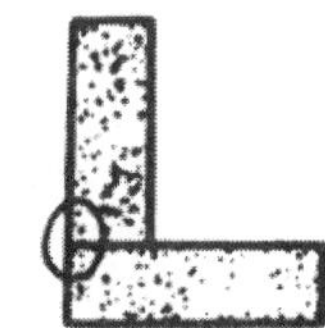

CLOSED CORNER
(FLUSH) JOINT

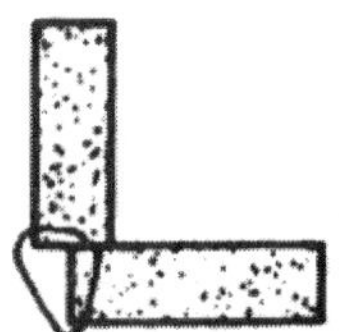

HALF OPEN
CORNER JOINT

WELDING POSITIONS

FLAT (F)

HORIZONTAL (H)

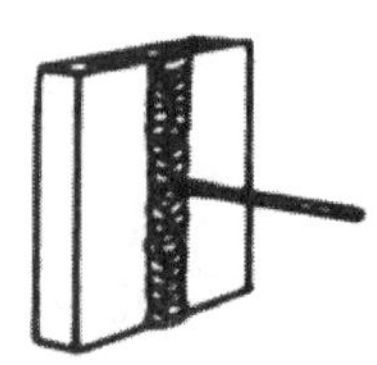

VERTICAL (V)

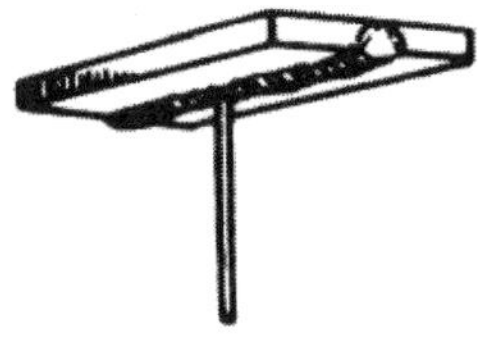

OVERHEAD (OH)

BOLTS IN COMMON USAGE

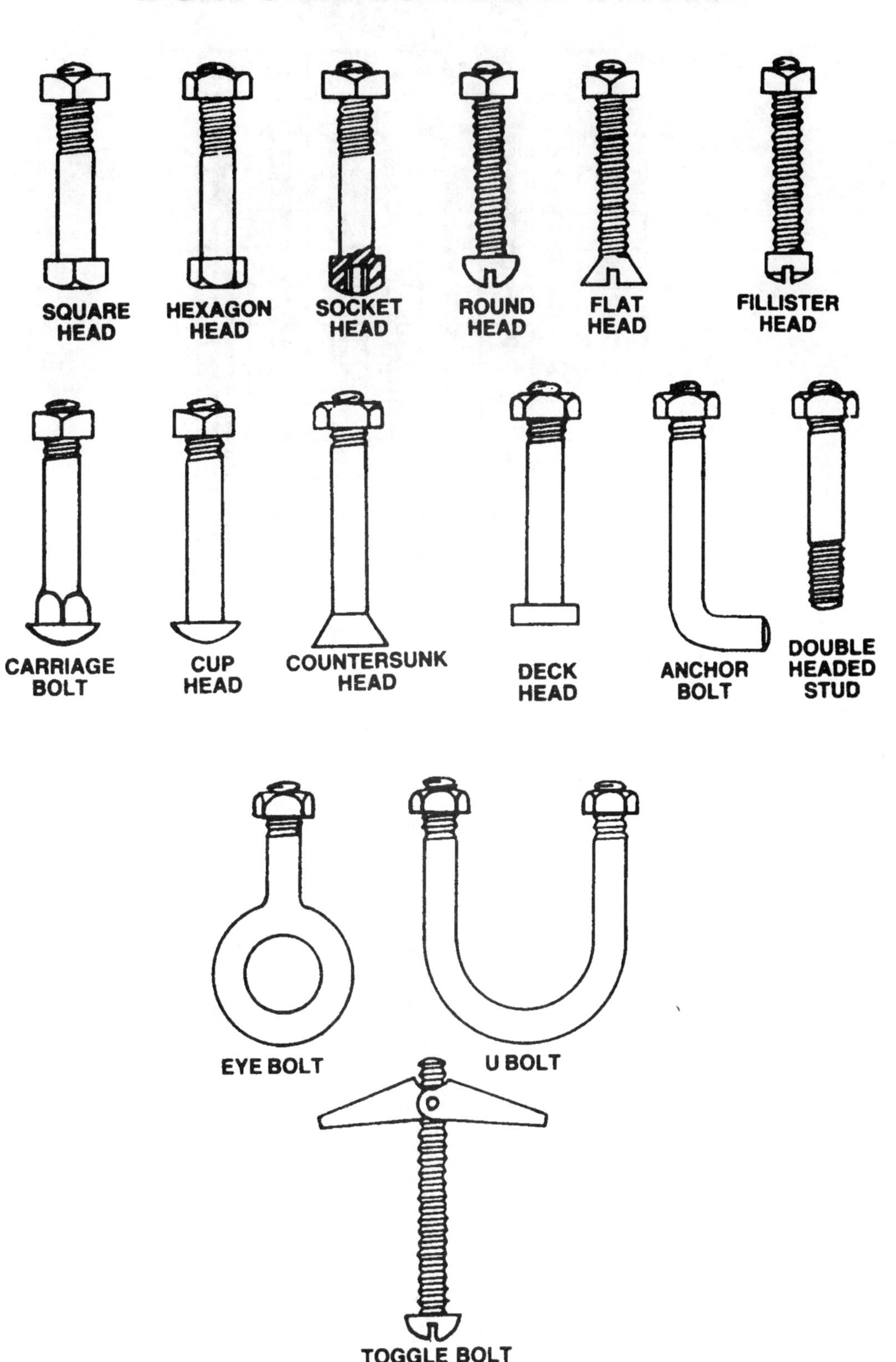

COMMON WIRE NAILS (ACTUAL SIZE)

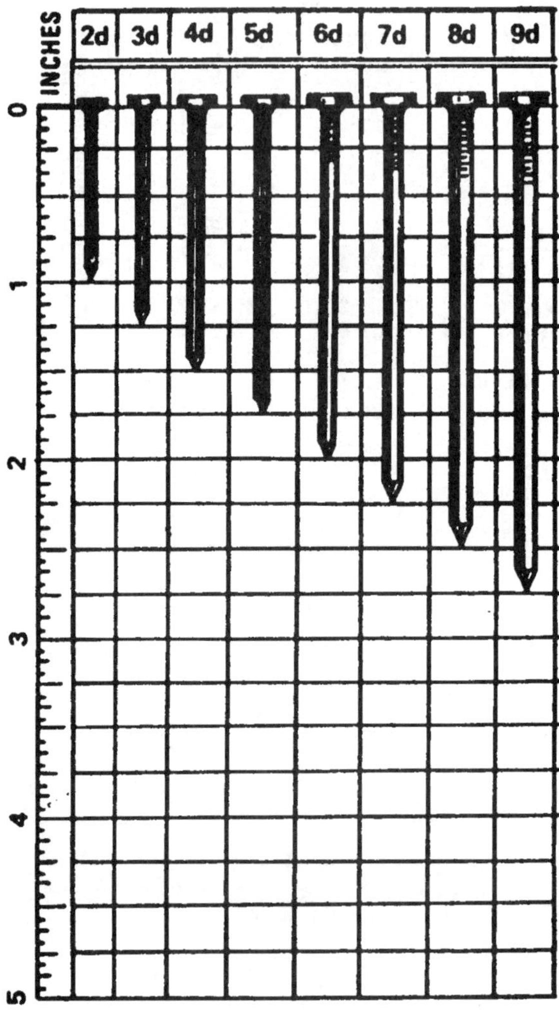

Cut Nails. Cut nails are angular-sided, wedge-shaped with a blunt point.

Wire Nails. Wire nails are round shafted, straight, pointed nails, and are used more generally than cut nails. They are stronger than cut nails and do not buckle as easily when driven into hard wood, but usually split wood more easily than cut nails. Wire nails are available in a variety of sizes varying from two penny to sixty penny.

Nail Finishes. Nails are available with special finishes. Some are galvanized or cadmium plated to resist rust. To increase the resistance to withdrawal, nails are coated with resins or asphalt cement (called cement coated). Nails which are small, sharp-pointed, and often placed in the craftsman's mouth (such as lath or plaster board nails) are generally blued and sterilized.

COMMON WIRE NAILS (ACTUAL SIZE) (Cont.)

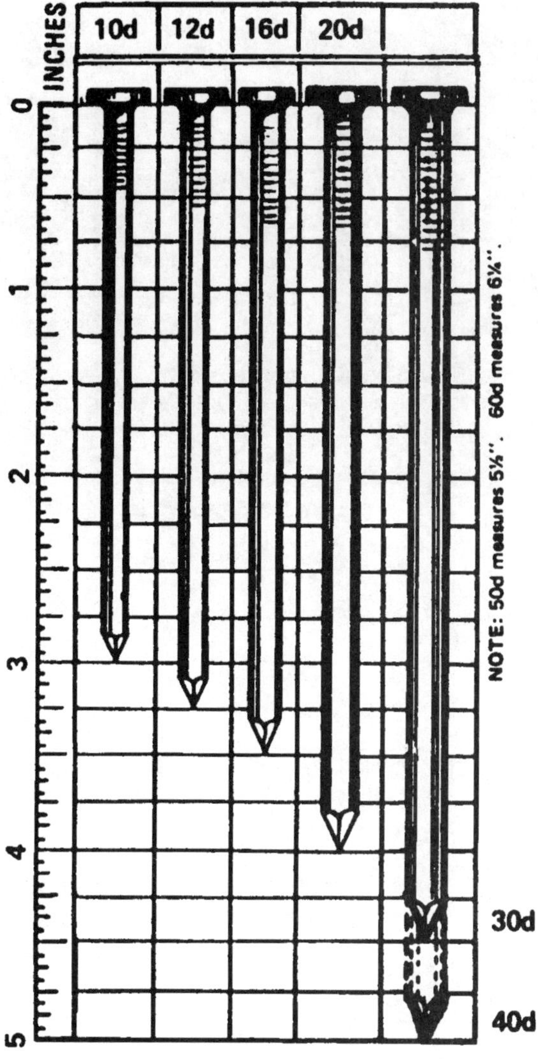

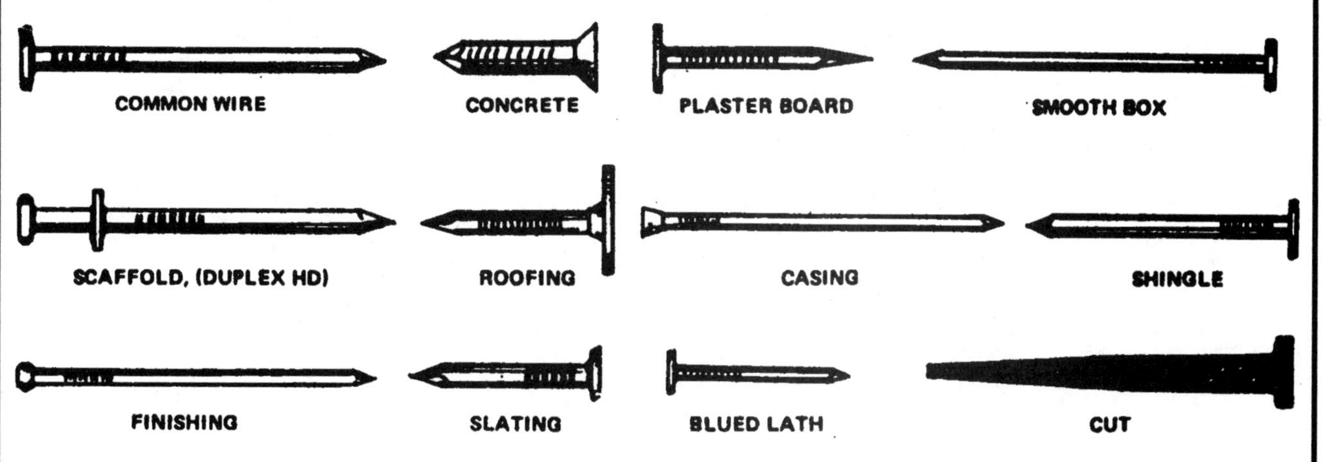

PLYWOOD — BASIC GRADE MARKS
AMERICAN PLYWOOD ASSOCIATION (APA)

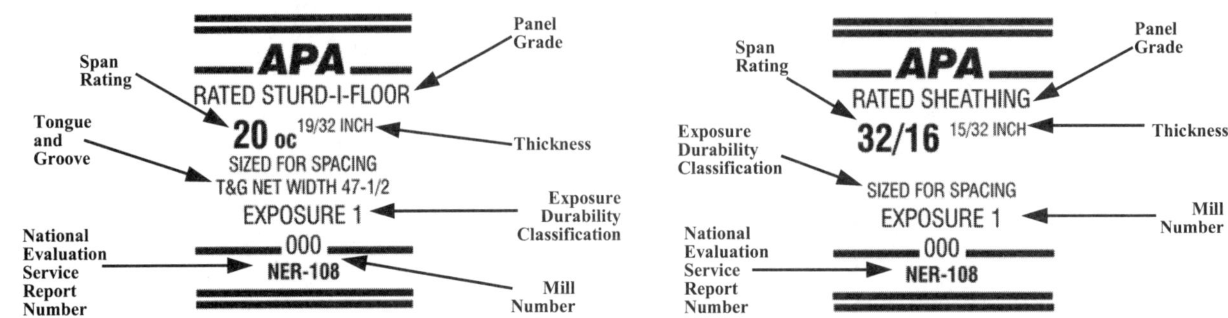

The American Plywood Association's trademarks appear only on products manufactured by APA member mills. The marks signify that the product is manufactured in conformance with APA performance standards and/or U.S. Product Standard PS 1-83 for Construction and Industrial Plywood.

APA A-C

For use where appearance of one side is important in exterior applications such as soffits, fences, structural uses, boxcar and truck linings, farm buildings, tanks, trays, commercial refrigerators, etc. **Exposure Durability Classification: Exterior. Common Thicknesses:** ¼, 11/32, ¾, 15/32, ½, 19/32, 5/8, 23/32, ¾.

APA A-D

For use where appearance of only one side is important in interior applications, such as paneling, built-ins, shelving, partitions, flow racks, etc. **Exposure Durability Classifications: Interior, Exposure 1. Common Thicknesses:** ¼, 11/32, 3/8, 15/32, ½, 19/32, 5/8, 23/32, ¾.

APA B-C

Utility panel for farm service and work buildings, boxcar and truck linings, containers, tanks, agricultural equipment, as a base for exterior coatings and other exterior uses. **Exposure Durability Classification: Exterior. Common Thicknesses:** ¼, 11/32, ¾, 15/32, ½, 19/32, 5/8, 23/32, ¾.

APA B-D

Utility panel for backing, sides or built-ins, industry shelving, slip sheets, separator boards, bins and other interior or protected applications. **Exposure Durability Classifications: Interior, Exposure 1. Common Thicknesses:** ¼, 11/32, 3/8, 15/32, ½, 19/32, 5/8, 23/32, ¾.

APA proprietary concrete form panels designed for high reuse. Sanded both sides and mill-oiled unless otherwise specified. Class I, the strongest, stiffest and more commonly available, is limited to Group 1 faces, Group 1 or 2 crossbands, and Group 1, 2, 3 or 4 inner plies. Class II is limited to Group 1 or 2 faces (Group 3 under certain conditions) and Group 1, 2, 3 or 4 inner plies. Also available in HDO for very smooth concrete finish, in Structural I, and with special overlays. **Exposure Durability Classification: Exterior. Common Thicknesses:** 19/32, ¾, 23/32, ¼.

Plywood panel manufactured with smooth, opaque, resin-treated fiber overlay providing ideal base for paint on one or both sides. Excellent material choice for shelving, factory work surfaces, paneling, built-ins, signs and numerous other construction and industrial applications. Also available as a 303 Siding with texture-embossed or smooth surface on one side only and Structural I. **Exposure Durability Classification: Exterior. Common Thicknesses:** 11/32, 3/8, 19/32, ½, 19/32, 5/8, 23/32, ¾.

SPECIALTY PANELS

> HDO • A • A • G-1 • EXT APA • 000 • PS1 83

Plywood panel manufactured with a hard, semi-opaque resin-fiber overlay on both sides. Extremely abrasion resistant and ideally suited to scores of punishing construction and industrial applications, such as concrete forms, industrial tanks, work surfaces, signs, agricultural bins, exhaust ducts, etc. Also available with skid-resistant screen-grid surface and in Structural I. *Exposure Durability Classification:* **Exterior.** *Common Thicknesses:* **3/8, ½, 5/8, 3/4**

> MARINE• A • A • EXT APA • 000 • PS1 83

Specialty designed plywood panel made only with Douglas fir or western larch, solid jointed cores, and highly restrictive limitations on core gaps and faces repairs. Ideal for both hulls and other marine applications. Also available with HDO or MDO faces. *Exposure Durability Classification:* **Exterior.** *Common Thicknesses:* **1/4, 3/8, ½, 5/8, 3/4.**

Unsanded and touch-sanded panels, and panels with "B" or better veneer on one side only, usually carry the APA trademark on the panel back. Panels with both sides of "B" or better veneer, or with special overlaid surfaces (such as Medium Density Overlay), carry the APA trademark on the panel edge, like this:

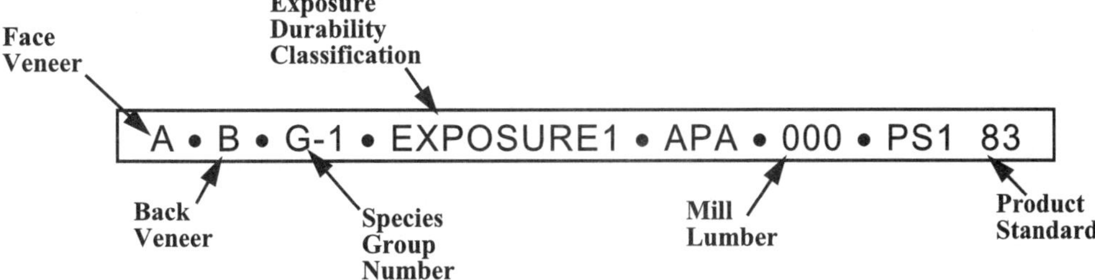

GLOSSARY OF TERMS
Some of the words and terms used in the grading of lumber follow:

Bow. A deviation flatwise from a straight line drawn from end to end of the piece. It is measured at the point of greatest distance from the straight line.

Checks. A separation of the wood which normally occurs across the annual rings and usually as a result of seasoning.

Crook. A deviation edgewise from a straight line drawn from end to end of the piece. It is measured at the point of greatest distance from the straight line.

Cup. A deviation from a straight line drawn across the piece from edge to edge. It is measured at the point of greatest distance from the straight line.

Flat Grain. The annual growth rings pass through the piece at an angle of less than 45 degrees with the flat surface of the piece.

Warp. Any deviation from a true or plane surface, including crook, cup, bow or any combination thereof.

Mixed Grain. The piece may have vertical grain, flat grain or a combination of both vertical and flat grain.

Pitch. An accumulation of resin which occurs in separations in the wood or in the wood cells themselves.

Shake. A separation of the wood which usually occurs between the rings of annual growth.

Splits. A separation of the wood due to tearing apart of the wood cells.

Vertical Grain. The annual growth rings pass through the piece at an angle of 45 degrees or more with the flat surface of the piece.

Wane. Bark or lack of wood from any cause, except eased edges (rounded) on the edge or corner of a piece of lumber.

LUMBER GRADING
GRADING-MARK ABBREVIATIONS

GRADES

(Listed alphabetically — not by quality)

COM	Common
CONST	Construction
ECON	Economy
No. 1	Number One
SEL-MER	Select Merchantable
SEL-STR	Select Structural
STAN	Standard
UTIL	Utility

ALSC TRADEMARKS

CLIS	California Lumber Inspection Service
NELMA	Northeastern Lumber Mfrs. Assoc., Inc.
NH&PMA	Northern Hardwood & Pine Mfrs. Assoc., Inc.
PLIB	Pacific Lumber Inspection Bureau
RIS	Redwood Inspection Service
SPIB	Southern Pine Inspection Bureau
TP	Timber Products Inspection
WCLB	West Coast Lumber Inspection Bureau
WWP	Western Wood Products Association

SPECIES GROUPINGS

AF	Alpine Fir
DF	Douglas Fir
HF	Hem Fir
SP	Sugar Pine
PP	Ponderosa Pipe
LP	Lodgepole Pine
IWP	Idaho White Pine
ES	Engelmann Spruce
WRC	Western Red Cedar
INC CDR	Incense Cedar
L	Larch
LP	Lodgepole Pine
MH	Mountain Hemlock
WW	White Wood

MOISTURE CONTENT

S-GRN	Surfaced at a moisture content of more than 19%.
S-DRY	Surfaced at a moisture content of 19% or less.
MC-15	Surfaced at a moisture content of 15% or less.

FRAMING ESTIMATING RULES OF THUMB

For 16" O.C. stud partitions figure 1 stud for every L.F. of wall; add for top and bottom plates.

For any type of framing, the quantity of basic framing members (in L.F.) can be determined based on spacing and surface area (S.F.):

12" O.C.	1.2 L.F./S.F.
16" O.C.	1.0 L.F./S.F.
24" O.C.	0.8 L.F./S.F.

(Doubled-up members, bands, plates, framed openings, etc., must be added.)

Framing accessories, nails, joist hangers, connectors, etc., should be estimated as separate material costs. Installation should be included with framing. Rule of thumb allowance is 0.5 to 1.5% of lumber cost for rough hardware. Another is 30 to 40 pounds of nails per M.B.F.

BOARD FEET/LINEAR FEET FOR LUMBER

Nominal Size	Actual Size	Board Feet Per Linear Foot	Linear Feet Per 1000 Board Feet
1 x 2	¾ x 1 ½	.167	6000
1 x 3	¾ x 2 ½	.250	4000
1 x 4	¾ x 3 ½	.333	3000
1 x 6	¾ x 5 ½	.500	2000
1 x 8	¾ x 7 ¼	.666	1500
1 x 10	¾ x 9 ¼	.833	1200
1 x 12	¾ x 11 ¼	1.0	1000
2 x 2	1 ½ x 1 ½	.333	3000
2 x 3	1 ½ x 2 ½	.500	2000
2 x 4	1 ½ x 3 ½	.666	1500
2 x 6	1 ½ x 5 ½	1.0	1000
2 x 8	1 ½ x 7 ¼	1.333	750
2 x 10	1 ½ x 9 ¼	1.666	600
2 x 12	1 ½ x 11 ¼	2.0	500

BOARD FEET CONVERSION TABLE

Nominal Size (In.)	ACTUAL LENGTH IN FEET								
	8	**10**	**12**	**14**	**16**	**18**	**20**	**22**	**24**
1 x 2		1 2/3	2	2 1/3	2 2/3	3	3 ½	3 2/3	4
1 x 3		2 ½	3	3 ½	4	4 ½	5	5 ½	6
1 x 4	2 ¾	3 1/3	4	4 2/3	5 1/3	6	6 2/3	7 1/3	8
1 x 5		4 1/6	5	5 5/6	6 2/3	7 ½	8 1/3	9 1/6	10
1 x 6	4	5	6	7	8	9	10	11	12
1 x 7		5 5/8	7	8 1/6	9 1/3	10 ½	11 2/3	12 5/6	14
1 x 8	5 1/3	6 2/3	8	9 1/3	10 2/3	12	13 1/3	14 2/3	16
1 x 10	6 2/3	8 1/3	10	11 2/3	13 1/3	15	16 2/3	18 1/3	20
1 x 12	8	10	12	14	16	18	20	22	24
1¼ x 4		4 1/6	5	5 5/6	6 2/3	7 ½	8 1/3	9 1/6	10
1¼ x 6		6 ¼	7 ½	8 ¾	10	11 ¼	12 ½	13 ¾	15
1¼ x 8		8 1/3	10	11 2/3	13 1/3	15	16 2/3	18 1/3	20
1¼ x 10		10 5/12	12 ½	14 7/12	16 2/3	18 ¾	20 5/6	22 11/12	25
1¼ x 12		12 ½	15	17 ½	20	22 ½	25	27 ½	30
1½ x 4	4	5	6	7	8	9	10	11	12
1½ x 6	6	7 ½	9	10 ½	12	13 ½	15	16 ½	18
1½ x 8	8	10	12	14	16	18	20	22	24
1½ x 10	10	12 ½	15	17 ½	20	22 ½	25	27 ½	30
1½ x 12	12	15	18	21	24	27	30	33	36
2 x 4	5 1/3	6 2/3	8	9 1/3	10 1/3	12	13 1/3	14 2/3	16
2 x 6	8	10	12	14	16	18	20	22	24
2 x 8	10 2/3	13 1/3	16	18 2/3	21 1/3	24	26 2/3	29 1/3	32
2 x 10	13 1/3	16 2/3	20	23 1/3	26 2/3	30	33 1/3	36 2/3	40
2 x 12	16	20	24	28	32	36	40	44	48
3 x 6	12	15	18	21	24	27	30	33	36
3 x 8	16	20	24	28	32	36	40	44	48
3 x 10	20	25	30	35	40	45	50	55	60
3 x 12	24	30	36	42	48	54	60	66	72
4 x 4	10 2/3	13 1/3	16	18 2/3	21 1/3	24	26 2/3	29 1/3	32
4 x 6	16	20	24	28	32	36	40	44	48
4 x 8	21 1/3	26 2/3	32	37 1/3	42 2/3	48	53 1/3	58 2/3	64
4 x 10	26 2/3	33 1/3	40	46 2/3	53 1/3	60	66 2/3	73 1/3	80
4 x 12	32	40	48	56	64	72	80	88	96

SLOPE AREA CALCULATIONS

Rise and Run	Multiply Flat Area by	LF of Hips or Valleys per LF of Common Run
2 in 12	1.014	1.424
3 in 12	1.031	1.436
4 in 12	1.054	1.453
5 in 12	1.083	1.474
6 in 12	1.118	1.500
7 in 12	1.158	1.530
8 in 12	1.202	1.564
9 in 12	1.250	1.600
10 in 12	1.302	1.641
11 in 12	1.357	1.685
12 in 12	1.413	1.732

DOWNSPOUT/VERTICAL LEADER CALCULATIONS

Roof Type	Slope	S.F. Roof/ Sq. In. Leader
Gravel	Less than ¼" per foot	300
Gravel	Greater than ¼" per foot	250
Metal or Shingle	Any	200

Alternate calculations:

$$\text{Diameter of downspout/leader} = 1.128 \sqrt{\frac{\text{Area of drainage}}{\text{SF Roof/Sq. Inch}}}$$

TYPICAL MINIMUM SIZE OF VERTICAL CONDUCTORS AND LEADERS

Size of leader or conductor (Inches)	Maximum projected roof area (Square feet)
2	544
2 ½	987
3	1,610
4	3,460
5	6,280
6	10,200
8	22,000

TYPICAL MINIMUM SIZE OF ROOF GUTTERS

Diameter gutter (Inches)	MAXIMUM PROJECTED ROOF AREA FOR GUTTERS OF VARIOUS SLOPES			
	1/16 in. Ft. slope (Sq. ft.)	1/8 in. per Ft. slope (Sq. ft.)	¼ in. per Ft. slope (Sq. ft.)	1/2 in. per Ft. /slope (Sq. ft.)
3	170	240	340	480
4	360	510	720	1,020
5	625	880	1,250	1,770
6	960	1,360	1,920	2,770
7	1,380	1,950	2,760	3,900
8	1,990	2,800	3,980	5,600
10	3,600	5,100	7,200	10,000

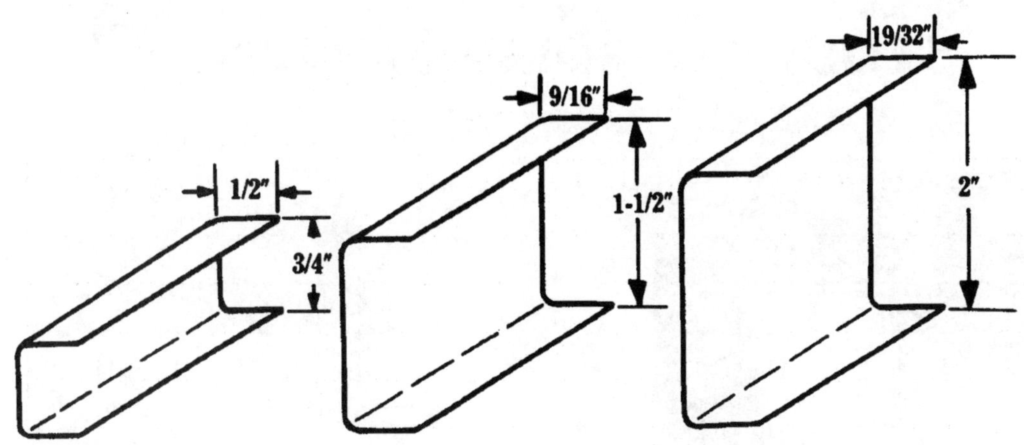

COLD ROLLED CHANNELS

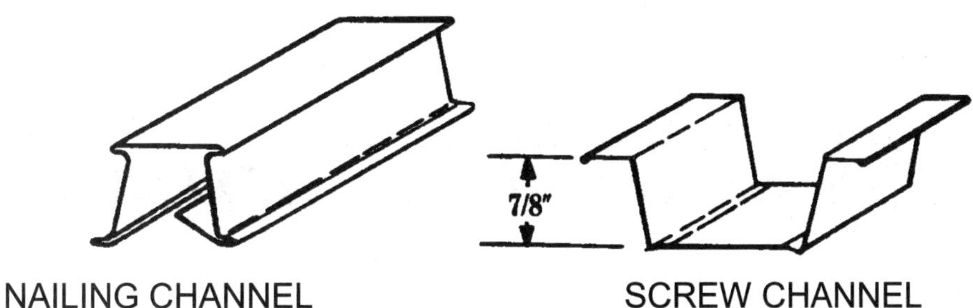

NAILING CHANNEL   **SCREW CHANNEL**

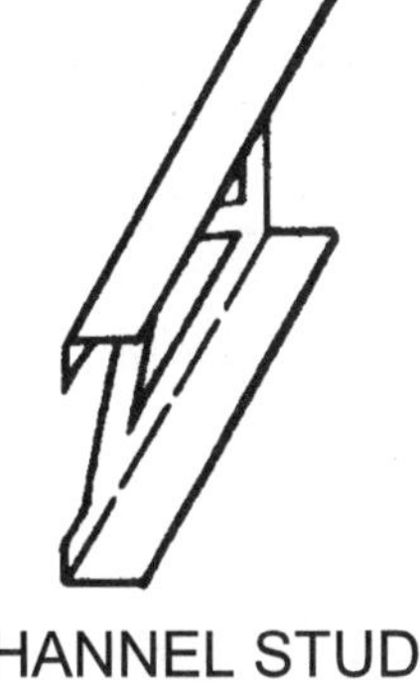

CHANNEL STUD

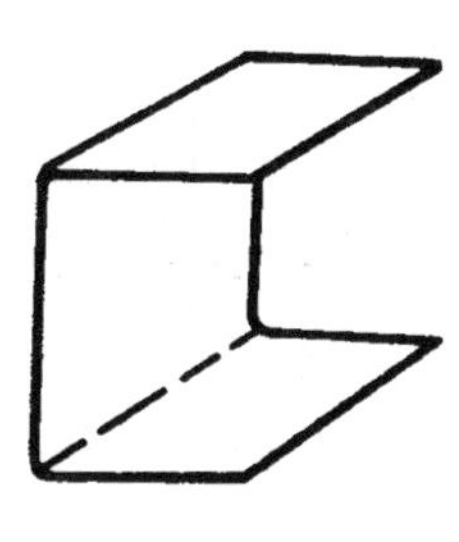

WIDE FLANGE CHANNEL

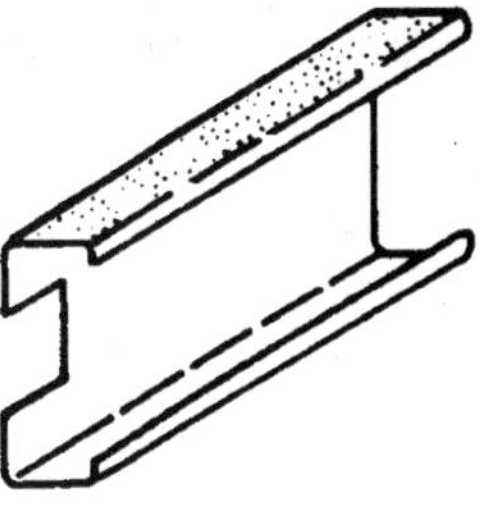

CEE STUD

STUDLESS SOLID PARTITION

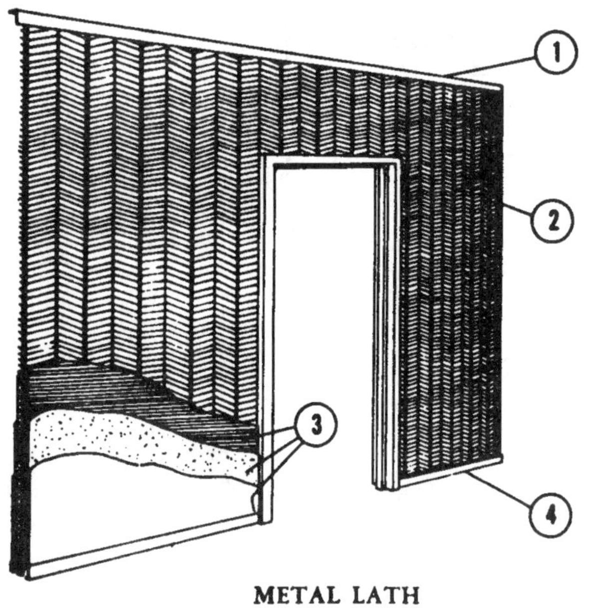
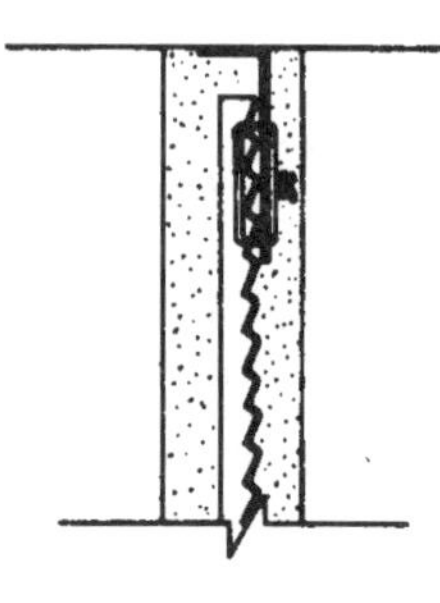
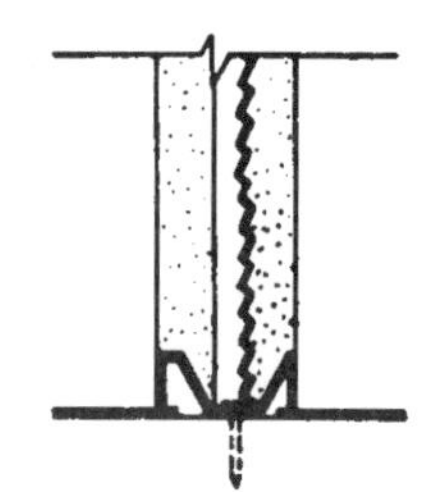

(1) Ceiling Runner
(2) Rib Metal Lath
(3) Plaster
(4) Combination Floor
    Runner and Screed

METAL LATH

STUDLESS SOLID PARTITION

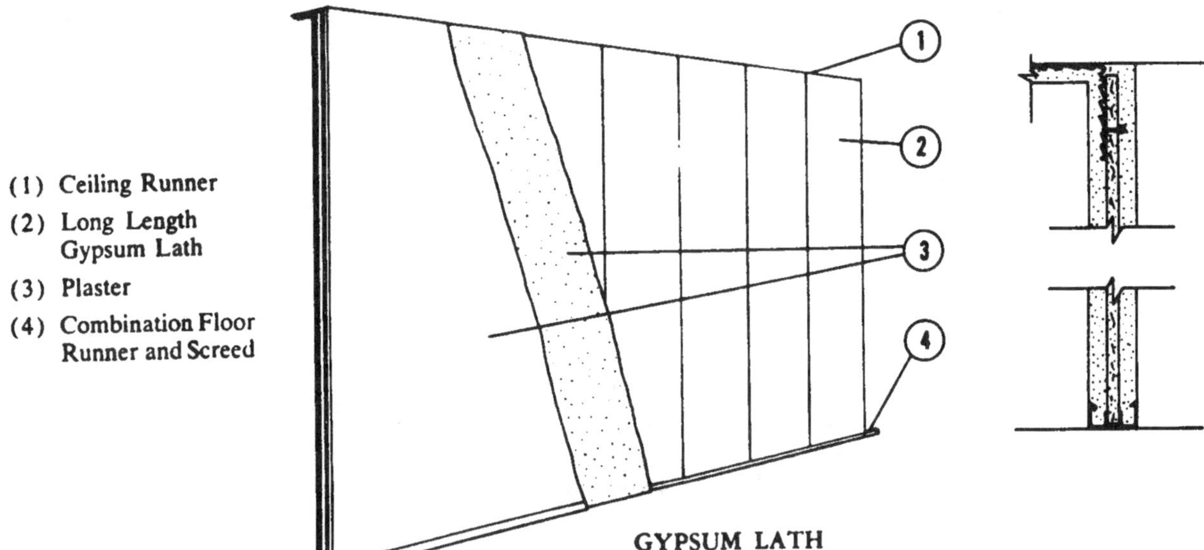

(1) Ceiling Runner
(2) Long Length
    Gypsum Lath
(3) Plaster
(4) Combination Floor
    Runner and Screed

GYPSUM LATH

STEEL STUD
Hollow Partition

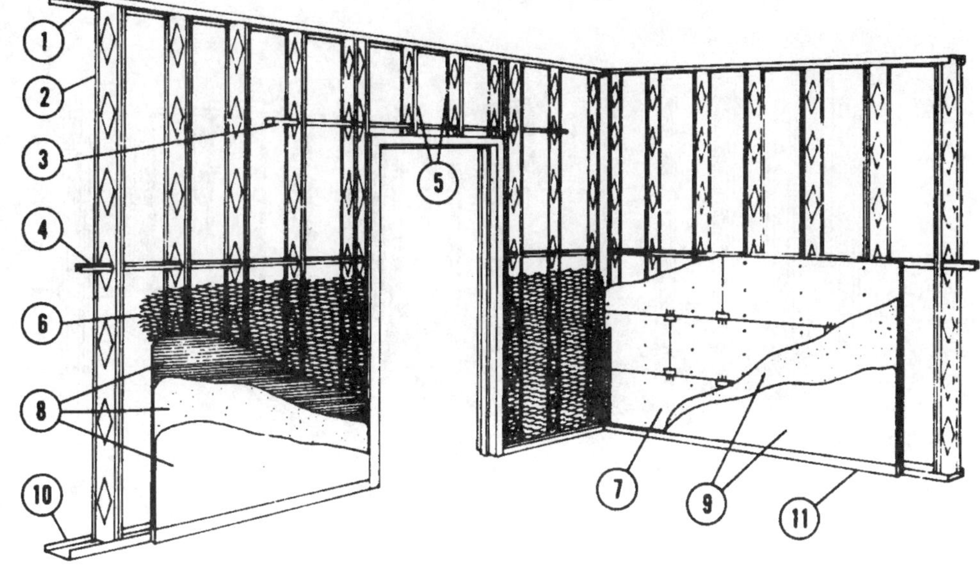

SCREW STUD
Hollow Partition

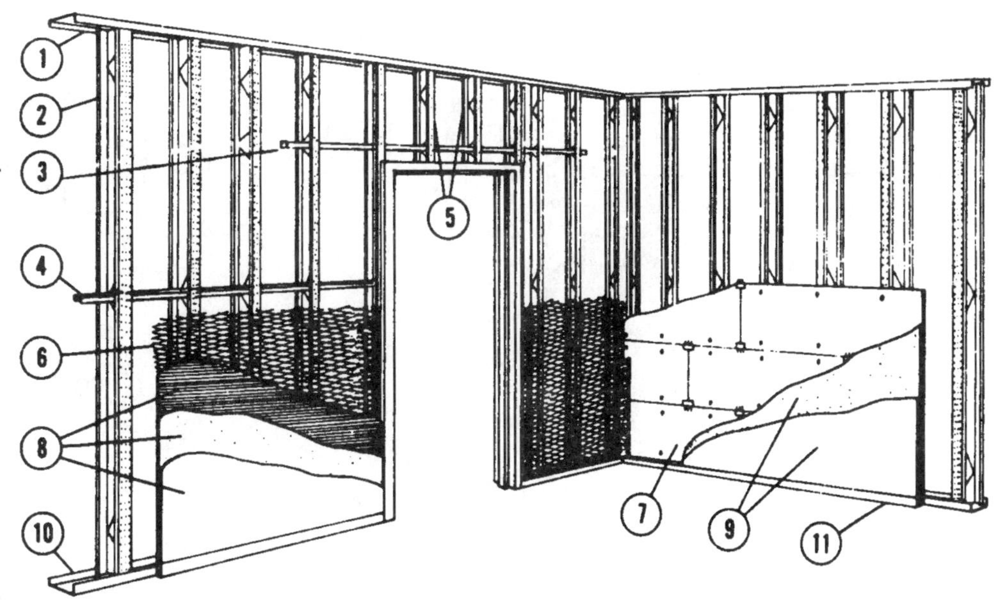

LOAD-BEARING HOLLOW PARTITION
Structural Stud

(1) Ceiling Runner Track
(2) Structural Studs (prefabricated)
(3) Structural Stud (nailable)
(4) Jack Studs
(5) Partition Stiffener (bridging)
(6) Metal or Wire Fabric Lath (wired-tied, nailed or stapled)
(7) Gypsum Lath (nailed or stapled)
(8) Three Coats of Plaster (Scratch, Brown, Finish)
(9) Two Coats of Plaster (Brown, Finish)
(10) Floor Runner Track
(11) Flush Metal Base

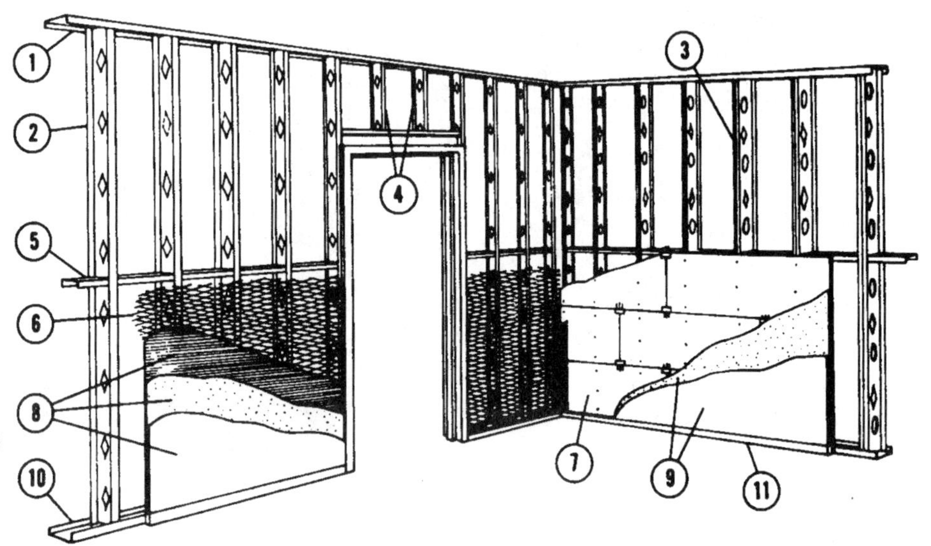

VERTICAL FURRING
With Studs

(1) Ceiling Runner Track
(2) Channel Studs
(3) Horizontal Stiffener
(4) Floor Runner
(5) Metal or wire Fabric Lath
(6) Gypsum Lath
(7) Bracing
(8) Three Coats of Plaster (Scratch, Brown, Finish)
(9) Two Coats of Plaster (Brown, Finish)
(10) Screw Channel Studs

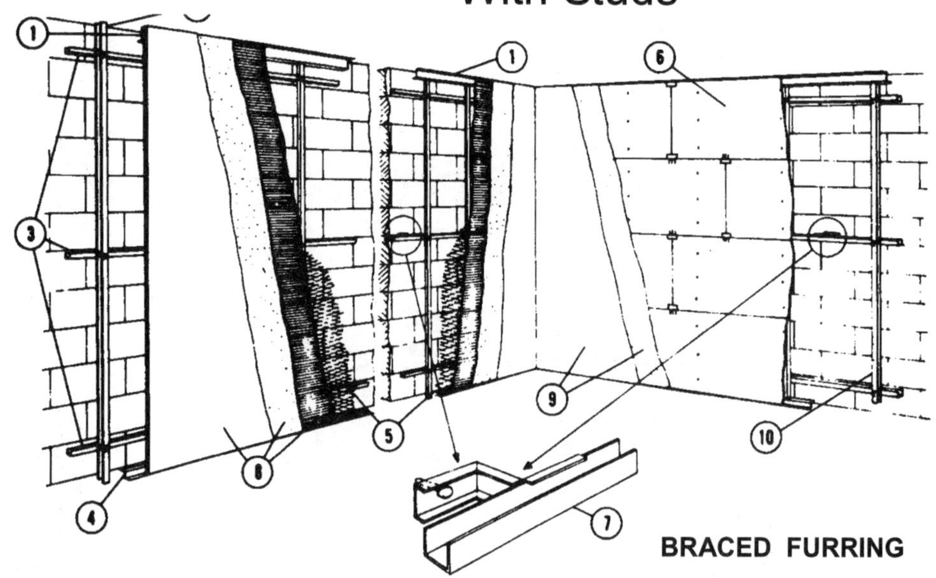

DOUBLE CHANNEL STUD
Hollow Partition

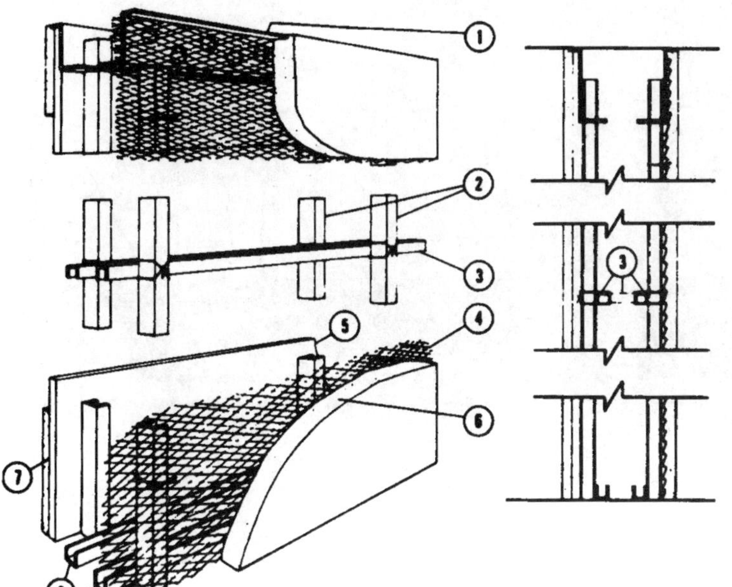

(1) Ceiling Runner
(2) Channel Studs
(3) Partition Stiffener and Channel Spacer
(4) Metal or Wire Fabric Lath (wire-tied)
(5) Gypsum Lath (clipped on)
(6) Three Coats of Plaster (Scratch, Brown, Finish)
(7) Two Coats of Plaster (Brown, Finish)
(8) Floor Runners (channel)

SINGLE CHANNEL STUD
Solid Partition

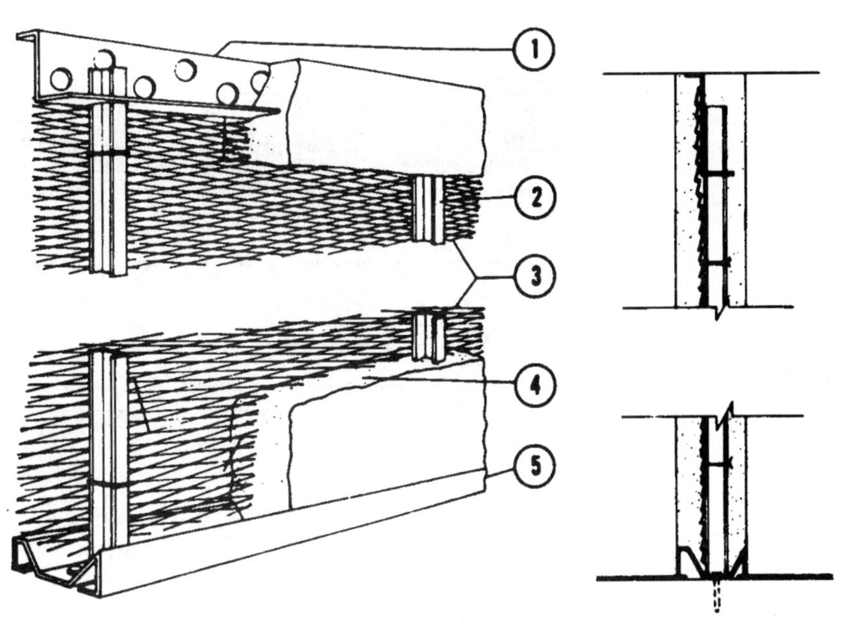

(1) Ceiling Runner

(2) Channel Stud

(3) Metal or Wire Fabric Lath (wire-tied)

(4) Plaster

(5) Combination Floor Runner and Screed

CEILINGS

STEEL JOISTS   CONCRETE JOISTS

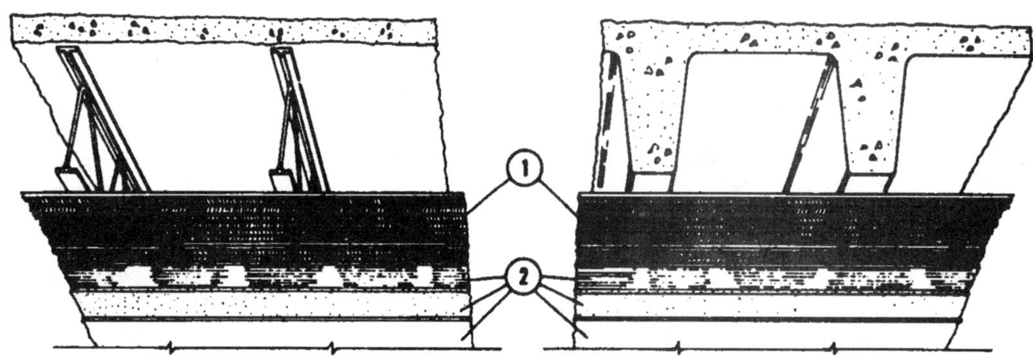

WOOD JOISTS

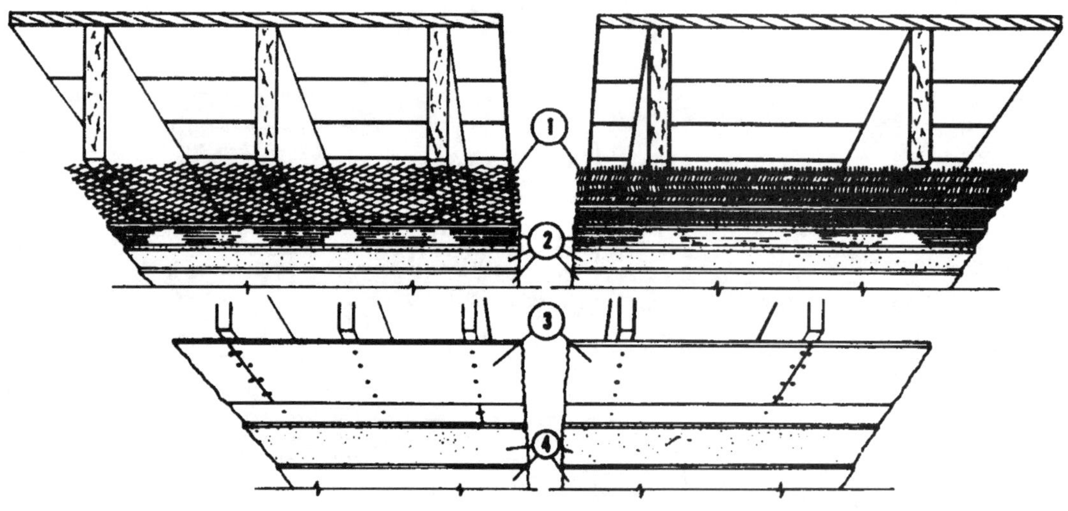

(1) Metal or Wire Fabric Lath

(2) Plaster

(3) Gypsum Lath

(4) Plaster

SUSPENDED CEILINGS

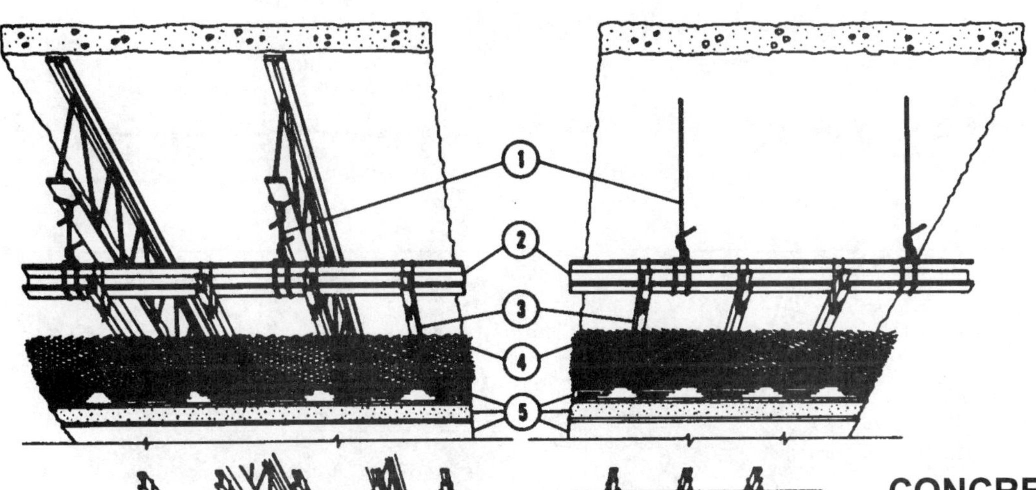

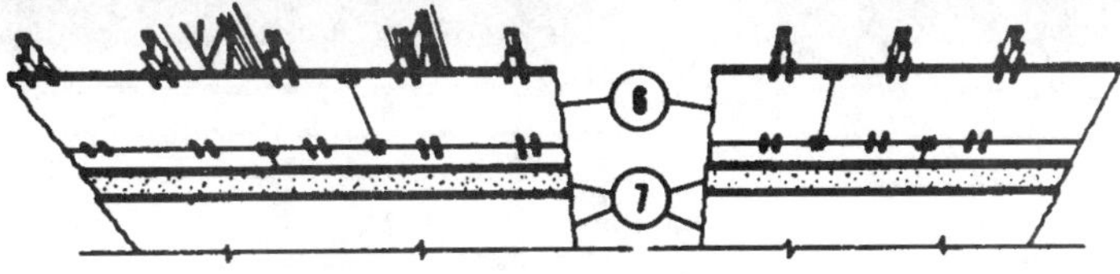

(1) Hanger
(2) Main Runner Channel
(3) Furring Channel
(4) Metal or Wire Fabric Lath
(5) Plaster
(6) Gypsum Lath
(7) Plaster

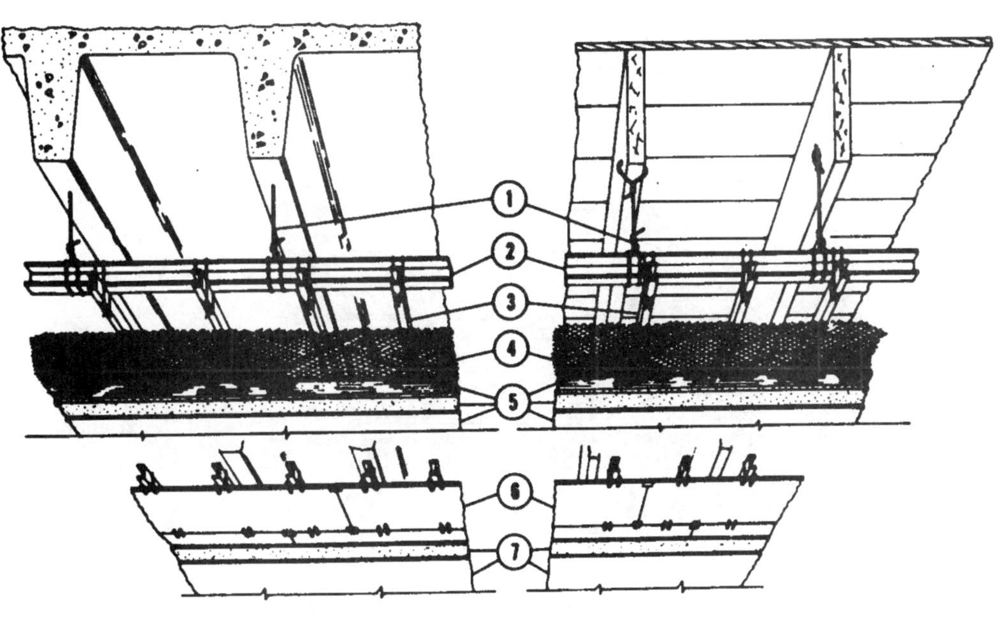

FURRED CEILINGS

STEEL JOISTS   CONCRETE JOISTS

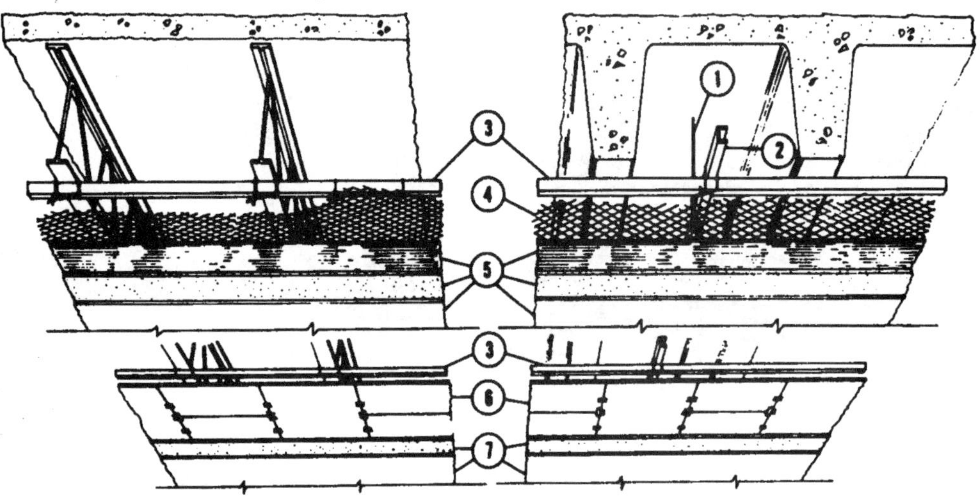

WOOD JOISTS

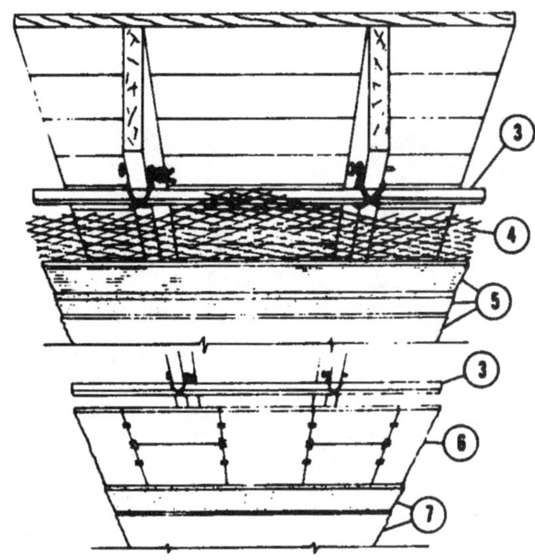

(1) Hanger

(2) ¾-inch Channel

(3) Cross Furring

(4) Metal or Wire Fabric Lath

(5) Plaster

(6) Gypsum Lath

(7) Plaster

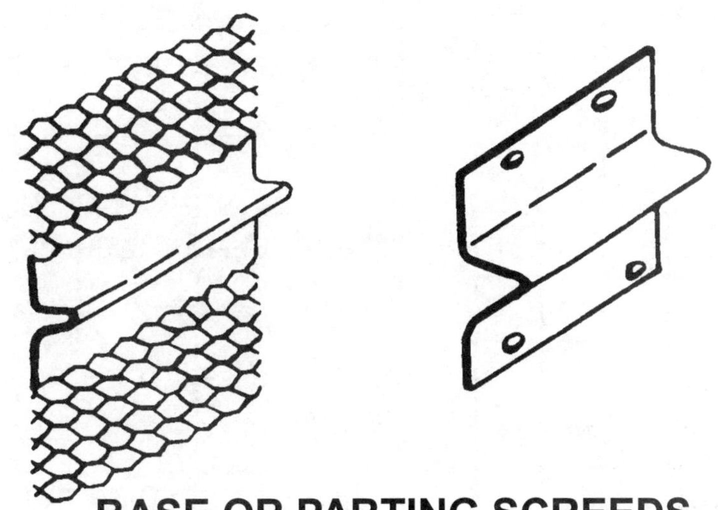

BASE OR PARTING SCREEDS

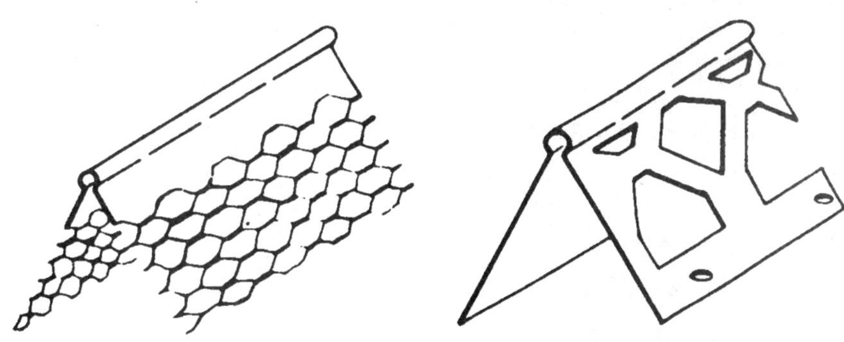

SMALL NOSE CORNER BEADS

WIRE BULL NOSE CORNER BEADS

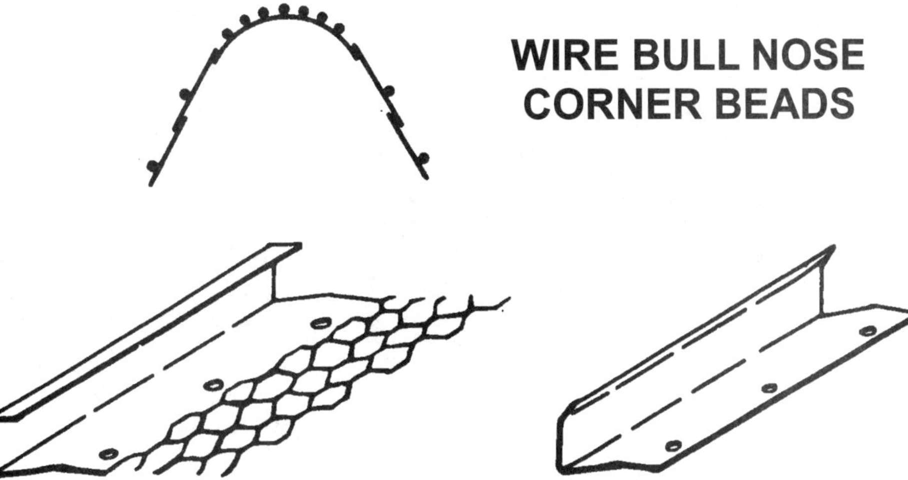

SQUARE CASING BEADS

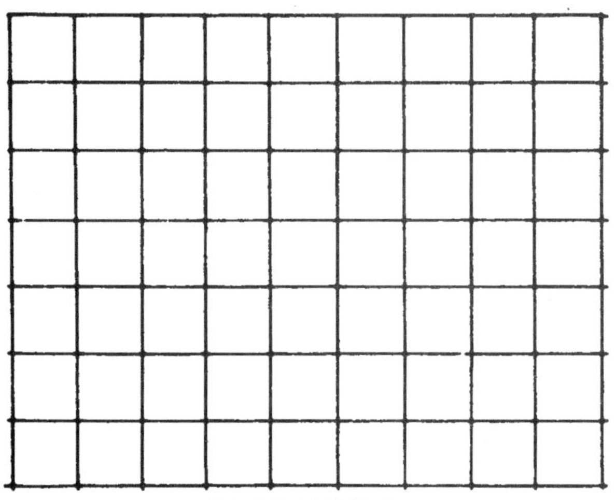

PLAIN WIRE
FABRIC LATH

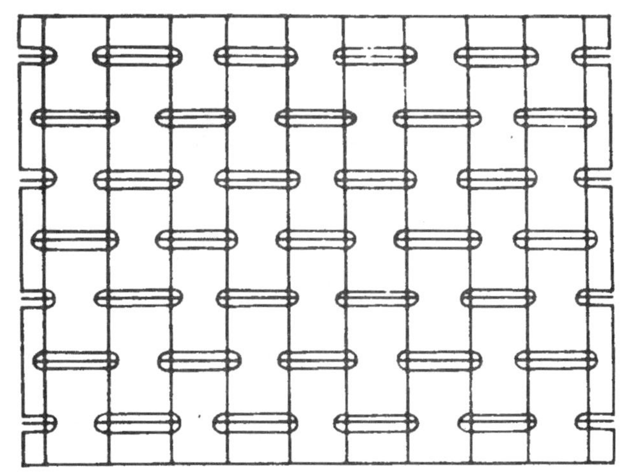

SELF-FURRING
WIRE FABRIC LATH

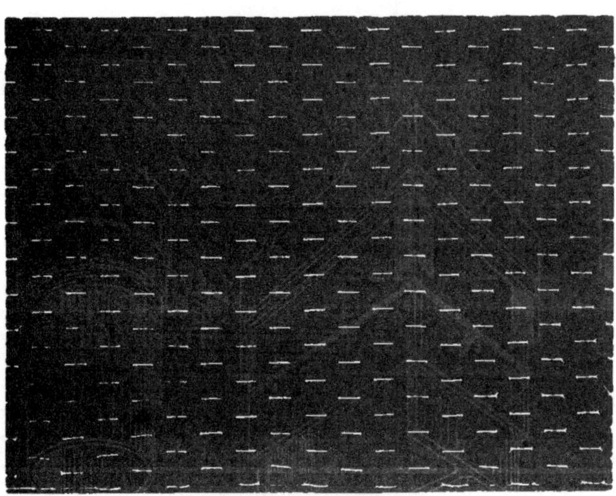

PAPER BACKED
WOVEN WIRE
FABRIC LATH

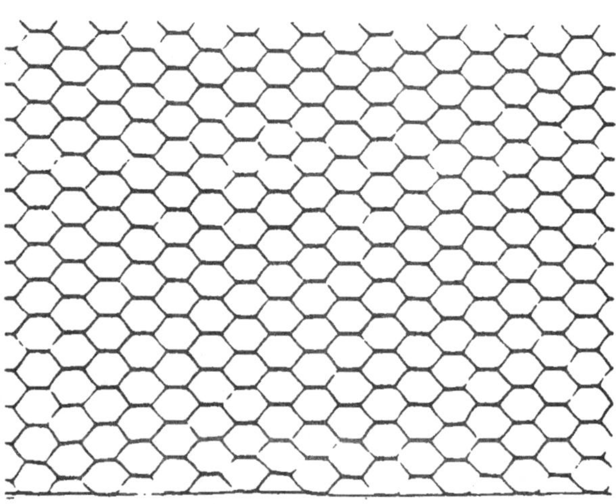

WOVEN WIRE
FABRIC LATH
(Also Available Self-Furred)

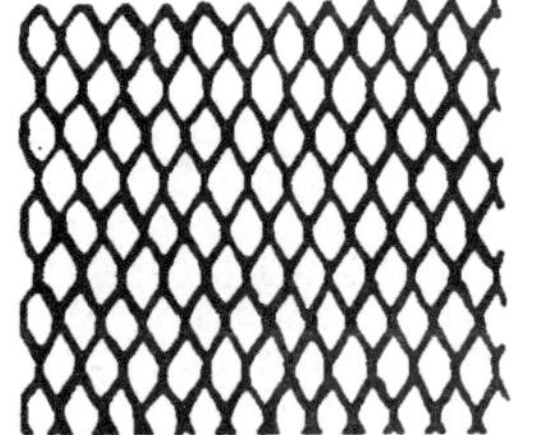

FLAT
DIAMOND MESH
METAL LATH

SELF-
FURRING
METAL LATH

FLAT RIB
METAL LATH

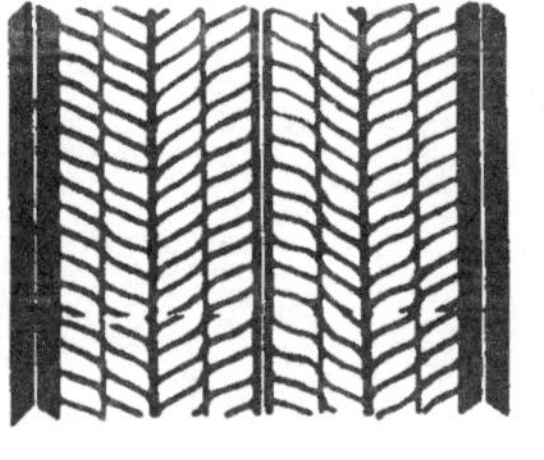

RIB
METAL LATH

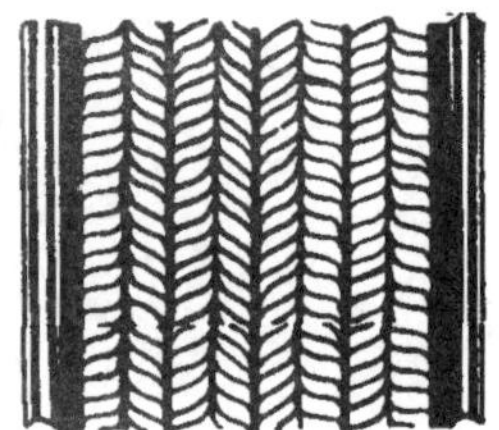

RIB
METAL LATH

VERTICAL FURRING

Studless

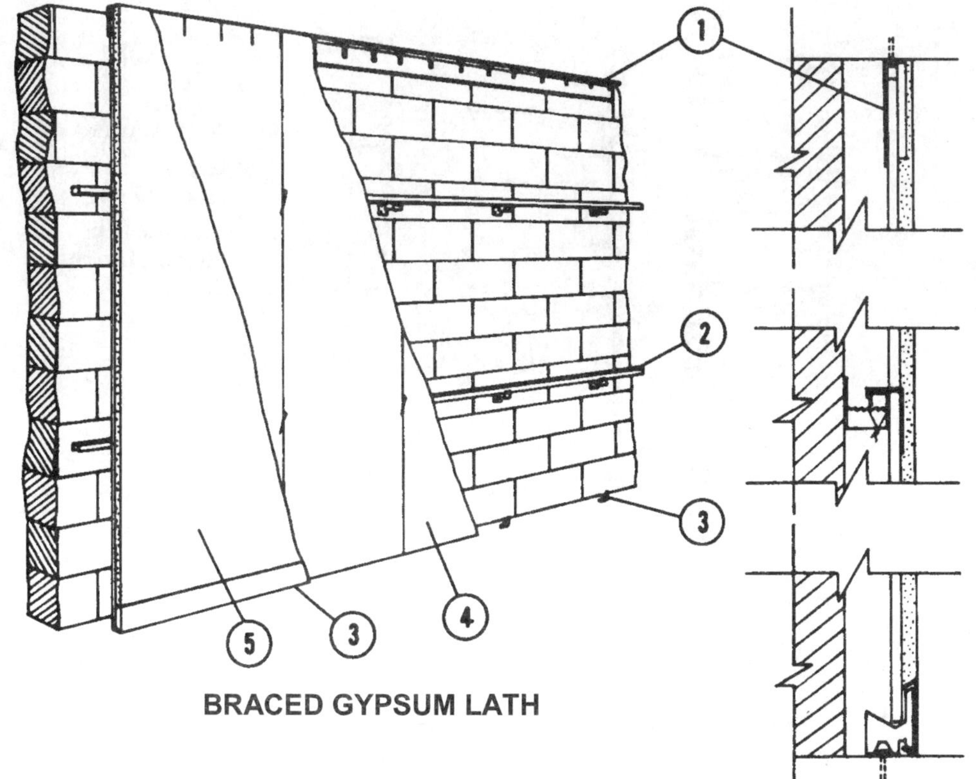

(1) Ceiling Runner

(2) Horizontal Stiffener (secured to bracing attachment)

(3) Metal Base and Clips

(4) Gypsum Lath

(5) Three Coats of Plaster (Scratch, Brown, Finish) (Minimum plaster thickness is ¾ inch.)

BRACED GYPSUM LATH

COLUMN FURRING

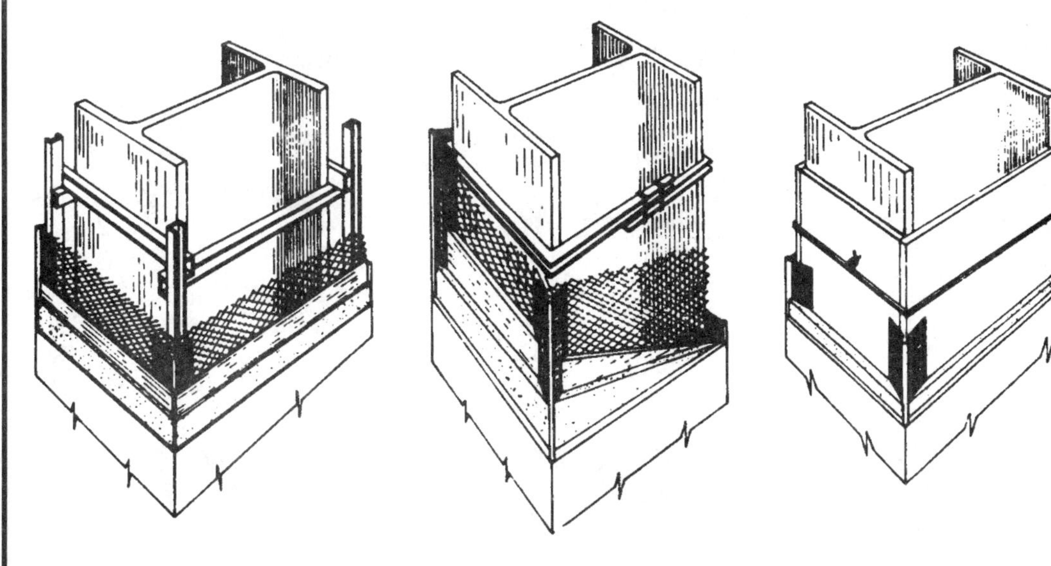

CONTACT FURRING

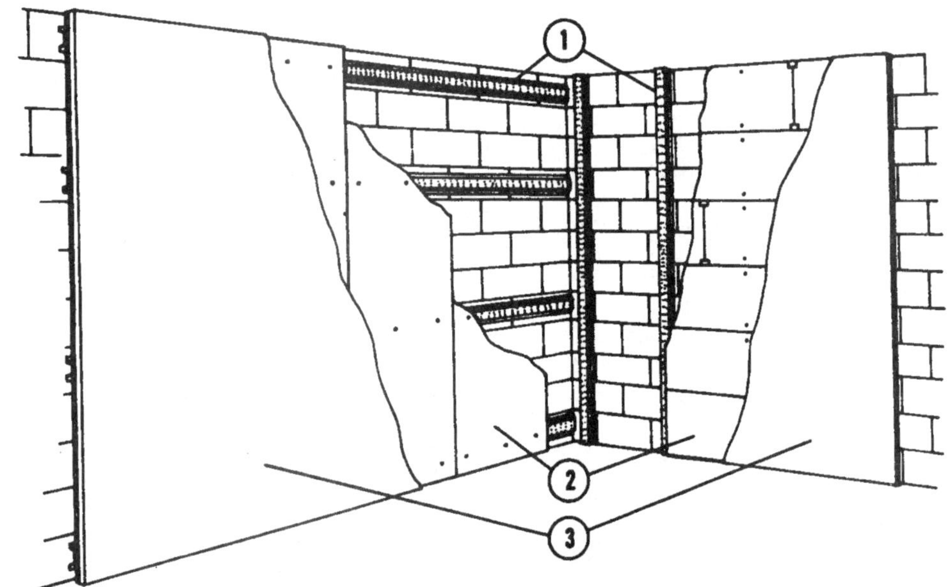

FALSE BEAMS

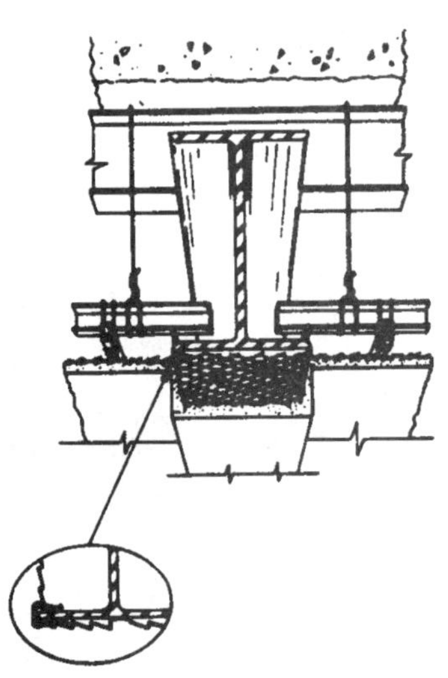

LATH DIRECT TO UNDERSIDE OF BEAM

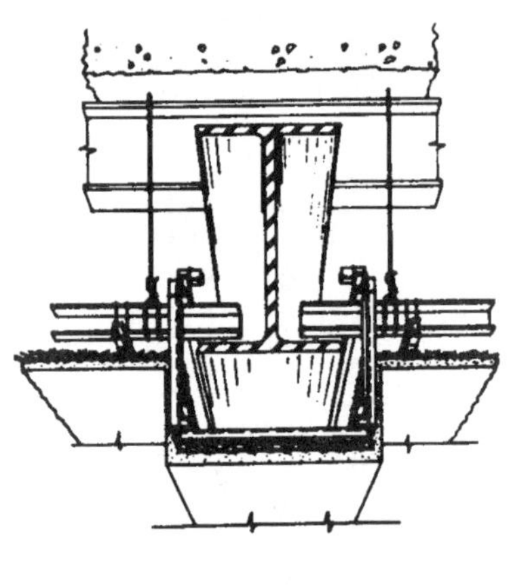

CHANNEL BRACKETS TO MAIN RUNNERS

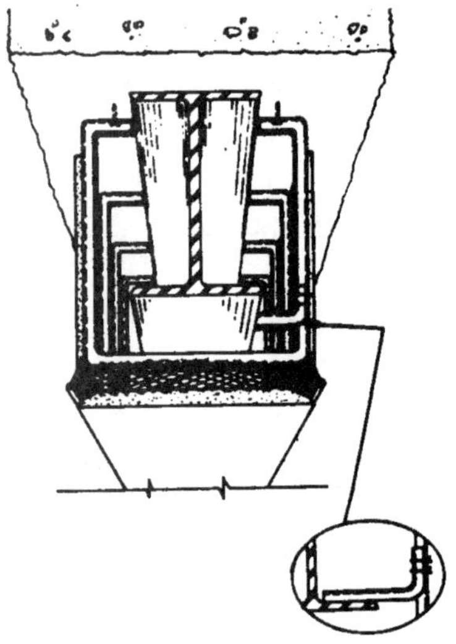

CHANNEL BRACKETS TO SLAB

MAXIMUM SPACING OF SUPPORTS FOR METAL LATH (Inches)

Type Of Lath	Weight of Lath Lb. Per Sq. Yd.	WALLS AND PARTITIONS			CEILINGS	
		Wood Studs	Solid Partitions	Steel Studs Wall Furring, Etc.	Wood or Concrete	Metal
Diamond Mesh (flat expanded)	2.5	16	16	13 1/2	12	12
	3.4	16	16	16	16	
Flat Rib	2.75	16	16	16	16	16
	3.4	19	24 (3)	19	19	19
3/8" Rib (1) (2)	3.4	24	(4)	24	24	24
	4.0	24	(4)	24	24	24
3/4" Rib	5.4	-	(4)	24 (5)	36 (6)	36 (6)
Sheet Lath	4.5	24	(4)	24	24	24

NOTE: Weights are exclusive of paper, fiber or other backing.

(1) 3.4 lb. 3/8" Rib Lath is permissible under Concrete Joists at 27" c.c.

(2) These spacings are based on a narrow bearing surface for the lath. When supports with a relatively wide bearing surface are used, these spacings may be increased accordingly, and still assure satisfactory work.

(3) This spacing permissible for Solid Partitions not exceeding 16' in height. For greater heights, permanent horizontal stiffener channels or rods must be provided on channel side of partitions, every 6' vertically, or else spacing shall be reduced 25%.

(4) For studless solid partitions, lath erected vertically.

(5) For interior wall furring or for application over solid surfaces for stucco.

(6) For contact or ceilings only.

TYPES OF LATH-ATTACHMENT TO WOOD AND METAL SUPPORTS

TYPE OF LATH	NAILS Type & Size	MAXIMUM SPACING Vertical (In Inches)	Horizontal	SCREWS MAXIMUM SPACING Vertical (In Inches)	Horizontal	Wire Gauge No.	Crown	Leg	STAPLES Round of Flattened Wire MAXIMUM SPACING Vertical (In Inches)	Horizontal
1. Diamond Mesh Expanded Metal Lath and Flat Rib Metal Lath	4d blued smooth box 1 ½ No. 14 gauge 7/32" head (clinched) 1" No.11 gauge 7/16" head, barbed 1 ½" No.11 gauge 7/16" head, barbed	6 6 6	- - 6	6	6	16	¾	7/8	6	6
2. 3/8" Rib Metal Lath and Sheet Lath	1 ½" No. 11 ga. 7/16" head, barbed	6	6	6	6	16	¾	1 ½	At Ribs	At Ribs
3. ¾" Rib Metal Lath	4d common 1 ½" No.12 ½ gauge ¼" head 2" No.11 gauge 7/16" head, barbed	At Ribs	- At Ribs	At Ribs	At Ribs	16	¾	1 5/8	At Ribs	At Ribs
4. Wire Fabric Lath	4d blued smooth box (clinched) 1" No.11 gauge 7/16" head, barbed	6 6	- -	6	6	16	¾	7/8	6	6
	1 ½" No.11 gauge 7/16" head, barbed 1 ¼" No. 12 ga. 3/8" head, furring 1" No.12 gauge 3/8" head	6 6 6	6 6			16	7/16	7/8	6	6
5. 3/8" Gypsum Lath	1 1/8" No.13 gauge 12/61" head, blued	8	8	8	8	16	¾	7/8	8	8
6. ½" Gypsum Lath	1 ¼" No.13 gauge 12/61" head, blued	8	8 6	8	8 6	16	¾	1 1/8	8	8 6

STRESS RELIEF (CONTROL JOINTS)

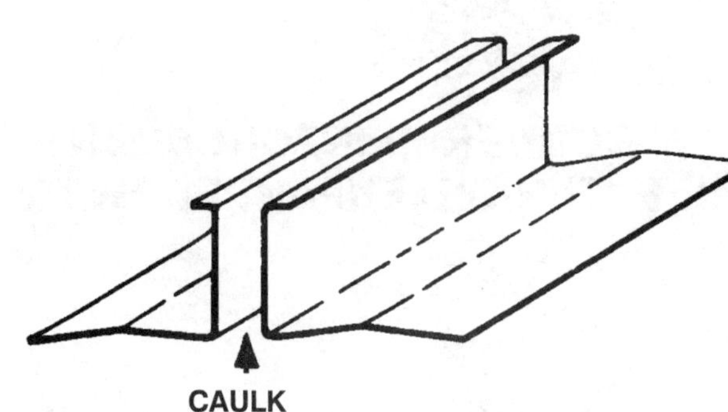

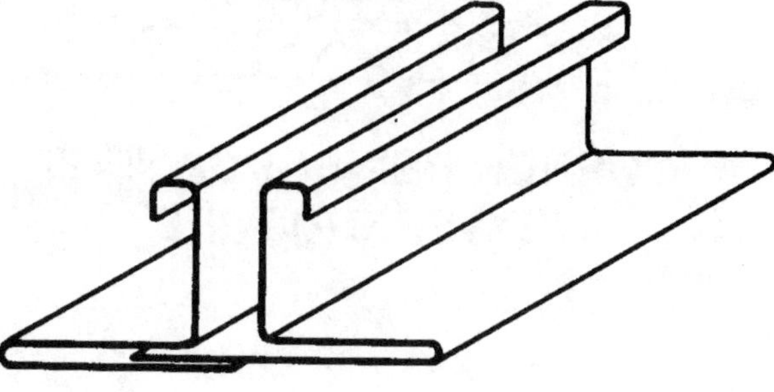

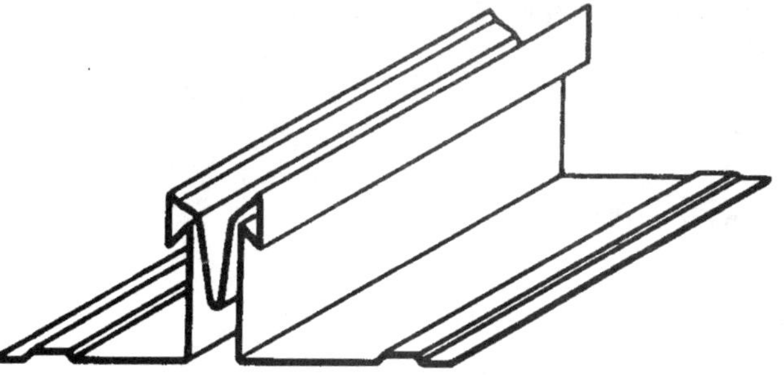

REVEALS

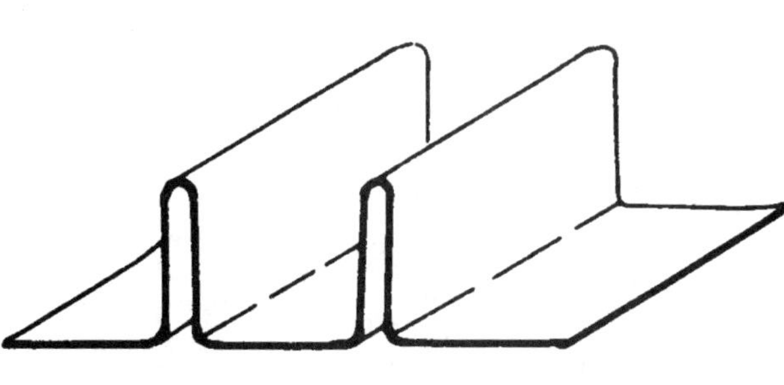

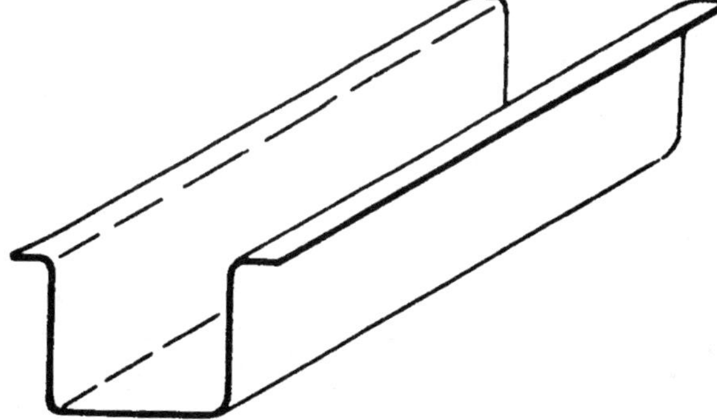

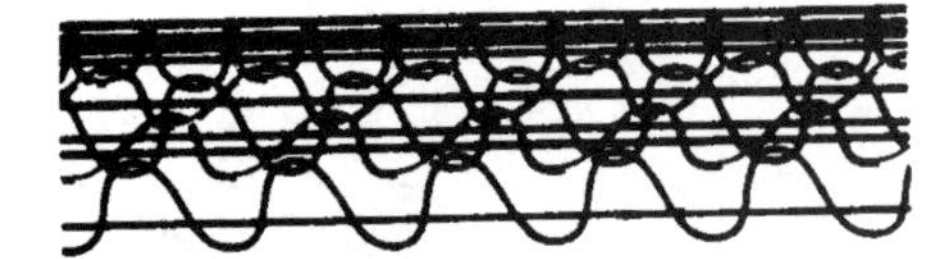

CORNER REINFORCEMENT (EXTERIOR) WIRE

CORNER REINFORCEMENT (EXTERIOR) EXPANDED METAL

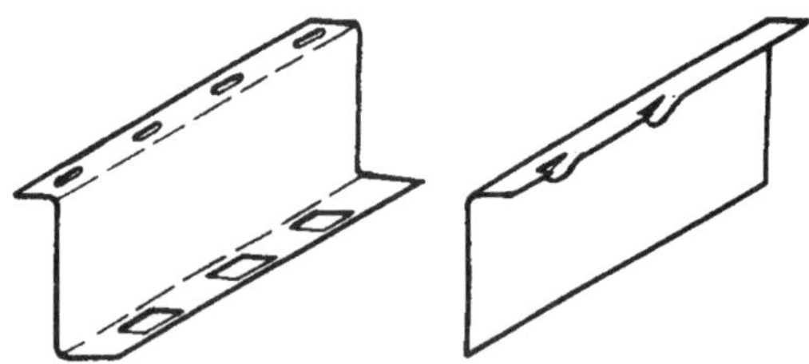

PARTITION RUNNERS (Z AND L SHAPE)

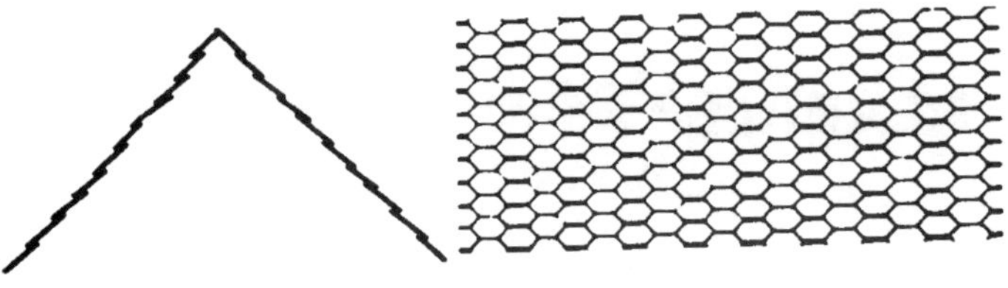
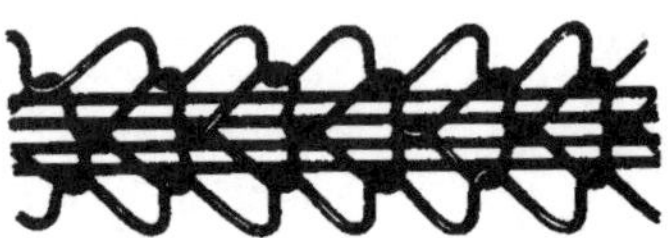

EXPANDED METAL CORNERITE

WIRE CORNERITE

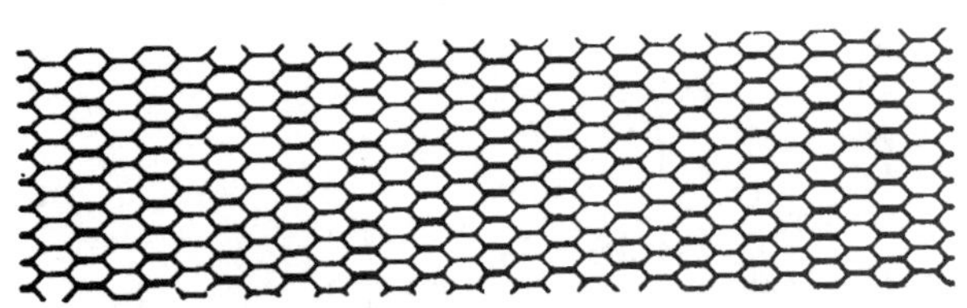
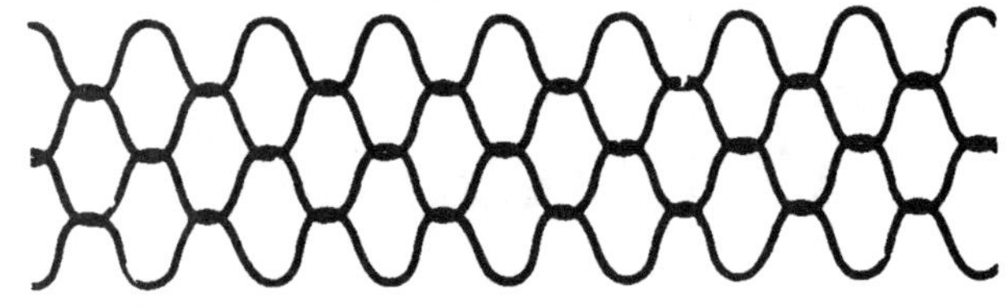

STRIP REINFORCEMENT (EXPANDED METAL)

STRIP REINFORCEMENT (WIRE)

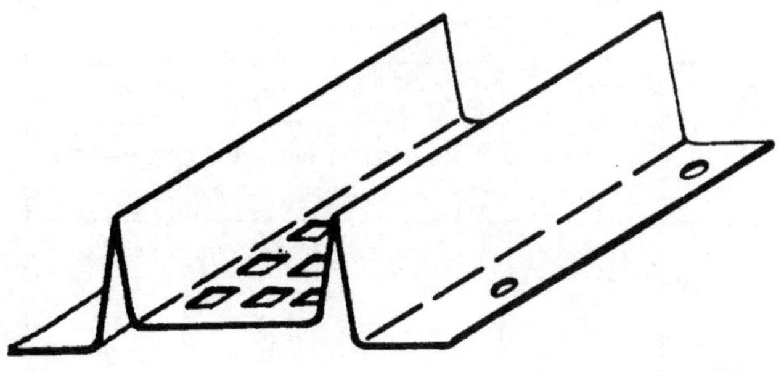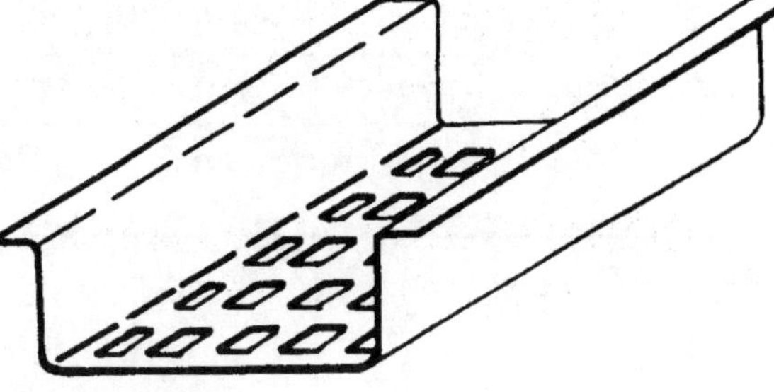

VENTILATING SCREEDS

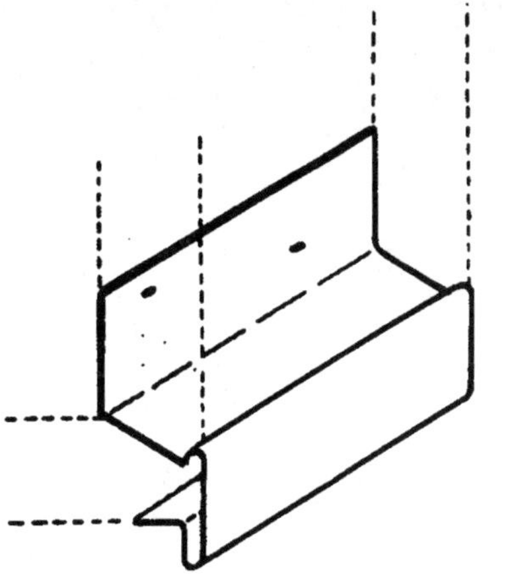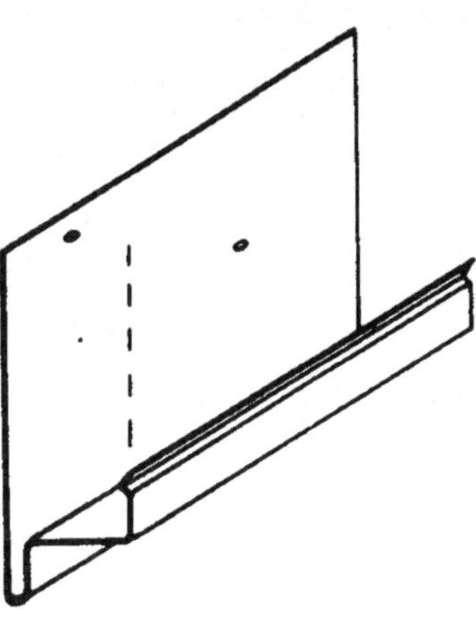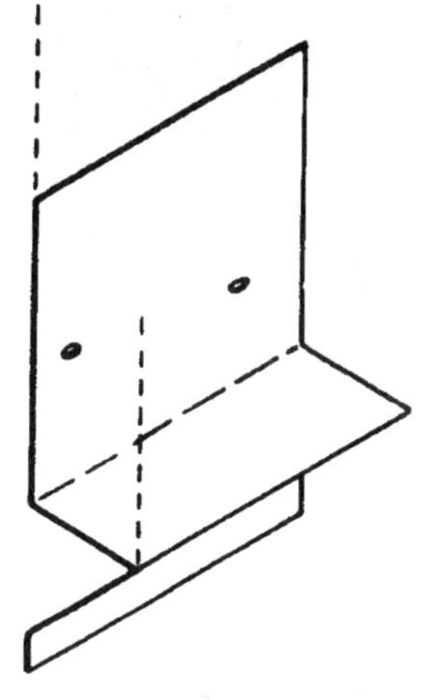

DRIP SCREEDS

WEEP SCREED
(Also Available with Perforations)

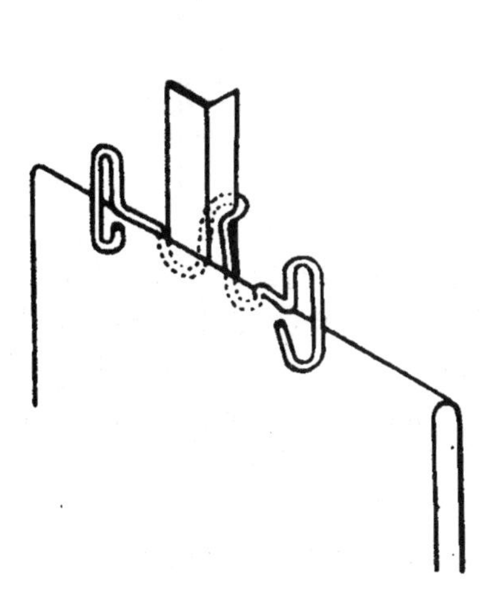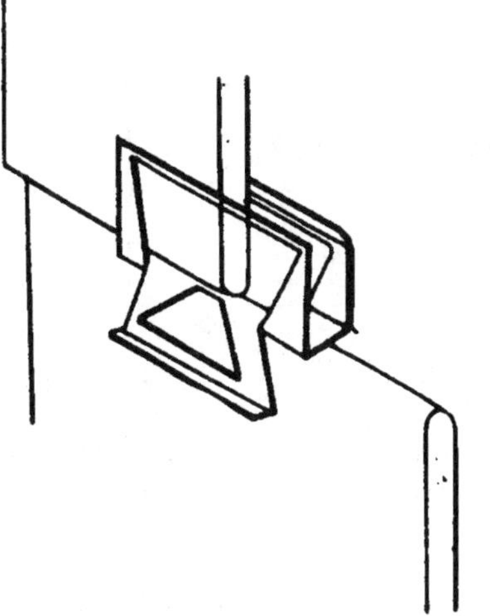

GYPSUM LATH ATTACHMENTS CLIPS

VERTICAL FURRING

VERTICAL FURRING MEMBER	UNBRACED				BRACED			
	STUD SPACING				STUD SPACING			
	24"	19"	16"	12"	24"	19"	16"	12"
	Maximum Furring Heights				Maximum Vertical Distance Between Braces			
3/4" Channel	6'	7'	8'	9'	5'	5'	6'	7'
1 1/2" Channel	8'	9'	10'	12'	6'	7'	8'	9'
2" Channel	9'	10'	11'	13'	7'	8'	9'	10'
2" Prefab. Stud	8'	9'	10'	11'	6'	7'	8'	9'
2 1/2" Prefab. Stud	10'	11'	12'	14'	8'	9'	10'	11'
3 1/4" Prefab. Stud	14'	16'	17'	20'	11'	13'	14'	16'

TYPES OF LATH—MAXIMUM SPACING OF SUPPORTS

TYPE OF LATH	Minimum Weight (psy), Gauge & Mesh Size	VERTICAL			HORIZONTAL	
		WOOD	METAL		Wood or Concrete	Metal
			Solid Plaster Partitions	Other		
Expanded Metal Lath (Diamond Mesh)	2.5 3.4	16" 16"	16" 16"	12" 16"	12" 16"	12" 16"
Flat rib Expanded Metal Lath	2.75 3.4	16" 19"	16" 24"	16" 19"	16" 19"	16" 19"
Stucco Mesh Expanded Metal Lath	1.8 and 3.6	16"	-	-	-	-
3/8" Rib Expanded Metal Lath	3.4 4.0	24" 24"	-	24" 24"	24" 24"	24" 24"
Sheet Lath	4.5	24"	-	24"	24"	24"
3/4" Rib Expanded Metal Lath (Not manufactured in West)	5.4	-	-	-	36"	36"
Wire Fabric Lath — Welded	1.95 lbs.,11 ga.,2"x2" 1.4 lbs.,16 ga.,2"x2" 1.4 lbs.,18 ga.,1"x1"	24" 16" 16"	24" 16" -	24" 16" -	24" 16" -	24" 16" -
Wire Fabric Lath — Woven	1.4 lbs.,17 ga.,1 1/2" Hex. 1.4 lbs.,18 ga.,1" Hex.	24" 24"	16" 16"	16" 16"	24" 24"	16" 16"
3/8" Gypsum Lath (plain)	-	16"	-	16"	16"	16"
(Large Size)	-	16"	-	16"	16"	16"
1/2" Gypsum Lath (plain)	-	24"	-	24"	24"	24"
(Large Size)	-	24"	No supports; Erected vertically	24"	24"	16"
5/8" Gypsum Lath (Large Size)	-	24"	No supports; Erected vertically	24"	24"	16"

PLASTERING TABLES

THICKNESS OF PLASTER

PLASTER	FINISHED THICKNESS OF PLASTER FROM FACE OF LATH, MASONRY, CONCRETE	
	Gypsum Plaster	Portland Cement Plaster
Expanded Metal Lath	5/8" minimum	5/8" minimum
Wire Fabric Lath	5/8" minimum	3/4" minimum (interior)
		7/8" minimum (exterior)
Gypsum Lath	1/2" minimum	
Gypsum Veneer Base	1/16" minimum	1/2" minimum
Masonry Walls	1/2" minimum	7/8" maximum
Monolithic Concrete Walls	5/8" maximum	1/2" maximum
Monolithic Concrete Ceilings	3/8" maximum	

GYPSUM PLASTER PROPORTIONS

NUMBER OF COATS	COAT	PLASTER BASE OR LATH	MAXIMUM VOLUME AGGREGATE PER 100# NEAT PLASTER (CUBIC FEET)	
			Damp Loose Sand	Perlite or Vermiculite
Two-Coat Work	Basecoat	Gypsum Lath	2 1/2	2 1/2
	Basecoat	Masonry	3	3
Three-Coat Work	First Coat	Lath	2	2
	Second Coat	Lath	3	3
	First & Second Coat	Masonry	3	3

PORTLAND CEMENT PLASTER

COAT	VOLUME CEMENT	MAXIMUM WEIGHT (OR VOLUME) LIME PER VOLUME CEMENT	MAXIMUM VOLUME SAND PER VOLUME CEMENT	APPROXIMATE MINIMUM THICKNESS	MINIMUM PERIOD MOIST CURING	MINIMUM INTERVAL BETWEEN COATS
First	1	20 lbs.	4	3/8"	48 Hours	48 Hours
Second	1	20 lbs.	5	1st & 2nd Coats total 3/4"	48 Hours	7 Days
Finish	1	1	3	1st, 2nd & Finish Coats total 7/8"	-	

PORTLAND CEMENT - LIME PLASTER

COAT	VOLUME CEMENT	MAXIMUM WEIGHT (OR VOLUME) LIME PER VOLUME CEMENT	MAXIMUM VOLUME SAND PER VOLUME CEMENT	APPROXIMATE MINIMUM THICKNESS	MINIMUM PERIOD MOIST CURING	MINIMUM INTERVAL BETWEEN COATS
First	1	1	4	3/8"	48 Hours	48 Hours
Second	1	1	4 1/2	1st & 2nd Coats total 3/4"	48 Hours	7 Days
Finish	1	1	3	1st, 2nd & Finish Coats total 7/8"	-	

METAL STUD CONSTRUCTION

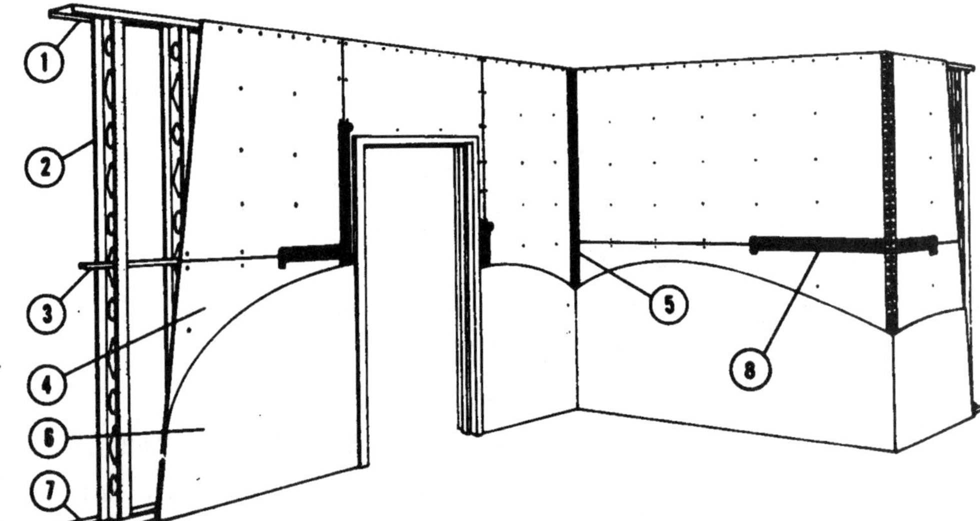

(1) Ceiling Runner Track

(2) Metal Stud (nailable or screw)

(3) Horizontal Stiffener

(4) Large Size Lath

(5) Angle Reinforcement

(6) Veneer Plaster 1/16 to 1/8 inch thick)

(7) Floor Runner Track

(8) Joint Reinforcement

WOOD STUD CONSTRUCTION

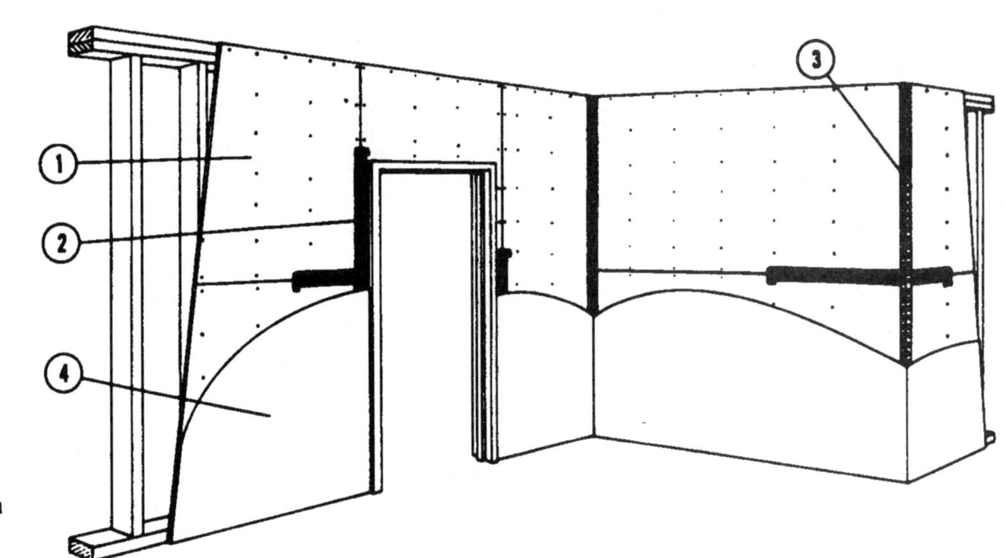

(1) Large Size Lath

(2) Joint Reinforcement

(3) Corner Bead

(4) Veneer Plaster (1/16 to 1/8 inch thick)

EXTERIOR LATH AND PLASTER

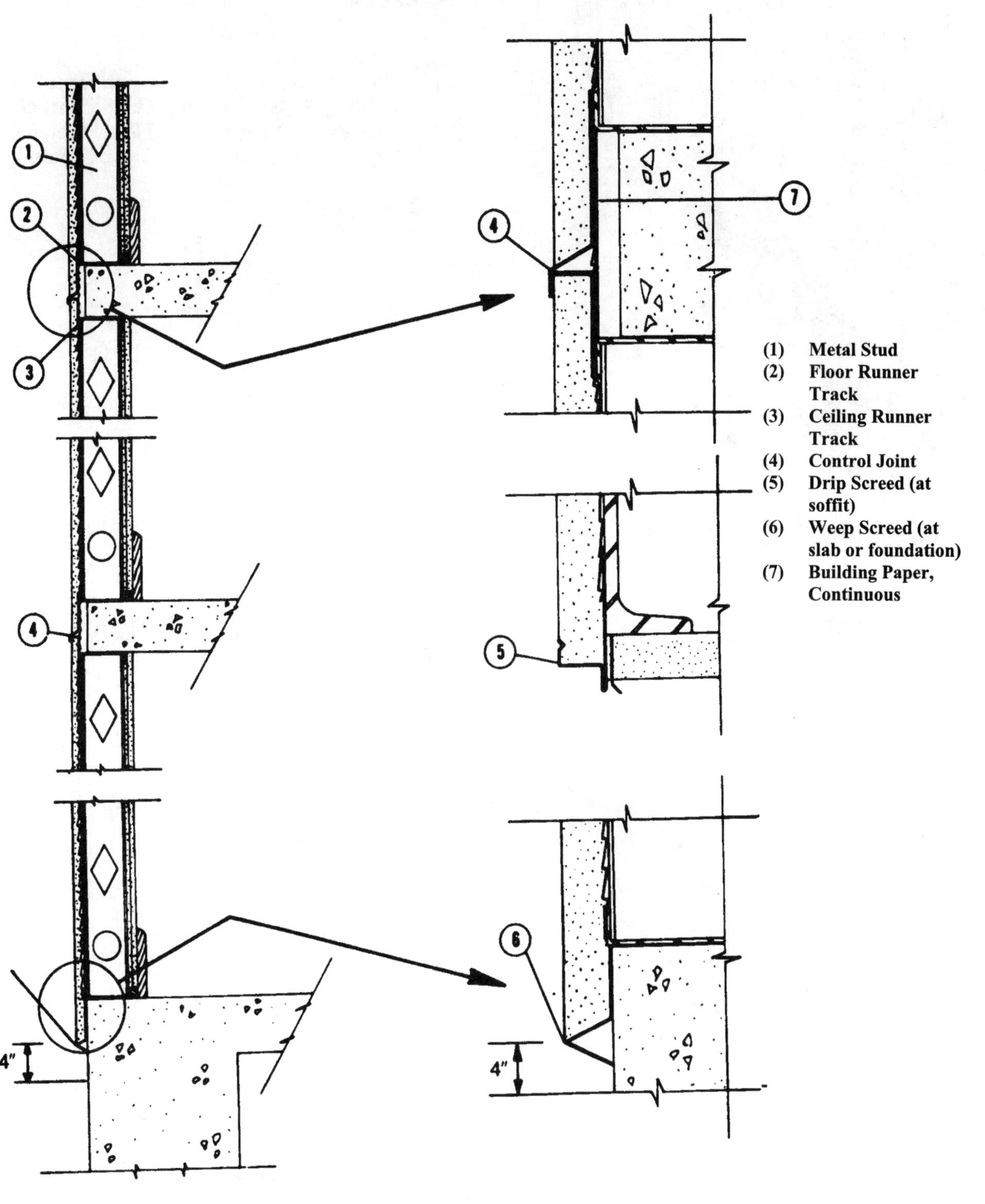

EXTERIOR LATH AND PLASTER

OPEN WOOD FRAME CONSTRUCTION

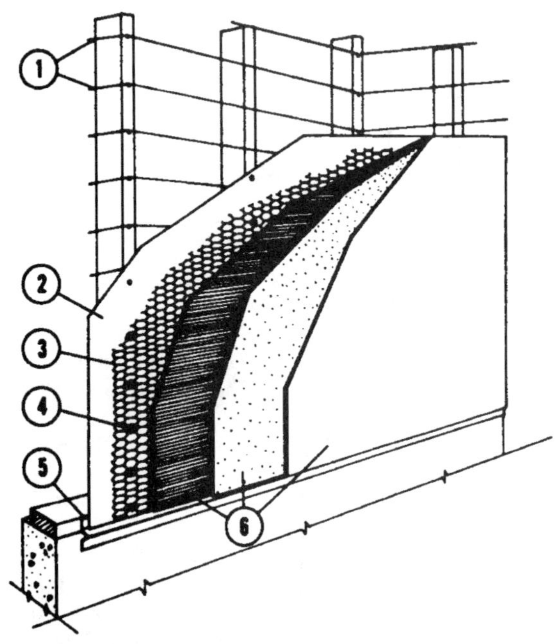

(1) Wire Backing
(2) Building Paper
(3) Wire Fabric Lath
(4) Approved Fasteners

(5) Weep Screed
(6) Three Coats of Plaster (Scratch, Brown, Finish)

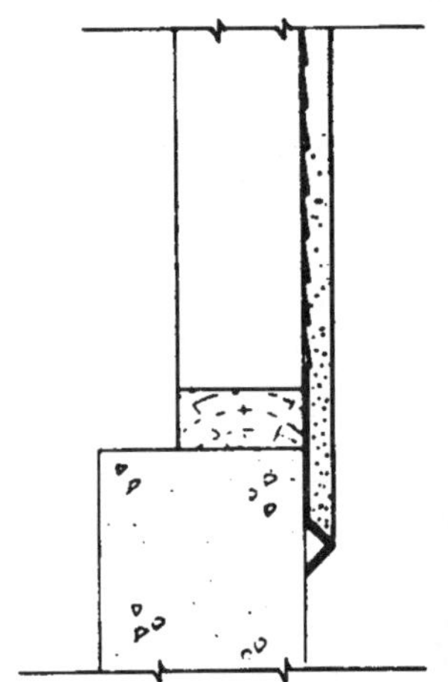

SHEATHED WOOD FRAME CONSTRUCTION

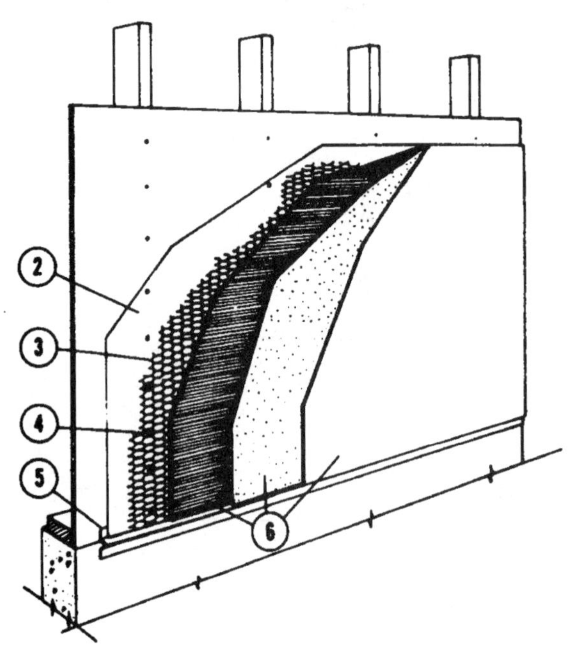

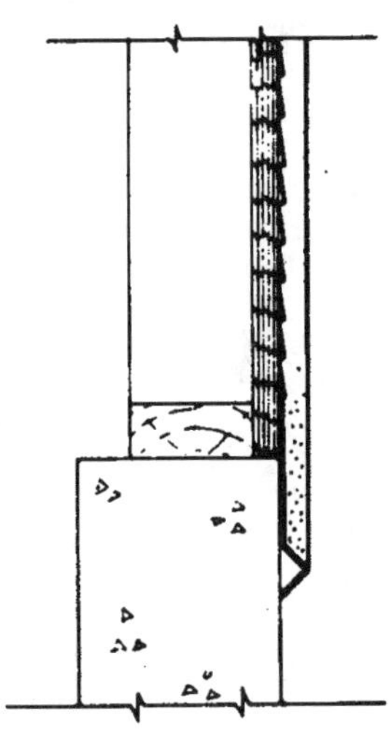

Paint Failure. There is a cause for every paint failure, and in most instances the failure can be prevented by observing specific precautions. The most often observed paint failures are: cracking flaking, scaling, alligatoring, bleeding, excessive chalking, blistering, fading, spotting, washing and discoloration. The probable causes, preventive measures and corrective measures are discussed in the following subsections:

Alligatoring. When a rupturing of the top paint coat causes the surface to break up into irregular areas separated by wide cracks in "alligator hide" fashion, the condition is called alligatoring. Its cause is due to applying paint to unseasoned wood or from applying a heavy coat of paint over a relatively soft undercoat before it has thoroughly dried. To correct, remove the entire paint film and repaint.

Bleeding. When the color of a previous coat is absorbed into the top coat, this condition is called bleeding. It is caused by the partial solubility of the pigment of the undercoat in the vehicle of the top coat. Asphalt or bituminous undercoat are particularly susceptible to bleeding. In order to prevent or correct bleeding, care must be exercised to select a vehicle for the top coats which is not a solvent for the undercoat pigments.

Blistering. Blistering is evidenced by blister like irregularities on the film of a painted surface. The most common cause of blistering is the application of paint over a damp or wet surface. Under the action of the sun's rays, the moisture is drawn out of the wood (or vaporized on a metallic surface), raising the paint coating with it in the form of blisters. Blistering can also be caused by using too much dryer in the undercoat. To avoid blistering, do not paint damp or wet surfaces, green lumber or greasy spots. Avoid excessive amounts of dryers in undercoat. To correct, remove the paint film in the blistered area, let the surface dry and repaint.

Blushing. A surface in which blushing has occurred is characterized by a white discoloration in the paint film or a precipitation of an ingredient. It is caused by the condensation of moisture on the paint film or by improper composition of the vehicle or solvent. By applying paint under conditions which do not permit moisture to condense on the applied film, blushing can be avoided. Do not paint cold metal under humid conditions. Blushing can be corrected only by removing the paint film in the affected areas and repainting.

Chalking. Chalking of a painted surface can be detected by gentle rubbing which will disclose loose powder on the paint film. Slow, uniform chalking is a desirable quality in the paint, but excessive chalking is undesirable and indicates a paint film failure. It is caused by excessive rain, fog or high humidity during the application of the paint or during the drying period. Paints low in binder content or high in inert pigments have a tendency toward early and excessive chalking.

To prevent chalking, use a paint which is neither deficient in oil content nor high in inert pigment content and apply only under dry conditions. To correct chalked surfaces, remove all loose chalked substance from the surface with a wire brush and repaint the surface with a paint of good quality.

Checking. Checking of a painted surface can be detected by the appearance in the topcoat of small openings or ruptures which divide the surface into small irregular areas, leaving the undercoat visible through the breaks in the topcoat. It is usually caused by too soft an undercoat or by applying a coat over an underlying coat which has not thoroughly dried. To correct, wire brush or sandpaper to remove the loose film and apply a new coat of paint, making sure it is as elastic as the previous coat.

Cracking. Cracks are breaks in a paint film which extend through to the surface to which the paint was applied. Where cracking is severe, flaking, scaling or peeling usually follows. It is usually caused by improper compounding of the paint, resulting in a hard and brittle paint film lacking in elasticity. Low grade paints are often inelastic because they are deficient in oils or contain too much inert material. Cracking will also result from painting a surface upon which too many coats of paint already exist. To avoid cracking, use paint of proper consistency and elasticity, and remove heavy coats of paint from the surface to be painted prior to repainting.

Crawling or Creeping. This defect can be distinguished by the little drops or islands which are formed by the paint film. It often occurs when paint is applied on an oily, waxy or greasy surface or on a very smooth surface such as glass and polished metal. To correct, remove all greasy or oily spots from the surface. On glossy surfaces, wash with a mild soda ash solution or sand with fine sandpaper.

Dulling. Dulling is characterized by a loss in gloss which develops in high gloss paints and enamels and is caused by improper compounding, use of very old stocks of paints and enamels or through the use of too much turpentine as a thinner. To avoid dulling, use fresh paint stocks of good quality. To correct, sand the dulled coat with fine sandpaper and repaint.

Flaking. Flaking is the dropping off of small pieces of a paint coat which has generally started with checking, developing into cracking, and finally the small cracked sections fail away from the surface. The causes and corrective measures are the same as for cracking.

Mildew. Mildew, a form of plant life, is a fungus frequently found on exposed surfaces in damp, warm locations, particularly on soft paint films. It is caused by soft paint films becoming sticky in the warm location and windblown spores and decayed and dried vegetation adhering to the surfaces. The oil in the paint becomes infected, and breeding of the mildew spores takes place in the damp, warm environment. Mildew can be prevented by using a hard drying paint, applied under dry conditions. Paints which contain zinc oxide are resistant to mildew infections. To correct, remove the paint coat with a blowtorch, wash the area with an abrasive soap and water, or with a water solution of trisodium phosphate, rinsing the surface with clear water and allowing it to dry. The addition of one half ounce of mercuric chloride per gallon of paint or the use of less oils and more turpentine is advisable where mildew is likely to occur. Exercise extreme caution in handling of paints with mercuric chloride or other fungicides to prevent poisoning of the skin.

Peeling. Peeling of a paint film is evidenced by large scales of the film curling and peeling off the painted surface. It is usually caused by the application of paint in the presence of moisture or from being applied over a faulty priming coat. It can be caused by moisture getting behind the paint film at a corner, or in the case of wood, at knotholes which are not properly sealed. Painting unseasoned lumber will also cause peeling. To avoid peeling, seal knot holes and end grains of lumber with shellac before painting and insure that the edges and corners of steel shapes are thoroughly covered. To correct, remove all loose, peeled paint by wire brush or blowtorch, clean the surface of dust film, and repaint on a dry surface only.

Runs and Sags. Ripples, runs and sags on a vertical surface are caused by the application of too heavy a coat of paint, or by using a paint which has been thinned excessively. Other contributing causes are the use of stiff, inflexible brushes or incomplete brushing of the film. It can be prevented by applying a uniform coat of paint of the correct consistency, using a flexible brush and properly brushing out the paint. When runs or sags appear on spray painted surfaces, it is also indicative that too heavy a coat was applied, or that the paint was excessively thinned. To correct, remove the runs or sags with sandpaper and repaint the area.

Scaling. Scaling is an aggravated form of flaking. It is evidenced by large sections of paint coming loose and falling from the painted surface. Scaling is usually preceded by cracking, and like cracking, is usually caused by the paint drying hard and brittle and being unable to expand or contract as the painted object does with changes in temperature and moisture. Scaling frequently occurs when unseasoned lumber is painted. In some cases, previous coats of paint may have lost their elasticity and become lifeless. This causes poor adhesion in the old coat and, in drying, the new coat shrinks and pulls the old film loose from the painted surface. To avoid, do not paint unseasoned lumber, or over a paint which dries hard and brittle, or over old lifeless paint films. Remove the paint with commercial paint removers, scrapers or blowtorch. Remove dust from the surface and repaint only when the surface is thoroughly dry, using a paint which dries with an elastic film.

Slow Drying. The drying time of a paint film varies with the inherent characteristics of the pigment or of the vehicle. However, certain faulty conditions may prolong the drying period and cause the condition to be termed a paint fault. Paints, which under normal drying conditions are tacky or sticky for 12 hours or longer after application, are likely to catch dust and dirt, promote mildew formations or fail by checking or alligatoring. It is usually caused by using paints to which a small amount of mineral oil has been added through negligence or error. Such paint may never dry thoroughly. Old linseed oil that has become fatty by exposure or inferior dryers and thinners frequently contribute to slow drying of paint coats. This condition can be avoided by painting when the temperature is above 50°F or by applying thin coats of paint in colder weather. Use less oil and allow ample time for each coat to dry before applying subsequent coats. Once paint is applied and fails to dry, the condition can be corrected by allowing sufficient additional time for drying, or by removing the paint with commercial paint removers or by scraping, and then repainting with the right type of paint under proper conditions.

Spotting. The appearance of discolored spots or craters in a painted surface is called spotting. It is usually caused by too few coats, such as painting on new work using only two coats, or painting old work with one coat. The lack of controlled penetration causes uneven fading in the pigments. Sap in wood may affect the paint and cause spotting. Spotting may be caused by improperly sealed nail heads, or rain or hail on a freshly painted surface. On plastered surfaces, an inferior primer or sealer may cause spotting as the alkali if the plaster burns through. To prevent spotting, apply sufficient coats of paint and avoid painting when rain is imminent. Insure that all old paint, plaster surfaces and nail heads are properly sealed with quality paint films. To correct, apply an additional coat of paint. Paint surfaces spotted from rain or hail must be sanded smooth before repainting.

Washing. Washing of paint is evidenced by streaks in the paint surface, the discoloration generally accumulating at the lower edges of boards, panels, girders, etc. Washing is caused by using paints having pigments which form water soluble compounds or by painting under damp conditions. This defect often occurs when paints of inferior quality are used. To prevent washing, use only good quality paints. To correct, remove the paint film completely and repaint.

Wrinkling. Wrinkling of a paint coat is evidenced by the paint film gathering in small wrinkles. Wrinkling may be caused by the application of an excessively thick coat, or by failure to brush out the paint properly. It is frequently caused by too much dryer in the paint or by using paints which have been excessively thinned with oil and applied thick. To avoid wrinkling, do not apply thick coats or use excessive amounts of dryer or oil. To correct, remove the wrinkles with sandpaper and repaint with properly thinned paint which does not have excessive amount of dryer or oil in it. If the coat of paint is excessively wrinkled, strip off the old coat and repaint.

Painting Checklist.

(1) Is all paint material tested or approved?
(2) Should surface preparation samples for sandblasting be required?
(3) Are color samples required?
(4) Is the contractor required to submit a list of paint materials for approval?
(5) Are delivered paint materials as specified?
(6) Prior to painting, check each surface with contractor for preparation, type of paint, number of coats and color.
(7) Is pretreatment required?
(8) Is paint properly mixed? Not excessively thinned?
(9) Is coverage satisfactory? Surface appearance and texture? Color?
(10) Is adequate ventilation provided in confined space? Lighting conditions satisfactory?
(11) Are weather conditions satisfactory for painting?

PIPE WEIGHTS

CAST IRON PIPE

SERVICE WEIGHT

Size, Inches	Weight of Pipe	Weight of Water	Total Weight— Lbs.
2"	3.8	1.45	5.3
3"	5.6	3.2	8.8
4"	7.5	5.5	13.0
5"	9.8	8.7	18.5
6"	12.4	12.5	24.9
8"	18.5	21.7	40.2

EXTRA HEAVY

Size, Inches	Weight of Pipe	Weight of Water	Total Weight — Lbs.
2"	4.3	1.45	5.8
3"	8.3	3.2	11.5
4"	10.8	5.5	16.3
5"	13.3	8.7	22.0
6"	16.0	12.5	28.5
8"	26.5	21.7	48.2

STEEL PIPE

Pipe Size	W/40	H_2O/Lbs.	Total Lbs./L.F.
2"	3.65	1.45	5.1
2 1/2"	5.79	2.07	7.86
3"	7.57	3.2	10.77
3 1/2"	9.11	4.28	13.39
4"	10.8	5.51	16.31
5"	14.6	8.66	23.26
6"	18.0	12.5	30.5
8"	28.6	21.66	50.26
10"	40.5	34.15	74.65

SPRINKLER AREA CALCULATIONS

Typical maximum floor area allowed per system riser:

Light Hazard	**Ordinary Hazard**	**Extra Hazard**
52,000 S.F.	40,000-52,000 S.F.	25,000 S.F.

Typical maximum floor area coverage allowed per sprinkler head:

Light Hazard	**Ordinary Hazard**	**Extra Hazard**
130-200 S.F.	100-130 S.F.	90 S.F.

Typical maximum spacing between lines and sprinkler heads:

Light Hazard	**Ordinary Hazard**	**Extra Hazard**
12-15 feet	12-15 feet	12 feet

Note: This data is for estimating purposes only. Check all applicable codes and regulations for specific requirements.

SPRINKLER HEAD CALCULATIONS
Typical maximum quantity of sprinkler heads allowed by pipe size.

Light Hazard:

For sprinklers below ceiling:

Steel		Copper	
1 in. pipe	2 sprinklers	1 in. tube	2 sprinklers
1 ¼ in. pipe	3 sprinklers	1 ¼ in. tube	3 sprinklers
1 ½ in. pipe	5 sprinklers	1 ½ in. tube	5 sprinklers
2 in. pipe	10 sprinklers	2 in. tube	12 sprinklers
2 ½ in. pipe	30 sprinklers	2 ½ in. tube	40 sprinklers
3 in. pipe	60 sprinklers	3 in. tube	65 sprinklers
3 ½ in. pipe	100 sprinklers	3 ½ in. tube	115 sprinklers

For sprinklers above and below ceiling:

Steel		Copper	
1 in.	2 sprinklers	1 in.	2 sprinklers
1 ¼ in.	4 sprinklers	1 ¼ in.	4 sprinklers
1 ½ in.	7 sprinklers	1 ½ in.	7 sprinklers
2 in.	15 sprinklers	2 in.	18 sprinklers
2 ½ in.	50 sprinklers	2 ½ in.	65 sprinklers

Ordinary Hazard:

For sprinklers above ceiling:

Steel		Copper	
1 in. pipe	2 sprinklers	1 in. tube	2 sprinklers
1 ¼ in. pipe	3 sprinklers	1 ¼ in. tube	3 sprinklers
1 ½ in. pipe	5 sprinklers	1 ½ in. tube	5 sprinklers
2 in. pipe	10 sprinklers	2 in. tube	12 sprinklers
2 ½ in. pipe	20 sprinklers	2 ½ in. tube	25 sprinklers
3 in. pipe	40 sprinklers	3 in. tube	45 sprinklers
3 ½ in. pipe	65 sprinklers	3 ½ in. tube	75 sprinklers
4 in. pipe	100 sprinklers	4 in. tube	115 sprinklers
5 in. pipe	160 sprinklers	5 in. tube	180 sprinklers
6 in. pipe	275 sprinklers	6 in. tube	300 sprinklers

For sprinklers above and below ceiling:

Steel		Copper	
1 in.	2 sprinklers	1 in.	2 sprinklers
1 ¼ in.	4 sprinklers	1 ¼ in.	4 sprinklers
1 ½ in.	7 sprinklers	1 ½ in.	7 sprinklers
2 in.	15 sprinklers	2 in.	18 sprinklers
2 ½ in.	30 sprinklers	2 ½ in.	40 sprinklers
3 in.	60 sprinklers	3 in.	65 sprinklers

Extra Hazard:

For sprinklers below ceiling:

Steel		Copper	
1 in. pipe	1 sprinkler	1 in. tube	1 sprinkler
1 ¼ in. pipe	2 sprinklers	1 ¼ in. tube	2 sprinklers
1 ½ in. pipe	5 sprinklers	1 ½ in. tube	5 sprinklers
2 in. pipe	8 sprinklers	2 in. tube	8 sprinklers
2 ½ in. pipe	15 sprinklers	2 ½ in. tube	20 sprinklers
3 in. pipe	27 sprinklers	3 in. tube	30 sprinklers
3 ½ in. pipe	40 sprinklers	3 ½ in. tube	45 sprinklers
4 in. pipe	55 sprinklers	4 in. tube	65 sprinklers
5 in. pipe	90 sprinklers	5 in. tube	100 sprinklers
6 in. pipe	150 sprinklers	6 in. tube	170 sprinklers

Note: This data is for estimating purposes only. Check all applicable codes and regulations for specific requirements.

SPRINKLER HAZARD OCCUPANCIES

Typical Light Hazard Occupancies:

Churches
Clubs
Eaves and overhangs, if combustible construction with
  no combustible beneath
Educational
Hospitals
Institutional
Libraries, except large stack rooms
Museums

Nursing or convalescent homes
Office, including data processing
Residential
Restaurant seating areas
Theaters seating areas
Theaters and auditoriums excluding stages and pro-
  sceniums
Unused attics

Typical Ordinary Hazard Occupancies (Group 1):

Automobile parking garages
Bakeries
Beverage manufacturing
Canneries
Dairy products manufacturing and processing

Electronic plants
Glass and glass products manufacturing
Laundries
Restaurant service areas

Typical Ordinary Hazard Occupancies (Group 2):

Cereal mills
Chemical plants - ordinary
Cold Storage warehouses
Confectionery products
Distilleries
Leather goods mfg.
Libraries-large stack room areas

Mercantiles
Machine shops
Metal working
Printing and publishing
Textile mfg.
Tobacco Products mfg.
Wood product assembly

Typical Ordinary Hazard Occupancies (Group 3):

Feed mills
Paper and pulp mills
Paper process plants
Piers and wharves
Repair garages
Tire manufacturing

Warehouses (having moderate to higher combustibil-
  ity of content, such as paper, household furni-
  ture, paint, general storage, whiskey, etc.)[1]
Wood machining

Typical Extra Hazard Occupancies (Group 1):

Combustible hydraulic fluid use areas
Die casting
Metal extruding
Plywood and particle board manufacturing
Printing (using inks with below 100°F [37.8°C] flash
  points)
Rubber reclaiming, compounding, drying, milling, vul-
  canizing

Saw mills
Textile picking, opening, blending, garnetting, card-
  ing, combining of cotton, synthetics, wool
  shoddy or burlap
Upholstering with plastic foams

Typical Extra Hazard Occupancies (Group 2):

Asphalt saturating
Flammable liquids spraying
Flow coating
Mobile Home or Modular Building assemblies (where
  finished enclosure is present and has combustible
  interiors)

Open Oil quenching
Solvent cleaning
Varnish and paint dipping

TYPICAL MINIMUM SIZE OF HORIZONTAL BUILDING STORM DRAINS AND BUILDING STORM SEWERS

Diameter of drain	Maximum projected area in square feet for various slopes		
	1/8 inch per feet slope	1/4 inch per feet slope	1/2 inch per feet slope
3	822	1160	1644
4	1880	2650	3760
5	3340	4720	6680
6	5350	7550	10700
8	11500	16300	23000
10	20700	29200	41400
12	33300	47000	66600
15	59500	84000	119000

PLUMBING BUDGETS

Budget estimates for plumbing can be determined as a percentage of total building costs depending on building type. For example:

Apartments	9 to 12 percent
Assembly	4 to 7 percent
Banks	3 to 6 percent
Dormitories	7 to 10 percent
Factories	4 to 8 percent
Hospitals	8 to 12 percent
Motels	9 to 12 percent
Office Buildings	4 to 7 percent
Retail (small)	4 to 7 percent
Retail (large)	3 to 6 percent
Schools	3 to 6 percent
Warehouses	3 to 7 percent

MECHANICAL / PLUMBING VENTS   15410

TYPICAL SIZE & LENGTH OF PLUMBING VENTS

Diameter of soil or waste stack (in.)	Total fixture units connected to stack (dfu)	DIAMETER OF VENT PIPE										
		1-1/4	1	2	2-1/2	3	4	5	6	8	10	12
1-1/4	2	30										
1-1/2	8	50	150									
1-1/2	10	30	100									
2	12	30	75	200								
2	20	26	50	150								
2-1/2	42		30	100	300							
3	10		42	150	360	1040						
3	21		32	110	270	810						
3	53		27	94	230	680						
3	102		25	86	210	620						
4	43			35	85	250	980					
4	140			27	65	200	750					
4	320			23	55	170	640					
4	540			21	50	150	580					
5	190				28	82	320	990				
5	490				21	63	250	760				
5	940				18	49	210	670				
5	1400				16	33	190	590				
6	500					26	130	400	1000			
6	1100					22	100	310	780			
6	2000					20	84	260	660			
6	2900						77	240	600			
8	1800						31	95	240	940		
8	3400						24	73	190	720		
8	5600						20	62	160	610		
8	7600						18	56	140	560		

TYPICAL SIZES OF FIXTURE WATER SUPPLY PIPES

Fixture	Nominal pipe size (inches)
Bath tubs	1/2
Combination sink and tray	1/2
Drinking fountain	3/8
Dishwasher (domestic)	1/2
Kitchen sink, residential	1/2
Kitchen sink, commercial	3/4
Lavatory	3/8
Laundry tray, 1, 2 or 3 compartments	1/2
Shower (single head)	1/2
Sinks (service, slop)	1/2
Sinks flushing rim	3/4
Urinal (flash tank)	3/8
Urinal (direct flush valve)	1
Water closet (tank type)	3/8
Water closet (flush valve type)	1
Hose bibs	1/2
Wall hydrant	1/2

TYPICAL VENTILATION AIR REQUIREMENTS
FOR SPECIAL USES

Occupancy Classification	Required ventilation air in cfm per human occupant
Special areas	
Lockers	2 *
	(or 30 per locker)
Wardrobes	2 *
Public bathrooms	40 **
Private bathrooms	25 **
Swimming pools	15
	(per occupant)
Water closet (flush valve type)	1
Exitways and corridors	1 1/2 *

*Per square foot floor area.
**Per water closet or urinal.

TYPICAL VENTILATION AIR REQUIREMENTS
FOR RETAIL USES

Occupancy Classification	Required ventilation air in cfm per human occupant
Mercantile	
Sales floors and showrooms (basement & grade floors)	7
Sales and showrooms (upper floors)	7
Storage areas	5
Dressing rooms	7
Malls	7
Shipping areas	15
Elevators	7
Supermarkets	
Meat processing rooms	5
Drugs stores	
Pharmacists' work rooms	20
Specialty shops	
Pet shops	1.0*
Florists	5
Greenhouses	5

*cfm per sq. ft. floor area.

TYPICAL VENTILATION AIR REQUIREMENTS
FOR RESIDENTIAL USES

Occupancy Classification	Required ventilation air in cfm per human occupant
Residential	
General living areas	5
Bedrooms	5
Kitchens	20
Basements, utility rooms	5
Mobile homes	5
Hotels, motels	
Bedrooms (single, double)	7
Living rooms (suites)	10
Corridors	5
Lobbies	7
Conference rooms (small)	20
Assembly rooms (large)	15

TYPICAL VENTILATION AIR REQUIREMENTS
FOR STORAGE USES

Occupancy Classification	Required ventilation air in cfm per human occupant
Storage	
Garages, service stations, parking garages (enclosed)	1.5*
Auto repair shops	1.5**
Warehouses	
General	7

*cfm/s.f. floor area
**Must have positive engine exhaust system.

TYPICAL VENTILATION AIR REQUIREMENTS
FOR FACTORY AND INDUSTRIAL USES

Occupany Classification	Required ventilation air in cfm per human occupant
Factory and industrial	
Metalworking & finishing	35
Automotive engine test	
Paint spray booths	
Picking, etching & plating lines	*Require Special Exhaust Systems*
Degreasing booths	
Sandblasting booths	
Chemicals and pharmaceuticals	
Dusty operations	30
Rooms containing potential gas emitters	20
Drying oven rooms	15
Fermentation rooms	15
Pillmaking booths	10
Packaging areas	10
Utility rooms	7
Computer rooms	7
Textiles-clothes manufacturer	15
Electronics & aerospace circuit board & soldering rooms	20
Wood products, papermaking	20
Brewing, distilling, wineries, bottling	20*
Food processing	20
Tobacco processing	20
Power plants	
Control rooms	10
Boiler rooms	35
Generator rooms	20
Sewage treatment plants	
Control rooms	10
Compressor/blower motor rooms	20
Glass & ceramic manufacturer	20
Agricultural	20

TYPICAL VENTILATION AIR REQUIREMENTS
FOR BUSINESS USES

Occupany Classification	Required ventilation air in cfm per human occupant
Business	
Banks	
(see offices)	
Vaults	5
Barber, beauty and health services	
Beauty shops (hair dressers)	25
Reducing salons	25
Sauna baths, steam rooms	5
Barber shops	7
Photo studios	
Camera rooms, stages	5
Dark rooms	
Shoe repair shops	
Workrooms/trade areas	10
Offices	
General office space and showrooms	15
Conference rooms	25
Drafting/art rooms	7
Doctor's consultation rooms	10
Waiting rooms	10
Lithographing rooms	7
Diazo printing rooms	7
Computer rooms	5
Keypunch rooms	7
Communication	
TV/radio broadcasting booths, studios	30
Motion picture and TV stages	30
Pressrooms	15
Composing rooms	7
Engraving rooms	7
Telephone switchboard rooms (manual)	7
Telephone switchgear rooms (automatic)	7
Teletypewriter/facsimile rooms	5
Research institutes	
Laboratories:	
Light duty; non-chemical	15
Chemical	15
Heavy-duty	15
Radioisotope, chemical & biologically toxic	15
Machine shops	15
Dark rooms, spectroscopy rooms	10
Animal rooms	40
Veterinary hospitals	
Kennels, stalls	25
Operating rooms	25
Reception rooms	10

TYPICAL VENTILATION AIR REQUIREMENTS
FOR INSTITUTIONAL USES

Occupancy Classification	Required ventilation air in cfm per human occupant
Institutional	
Prisons	
Cell blocks	7
Eating halls..	15
Guard stations	7

AIR CONDITIONING
RECOMMENDED SHEET METAL GAUGES AND CONSTRUCTION FOR RECTANGULAR DUCT

LOW PRESSURE — LOW VELOCITY = 2" W.G. MAX

PLATE NO.	DIMENSION OF LONGEST SIDE OF DUCT	Steel Metal Gauges		AT JOINTS					Reinforcing Between Joints
		Steel	Aluminum	Plain "S" Slip (B) / Pocket Lock (K) / Drive Slip (A)	Hemmed "S" Slip (C) / Bar Slip (E) / Standing Seam (1)	Reinforced Bar Slip (G)	Angle Slip (H) / Alternate Bar Slip (F) / Angle RFD Pocket (L)	Companion Angles (M) / Angle Reinforced Standing Seam (J)	
6	Thru 12"	26	24 (.020)	A-B-K	———	———	———	———	
6	13" thru 18"	24	22 (.025)	A-B-K	———	———	———	———	
7 7A	19" thru 30"	24	22 (.025)	K @ 5' cc A	C-E- @ 5' cc C-E- @ 10' cc	———	———	———	1" x 1" x 1/8" @ 5' cc
8	31" thru 42"	22	20 (.032)	K @ 5' cc	E-G-K @ 5' cc E-G-K @ 10' cc	———	———	———	1" x 1" x 1/8" @ 5' cc
9	43" thru 54"	22	20 (.032)	K @ 4' cc K @ 8' cc	E-@ 4' cc E-@ 8' cc	G- @ 4' cc G- @ 8' cc	———	———	1½" x 1½" x 1/8" @ 4' cc
9	55" thru 60"	20	18 (.040)	K @ 4' cc K @ 8' cc	E-@ 4' cc E-@ 8' cc	G- @ 4' cc G- @ 8' cc	———	———	1½" x 1½" x 1/8" @ 4' cc
10	61" thru 84"	20	18 (.040)	———	———	G- @ 4' cc G- @ 5' cc	H- @ 4' cc F- @ 4' cc L- @ 4' cc H- @ 5' cc F- @ 5' cc L- @ 5' cc	J- @ 2' cc	1½" x 1½" x 1/8" @ 2' cc 1½" x 1½" x 1/8" @ 2'- 6" cc
11	85" thru 96"	18	16 (.051)	———	———	———	H- @ 4' cc L- @ 4' cc H- @ 5' cc L- @ 5' cc	M- @ 4' cc M- @ 5' cc J- @ 2' cc	1½" x 1½" x 3/16" @ 2' cc 1½" x 1½" x 3/16" @ 2'- 6" cc 1½" x 1½" x 3/16" @ 2' cc
12	Over 96"	18	16 (.051)	———	———	———	H- @ 4' cc L- @ 4' cc H- @ 5' cc L- @ 5' cc	M- @ 4' cc M- @ 5' cc J- @ 2' cc	2" x 2" x ¼ @ 2' cc 2" x 2" x ¼ @ 2'- 6" cc 2" x 2" x ¼ @ 2' cc

H (height dimension) —— up to 42" = 1"
H (height dimension) —— 43" to 96" = 1½"
H (height dimension) —— over 96" = 2"

AIR CONDITIONING
TYPICAL DUCT CONNECTIONS
CROSS JOINTS FOR SHEET METAL DUCTWORK
(NOT TO SCALE)

H - HEIGHT REFERRED TO IN DIMESIONS

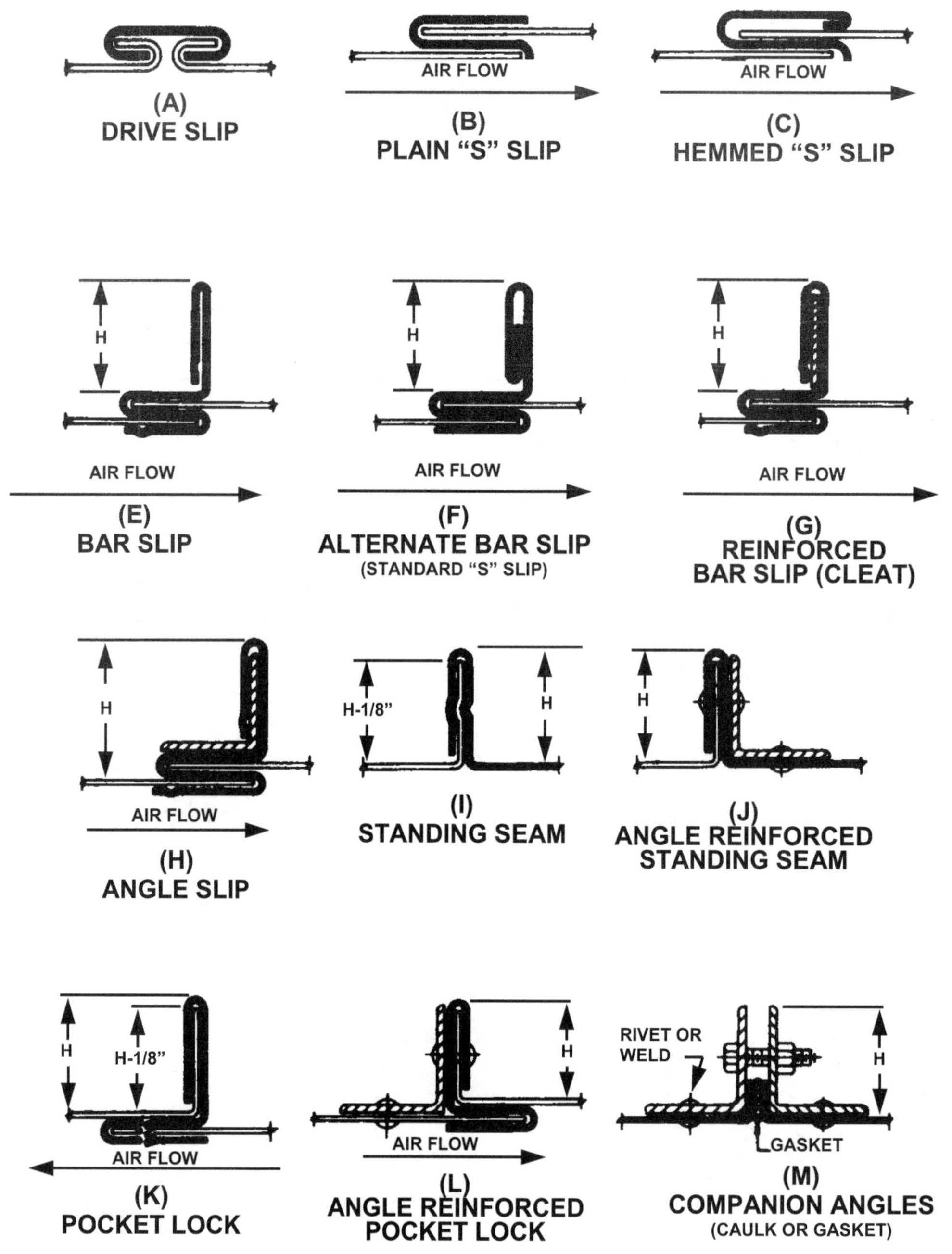

AIR CONDITIONING (Cont.)
LONGITUDINAL SEAMS
FOR SHEET METAL DUCTWORK

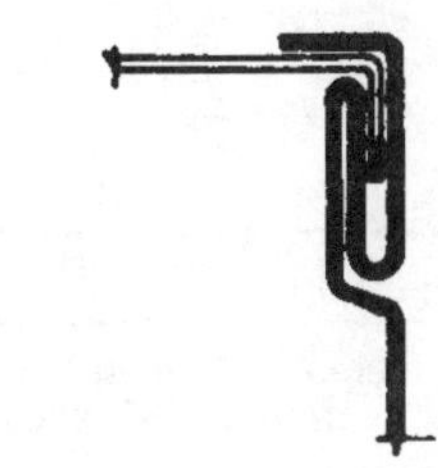

Fig. "N"
PITTSBURGH LOCK

Fig. "Z"
BUTTON PUNCH SNAP LOCK

Fig. "O"
ACME LOCK-GROOVED SEAM

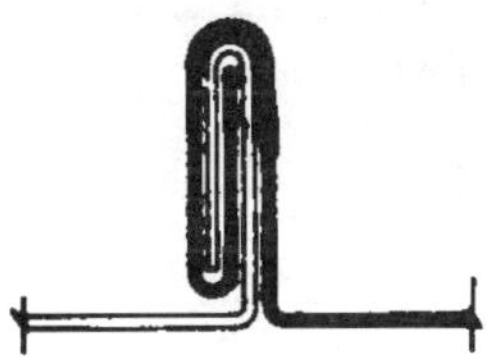

Fig. "T"
DOUBLE SEAM

Approximately 2" Spacing
Between "Buttons"

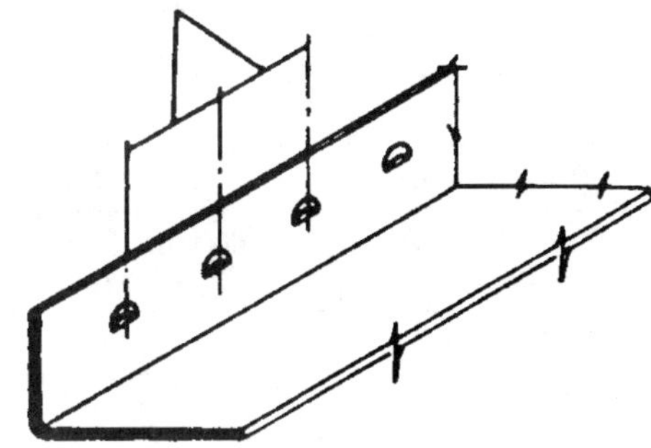

DETAIL NO.1
MALE PIECE-SNAP LOCK

TYPICAL DUCT CONSTRUCTION SHEET METAL GAGES IN ONE– AND TWO– FAMILY DWELLINGS

Metal Gauges (duct not enclosed in partitions)		
ROUND DUCTS		
Diameter, Inches	**Minimum thickness galvanized sheet gage**	**Minimum thickness aluminum B&S gage**
Less than 12	30	26
12-14..	28	26
15-18..	26	24
Over 18	24	22
RECTANGULAR DUCTS		
Width, Inches	**Minimum thickness galvanized sheet gage**	**Minimum thickness aluminum B&S gage**
Less than 14	28	24
14-24..	26	22
25-30..	24	22
Over 30	22	20
Metal Gauges (duct enclosed in partitions)		
Width, Inches	**Minimum thickness galvanized sheet gage**	**Minimum thickness aluminum B&S gage**
14 or less	30	26
Over 14	28	24

TYPICAL DUCT CONSTRUCTION SHEET METAL GLASS
(All uses except 1– and 2– family dwellings)

RECTANGULAR DUCTS		
Maximum side inches	**Steel min. Galv. Sheet Gage**	**Aluminum Min. B&S Gage**
Through 12	26 (0.022 in.)	24 (0.020 in.)
13 through 30	24 (0.028 in.)	22 (0.025 in.)
31 through 54	22 (0.034 in.)	20 (0.032 in.)
55 through 84	20 (0.040 in.)	18 (0.040 in.)
Over 84	18 (0.052 in.)	16 (0.051 in.)

ROUND DUCTS			
Diameter Inches	**Spiral seam duct**	**Longitudinal seam duct**	**Fittings**
	Steel min. Galv. Sht. Gage	**Steel min. Galv. Sht. Gage**	**Steel min. Galv. Sht. Gage**
Through 12	28 (0.019 in.)	26 (0.022 in.)	26 (0.022 in.)
13 through 18	26 (0.022 in.)	24 (0.028 in.)	24 (0.028 in.)
19 through 28	24 (0.028 in.)	22 (0.034 in.)	22 (0.034 in.)
29 through 36	22 (0.034 in.)	20 (0.040 in.)	20 (0.040 in.)
37 through 52	20 (0.040 in.)	18 (0.052 in.)	18 (0.052 in.)

APPLICATIONS FOR CONDUCTORS
USED FOR GENERAL WIRING

	AMBIENT TEMPERATURE								
	60°C 140°F	75°C 167°F	85°C 185°F	90°C 194°F	110°C 230°F	200°C 392°F	Dry	Dry or Wet	FEATURES
R	X						X		Code Rubber
RH		X					X		Heat Resistant
RHH				X			X		More Heat Resistant
RW	X							X	Moisture Resistant
RH-RW	X							X	Moisture and Heat Resistant
		X					X		Moisture and Heat Resistant
RHW		X						X	Moisture and Heat Resistant
RU	X						X		Latex Rubber
RUH		X					X		Heat Resistant
RUW	X							X	Moisture Resistant
T	X						X		Thermoplastic
TW	X							X	Moisture Resistant
THHN				X			X		Heat Resistant
THW		X						X	Moisture and Heat Resistant
THWN		X						X	Moisture and Heat Resistant
MI			X					X	Mineral Insulated Metal Sheathed
V			X				X		Varnished Cambric
AVA					X		X		With Asbestos
AVB				X			X		With Asbestos
AVL					X			X	With Asbestos

This table does not include special condition conductors, thickness of conductor insulation, or reference to all other protective coverings.

GENERAL CLASSIFICATION OF INSULATIONS:

A .... Asbestos
H .... Heat Resistant
MI .... Mineral Insulation
R .... Rubber

RU .... Latex Rubber
V .... Varnished Cambric
T .... Thermoplastic
W .... (Water) Moisture Resistant

WIRE AND SHEET METAL GAGES
(In Decimals of an Inch)

Name of Gage	American Wire Gage (A.W.G.) (Corresponds to Brown & Sharpe Gage)	Birmingham Iron Wire Gage (B.W.G.)	United States Standard Gage (U.S.S.G.)	
Principal Use	Electrical Wire & Non-Ferrous Sheet Metal	Iron or Steel Wire	Ferrous Sheet Metal	
Gage No.				Gage No.
00 00000				00 00000
0 00000	.5800			0 00000
00000	.5165	.500		00000
0000	.4600	.454		0000
000	.4096	.425		000
00	.3648	.380		00
0	.3249	.340		0
1	.2893	.300		1
2	.2576	.284		2
3	.2294	.259	23.91	3
4	.2048	.238	.2242	4
5	.1819	.220	.2092	5
6	.1620	.203	.1943	6
7	.1443	.180	.1793	7
8	.1285	.165	.1644	8
9	.1144	.148	.1495	9
10	.1019	.134	.1345	10
11	.0907	.120	.1196	11
12	.0808	.109	.1046	12
13	.0720	.095	.0897	13
14	.0641	.083	.0747	14
15	.0571	.072	.0673	15
16	.0508	.065	.0598	16
17	.0453	.058	.0538	17
18	.0403	.049	.0478	18
19	.0359	.042	.0418	19
20	.0320	.035	.0359	20
21	.0285	.032	.0329	21
22	.0253	.028	.0299	22
23	.0226	.025	.0269	23
24	.0201	.022	.0239	24
25	.0179	.020	.0209	25
26	.0159	.018	.0179	26
27	.0142	.016	.0164	27
28	.0126	.014	.0149	28
29	.0113	.013	.0135	29
30	.0100	.012	.0120	30
31	.0089	.010	.0105	31
32	.0080	.009	.0097	32
33	.0071	.008	.0090	33
34	.0063	.007	.0082	34
35	.0056	.005	.0075	35
36	.0050	.004	.0067	36
37	.0045		.0064	37
38	.0040		.0060	38
39	.0035			39
40	.0031			40

Geographic Cost Modifiers

The costs as presented in this book attempt to represent national averages. Costs, however, vary among regions, states and even between adjacent localities.

In order to more closely approximate the probable costs for specific locations throughout the U.S., this table of Geographic Cost Modifiers is provided. These adjustment factors are used to modify costs obtained from this book to help account for regional variations of construction costs and to provide a more accurate estimate for specific areas. The factors are formulated by comparing costs in a specific area to the costs as presented in the Costbook pages. An example of how to use these factors is shown below. Whenever local current costs are known, whether material prices or labor rates, they should be used when more accuracy is required.

Cost Obtained from Costbook Pages **X** Location Cost Adjustment Factor Divided by 100 **=** **Adjusted Cost**

For example, a project estimated to cost $125,000 using the Costbook pages can be adjusted to more closely approximate the cost in Los Angeles:

$$\$125,000 \text{ X } \frac{105}{100} = \$131,250$$

State	Metropolitan Areas	Multiplier
AK	ANCHORAGE	122
AL	ANNISTON	80
	AUBURN-OPELIKA	80
	BIRMINGHAM	79
	DOTHAN	77
	GADSDEN	77
	HUNTSVILLE	79
	MOBILE	82
	MONTGOMERY	77
	TUSCALOOSA	79
AR	FAYETTEVILLE-SPRINGDALE-ROGERS	72
	FORT SMITH	77
	JONESBORO	76
	LITTLE ROCK-NORTH LITTLE ROCK	78
	PINE BLUFF	77
	TEXARKANA	77
AZ	FLAGSTAFF	88
	PHOENIX-MESA	88
	TUCSON	87
	YUMA	89
CA	BAKERSFIELD	100
	FRESNO	102
	LOS ANGELES-LONG BEACH	105
	MODESTO	99
	OAKLAND	108
	ORANGE COUNTY	102
	REDDING	99
	RIVERSIDE-SAN BERNARDINO	100
	SACRAMENTO	102
	SALINAS	104
	SAN DIEGO	101
	SAN FRANCISCO	111
	SAN JOSE	109
	SAN LUIS OBISPO	98
	SANTA CRUZ-WATSONVILLE	104
	SANTA ROSA	105
	STOCKTON-LODI	101
	VALLEJO-FAIRFIELD-NAPA	104
	VENTURA	100
	SANTA BARBARA	103

GEOGRAPHIC COST MODIFERS

State	Metropolitan Areas	Multiplier
CO	BOULDER-LONGMONT	86
	COLORADO SPRINGS	91
	DENVER	91
	FORT COLLINS-LOVELAND	84
	GRAND JUNCTION	86
	GREELEY	84
	PUEBLO	88
CT	BRIDGEPORT	100
	DANBURY	100
	HARTFORD	99
	NEW HAVEN-MERIDEN	100
	NEW LONDON-NORWICH	98
	STAMFORD-NORWALK	103
	WATERBURY	99
DC	WASHINGTON	94
DE	DOVER	93
	WILMINGTON-NEWARK	94
FL	DAYTONA BEACH	83
	FORT LAUDERDALE	86
	FORT MYERS-CAPE CORAL	79
	FORT PIERCE-PORT ST. LUCIE	85
	FORT WALTON BEACH	87
	GAINESVILLE	81
	JACKSONVILLE	84
	LAKELAND-WINTER HAVEN	81
	MELBOURNE-TITUSVILLE-PALM BAY	88
	MIAMI	86
	NAPLES	88
	OCALA	85
	ORLANDO	85
	PANAMA CITY	75
	PENSACOLA	79
	SARASOTA-BRADENTON	80
	TALLAHASSEE	77
	TAMPA-ST. PETERSBURG-CLEARWATER	83
	WEST PALM BEACH-BOCA RATON	86
GA	ALBANY	76
	ATHENS	78
	ATLANTA	85
	AUGUSTA	74
	COLUMBUS	75
	MACON	78
	SAVANNAH	78

State	Metropolitan Areas	Multiplier
HI	HONOLULU	121
IA	CEDAR RAPIDS	87
	DAVENPORT	89
	DES MOINES	90
	DUBUQUE	85
	IOWA CITY	89
	SIOUX CITY	85
	WATERLOO-CEDAR FALLS	84
ID	BOISE CITY	88
	POCATELLO	87
IL	BLOOMINGTON-NORMAL	95
	CHAMPAIGN-URBANA	94
	CHICAGO	103
	DECATUR	93
	KANKAKEE	96
	PEORIA-PEKIN	95
	ROCKFORD	95
	SPRINGFIELD	93
IN	BLOOMINGTON	91
	EVANSVILLE	89
	FORT WAYNE	90
	GARY	97
	INDIANAPOLIS	93
	KOKOMO	89
	LAFAYETTE	90
	MUNCIE	90
	SOUTH BEND	90
	TERRE HAUTE	90
KS	KANSAS CITY	88
	LAWRENCE	84
	TOPEKA	83
	WICHITA	83
KY	LEXINGTON	84
	LOUISVILLE	86
	OWENSBORO	84
LA	ALEXANDRIA	79
	BATON ROUGE	82
	HOUMA	82
	LAFAYETTE	81
	LAKE CHARLES	82
	MONROE	79
	NEW ORLEANS	85
	SHREVEPORT-BOSSIER CITY	80

GEOGRAPHIC COST MODIFERS

01025

State	Metropolitan Areas	Multiplier
MA	BARNSTABLE-YARMOUTH	103
	BOSTON	105
	BROCKTON	102
	FITCHBURG-LEOMINSTER	99
	LAWRENCE	99
	LOWELL	98
	NEW BEDFORD	102
	PITTSFIELD	98
	SPRINGFIELD	98
	WORCESTER	98
MD	BALTIMORE	90
	CUMBERLAND	86
	HAGERSTOWN	87
ME	BANGOR	85
	LEWISTON-AUBURN	86
	PORTLAND	88
MI	ANN ARBOR	95
	DETROIT	96
	FLINT	92
	GRAND RAPIDS-MUSKEGON-HOLLAND	89
	JACKSON	91
	KALAMAZOO-BATTLE CREEK	86
	LANSING-EAST LANSING	91
	SAGINAW-BAY CITY-MIDLAND	90
MN	DULUTH	95
	MINNEAPOLIS-ST. PAUL	100
	ROCHESTER	95
	ST. CLOUD	97
MO	COLUMBIA	89
	JOPLIN	85
	KANSAS CITY	91
	SPRINGFIELD	86
	ST. JOSEPH	89
	ST. LOUIS	89
MS	BILOXI-GULFPORT-PASCAGOULA	78
	JACKSON	77
MT	BILLINGS	88
	GREAT FALLS	88
	MISSOULA	86
NC	ASHEVILLE	73
	CHARLOTTE	75

State	Metropolitan Areas	Multiplier
NC	FAYETTEVILLE	75
	GREENSBORO-WINSTON-SALEM-HIGH POINT	74
	GREENVILLE	74
	HICKORY-MORGANTON-LENOIR	71
	RALEIGH-DURHAM-CHAPEL HILL	74
	ROCKY MOUNT	71
	WILMINGTON	75
ND	BISMARCK	86
	FARGO	89
	GRAND FORKS	87
NE	LINCOLN	82
	OMAHA	86
NH	MANCHESTER	90
	NASHUA	88
	PORTSMOUTH	86
NJ	ATLANTIC-CAPE MAY	99
	BERGEN-PASSAIC	101
	JERSEY CITY	100
	MIDDLESEX-SOMERSET-HUNTERDON	97
	MONMOUTH-OCEAN	99
	NEWARK	102
	TRENTON	99
	VINELAND-MILLVILLE-BRIDGETON	98
NM	ALBUQUERQUE	86
	LAS CRUCES	83
	SANTA FE	90
NV	LAS VEGAS	97
	RENO	96
NY	ALBANY-SCHENECTADY-TROY	91
	BINGHAMTON	90
	BUFFALO-NIAGARA FALLS	97
	ELMIRA	85
	GLENS FALLS	84
	JAMESTOWN	92
	NASSAU-SUFFOLK	103
	NEW YORK	118
	ROCHESTER	95
	SYRACUSE	92
	UTICA-ROME	92

State	Metropolitan Areas	Multiplier
OH	AKRON	92
	CANTON-MASSILLON	90
	CINCINNATI	90
	CLEVELAND-LORAIN-ELYRIA	93
	COLUMBUS	90
	DAYTON-SPRINGFIELD	88
	LIMA	90
	MANSFIELD	87
	STEUBENVILLE	93
	TOLEDO	92
	YOUNGSTOWN-WARREN	87
OK	ENID	79
	LAWTON	80
	OKLAHOMA CITY	81
	TULSA	80
OR	EUGENE-SPRINGFIELD	93
	MEDFORD-ASHLAND	91
	PORTLAND	95
	SALEM	94
PA	ALLENTOWN-BETHLEHEM-EASTON	94
	ALTOONA	92
	ERIE	93
	HARRISBURG-LEBANON-CARLISLE	90
	JOHNSTOWN	93
	LANCASTER	91
	PHILADELPHIA	102
	PITTSBURGH	93
	READING	93
	SCRANTON-WILKES-BARRE-HAZLETON	91
	STATE COLLEGE	88
	WILLIAMSPORT	89
	YORK	92
RI	PROVIDENCE	98
SC	AIKEN	81
	CHARLESTON-NORTH CHARLESTON	76
	COLUMBIA	76
	FLORENCE	73
	GREENVILLE-SPARTANBURG-ANDERSON	76
	MYRTLE BEACH	81
SD	RAPID CITY	80
	SIOUX FALLS	80

State	Metropolitan Areas	Multiplier
TN	CHATTANOOGA	79
	JACKSON	78
	JOHNSON CITY	75
	KNOXVILLE	79
	MEMPHIS	83
	NASHVILLE	82
TX	ABILENE	77
	AMARILLO	78
	AUSTIN-SAN MARCOS	78
	BEAUMONT-PORT ARTHUR	79
	BROWNSVILLE-HARLINGEN-SAN BENITO	80
	BRYAN-COLLEGE STATION	78
	CORPUS CHRISTI	76
	DALLAS	83
	EL PASO	76
	FORT WORTH-ARLINGTON	82
	GALVESTON-TEXAS CITY	81
	HOUSTON	84
	LAREDO	70
	LONGVIEW-MARSHALL	75
	LUBBOCK	78
	MCALLEN-EDINBURG-MISSION	75
	ODESSA-MIDLAND	76
	SAN ANGELO	75
	SAN ANTONIO	80
	TEXARKANA	77
	TYLER	77
	VICTORIA	77
	WACO	77
	WICHITA FALLS	77
UT	PROVO-OREM	84
	SALT LAKE CITY-OGDEN	83
VA	CHARLOTTESVILLE	79
	LYNCHBURG	79
	NORFOLK-VIRGINIA BEACH-NEWPORT NEWS	81
	RICHMOND-PETERSBURG	83
	ROANOKE	76
VT	BURLINGTON	91
WA	BELLINGHAM	100
	BREMERTON	98
	OLYMPIA	97

GEOGRAPHIC COST MODIFERS

State	Metropolitan Areas	Multiplier
WA	RICHLAND-KENNEWICK-PASCO	95
	SEATTLE-BELLEVUE-EVERETT	102
	SPOKANE	97
	TACOMA	99
	YAKIMA	95
WI	APPLETON-OSHKOSH-NEENAH	90
	EAU CLAIRE	90
	GREEN BAY	91
	JANESVILLE-BELOIT	90
	KENOSHA	92
	LA CROSSE	89
	MADISON	89
	MILWAUKEE-WAUKESHA	96
	RACINE	92
	WAUSAU	90
WV	CHARLESTON	89
	HUNTINGTON	90
	PARKERSBURG	87
	WHEELING	89
WY	CASPER	87
	CHEYENNE	89

Square Foot Tables

The following Square Foot Tables list hundreds of actual projects for dozens of building types, each with associated building size, total square foot building cost and percentage of project costs for total mechanical and electrical components. This data provides an overview of construction costs by building type. These costs are for actual projects. The variations within similar building types may be due, among other factors, to size, location, quality and specified components, materials and processes. Depending upon all such factors, specific building costs can vary significantly and may not necessarily fall within the range of costs as presented. The data has been updated to reflect current construction costs.

SQUARE FOOT TABLES

COMMERCIAL

AUTO DEALERSHIP

Project Size Gross S.F.	Project Cost $/S.F.	% Cost Mechanical	% Cost Electrical
7,700	142.90	16.7	9.4
16,100	81.10	10.2	15.9
20,000	85.20	12.9	23.4
26,300	86.80	12.5	22.0
43,600	68.80	19.4	13.2
53,600	114.60	12.5	11.5

BUSINESS CENTER

3,900	85.60	12.0	9.2
9,900	82.80	9.1	7.6
54,400	55.40	3.4	12.2
135,000	60.20	8.2	1.5

CINEMA

18,000	195.70	10.9	6.7
22,500 A	120.40	6.6	4.2

MALL/PLAZA

9,700	60.60	15.0	13.3
10,500	96.80	8.0	13.5
16,300	88.20	9.4	10.6
26,900	87.90	15.0	7.0
36,000	74.40	10.0	11.0
36,300	84.50	12.4	8.2
44,720	88.20	18.3	12.4
59,100	86.40	9.8	9.5
60,000 R	88.60	10.0	9.5
64,100	89.10	22.4	18.4
66,000	102.00	13.5	11.5
67,400	74.20	21.0	15.0
73,500	107.20	14.9	6.9

MALL/PLAZA (Cont.)

Project Size Gross S.F.	Project Cost $/S.F.	% Cost Mechanical	% Cost Electrical
142,000	67.40	7.1	8.0
220,000	178.00	11.0	6.4
223,700	52.30	9.0	9.3
321,200	62.20	7.3	7.4
379,900	81.20	11.2	6.2
405,100	83.70	13.9	6.0
482,000	139.40	10.5	9.4
630,000	93.20	12.2	12.4

RESTAURANT

4,300 R	153.90	6.9	8.7
4,400 R	226.50	14.6	8.5
5,800	184.10	28.0	10.6
6,800 A	215.20	7.0	11.1
7,360 R	231.80	16.0	6.5
9,600	230.10	24.7	13.1
10,000 R	239.40	21.0	10.0
10,100	206.50	28.6	18.4
10,600	380.40	20.4	6.4
22,900 R	247.20	15.8	16.9

RETAIL STORE

1,000	231.80	12.8	6.7
3,000 R	207.40	14.3	10.5
12,300	284.30	14.0	10.0
30,000	143.70	15.6	26.2
61,300	74.40	13.3	13.0
115,000	102.80	14.6	11.3
154,700	145.70	11.2	12.4
314,700 R	122.90	13.8	9.4

A = Addition   R = Remodel

SQUARE FOOT TABLES

RESIDENTIAL

APARTMENTS

Project Size Gross S.F.	Project Cost $/S.F.	% Cost Mechanical	% Cost Electrical
3,700	86.10	14.9	4.4
13,900	117.90	7.9	7.2
19,200 R	187.30	45.3	7.5
19,700	109.70	7.4	10.4
23,700	120.50	10.6	4.3
26,500	114.10	25.2	12.8
35,100	93.30	16.4	5.6
54,000	161.50	23.3	13.1
62,700	117.40	17.0	9.0
67,300	107.20	13.8	8.4
70,600	60.40	18.1	7.8
75,300	133.20	13.2	8.1
75,600	129.60	14.5	8.9
72,200	112.30	18.4	10.9
77,600	145.30	26.9	14.3
88,100	134.90	15.3	9.3
89,500	126.50	10.7	11.1
94,100	76.20	8.5	6.8
96,000	95.20	17.0	13.3
102,000	165.10	17.8	12.5
103,200	75.10	12.1	8.9
103,600	120.90	19.1	9.0
105,200	153.90	14.9	8.9
106,200	110.80	12.8	9.3
110,900	106.50	15.6	8.4
111,800	151.20	17.3	7.4
115,900	113.40	12.5	8.6
117,200	70.90	12.9	8.0
119,000	97.30	15.6	8.4

APARTMENTS (Cont.)

Project Size Gross S.F.	Project Cost $/S.F.	% Cost Mechanical	% Cost Electrical
119,400	63.80	18.6	8.7
144,300	141.70	15.0	8.6
176,300	120.50	19.1	9.0
192,300	70.90	10.1	6.0
210,900	139.50	19.1	9.5
220,200	143.60	14.8	7.5
253,900	186.50	20.6	7.7
369,500	136.30	15.8	9.0

CONDOS/TOWNHOUSE

Project Size Gross S.F.	Project Cost $/S.F.	% Cost Mechanical	% Cost Electrical
8,600	122.30	8.5	4.4
16,700	98.60	13.2	6.5
18,000	207.40	14.6	9.7
18,400	106.40	12.9	5.8
74,800	96.50	13.8	5.2
111,700	121.70	9.2	7.1
150,300	136.00	15.9	7.9
278,800	212.50	14.1	7.9
1,109,900	97.90	9.8	4.7

SINGLE-FAMILY HOMES

Project Size Gross S.F.	Project Cost $/S.F.	% Cost Mechanical	% Cost Electrical
600 R	88.10	16.3	2.0
900	114.40	33.0	3.0
2,100	132.40	8.8	4.0
2,200	283.80	23.8	3.4
2,500	139.10	6.1	7.1
2,900	134.20	7.7	2.8
3,000	94.20	9.5	6.8
3,100	210.70	15.5	4.6
3,600	131.90	8.5	3.0
3,700	184.50	9.8	3.4

A = Addition   R = Remodel

SQUARE FOOT TABLES

RESIDENTIAL (Cont.)

SINGLE-FAMILY HOMES (Cont.)

Project Size Gross S.F.	Project Cost $/S.F.	% Cost Mechanical	% Cost Electrical
4,200	160.40	15.5	5.7
4,600	254.70	9.0	4.7
5,200	331.00	8.3	6.0
5,700	193.90	7.4	3.7
5,700	131.80	8.7	12.6
21,300*	94.00	26.0	4.0
22,700*	87.30	27.0	4.6
45,000*	132.90	7.8	2.5
51,458*	96.30	11.0	5.0

*TOWNHOUSES

EDUCATIONAL

ADMINISTRATION (OFFICES)

Project Size Gross S.F.	Project Cost $/S.F.	% Cost Mechanical	% Cost Electrical
53,700	226.50	15.4	9.0

ATHLETIC FACILITY

Project Size Gross S.F.	Project Cost $/S.F.	% Cost Mechanical	% Cost Electrical
38,100	224.70	16.8	9.8
44,100	207.40	16.8	8.4
100,000	141.70	19.2	5.3
160,000	256.80	13.7	7.9
247,500	215.40	11.2	9.0
271,000	202.40	13.2	7.6
283,100	243.80	14.6	6.0

AUDITORIUM/PERFORMING ARTS

Project Size Gross S.F.	Project Cost $/S.F.	% Cost Mechanical	% Cost Electrical
9,900	363.10	17.5	29.2
17,800	332.00	11.4	16.1
29,200	328.50	16.2	10.5
62,700	238.90	13.0	12.9

CLASSROOM

Project Size Gross S.F.	Project Cost $/S.F.	% Cost Mechanical	% Cost Electrical
35,400	329.40	10.2	6.8
70,000	148.70	24.3	13.1
78,900	276.70	13.3	14.3
80,100	224.70	16.9	10.5
100,000	223.90	21.1	9.8
166,000	154.70	16.1	11.2
298,400	145.10	11.3	11.9

COMPLETE COLLEGE FACILITIES

Project Size Gross S.F.	Project Cost $/S.F.	% Cost Mechanical	% Cost Electrical
95,300	233.40	19.4	11.7
450,000	278.50	18.6	10.8

ELEMENTARY SCHOOL

Project Size Gross S.F.	Project Cost $/S.F.	% Cost Mechanical	% Cost Electrical
18,000	170.80	19.1	13.8
30,800	157.10	16.0	10.0
31,600	131.40	12.1	11.3
35,700	192.00	17.3	8.7
40,000	166.90	18.5	13.4
40,500	146.30	22.1	11.3
57,000	126.40	22.1	7.5
69,700	197.10	20.6	9.3
91,400	158.90	18.2	7.6

HIGH SCHOOL

Project Size Gross S.F.	Project Cost $/S.F.	% Cost Mechanical	% Cost Electrical
116,400	174.50	18.1	12.9
133,000	150.30	17.8	10.7
184,000	264.70	23.0	10.0
217,200 R	142.60	26.4	10.6
254,000 R	110.00	17.2	14.6
431,700	167.70	13.1	9.6

A = Addition   R = Remodel

SQUARE FOOT TABLES

EDUCATIONAL (Cont.)

JUNIOR HIGH SCHOOL

Project Size Gross S.F.	Project Cost $/S.F.	% Cost Mechanical	% Cost Electrical
26,000	219.80	9.5	9.3
28,100	136.30	11.9	9.5
52,800	176.30	18.3	8.6
91,600	216.80	21.2	11.0
123,700	177.40	29.1	9.1

LABORATORY/RESEARCH

Project Size Gross S.F.	Project Cost $/S.F.	% Cost Mechanical	% Cost Electrical
9,200	380.40	25.4	4.5
80,300	299.10	20.2	15.5

LIBRARY

Project Size Gross S.F.	Project Cost $/S.F.	% Cost Mechanical	% Cost Electrical
6,900	199.40	17.9	10.3
8,200	183.00	18.8	10.6
12,000	237.30	19.4	15.9
15,000	207.40	15.7	18.7
16,300	175.70	11.3	7.4
28,600	157.40	15.7	9.0
30,100	213.50	13.0	11.0
37,700	167.70	14.6	8.0
43,500	141.70	16.6	6.7
47,900	248.40	29.7	9.0
51,400	253.80	13.1	12.4
63,400 A	190.90	13.5	8.3
64,000	178.70	10.9	11.4
74,000	202.40	17.2	8.0
75,600	268.60	12.5	16.7
176,000	148.80	11.2	9.8

SPECIAL NEEDS FUNCTION

Project Size Gross S.F.	Project Cost $/S.F.	% Cost Mechanical	% Cost Electrical
15,200	168.70	16.6	8.4
27,900	203.30	19.4	11.2

STUDENT CENTER/MULTIPURPOSE

Project Size Gross S.F.	Project Cost $/S.F.	% Cost Mechanical	% Cost Electrical
90,000	256.60	16.2	10.0
49,600	189.50	18.2	9.2
187,700	285.30	15.3	8.8
194,800	150.70	17.2	8.5

HOTEL/MOTEL

CONVENTION/CONFERENCE CENTER

Project Size Gross S.F.	Project Cost $/S.F.	% Cost Mechanical	% Cost Electrical
8,600 A	225.30	20.7	16.3
71,900	249.80	12.5	13.8
433,800	130.60	22.1	8.3

HOTEL

Project Size Gross S.F.	Project Cost $/S.F.	% Cost Mechanical	% Cost Electrical
19,900 A	128.80	16.8	5.8
25,875 R	115.50	11.8	10.5
48,400 A	230.90	23.6	8.2
64,300 R	311.20	20.3	10.1
104,200 A	144.60	15.0	8.8
108,040	123.50	13.0	7.0
110,100	159.90	18.4	10.5
132,000	298.80	16.4	5.4
135,900 A	184.70	15.2	7.7
144,100 A	207.40	19.3	11.5
231,000	224.30	15.2	8.4
449,800 A	129.60	13.1	7.0

HOTEL/INN

Project Size Gross S.F.	Project Cost $/S.F.	% Cost Mechanical	% Cost Electrical
57,400	149.90	13.0	10.0
73,000	98.20	24.4	18.1
75,900	130.60	16.5	7.6
162,000	157.10	17.5	8.0
197,000	150.40	15.7	7.7
277,900	110.40	18.8	9.4

A = Addition   R = Remodel

SQUARE FOOT TABLES

INDUSTRIAL

MANUFACTURING

Project Size Gross S.F.	Project Cost $/S.F.	% Cost Mechanical	% Cost Electrical
14,300	127.60	13.0	7.0
18,500	192.40	20.5	13.5
26,600	68.80	6.1	12.1
31,400	152.00	18.9	17.7
33,400	87.90	23.1	15.8
37,300	65.50	3.0	24.0
43,400	87.30	14.7	11.7
45,400	141.70	15.9	15.0
79,800	147.30	16.0	14.5
81,100	157.40	8.2	8.4
137,400	98.80	41.6	13.6
179,600	114.00	26.3	13.6
186,000	106.50	19.5	11.7

RESEARCH AND DEVELOPMENT

Project Size Gross S.F.	Project Cost $/S.F.	% Cost Mechanical	% Cost Electrical
89,140	203.50	20.8	9.0
100,400	269.60	36.1	25.1
114,200	264.00	21.4	9.8
125,000	179.50	20.8	8.4
140,000	250.90	20.1	11.2

WAREHOUSE W/OFFICE

Project Size Gross S.F.	Project Cost $/S.F.	% Cost Mechanical	% Cost Electrical
14,000	58.60	6.5	9.2
19,000	52.50	2.3	1.8
19,700	69.00	9.5	7.0
31,200	59.20	7.0	7.0
40,500	75.60	6.6	10.5
62,000	96.20	10.8	10.0
96,200	55.70	2.2	6.3
105,000	54.10	5.3	11.1
149,800	56.50	14.5	8.6

WAREHOUSE W/OFFICE (Cont.)

Project Size Gross S.F.	Project Cost $/S.F.	% Cost Mechanical	% Cost Electrical
168,600	62.20	11.4	6.3
209,600	57.40	4.9	10.3
402,400	78.00	14.9	8.3

MEDICAL

EDUCATION CENTER

Project Size Gross S.F.	Project Cost $/S.F.	% Cost Mechanical	% Cost Electrical
35,400	349.10	10.2	6.8

HOSPITALS

Project Size Gross S.F.	Project Cost $/S.F.	% Cost Mechanical	% Cost Electrical
9,300 R	344.90	29.5	12.0
15,900 A	254.50	27.4	15.3
16,600 R	91.70	14.7	9.6
22,000	579.20	31.6	17.5
39,100	288.60	23.3	7.9
63,800	232.50	23.4	7.8
98,000 A	334.50	20.5	17.0
100,200 A	533.60	22.2	9.6
103,900 A	295.70	31.1	11.7
109,300	293.00	20.7	15.7
148,700	230.20	25.2	11.8
154,700	373.90	19.3	10.5
165,484	325.00	21.5	14.8
165,700 A	317.10	26.2	17.7
179,400 A	193.20	28.9	20.3
182,800	287.60	19.1	14.0
265,000	343.10	31.2	14.1
281,100	328.70	24.1	14.6
435,000	340.50	21.5	13.7
694,300	188.60	28.9	9.7
772,300	347.50	31.5	13.4

A = Addition   R = Remodel

For more information subscribe to **Design Cost & Data**

SQUARE FOOT TABLES

MEDICAL (Cont.) | PUBLIC FACILITIES

MEDICAL OFFICES/CENTERS

Project Size Gross S.F.	Project Cost $/S.F.	% Cost Mechanical	% Cost Electrical
3,000	145.80	13.7	11.5
5,500	151.80	8.5	18.7
10,000	183.50	13.9	9.4
10,600	232.40	13.2	8.9
16,300	192.30	21.8	13.2
18,300	104.90	7.1	13.4
20,600	150.70	14.3	6.4
24,900	265.30	21.0	15.1
27,000	267.90	19.7	10.1
28,400	186.80	13.6	9.3
30,500	202.40	17.2	12.7
32.000	160.40	18.1	10.5
44,300	161.70	24.4	16.2
50,200	93.30	9.8	4.9
51,200	236.40	21.0	12.0
64,600	112.60	13.2	6.4
66,000	104.90	9.8	8.8
80,000	90.00	8.2	8.3
137,175	150.40	12.8	6.6

NURSING HOMES

Project Size Gross S.F.	Project Cost $/S.F.	% Cost Mechanical	% Cost Electrical
11,600 A	440.80	53.2	7.9
16,800	287.00	33.3	7.8
31,900 A	215.80	22.0	11.0
64,100	185.00	20.6	11.1
290,000	259.40	16.1	13.6

RESEARCH

Project Size Gross S.F.	Project Cost $/S.F.	% Cost Mechanical	% Cost Electrical
34,600	245.60	18.1	3.0

PUBLIC FACILITIES

ANIMAL CENTER

Project Size Gross S.F.	Project Cost $/S.F.	% Cost Mechanical	% Cost Electrical
20,000	311.20	20.7	4.6
39,100	232.50	22.9	6.8
44,300	221.60	8.1	5.8

AUTO DEALERSHIP

7,700	145.00	16.7	9.4

BROADCASTING

20,000	424.20	20.0	13.0
29,500	327.30	16.6	15.0
45,000 R	237.30	15.0	13.0

CIVIC CENTER

6,000	233.70	9.2	2.8
23,900	293.60	12.3	10.9
34,400	133.80	3.5	17.8
69,800 A	350.60	17.7	8.8
206,500	189.60	15.0	11.0

CORRECTION FACILITIES

44,600	283.50	20.9	13.1
66,000	173.00	15.0	23.0
257,800 A	299.30	20.9	10.7
360,000	201.60	32.7	13.2

FIRE STATION

6,900	253.00	12.4	9.7
7,600	194.40	16.0	9.5
8,430	267.30	12.8	9.6
9,600	233.10	13.5	11.8

A = Addition   R = Remodel

SQUARE FOOT TABLES

PUBLIC FACILITIES (Cont.)

GOVERMENT BUILDINGS

Project Size Gross S.F.	Project Cost $/S.F.	% Cost Mechanical	% Cost Electrical
12,300	181.40	13.5	10.7
23,500	297.20	11.1	15.6
27,300	234.60	19.5	9.3
31,600	264.20	27.1	11.3
46,600	248.60	17.4	13.8
72,100	270.70	24.4	10.7
78,200	224.60	19.7	15.2
332,900	217.90	16.2	14.2
364,100	285.40	14.6	12.9
771,000	315.40	17.6	11.3

MUSEUM

Project Size Gross S.F.	Project Cost $/S.F.	% Cost Mechanical	% Cost Electrical
27,600	217.70	17.8	14.1
30,100	250.10	18.6	7.9
43,264	233.10	11.1	8.3
63,000	252.20	8.8	18.1

PARKING GARAGE

Project Size Gross S.F.	Project Cost $/S.F.	% Cost Mechanical	% Cost Electrical
66,000	66.60	2.8	3.1
169,000	49.10	10.3	3.7
562,700	44.30	2.4	6.2

TRANSPORTATION

Project Size Gross S.F.	Project Cost $/S.F.	% Cost Mechanical	% Cost Electrical
7,300	400.20	15.1	3.2
14,300	321.60	13.3	16.6
23.000	222.20	9.6	13.7
35,500	185.00	1.0	19.0
49,100	275.90	35.5	11.6

TRANSPORTATION (Cont.)

Project Size Gross S.F.	Project Cost $/S.F.	% Cost Mechanical	% Cost Electrical
288,100	159.00	8.3	13.3
2,160,000	257.60	23.3	11.5

OFFICES

BANKS

Project Size Gross S.F.	Project Cost $/S.F.	% Cost Mechanical	% Cost Electrical
2,900	370.00	5.3	4.3
3,100	165.10	5.9	6.7
3,300	240.20	10.3	16.4
3,600	181.00	10.3	12.2
4,000	206.10	8.0	9.0
4,100	203.90	21.4	13.0
4,200	220.00	8.6	14.21
4,400	254.20	12.2	12.7
4,500	157.80	12.4	13.5
4,900	249.30	11.7	11.2
5,900	185.30	9.3	13.8
6,000	208.40	11.6	7.3
6,100	280.20	11.0	8.0
7,000	376.90	6.0	9.0
7,300	230.50	11.9	11.3
7,700	258.30	8.0	7.5
7,800	283.10	11.0	11.7
8,000	143.90	10.0	14.0
9,200	228.20	9.9	12.1
9,400	165.20	11.7	11.8
10,200	344.70	12.6	12.6
12,600	123.20	7.0	18.0

A = Addition   R = Remodel

SQUARE FOOT TABLES

OFFICES (Cont.)

BANKS (Cont.)

Project Size Gross S.F.	Project Cost $/S.F.	% Cost Mechanical	% Cost Electrical
13,300	217.10	9.5	8.3
13,800	194.40	10.00	9.7
15,000	166.40	15.4	12.4
15,200	133.00	9.2	12.9
15,500	169.60	9.8	10.3
16,000	111.40	13.4	23.1
20,100	104.80	13.0	11.0
21,700	221.60	8.8	11.3
44,800	189.30	13.0	8.2
53,200	341.90	14.9	7.2
62,100	196.30	10.1	7.9
95,100	258.20	13.5	4.3

OFFICE BUILDINGS

Project Size Gross S.F.	Project Cost $/S.F.	% Cost Mechanical	% Cost Electrical
2,600	232.20	17.2	9.4
3,400	189.80	10.5	11.3
3,800	186.30	16.4	11.8
4,400	184.10	12.8	8.5
4,500	158.30	13.0	7.0
5,100	115.10	16.6	10.2
5,200	150.90	8.0	5.7
6,700	224.70	16.8	10.4
7,500	241.90	10.1	8.0
7,900	188.40	17.4	9.0
8,100	291.20	10.6	11.0
10,600	168.40	9.9	9.7
10,900	96.90	13.8	10.8

OFFICE BUILDINGS (Cont.)

Project Size Gross S.F.	Project Cost $/S.F.	% Cost Mechanical	% Cost Electrical
11,300	191.30	17.0	6.0
13,000	130.30	15.0	9.0
14,400	170.90	19.8	12.9
14,500	131.00	17.3	12.5
17,000	187.00	14.7	7.9
17,800 A	101.10	10.4	11.0
18,100	195.90	22.5	11.6
19,300	106.40	11.1	8.1
24,600	98.30	18.5	14.1
27,700	118.20	19.6	5.5
27,800	214.50	12.7	5.1
27,800	115.50	17.8	10.3
32,500 R	197.70	13.9	6.4
35,400	115.40	15.0	12.0
36.500	99.90	10.2	10.2
42,300	124.00	10.3	7.7
44,400	168.90	23.5	14.7
44,400	88.80	11.0	5.0
44,500	122.30	11.5	3.1
45,400	105.60	19.1	13.2
47,300	108.70	18.5	8.0
49,700	182.50	22.1	7.2
50,000	174.00	19.4	15.6
50,400	188.70	23.2	7.8
52,200	122.60	18.3	7.8
52,900	142.90	4.4	3.9
53,700	218.70	15.4	9.0

A = Addition   R = Remodel

SQUARE FOOT TABLES

OFFICES (Cont.)

OFFICE BUILDINGS (Cont.)

Project Size Gross S.F.	Project Cost $/S.F.	% Cost Mechanical	% Cost Electrical
54,000	88.50	14.4	2.6
56,000	100.30	10.6	6.2
56,500	138.10	19.1	11.0
72,000 R	48.50	12.9	20.2
74,000	96.20	13.5	6.8
80,800	97.20	12.0	6.0
81,800	133.80	21.1	9.5
81,900	112.60	19.9	9.0
82,000	156.50	14.0	4.2
83,100	184.50	16.0	8.6
85,400	161.70	18.7	8.6
86,200	126.70	14.7	11.0
99,900	146.60	16.6	6.6
100,000	163.80	10.6	12.1
100,000	96.60	16.3	10.4
116,400	154.70	12.6	10.0
134,500	315.60	14.1	12.4
140,000	244.60	20.1	11.2
155,700	304.70	11.1	6.1
171,000	142.80	24.0	7.7
174,300	175.80	12.3	9.1
203,300	237.70	19.0	13.2
265,800	329.00	11.0	10.0
287,300	91.10	10.7	4.1
319,800	142.80	13.4	5.6
350,000	211.00	20.0	13.0
360,900	127.70	14.2	11.1
394,000	101.50	22.5	8.7

OFFICE BUILDINGS (Cont.)

Project Size Gross S.F.	Project Cost $/S.F.	% Cost Mechanical	% Cost Electrical
430,000	152.00	21.8	9.3
490,000	161.60	11.2	7.5
588,400	332.30	15.0	9.0
606,000	132.90	9.4	8.8
620,000	366.50	12.6	12.8
733,500	101.50	10.8	4.2

RECREATIONAL

ARENA

Project Size Gross S.F.	Project Cost $/S.F.	% Cost Mechanical	% Cost Electrical
315,200	380.40	10.4	8.00
385,800	217.70	13.2	8.80
727,000	196.90	14.9	7.40

HEALTH CLUB

Project Size Gross S.F.	Project Cost $/S.F.	% Cost Mechanical	% Cost Electrical
15,900	106.40	8.7	9.7
21,800	150.60	10.3	10.3
30,100	232.80	11.4	19.4
66,400	114.60	10.7	8.5

RECREATIONAL CENTER

Project Size Gross S.F.	Project Cost $/S.F.	% Cost Mechanical	% Cost Electrical
9,900	206.80	6.6	9.8
14,000	134.90	11.2	5.0
14,000	161.70	11.3	16.7
15,700	123.20	17.8	14.3
20,000	279.30	19.1	8.9
21,200	157.20	16.0	9.6

A = Addition   R = Remodel

SQUARE FOOT TABLES

RECREATIONAL (Cont.)

RECREATIONAL CENTER (Cont.)

Project Size Gross S.F.	Project Cost $/S.F.	% Cost Mechanical	% Cost Electrical
26,000	183.00	9.3	7.0
53,400 A	219.00	11.7	6.4
69,800	355.00	17.7	8.8

RELIGIOUS

CHURCH

Project Size Gross S.F.	Project Cost $/S.F.	% Cost Mechanical	% Cost Electrical
4,100	249.30	8.8	13.5
10,400 R	339.30	12.8	8.1
11,100	149.50	10.9	12.7
13,400	223.90	5.5	6.0
14,500	152.90	12.8	7.4
15,200	198.10	18.3	8.0
15,700	172.40	14.4	8.4

CHURCH (Cont.)

Project Size Gross S.F.	Project Cost $/S.F.	% Cost Mechanical	% Cost Electrical
16,000	280.60	14.1	9.6
20,900	175.00	16.0	14.0
21,500	192.10	7.3	8.0
22,900	173.50	12.1	9.5
30,600	131.10	15.5	8.0
42,700	148.10	18.6	7.7

MULTI-PURPOSE

Project Size Gross S.F.	Project Cost $/S.F.	% Cost Mechanical	% Cost Electrical
4,400	153.20	11.5	17.5
5,800 A	195.30	8.9	7.5
6,400	266.60	15.6	15.8
9,000	130.60	7.7	5.9
9,000	185.30	16.0	6.6
10,100	124.30	11.1	12.0
12,000	245.80	13.1	10.7
18,400	134.60	10.8	10.1
19,500	237.30	16.0	17.3

A = Addition   R = Remodel

For more information subscribe to **Design Cost & Data**

BNi Building News

Notes

Notes

Notes

Notes

Notes

Notes